A SOURCE BOOK IN MATHEMATICS, 1200–1800

A SOURCE BOOK IN MATHEMATICS, 1200–1800

EDITED BY D. J. STRUIK

PRINCETON UNIVERSITY PRESS, PRINCETON, NEW JERSEY

Published by Princeton University Press, 41 William Street, Princeton, New Jersey 08540
In the United Kingdom: Princeton University Press, Oxford

First Princeton Paperback printing, 1986
Third printing, 1990
LCC 85-43381
ISBN 0-691-08404-1
ISBN 0-691-02397-2 (pbk.)

Printed in the United States of America by Princeton University Press,
Princeton, New Jersey

9 8 7 6 5 4 3

TO THE MEMORY OF GEORGE SARTON

ERRATA ET ADDENDA

Page	Line f.b. = from below	
12	3	Macdonald's book was reprinted in 1966.
69	note	The Witmer translation was published in 1968, title: *The Great Art*.
75	14	H. Guericke and K. Reich published a German translation (München, 1973).
152	10 f.b.	Between "diminished by" and "the square of z" add: "a multiplied by z, or the cube of z is equal to a multiplied."
168	11	For "it was written . . . after 1676" read "it was first written in 1667 or 1668 and rewritten later."
189	19	After "179-388" add "and Margaret E. Baron, *The origins of the infinitesimal calculus* (Oxford etc., 1969)."
215	14 of note	Read: $\Sigma xy = \Sigma(\frac{1}{4}a^2 - z^2) = \frac{1}{4}\Sigma a^2 - \Sigma z^2$.
217	note A	Read: $CD = \underline{a}$
223	9 f.b.	For "2ae+e" read "2ae+e^2."
236	9 f.b.	Read: $DS_2 = \sin(\text{quartare}-FR_2)$.
246	2, 3 of note 2	For "and maybe medieval," read "and are medieval, e.g. Bradwardine."
253	12 f.b.	For "His most famous disciple was," read "Among the attendants at his lectures may have been."
255	note 3	Read: "The theorem that FT is tangent to the curve VIF."
262	note 10	Read in the first formula $\int_0^y y\frac{dy}{ds}dx$ for $\int_0^x y\frac{dy}{dx}dx$
270	5 f.b.	It is J. O. Fleckenstein.
271	19 f.b.	For "III(1863" read "I(1858)"; for "Band VII" read "Band V."

Page	Line	
280	end text	After "213-222," add "and H. J. M. Bos, *Differentials, higher order differentials and the derivative in the Leibnizian calculus*, Arch. Hist. Exact Sc. 14 (1974), 1-90.
281	13	For "Band III" read "Band I."
284	15 f.b.	Omit "under Barrow."
285	14 f.b.	For "(1959) vol 1," read "(1960) vol. 2."
291	8, 9 f.b.	Hofmann's book now exists in a (revised) English translation, *Leibniz in Paris 1672-1676* (Cambridge University Press, 1974). See also *Leibniz à Paris*, a symposium of 1976 reported in *Historia Mathematica* 9 (1982), 113-123.
311	16	For "1930" read "(1938)."
317	18	For $\frac{nn-n}{n}$ read $\frac{nn-n}{2}$
	note 7	For $1+2+\cdots+n$ read $1+2+---+(n-1)$. After "before" read "the then well-known fact."
324	8	For "305-558" read "385-558."
329	10	For "$\Delta x+2\Delta x+\Delta^3 x$" read "$\Delta x+2\Delta^2 x+\Delta^3 x$." For "$3\Delta^2 x+\Delta^4 x$" read "$3\Delta^3 x+\Delta^4 x$."
348	3 f.b.	For $(\sin)^2$ read $(\sin 7)^3$.
349	10 f.b.	For "infinitely small" read "infinitely large."
351	13	After "vol 6" add "(1809)."
355	15	For "$\Psi(s)$ is an even" read "$\Psi(s)$ is an odd."
368		On the history of the function concept consult also sections of M. Kline, *Mathematical thought from ancient to modern times* (Oxford University Press, 1972).

Furthermore:
English translations have appeared
1) Of *Witelonis Perspectivae Liber Primus: Perspectiva Book I* by Witelo (ca. 1230/35 — after ca. 1275), Studia Copernicana 15 (1977), ed. S. Unguru (Polish Academy of Sciences).

2) Of a section on divergent series by Euler, part of *De seriebus divergentibus*, Novi Comm. Acad. Sc. Petrop. 5 (1754/55), 205-237, Opera Omnia (1) 14, 585-617; see *Historia Matematica* 3 (1976), 141-160.

Some work of Lagrange, Legendre, and Laplace belonging to our period 1200-1800 is found in English translation in the *Source book in classical analysis* (Harvard University Press, 1973), ed. G. Birkhoff, with Uta Merzbach.

On the authors of the selections included in our book we can find usually excellent articles on life and work in the *Dictionary of scientific biography*, 16 vols. (Scribner, New York, 1970-1980).

PREFACE

This Source Book contains selections from mathematical writings of authors in the Latin world, authors who lived in the period between the thirteenth and the end of the eighteenth century. By Latin world I mean that there are no selections taken from Arabic or other Oriental authors, unless, as in the case of Al-Khwārizmī, a much-used Latin translation was available. The choice was made from books and from shorter writings. Usually only a significant part of the document has been taken, although occasionally it was possible to include a complete text. All selections are presented in English translation. Reproductions of the original text, desirable from a scientific point of view, would have either increased the size of the book far too much, or made it necessary to select fewer documents in a field where even so there was an *embarras du choix*. I have indicated in all cases where the original text can be consulted, and in most cases this can be done in editions of collected works available in many university libraries and in some public libraries as well.

It has not often been easy to decide to which selections preference should be given. Some are fairly obvious; parts of Cardan's *Ars magna*, Descartes's *Géométrie*, Euler's *Methodus inveniendi*, and some of the seminal work of Newton and Leibniz. In the selection of other material the editor's decision whether to take or not to take was partly guided by his personal understanding or feelings, partly by the advice of his colleagues. It stands to reason that there will be readers who miss some favorites or who doubt the wisdom of a particular choice. However, I hope that the final pattern does give a fairly honest picture of the mathematics typical of that period in which the foundations were laid for the theory of numbers, analytic geometry, and the calculus.

The selection has been confined to pure mathematics or to those fields of applied mathematics that had a direct bearing on the development of pure mathematics, such as the theory of the vibrating string. The works of scholastic authors are omitted, except where, as in the case of Oresme, they have a direct connection with writings of the period of our survey. Laplace is represented in the Source Book on nineteenth-century calculus.

Some knowledge of Greek mathematics will be necessary for a better understanding of the selections: Diophantus for Chapters I and II, Euclid for Chapter III, and Archimedes for Chapter IV. Sufficient reference material for this purpose is found in M. R. Cohen and I. E. Drabkin, *A source book in Greek science* (Harvard University Press, Cambridge, Massachusetts, 1948). Many of the classical authors are also easily available in English editions, such as those of Thomas Little Heath.

It was often a difficult task to decide on how much of the archaic flavor should be preserved, especially in notation. I have tried to find a middle road by keeping some archaic forms, explaining them in notes, and modernizing other places while indicating what the original text looked like. Some of the illustrations may give additional insight.

The editor takes full responsibility for the translation. Where already existing English translations have been utilized, they have been checked with the original. Remarks by the editor in the Selections are indicated by square brackets.

Those in search of English translations of older mathematical writings, not included in this Source Book, should consult D. E. Smith, *A source book in mathematics* (McGraw-Hill, New York, 1929; Dover, New York, 1959), Henrietta O. Midonick, *The treasury of mathematics* (Philosophical Library, New York, 1965), and further D. J. Struik, "A selected list of mathematical books and articles published after 1200 and translated into English," *Scripta Mathematica 15* (1949), 115–131.

The following institutions have permitted the use of certain texts for our selections: the Stevin Committee of the Royal Netherlands Academy of Sciences for Selections I.3 and IV.1; Akademie Verlag, Berlin, for Selection I.11; the University of California Press, Berkeley, California, for Selection I.14; E. J. Brill, Ltd., Leiden, for Selection III.1; the University of Wisconsin Press, Madison, Wisconsin, for Selection III.2; Northwestern University Press, Evanston, Illinois, for Selections IV.1 and IV.2; Teachers College, Columbia University, New York, for Selection IV.10; the Open Court Publishing Co., La Salle, Illinois, for Selection IV.14; the Royal Society of Edinburgh, the Royal Society, London, the University Library, Cambridge, and the British Museum, London, for Selection V.4.

Thanks for encouragement and advice are due to many colleagues, in particular to C. B. Boyer, I. B. Cohen, J. E. Hofmann, J. F. Scott, C. J. Scriba, Maria Spoglianti, R. Taton, D. T. Whiteside, and A. B. Yuškevič; and especially to Ruth Ramler Struik, who also contributed to the translations from the Latin. For the drawings I am obliged to Alfonso Vera Mackintosh, Jaime Barcena Armento, Federico A. Grageda Venegas, Mrs. Joanna Muckenhoupt Enzmann, and W. Minty, and for generous editorial care to Mr. J. D. Elder of the Harvard University Press.

D. J. Struik

Belmont, Massachusetts
November 1967

CONTENTS

CHAPTER I ARITHMETIC

CHAPTER II ALGEBRA

CHAPTER III GEOMETRY

CHAPTER IV ANALYSIS BEFORE NEWTON AND LEIBNIZ

CHAPTER V NEWTON, LEIBNIZ, AND THEIR SCHOOL

ABBREVIATIONS OF TITLES

Jakob Bernoulli, *Opera = Jacobi Bernoulli Opera*, ed. G. Cramer, 2 vols., Geneva, 1744.

Johann Bernoulli, *Opera omnia = Johannis Bernoulli Opera omnia*, ed. G. Cramer, 4 vols., Lausanne, 1742.

Cantor, *Geschichte* = M. Cantor, *Vorlesungen über Geschichte der Mathematik*, Teubner, Leipzig; vol. I (3rd ed., 1907), vol. II (2nd ed., 1900), vol. III (2nd ed., 1901), vol. IV (1908).

Descartes, *Oeuvres = Oeuvres de Descartes*, ed. C. Adam and P. Tannery, 12 vols., Cerf, Paris, 1897–1918.

Euler, *Opera omnia = Leonhard Euler Opera omnia*, ed. Soc. Scient. Natur. Helveticae, Teubner, Leipzig and Berlin, 1911–1938, Füssli, Zürich; three series.

Fermat, *Oeuvres = Oeuvres de Fermat*, ed. P. Tannery and C. Henry, 4 vols., Gauthier-Villars, Paris, 1891–1912.

Gauss, *Werke* = Carl Friedrich Gauss, *Werke*, ed. Königliche Gesellschaft für Wissenschaften, Göttingen, 12 vols., Leipzig and Berlin, 1863–1929.

Heath, *Euclid's Elements = Euclid's Elements*, translated with introduction and commentary by Sir T. L. Heath, 3 vols., Cambridge University Press, Cambridge, England, 1925; Dover, New York, 1956.

Huygens, *Oeuvres complètes = Oeuvres complètes de Christiaan Huygens*, ed. Société Hollandaise des Sciences, 22 vols., Nyhoff, The Hague, 1888–1950.

Lagrange, *Oeuvres = Oeuvres de Lagrange*, ed. G. Darboux and J. A. Serret, 14 vols., Gauthier-Villars, Paris, 1867–1892.

Leibniz, *Mathematische Schriften = Leibnizens Mathematische Schriften*, ed. C. I. Gerhardt, 7 vols. in two Abtheilungen, Schmidt, Halle, 1849–1863.

Smith, *History of mathematics* = D. E. Smith, *History of Mathematics*, 2 vols., Ginn, Boston, 1923, 1925; Dover, New York, 1957.

Smith, *Source book* = D. E. Smith, *A Source book in mathematics*, McGraw-Hill, New York, 1929; Dover, New York, 1959.

Tropfke, *Geschichte* = J. Tropfke, *Geschichte der Elementar-Mathematik in systematischer Darstellung*, 7 vols., De Gruyter, Berlin and Leipzig, 1921–1924; third editions exist of vol. I (1930), vol. II (1933), vol. IV (1940).

The *Mémoires* of the Paris Academy of Sciences are quoted under their title: *Histoire de l'Académie Royale, Paris*, short for *Histoire de l'Académie Royale des Sciences, Paris*.

The *Mémoires* of the Berlin Academy of Sciences are quoted under their title: *Histoire de l'Académie Royale, Berlin*, short for *Histoire de l'Académie Royale des Sciences et des Belles Lettres de Berlin*. After 1772 they were published under the title of *Nouveaux Mémoires de l'Académie Royale* ... When, in quoting these *Mémoires* in both cases, years such as 1745 (1746) are mentioned, it means that the volume is for the year 1745 and was published in 1746.

A SOURCE BOOK IN MATHEMATICS, 1200–1800

CHAPTER I ARITHMETIC

The study of mathematics in medieval Latin Europe,[1] after the eleventh century, was stimulated by Latin translations from the Greek and, especially, from the Arabic. They were prepared in those places where the contact between the Christian and the Islamic civilizations was the most intimate, notably in Sicily, Southern Italy, and Spain. Some of the most prolific translators were Adelard of Bath, Robert of Chester, and Gerhard of Cremona in the twelfth century and Johannes Campanus in the thirteenth. In this way Latin Europe became acquainted with the geometry of Euclid (*c.* 300 B.C.), originally composed in Greek (and previously known only through secondhand abstracts), and the arithmetic and algebra developed in the countries of Islam. Here the author with perhaps the greatest influence was Mohammed Al-Khwārizmī, who worked at Bagdad and wrote his Arabic in the early part of the ninth century. His name is preserved in our term "algorithm," and our word "algebra" is derived from the Arabic title of Mohammed's book on equations (see Chapter II).

It was through Islamic channels that Latin Europe became acquainted with our present decimal position system of numbers, based on the ten symbols which we now write 1, 2, 3, 4, 5, 6, 7, 8, 9, 0. Their gradual penetration into Europe may have come about, in ways only partially traceable, along the trade routes that connected the Christian and the Islamic worlds. It was also accomplished through scholarly manuscripts, among which those by Al-Khwārizmī and the learned merchant Leonardo of Pisa had a considerable reputation. Leonardo's *Liber abaci* (1202) begins with the introduction of the ten symbols and then develops an arithmetic based on these symbols, followed by a theory of equations.

With Fermat, in the first half of the seventeenth century, begins the study of the abstract theory of numbers. But it was not until a century later that a first-class mathematician accepted Fermat's challenge. Euler's work, starting around 1730, marks the beginning of the period of continual research in number theory. But for many years Euler worked almost alone in this field. Only in the second half of the eighteenth century was he joined by some others, such as Lagrange and Legendre. This phase can be said to have ended with the great texts by Legendre (1797–98) and Gauss (1801). The work of the young Gauss inaugurates a new phase in the history of number theory.

---

[1] We use the term Latin Europe to denote those parts of Europe in which Latin was the principal language employed by the literate for communication. It comprised chiefly Italy, Christian Spain, France, Germany (with the Low Countries), England, and Scandinavia.

In the choice of the selections, especially those of Euler, we had to be fairly arbitrary. We decided to concentrate on power residues. The reader who wants to know more about eighteenth-century number theory should consult M. Cantor, *Vorlesungen über Geschichte der Mathematik*, vol. III (Teubner, Leipzig, 1898) and especially, in vol. IV (1908), the article by F. Cajori. Further details can be found in the three volumes of L. E. Dickson, *History of the theory of numbers* (Carnegie Institution, Washington, D.C., 1919–1923; 2d ed., 1934).

Among the problems solved in the *Liber abaci* one has acquired considerable fame. It may well have been invented by Leonardo, and with it we open our collection of texts.

1 LEONARDO OF PISA. THE RABBIT PROBLEM

In Leonardo of Pisa, also called Fibonacci, we meet the first outstanding mathematician of the Latin Middle Ages. He was a merchant of Pisa who traveled widely in the world of Islam, and took the opportunity of studying Arabic mathematical writings. His work is in the spirit of the Arabic mathematics of his day, but also reveals his own position as an independent thinker. Leonardo's *Liber abaci* (1202, revised 1228) circulated widely in manuscript, but was published only in 1857: *Scritti di Leonardo Pisano* (pubbl. da B. Boncompagni, Tipografia delle scienze matematiche e fisiche, Rome; 2 vols., 459 pp.). It is a voluminous compendium on arithmetic and its mercantile practice (even finger counting), the theory of linear, quadratic, and simultaneous sets of equations, square and cube roots. One of the principal features of the book is that, from the first page on, Leonardo introduces and uses the decimal position system. The first chapter opens with the sentence: "These are the nine figures of the Indians

$$9\ 8\ 7\ 6\ 5\ 4\ 3\ 2\ 1$$

With these nine figures, and with this sign 0 which in Arabic is called zephirum,[1] any number can be written, as will below be demonstrated."

We confine ourselves here to presenting, in translation from the Latin, two interesting sections from the *Liber abaci* that may very well be original contributions. Since the first introduces *paria coniculorum*, we know it as the rabbit problem. It stands by itself (vol. I, 283–284), sandwiched in between other problems; the one before it deals with the so-called perfect numbers 6, 28, 496, . . . , and the one after with the solution of a system of four linear equations with four unknowns.

How many pairs of rabbits can be bred from one pair in one year?

A man has one pair of rabbits at a certain place entirely surrounded by a wall. We wish to know how many pairs can be bred from it in one year, if the nature of these rabbits is such that they breed every month one other pair and begin to

---

[1] *Zephirum, zephyr*, from Arabic *as-sifr*, literal translation of Sanskrit *sunya* = empty, has in its turn led to French *chiffre*, German *Ziffer*, and English *zero* and *cipher*. The meaning of these terms is either zero or number digit in general, which shows the importance attached to the 0 in the understanding of the decimal position system, when in the later Middle Ages it gradually penetrated Latin Europe.

breed in the second month after their birth. Let the first pair breed a pair in the first month, then duplicate it and there will be 2 pairs in a month. From these pairs one, namely the first, breeds a pair in the second month, and thus there are 3 pairs in the second month. From these in one month two will become pregnant, so that in the third month 2 pairs of rabbits will be born. Thus there are 5 pairs in this month. From these in the same month 3 will be pregnant, so that in the fourth month there will be 8 pairs. From these pairs 5 will breed 5

	pairs
	1
first	2
second	3
third	5
fourth	8
fifth	13
sixth	21
seventh	34
eighth	55
ninth	89
tenth	144
eleventh	233
twelfth	377

other pairs, which added to the 8 pairs gives 13 pairs in the fifth month, from which 5 pairs (which were bred in that same month) will not conceive in that month, but the other 8 will be pregnant. Thus there will be 21 pairs in the sixth month. When we add to these the 13 pairs that are bred in the 7th month, then there will be in that month 34 pairs ... [and so on, 55, 89, 144, 233, 377, ...]. Finally there will be 377. And this number of pairs has been born from the first-mentioned pair at the given place in one year. You can see in the margin how we have done this, namely by combining the first number with the second, hence 1 and 2, and the second with the third, and the third with the fourth ... At last we combine the 10th with the 11th, hence 144 with 233, and we have the sum of the above-mentioned rabbits, namely 377, and in this way you can do it for the case of infinite numbers of months.[2]

Here is a section (I, 24) in which Leonardo introduces a kind of continued fraction, which he writes $\dfrac{eca}{fdb}$, or in our notation:

$$\frac{eca}{fdb} = a + \cfrac{c + \cfrac{e}{f}}{d} \Big/ b = \frac{adf + cf + e}{bdf} = \frac{a}{b} + \frac{c}{d}\cdot\frac{1}{b} + \frac{e}{f}\cdot\frac{1}{b}\cdot\frac{1}{d}.$$

Below some line [branchlet, *virgula*] let there be 2, 6, 10, and above the 2 be 1, above 6 be 5, and above 10 be 7, which appears as $\frac{1}{2}\frac{5}{6}\frac{7}{10}$. The 7 above 10 at the head of the line represents seven-tenths and the 5 above 6 denotes five-sixths

---

[2] This sequence of numbers, 1, 2, 3, 5, 8, ..., u_n, ..., with the property that $u_n = u_{n-1} + u_{n-2}, u_0 = 1, u_1 = 1$, is called a *Fibonacci series*. It has been the subject of many investigations, and is closely connected with the golden section, that is, the division of a line segment AB by a point P such that $AP : AB = PB : AP$. See, for example, R. C. Archibald, "The golden section," *American Mathematical Monthly 25* (1918), 232–238; D'Arcy W. Thompson, *On growth and form* (Cambridge University Press, New York, 1942), 912–933; H. S. M. Coxeter, "The golden section, phyllotaxis, and Wythoff's game," *Scripta Mathematica 19* (1950), 135–143; E. B. Dynkin and W. A. Uspenski, *Mathematische Unterhaltungen*, II (Deutscher Verlag der Wissenschaften, Berlin, 1956); and N. N. Vorob'ev, *Fibonacci numbers*, trans. by H. Mors (Blaisdell, New York, London, 1961).

of the decimal part and 1 above 2 denotes half of the sixth of the decimal part...[3] We shall say that fractions which are on a branch [*virga*] will be in grades, so that the first grade of them is the fraction which is at the head of the branch to the right. The second grade is the fraction that follows towards the left. For instance, in the branch above, hence in $\frac{1}{2}\frac{8}{8}\frac{7}{10}$, the $\frac{7}{10}$ are in the first grade of this branch, $\frac{8}{8}$ in the second one...

---

2 RECORDE. ELEMENTARY ARITHMETIC

With the growing interest in mercantile reckoning and the spread of printing, the number of textbooks of elementary arithmetic increased rapidly from the latter half of the fifteenth century onward. Their character has been described in L. C. Karpinski, *The history of arithmetic* (Rand McNally, Chicago, 1925) and in Smith, *History of mathematics*, II, chaps. 1–3, with many illustrations. Smith, *Source book*, 1–12, has an English translation of a section of the so-called *Treviso arithmetic* of 1478 (the first printed arithmetic). Typical of all is their introduction to the art of reckoning with the aid of the decimal position system, with digits almost or exactly the same as those we use. Many books have chapters on finger reckoning and on the use of counters for computation on an abacus. Notations for addition, subtraction, multiplication, and division still vary, though the use of + and − for addition and subtraction is fairly common. As an example we present here, in facsimile, some pages of the first arithmetic printed in the English language, *The ground of artes* by Robert Recorde (*c.* 1510–1558). Recorde, a Cambridge M.D. and physician to Edward VI and Mary Tudor, wrote several books on mathematics and astronomy that were long in use in England. *The ground of artes*, first published in London between 1540 and 1542 (the oldest extant edition has the date 1543), was regularly reprinted and reedited; there exists an edition of 1699.

In the pages that we reproduce (Figs. 1, 2, 3) we see how Recorde performed division in Arabic numerals, and how he taught addition by means of counters, which have long been in use and are still popular in Russia, Japan, and China. In the United States they are used by Chinese laundrymen and restaurant workers, and on baby pens. Recorde used the + and − signs, and in his algebra, *The whetstone of witte* (London, 1557), he introduced our sign for equality:

I will sette as I doe often in woorke use, a paire of paralleles, or Gemowe lines of one lengthe, thus: $=\!\!=\!\!=$, bicause noe.2. thynges, can be moare equalle.

In his use of the strange word "Gemowe" we see an example of Recorde's attempt to substitute English technical words for the current Latin ones. Stevin tried the same in

---

[3] Hence $\frac{1}{2}\frac{8}{8}\frac{7}{10} = \frac{7}{10} + \frac{8}{8}\cdot\frac{1}{10} + \frac{1}{2}\cdot\frac{1}{8}\cdot\frac{1}{10}$. On connections with an Islamic and perhaps pre-Islamic calculus of fractions see Cantor, *Geschichte*, I, 813.

Fig. 1

Diuision

that wpll be the nomber that shall a-
mounte, therfore is this wape most
easper. S. So is it, and also most
certapner, for such as I am, p mpght
qupckelp erre in multiplpenge, espe-
cpallp bepng smallp practised therin.
M. Then proue in some brefe example whether pou can do it, and so wpll
we make an ende. S. I wolde diuide
38468 bp 24, therfore fprst I sette the

table thus,	24	1
Then set I the two sumes	48	2
of diuision thus. 38468	72	3
And ouer the di- 24	96	4
uisor I fpnde 38, whiche I	120	5
seke in the table, and fpnde	144	6
it not, therfore take I the	168	7
nexte benethe it, which the	182	8
tabic hath, and that is 24,	216	9

the diuisor it selfe, against whiche is
set 1, whiche I take for the quotient,
whiche I set in his place. And now I
nede not to multiplp the diuisor bp it,
but onlp to withdrawe the diuisor out
of the 38 that is ouer it, z so remapneth

Diuision.   85

neth 14, as thus.   14
Then set I forwarde   38468 (1
the diuisor, and fpnd
ouer it 144, as appereth: then seke I
that nomber in p ta-   14
ble, and fpnde it, and   38468 (
agapnst it is 6, ther-   xxx4
fore I set 6 before 1,
for mp quotient, and I take that 144
for the iuste multiplication of the diuisor bp that quotient, and therfore
without anp newe multiplication, I
do subtracte that 144 from the other
144, and there resteth nothpnge, as
here pou map se.   xx
Therfore I set for-   38468 (16
warde p diuisor, but   xxx
sepng it wpll not be
in p nexte place (for then ouer 2 wold
be nothpnge) I set it forwarde twpse,
as pou se here.   xx
And for bpcause that   38468 (16
I coulde not set it in   xxx4
p nexte place folow-   x   2
pnge, therfore I sette a cpphet in the
     M.iiii.   quo-

Fig. 2

Accomptynge

in the same lpne in p second sume, to p
one p is there all redp: z then wpll the
hole sume appere (as
pou map wel se) to be
8746, which was p
fprst grosse summe, z
therfore I do per-
ceaue, that I hadde
well subtracted be-
fore. And thus pou map se how subtraction mape be trped bp Addition.
S. I perceaue the same order here w
couters, p I lernd before in figures.
M. Then let me se howe can pou trpe
Addition bp Subtraction. S. fprst
I wpl set forth this exaple of Addtion
where I haue added 2189 to 4583, z
the hole sume appereth to be 7177,

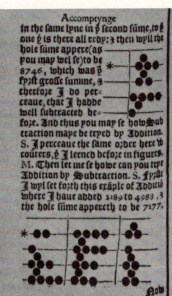

by counters.   119

Nowe to trpe whether that sume be
well added or no, I wpll subtract one
of the fprst two sumes from the thprd,
and pf I haue well done p remapner
wpll be lpke that other sume. As for
eraumple: I wpll subtracte the fprste
summe from the thprde, whiche I set
thus in thep
order.
Then do I
subtract 2000
of the fprste
summe fro p
second sume,
and then remapneth there 5000 thus.
Then in the thprd
lpne, I subtract p
100 of the fprste
summe, fro the se-
cond sume, where
is onelp 100 also,
z then in p thprde
lpne resteth nothpng. Then in the se-
conde lpne with his space ouer hpm,
I fpnde 80, which I shuld subtracte
     R.iiii.   from

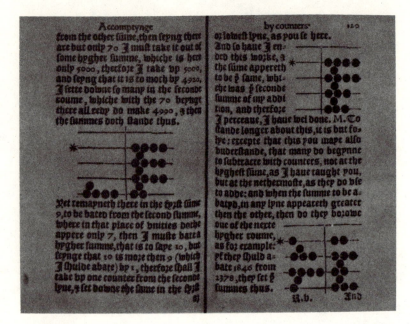

Fig. 3

Dutch, Kepler in German. The only one who was partly successful was Stevin. "Gemowe," also "gemew," means twin (French *gémeaux*, Latin *gemini*).

On Recorde see F. M. Clarke, "New light on Robert Recorde," *Isis 8* (1926), 50–70; see also L. D. Patterson, *Isis 42* (1951), 208–218.

For those who find it difficult to read the text, we transcribe here, in slightly modernized form, page 84ᵛ, beginning with the third line. It is a discussion between M, the master, and S, the scholar:

S. So is it, and also more certainer, for such as I am, that might quickly err in multiplying, especially being smally practised therein. *M.* Then prove in some brief example whether you can do it, and so will we make an end. *S.* I would divide 38468 by 24, therefore first I set the table thus. Then set I the two sums of division thus. And over the divisor I find 38, which I seek in the table, and find it not, therefore take I the next beneath it, which the table has, and that is 24, the divisor itself, against which is set 1, which I take for the quotient, which I set in his place. And now I need not to multiply the divisor by it, but only to withdraw the divisor out of the 38 that is over it, and so remains 14, as thus.

---

3 STEVIN. DECIMAL FRACTIONS

The introduction of decimal fractions as a common computational practice can be dated back to the Flemish pamphlet *De Thiende*, published at Leyden in 1585, together with a French translation, *La Disme*, by the Flemish mathematician Simon Stevin (1548–1620), then settled in the Northern Netherlands. It is true that decimal fractions were used by the Chinese many centuries before Stevin and that the Persian astronomer Al-Kāshī used both decimal and sexagesimal fractions with great ease in his *Key to arithmetic* (Samarkand, early fifteenth century).[1] It is also true that Renaissance mathematicians such as Christoff Rudolff (first half sixteenth century) occasionally used decimal fractions, in different types of notation. But the common use of decimal fractions, at any rate in European mathematics, can be directly traced to *De Thiende*, especially after John Napier (see p. 13) had modified Stevin's notation into the present one with the decimal point or comma.

Stevin's notation strikes us as clumsy, showing an unnecessary relation to the notation of sexagesimal fractions. However, for beginners in the difficult arts of multiplication and division, his method may have had a certain advantage. See further the introduction to the edition of *De Thiende* in *The principal works of Simon Stevin*, IIA (Swets-Zeitlinger, Amsterdam, 1958), 373–385. We take from this edition the English translation, based on that of Richard Norton and published in 1608. Another English translation, by V. Sanford, can be found in Smith, *Source book*, 20–34.

THE FIRST PART
Of the Definitions of the Dime.

THE FIRST DEFINITION

Dime is a kind of arithmetic, invented by the tenth progression, consisting in characters of ciphers, whereby a certain number is described and by which also all accounts which happen in human affairs are dispatched by whole numbers, without fractions or broken numbers.

Explication. Let the certain number be one thousand one hundred and eleven, described by the characters of ciphers thus 1111, in which it appears that each 1 is the 10th part of his precedent character 1; likewise in 2378 each unity of 8 is the tenth of each unity of 7, and so of all the others. But because it is convenient that the things whereof we would speak have names, and that this manner of computation is found by the consideration of such tenth or dime progression, that is that it consists therein entirely, as shall hereafter appear, we call this treatise fitly by the name of *Dime*, whereby all accounts happening in the affairs of man may be wrought and effected without fractions or broken numbers, as hereafter appears.

THE SECOND DEFINITION

Every number propounded is called COMMENCEMENT, whose sign is thus ⓪.
Explication. By example, a certain number is propounded of three hundred

---

[1] P. Luckey, *Die Rechenkunst bei Ğamšīd b. Mas'ūd al-Kāšī* (Steiner, Wiesbaden, 1951).

sixty-four: we call them the 364 *commencements*, described thus 364⓪, and so of all other like.

And each tenth part of the unity of the COMMENCEMENT we call the PRIME, whose sign is thus ①, and each tenth part of the unity of the prime we call the SECOND, whose sign is ②, and so of the other: each tenth part of the unity of the precedent sign, always in order one further.

Explication. As 3① 7② 5③ 9④, that is to say: 3 *primes*, 7 *seconds*, 5 *thirds*, 9 *fourths*, and so proceeding infinitely, but to speak of their value, you may note that according to this definition the said numbers are $\frac{3}{10}$, $\frac{7}{100}$, $\frac{5}{1000}$, $\frac{9}{10000}$, together $\frac{3759}{10000}$, and likewise 8⓪ 9① 3② 7③ are worth 8, $\frac{9}{10}$, $\frac{3}{100}$, $\frac{7}{1000}$, together $8\frac{937}{1000}$, and so of other like. Also you may understand that in this *dime* we use no fractions, and that the multitude of signs, except ⓪, never exceed 9, as for example not 7① 12②, but in their place 8① 2②, for they value as much.

The numbers of the second and third definitions beforegoing are generally called DIME NUMBERS.

The End of the Definitions

THE SECOND PART OF THE DIME.
Of the Operation or Practice.

Dime numbers being given, how to add them to find their sum.

The Explication Propounded: There are 3 orders of dime numbers given, of which the first 27⓪, 8①, 4②, 7③, the second 37⓪, 6①, 7②, 5③, the third 875⓪, 7①, 8②, 2③.

The Explication Required: We must find their total sum.

Construction. The numbers given must be placed in order as here adjoining, adding them in the vulgar manner of adding of whole numbers in this manner. The sum (by the first problem of our French Arithmetic[2]) is 941304, which are (that which the signs above the numbers do show) 941⓪ 3① 0② 4③. I say they are the sum required.

⓪	①	②	③
2	7	8	4	7
3	7	6	7	5
8 7	5	7	8	2
9 4	1	3	0	4

Demonstration. The 27⓪ 8① 4② 7③ given make by the 3rd definition before 27, $\frac{8}{10}$, $\frac{4}{100}$, $\frac{7}{1000}$, together $27\frac{847}{1000}$ and by the same reason the 37⓪ 6① 7② 5③ shall make $37\frac{675}{1000}$ and the 875⓪ 7① 8② 2③ will make $875\frac{782}{1000}$, which three numbers make by common addition of vulgar arithmetic $941\frac{304}{1000}$. But so much is the sum 941⓪ 3① 0② 4③; therefore it is the true sum to be demonstrated. Conclusion: Then dime numbers being given to be added, we have found their sum, which is the thing required.

---

[2] *L'Arithmétique de Simon Stevin de Bruges* (Leyden, 1585); see Stevin, *The principal works* (Swets-Zeitlinger, Amsterdam), vol. IIB (1958). Problem I (p. 81) is: "Given two arithmetical integer numbers. Find their sum."

Note that if in the number given there want some signs of their natural order, the place of the defectant shall be filled. As for example, let the numbers given be 8⓪ 5① 6② and 5⓪ 7②, in which the latter wanted the sign of ①; in the place thereof shall 0① be put. Take then for that latter number given 5⓪ 0① 7②, adding them in this sort.

	⓪	①	②
	8	5	6
	5	0	7
	1 3	6	3

This advertisement shall also serve in the three following propositions, wherein the order of the defailing figures must be supplied, as was done in the former example.

THE SECOND PROPOSITION: OF SUBTRACTION

A dime number being given to subtract, another less dime number given: out of the same to find their rest.

Explication Propounded: Be the numbers given 237⓪ 5① 7② 8③ and 59⓪ 7① 3② 9③.

The Explication Required: To find their rest.

Construction. The numbers given shall be placed in this sort, subtracting according to vulgar manner of subtraction of whole numbers, thus.

	⓪	①	②	③
	2 3 7	5	7	8
	5 9	7	3	9
	1 7 7	8	3	9

The rest is 177839, which values as the signs over them do denote 177⓪ 8① 3② 9③, I affirm the same to be the rest required.

Demonstration. The 237⓪ 5① 7② 8③ make (by the third definition of this Dime) $237\frac{5}{10}$, $\frac{7}{100}$, $\frac{8}{1000}$, together $237\frac{578}{1000}$, and by the same reason the 59⓪ 7① 3② 9③ value $59\frac{739}{1000}$, which subtracted from $237\frac{578}{1000}$, there rests $177\frac{839}{1000}$, but so much doth 177⓪ 8① 3② 9③ value; that is then the true rest which should be made manifest.

Conclusion. A dime being given, to subtract it out of another dime number, and to know the rest, which we have found.

THE THIRD PROPOSITION: OF MULTIPLICATION

A dime number being given to be multiplied, and a multiplicator given: to find their product.

The Explication Propounded: Be the number to be multiplied 32⓪ 5① 7②, and the multiplicator 89⓪ 4① 6②.

The Explication Required: To find the product.

Construction. The given numbers are to be placed as here is shown, multiplying according to the vulgar manner of multiplcation by whole numbers, in this manner, giving the product 29137122. Now to know how much they value, join the two last signs together as the one ② and the other ② also, which together make ④, and say that the last sign of the product shall be ④, which being known, all the rest are also known by their continued order. So that the product required is 2913⓪ 7① 1② 2③ 2④.

				⓪	①	②
				3	2	5 7
				8	9	4 6
			1	9	5	4 2
		1	3	0	2	8
	2	9	3	1	3	
	2 6	0	5	6		
	2 9	1	3	7	1	2 2
			⓪	①	②	③ ④

Demonstration. The number given to be multiplied, 32⓪ 5① 7② (as appears by the third definition of this Dime), $32\frac{5}{10}, \frac{7}{100}$, together $32\frac{57}{100}$; and by the same reason the multiplicator 89⓪ 4① 6② value $89\frac{46}{100}$ by the same, the said $32\frac{57}{100}$ multiplied gives the product $2913\frac{7122}{10000}$. But it also values 2913⓪ 7① 1② 2③ 2④.

It is then the true product, which we were to demonstrate. But to show why ② multiplied by ② gives the product ④, which is the sum of their numbers, also why ④ by ⑤ produces ⑨, and why ⓪ by ③ produces ③, etc., let us take $\frac{2}{10}$ and $\frac{3}{100}$, which (by the third definition of this Dime) are 2① 3②, their product is $\frac{6}{1000}$, which value by the said third definition 6③; multiplying then ① by ②, the product is ③, namely a sign compounded of the sum of the numbers of the signs given.

Conclusion. A dime number to multiply and to be multiplied being given, we have found the product, as we ought.

Note: If the latter sign of the number to be multiplied be unequal to the latter sign of the multiplicator, as, for example, the one 3④ 7⑤ 8⑥, the other 5① 4②, they shall be handled as aforesaid, and the disposition thereof shall be thus.

```
                        ④  ⑤  ⑥
                        3   7   8
                            5   4  ②
                    ─────────────────
                        1   5   1   2
                1   8   9   0
            ─────────────────────────
                2   0   4   1   2
                ④  ⑤  ⑥  ⑦  ⑧
```

THE FOURTH PROPOSITION: OF DIVISION

A dime number for the dividend and divisor being given: to find the quotient.

Explication Proposed: Let the number for the dividend be 3⓪ 4① 4② 3③ 5④ 2⑤ and the divisor 9① 6②.

Explication Required: To find their quotient.

Construction. The numbers given divided (omitting the signs) according to the vulgar manner of dividing of whole numbers, gives the quotient 3587; now to know what they value, the latter sign of the divisor ② must be subtracted from the latter sign of the dividend, which is ⑤, rests ③ for the latter are also manifest by their continued order, thus 3⓪ 5① 8② 7③ are the quotient required.

```
        I
        18
        5164
        7617      ⓪ ① ② ③
        344352   (3  5  8  7
        96666
        999
```

Demonstration. The number dividend given 3⓪ 4① 4② 3③ 5④ 2⑤ makes (by the third definition of this Dime) $3, \frac{4}{10}, \frac{4}{100}, \frac{3}{1000}, \frac{5}{10000}, \frac{2}{100000}$, together $3\frac{44352}{100000}$, and by the same reason the divisor 9① 6② values $\frac{96}{100}$, by which $3\frac{44352}{100000}$ being divided, gives the quotient $3\frac{587}{1000}$; but the said quotient values 3⓪ 5① 8② 7③, therefore it is the true quotient to be demonstrated.

Conclusion. A dime number being given for the dividend and divisor, we have found the quotient required.

Note: If the divisor's signs be higher than the signs of the dividend, there may be as many such ciphers 0 joined to the dividend as you will, or as many as shall be necessary: as for example, 7② are to be divided by 4⑤, I place after the 7 certain ⓪, thus 7000, dividing them as afore said, and in this sort it gives for the quotient 1750⓪.

```
        3   2
        7   0   0   0   (1  7  5  0  ⓪
        4   4   4   4
```

It happens also sometimes that the quotient cannot be expressed by whole numbers, as 4① divided by 3② in this sort, whereby appears that there will infinitely come 3's, and in such a case you may come so near as the thing requires, omitting the remainder.

$$
\begin{array}{llll}
\mathit{1} & \mathit{1} & \mathit{1} & (1 \\
\mathit{4} & \mathit{0} & \mathit{0} & \mathit{0} \quad 0 \quad 0 \quad 0 \quad (1 \\
\mathit{3} & \mathit{3} & \mathit{3} & \mathit{3}
\end{array}
\qquad
\begin{array}{lll}
① & ① & ② \\
3 & 3 & 3
\end{array}
$$

It is true, that 13⓪ 3① 3⅓② , or 13⓪ 3① 3② 3⅓③ etc. shall be the perfect quotient required. But our invention in this Dime is to work all by whole numbers. For seeing that in any affairs men reckon not of the thousandth part of a mite, es, grain, etc., as the like is also used of the principal geometers and astronomers in computations of great consequence, as Ptolemy and Johannes Montaregio,[3] have not described their tables of arcs, chords, or sines in extreme perfection (as possibly they might have done by multinomial numbers), because that imperfection (considering the scope and end of those tables) is more convenient than such perfection.

Note 2. The extraction of all kinds of roots may also be made by these dime numbers; as, for example, to extract the square root of 5② 2③ 9④ , which is performed in the vulgar manner of extraction in this sort, and the root shall be 2① 3② , for the moiety or half of the latter sign of the numbers given is always the latter sign of the root; wherefore, if the latter sign given were of a number impair, the sign of the next following shall be added, and then it shall be a number pair; and then extract the root as before. Likewise in the extraction of the cubic root, the third part of the latter sign given shall be always the sign of the root; and so of all other kinds of roots.

$$
\begin{array}{l}
\mathit{1} \\
\mathit{5} \quad \mathit{2} \quad \mathit{9} \\
\overline{\mathit{2} \qquad \mathit{3}} \\
\overline{\mathit{4}}
\end{array}
$$

THE END OF THE DIME

After this follows an Appendix in which different applications of the decimal method of counting to surveying, cloth measuring, wine gauging, and other trades and professions are described. The decimal division of weights and measures was not systematically introduced until the French Revolution. As to its introduction (and nonintroduction) into the United States, see C. D. Hellman, "Jefferson's efforts towards the decimalization of U.S. weights and measures," *Isis 16* (1931), 266–314.

4   NAPIER. LOGARITHMS

John Napier (or Neper, 1550–1617), a Scottish baron, computed a table of what he called logarithms, using the correspondence between an arithmetic and a geometric progression. He published his invention first in the *Mirifici logarithmorum canonis descriptio* (Edinburgh,

---

[3] Johannes Montaregio (1436–1476) is best known under his latinized name, Johannes Regiomontanus. This craftsman, humanist, astronomer, and mathematician of Nuremberg influenced the development of trigonometry by means of his widely used book *De triangulis omnimodis* (written *c.* 1464, printed in Nuremberg in 1533). The sines, for Regiomontanus as well as for Stevin, were half chords; see our note to the following text on Napier and Selection III.2.

1614), followed by the *Mirifici logarithmorum canonis constructio* (published posthumously at Edinburgh in 1619). Of the second work we give an extract here, from the translation by W. R. Macdonald, *The construction of the wonderful canon of logarithms* (Blackwood, Edinburgh, London, 1889). Napier's logarithms were not yet those we use at present; for instance, his logarithm of 10^7 is zero. The first table based on the decimal system such that $\log 1 = 0$, $\log 10 = 1$ was published by Napier's admirer, the London professor Henry Briggs (1561–1630), in his *Arithmetica logarithmica* (London, 1624). With the work of Stevin, Napier, and Briggs the application of the decimal system to computation was in principle completed. The relation between exponentials and logarithms and the systematic use of natural logarithms had to wait until later. See our text on this subject by Euler.

The idea of the logarithms also occurred to Napier's contemporary, the Swiss instrument maker Jost Bürgi (1552–1632), whose *Progress Tabulen* (Prague, 1620) contain a certain type of antilogarithms.

Curiously enough, though the word "logarithm" occurs in the title of the *Construction*, the text uses *numerus artificialis* instead, which we translate by *logarithm*.

The *Constructio* has an appendix on spherical trigonometry, where Napier uses the rules known as *rules of Napier* on spherical triangles (expressed in words).

The *Descriptio* opens with the following verse:

> Hic liber est minimus, si spectes verba, sed usum
> Si spectes, Lector, maximus hic liber est
> Disce, scies parvo tantum debere libello
> Te, quantum magnis mille voluminibus,

which we freely translate as follows:

> The use of this book is quite large, my dear friend,
> No matter how modest it looks,
> You study it carefully and find that it gives
> As much as a thousand big books.

The author signs his name as Andres Junius, Phil. prof. in Acad. Edinb. On Napier and his logarithms, see *Napier tercentenary memorial volume*, ed. C. G. Knott (Royal Society of Edinburgh, London, 1915), and J. F. Scott, *A history of mathematics* (Taylor and Francis, London, 1958), chap. IX.

Here follows part of the *Construction of the wonderful canon of logarithms*.

---

1. A logarithmic table [*tabula artificialis*] is a small table by the use of which we can obtain a knowledge of all geometrical dimensions and motions in space, by a very easy calculation... It is picked out from numbers progressing in continuous proportion.

2. Of continuous progressions, an arithmetical is one which proceeds by equal intervals; a geometrical, one which advances by unequal and proportionally increasing or decreasing intervals...

3. In these progressions we require accuracy and ease in working. Accuracy is obtained by taking large numbers for a basis; but large numbers are most easily made from small by adding cyphers.[1]

---

[1] Cyphers = zeros (see Selection I.1, Leonardo of Pisa).

Thus instead of 1000000, which the less experienced make the greatest sine,[2] the more learned put 10000000, whereby the difference of all sines is better expressed. Wherefore also we use the same for radius and for the greatest of our geometrical proportionals.

4. In computing tables, these large numbers may again be made still larger by placing a period after the number and adding cyphers...

5. In numbers distinguished thus by a period in their midst, whatever is written after the period is a fraction [*quicquid post periodam notatur fractio*], the denominator of which is unity with as many cyphers after it as there are figures after the period.[3]

Thus 10000000.04 is the same as $10000000\frac{4}{100}$; also 25.803 is the same as $25\frac{803}{1000}$; also 9999998.0005021 is the same as $9999998\frac{5021}{10000000}$, and so of others.

6. When the tables are computed, the fractions following the period may then be rejected without any sensible error. For in our large numbers, an error which does not exceed unity is insensible and as if it were none...

Then follow in Arts. 7–15 some rules for accurate counting with large numbers.

---

16. Hence, if from the radius with seven cyphers added you subtract its 10000000th part, and from the number thence arising its 10000000th part, and so on, a hundred numbers may very easily be continued geometrically in the proportion subsisting between the radius and the sine less than it by unity, namely between 10000000 and 9999999; and this series of proportionals we name the First table.

[2] Sin 90° = 100.0000, hence the radius R of the circle is 10^6. The sine of an angle was always defined as half the chord belonging to the double angle, hence sin $\alpha = CB = \frac{1}{2}$ chord $2\alpha = \frac{1}{2}CD$ [Fig. 4]. The numerical values of the sines therefore depended on the choice of R. Euler introduced dimensionless sines and tangents by consistently writing $R = 1$ (1748, see our extract of the *Introductio*, Selection V.15).

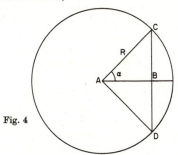

Fig. 4

[3] The clumsy notation for decimal fractions of Stevin is here replaced by the method of the decimal point. Napier's authority made this method generally accepted.

First table.

10000000.0000000	
1.0000000	
9999999.0000000	
.9999999	
9999998.0000001	
.9999998	
9999997.0000003	
to be continued up to	
9999900.0004950	

Thus from radius, with seven cyphers added for greater accuracy, namely, 10000000.0000000, subtract 1.0000000 you get 9999999.0000000; from this subtract .9999999, you get 9999998.0000001; and proceed in this way... until you create a hundred proportionals, the last of which, if you have computed rightly, will be 9999900.0004950.

17. The Second table proceeds from radius with six cyphers added, through fifty other numbers decreasing proportionally in the proportion which is easiest, and as near as possible to that subsisting between the first and last numbers of the First table.

Second table.

10000000.000000	
100.000000	
9999900.000000	
99.999000	
9999800.001000	
to be continued up to	
9995001.222927	

Thus the first and last numbers of the First table are 10000000.0000000 and 9999900.0004950, in which proportion it is difficult to form fifty proportional numbers. A near and at the same time an easy proposition is 100000 to 99999, which may be continued with sufficient exactness by adding six cyphers to radius and continually subtracting from each number its own 100000th part... and this table contains, besides radius which is the first, fifty other proportional numbers, the last of which, if you have not erred, you will find to be 9995001.222927.[4]

Article 18 has a Third table of 69 columns, from 10^{12} down by 2000th parts to 9900473.57808.

---

19. The first numbers of all the columns must proceed from radius with four cyphers added, in the proportion easiest and nearest to that subsisting between the first and the last numbers of the first column.

As the first and the last numbers of the first column are 10000000.0000 and 9900473.5780, the easiest proportion very near to this is 100 to 99. Accordingly sixty-eight numbers are to be continued from radius in the ratio of 100 to 99 by subtracting from each one of them its hundredth part.

20. In the same proportion a progression is to be made from the second number of the first column through the second numbers in all the columns, and from the third through the third, and from the fourth through the fourth, and from the others respectively through the others.

[4] This should be 9995001.224804.

Thus from any number in one column, by subtracting its hundredth part, the number of the same rank in the following column is made, and the numbers should be placed in order as follows.

Here follows a table of "Proportionals of the Third Table," with 69 columns, the last number in the sixty-ninth column being 4998609.4034, roughly half the original number 10000000.0000.

---

21. Thus, in the Third table, between radius and half radius, you have sixty-eight numbers interpolated, in the proportion of 100 to 99, and between each two of these you have twenty numbers interpolated in the proportion of 10000 to 9995; and again, in the Second table, between the first two of these namely between 10000000 and 9995000, you have fifty numbers interpolated in the proportion of 100000 to 99999; and finally, in the First table, between the latter, you have a hundred numbers interpolated in the proportion of radius or 10000000 to 9999999; and since the difference of these is never more than unity, there is no need to divide it more minutely by interpolating means, whence these three tables, after they have been completed, will suffice for computing a Logarithmic table.

Hitherto we have explained how we may most easily place in tables sines or natural numbers progressing in geometrical proportion.

22. It remains, in the Third table at least, to place beside the sines or natural numbers decreasing geometrically their logarithms or artificial numbers increasing arithmetically.

Articles 23 and 24 represent arithmetic increase and geometric decrease by points on a line.

---

25. Whence a geometrically moving point approaching a fixed one has its velocities proportionate to its distances from the fixed one.

Thus referring to the preceding figure [Fig. 1], I say that when the geometrically moving point G is at T, its velocity is as the distance TS, and when

Fig. 1

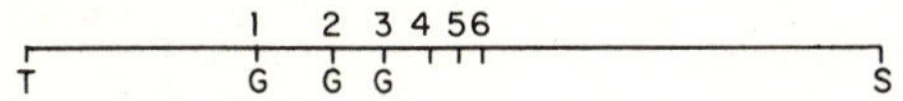

G is at 1 its velocity is as $1S$, and when at 2 its velocity is as $2S$, and so of the others. Hence, whatever be the proportion of the distances TS, $1S$, $2S$, $3S$, $4S$, etc., to each other, that of the velocities of G at the points T, 1, 2, 3, 4, etc., to one another, will be the same.

For we observe that a moving point is declared more or less swift, according as it is seen to be borne over a greater or less space in equal times. Hence the ratio of the spaces traversed is necessarily the same as that of the velocities. But the ratio of the spaces traversed in equal times, $T1$, 12, 23, 34, 45, etc., is that of the distances TS, $1S$, $2S$, $3S$, $4S$, etc. Hence it follows that the ratio to one another of the distances of G from S, namely TS, $1S$, $2S$, $3S$, $4S$, etc., is the same as that of the velocities of G at the points T, 1, 2, 3, 4, etc., respectively.

26. The logarithm of a given sine is that number which has increased arithmetically with the same velocity throughout as that with which radius began to decrease geometrically, and in the same time as radius has decreased to the given sine.

Let the line TS [Fig. 2] be the radius, and dS a given sine in the same line; let g move geometrically from T to d in certain determinate moments of time. Again, let bi be another line, infinite towards i, along which, from b, let a move arithmetically with the same velocity as g had at first when at T; and from the fixed point b in the direction of i let a advance in just the same moments of time up to the point c. The number measuring the line bc is called the logarithm of the given sine dS.[5]

Fig. 2

27. Whence nothing is the logarithm of radius [*Unde sinus totius nihil est pro artificiali*] . . .

28. Whence also it follows that the logarithm of any given sine is greater than the difference between radius and the given sine, and less than the difference between radius and the quantity which exceeds it in the ratio of radius to the given sine. And these differences are therefore called the limits of the logarithm.

Thus, the preceding figure being repeated [Fig. 3], and ST being produced beyond T to o, so that oS is to TS as TS to dS, I say that bc, the logarithm of the sine dS, is greater than Td and less than oT. For in the same time that g is borne from o to T, g is borne from T to d, because (by 24) oT is such a part of oS as Td is of TS, and in the same time (by the definition of a logarithm) is a borne from b to c; so that oT, Td, and bc are distances traversed in equal times. But since g when moving between T and o is swifter than at T, and

---

[5] In the language of the calculus: let $TS = a$ ($= 10^7$), $dS = y$; then the initial velocity ($t = c$) at g is a (see Art. 25), hence the velocity of g at d is $(d/dt)(a - y) = -dy/dt = y$, hence $y = a e^{-t}$. When $bc = x$, then $x = at = $ Nap log y. Hence Nap log $y = a \ln a/y$, so that (by Art. 27) for $y = a$, Nap log $a = 0$, where $\ln = \log_e$, the natural logarithm. The familiar rules for logarithmic computation do not apply:

$$\text{Nap log } xy = a(\ln a - \ln x - \ln y).$$

We should not be confused by the terms "radius" and "sine"; what is meant is a line segment TS and a section $dS \leqq TS$. When $a = 1$ the Nap log and the ln differ only in sign; this may have caused the confusion in some textbooks, which insist on calling the natural logarithms Napierian or Neperian logarithms.

between T and d slower, but at T is equally swift with a (by 26); it follows that oT the distance traversed by g moving swiftly is greater, and Td the distance traversed by g moving slowly is less, than bc the distance traversed by the point a with its medium motion, in just the same moments of time; the latter is, consequently, a certain mean between the two former. Therefore oT is called the greater limit, and Td the less limit of the logarithm which bc represents.

Fig. 3

29. Therefore to find the limits of the logarithm of a given sine.

By the preceding it is proved that the given sine being subtracted from radius the less limit remains, and that radius being multiplied into the less limit and the product divided by the given sine, the greater limit is produced, as in the following example.

30. Whence the first proportional of the First table, which is 9999999, has its logarithm between the limits 1.0000001 and 1.0000000...

31. The limits themselves differing insensibly, they or anything between them may be taken as the true logarithm...

32. There being any number of sines decreasing from radius in geometrical proportion, of one of which the logarithm or its limits is given, to find those of the others.

This necessarily follows from the definitions of arithmetical increase, of geometrical decrease, and of a logarithm... So that, if the first logarithm corresponding to the first sine after radius be given, the second logarithm will be double of it, the third triple, and so of the others; until the logarithms of all the sines be known...

33. Hence the logarithms of all the proportional sines of the First table may be included between near limits, and consequently given with sufficient exactness...

34. The difference of the logarithms of radius and a given sine is the logarithm of the given sine itself...

35. The difference of the logarithms of two sines must be added to the logarithm of the greater that you may have the logarithm of the less, and subtracted from the logarithm of the less that you may have the logarithm of the greater...

36. The logarithms of similarly proportioned sines are equidifferent.

This necessarily follows from the definitions of a logarithm and of the two motions... Also there is the same ratio of equality between the differences of the respective limits of the logarithms, namely as the differences of the less among themselves, so also of the greater among themselves, of which logarithms the sines are similarly proportioned.

37. Of three sines continued in geometrical proportion, as the square of the mean equals the product of the extremes, so of their logarithms the double of

the mean equals the sum of the extremes. Whence any two of these logarithms being given, the third becomes known . . .

38. Of four geometrical proportionals, as the product of the means is equal to the product of the extremes; so of their logarithms, the sum of the means is equal to the sum of the extremes. Whence any three of these logarithms being given, the fourth becomes known . . . [6]

39. The difference of the logarithms of two sines lies between two limits; the greater limit being to the radius as the difference of the sines to the less sine, and the less limit being to radius as the difference of the sines to the greater sine . . . [7]

Articles 40–46 show how to find logarithms.

---

47. In the Third table, beside the natural numbers, are to be written their logarithms; so that the Third table, which after this we shall always call the Radical table, may be made complete and perfect . . .

The Radical Table

First Column			Second Column			69th Column	
Natural Numbers	Logarithms		Natural Numbers	Logarithms		Natural Numbers	Logarithms
10000000.0000	.0		9900000.0000	100503.3		5048858.8900	6834225.8
9995000.0000	5001.2		9895050.0000	105504.6	. . .	5046334.4605	6839227.1
9990002.50000	10002.5		9890102.4750	110505.8		5043011.2932	6844228.3
:	:		:	:	. . .	:	:
9900473.5700	100025.0		9801468.8423	200528.2		4998609.4034	6934250.8

48. The Radical table being now completed, we take the numbers for the logarithmic table from it alone.

For as the first two tables were of service in the formation of the third, so this third Radical table serves for the construction of the principal Logarithmic table, with great ease and no sensible error.

49. To find most easily the logarithms of sines greater than 9996700.

This is done simply by the subtraction of the given sine from radius. For (by 29) the logarithm of the sine 9996700 lies between the limits 3300 and 3301; and these limits, since they differ from each other by unity only, cannot

[6] The modern theorem for the logarithm of a product does not hold, since the logarithm of unity is not zero. Hence Arts. 37 and 38, to express special cases.

[7] This is proved by the principle of proportion and of Article 36. This rule is used first in Arts. 40 and 41 as an illustration to find the logarithm of 9999975.5 from that of the nearest sine in the First table, 9999975.0000300, noting that the limits of the logarithms of the latter number are 25.0000025 and 25.000000, that the difference of the logarithms of the two numbers by the rule just given is .4999712, and that the limits for the logarithm of 9999975.5 are therefore 24.5000313 and 24.5000288, whence Napier lists the logarithm as 24.5000300.

Articles 41 to 45 illustrate the fact that one may now calculate the logarithms of all the "proportionals" in the First, Second, and Third tables, as well as of the sines or natural numbers not proportionals in these tables but near or between them.

differ from their true logarithm by any sensible error, that is to say, by an error greater than unity. Whence 3300, the less limit, which we obtain simply by subtraction, may be taken for the true logarithm. The method is necessarily the same for all sines greater than this.

50. To find the logarithms of all sines embraced within the limits of the Radical table.

Multiply the difference of the given sine and table sine nearest it by radius. Divide the product by the easiest divisor, which may be either the given sine or the table sine nearest it, or a sine between both, however placed. By 39 there will be produced either the greater or less limit of the difference of the logarithms, or else something intermediate, no one of which will differ by a sensible error from the true difference of the logarithms on account of the nearness of the numbers in the table. Wherefore (by 35), add the result, whatever it may be, to the logarithm of the table sine, if the given sine be less than the table sine; if not, subtract the result from the logarithm of the table sine, and there will be produced the required logarithm of the given sine.

Two examples are given. In the first the given sine is 7489557, the table sine of which nearest to it is 7490786.6119. The computation gives 2890752 for the logarithm.

---

51. All sines in the proportion of two to one have 6931469.22 for the difference of their logarithms [because this number is the logarithm of sine 5000000].

52. All sines in the proportion of ten to one have 23025842.34 for the difference of their logarithms.

Article 53 contains a short table of given proportions of sines and corresponding differences of logarithms; Art. 54 deals with the logarithms of all sines outside the limits of the Radical table.

---

55. As half radius is to the sine of half a given arc, so is the sine of the complement of the half arc to the sine of the whole arc ...[8]

56. Double the logarithm of an arc of 45 degrees is the logarithm of half radius ...

[8] Only here does Napier begin to introduce angles into the construction of his tables. Napier proves Arts. 55–57 by geometric principles and the preceding theorems concerning logarithms. He then often speaks of the logarithms of the arcs, meaning logarithms of the corresponding sines.

57. The sum of the logarithms of half radius and any given arc is equal to the sum of the logarithms of half the arc and the complement of the half arc. Whence the logarithm of the half arc may be found if the logarithms of the other three be given...

Article 58 deals with the logarithms of all arcs not less than 45 degrees.

---

59. To form a logarithmic table.

Here follows a description of the construction of a table of 45 pages, each page devoted to one degree divided into minutes.

Napier's table is constructed in quite the same form as that used at present, except that the second (sixth) column gives sines for the number of degrees indicated at the top (bottom) and of minutes in the first (seventh) column, the third (fifth) column gives the corresponding logarithm, and the fourth column gives the *differentiae* between the logarithms in the third and fifth columns, these being therefore essentially logarithmic tangents or cotangents. A few entries follow.

0° min	sines	logarithm	+/− differentiae	logarithm	sines	
0	0	infinitum	infinitum	0	10000000	69
1	2909	81425681	81425680	1	10000000	59
2	5818	74494213	74494211	2	9999998	58
3	8727	70439560	70439560	4	9999998	57

. . . .

30° min	sines	logarithm	+/− differentiae	logarithm	sines	
0	5000000	6931469	5493059	1483410	8660254	60
1	5002519	6926432	5486342	1440090	8658799	59
2	5005038	6921399	5479628	1441771	8657344	58

. . . .

44° min	sines	logarithm		logarithm	sines	
59	7069011	3468645	5818	3462827	7071068	1
60	7071068	3465735	0	3465735	7071068	0

min 45°

Hence log sin 3' = log 8727 = 70439560,

log sin 30° 1' = log 5002519 = 6926432,

log sin 45° = log 7071068 = 3465735; (half of log sin 30°, Art. 56),

also log sin 90° = log 10000000 = 0.

5 PASCAL. THE PASCAL TRIANGLE

The so-called Pascal triangle appears in a treatise by Blaise Pascal (1623–1662), published posthumously under the title *Traité du triangle arithmétique, avec quelques autres petits traités sur la même manière* (Paris, 1665). This treatise is important not only because of its careful examination of the properties of the binomial coefficients, but also because of their application to problems in games of chance. At one place Pascal expresses with clarity the principle of complete induction.

The Pascal triangle appears for the first time (so far as we know at present) in a book of 1261 written by Yang Hui, one of the mathematicians of the Sung dynasty in China.[1] The properties of binomial coefficients were discussed by the Persian mathematician Jamshid Al-Kāshī in his *Key to arithmetic* of c. 1425.[2] Both in China and in Persia the knowledge of these properties may be much older. This knowledge was shared by some of the Renaissance mathematicians, and we see Pascal's triangle on the title page of Peter Apian's German arithmetic of 1527. After this we find the triangle and the properties of binomial coefficients in several other authors.[3]

Pascal wrote his treatise probably by the end of 1654. It can be found in the *Oeuvres*, ed. L. Brunschvicg and P. Boutroux, III (Hachette, Paris, 1909), 456 seq., and in other editions of Pascal's work. A paraphrase of certain theorems can be found in H. Meschkowski, *Ways of thought of great mathematicians* (Holden-Day, San Francisco, 1964), 36–43.

TREATISE ON THE ARITHMETIC TRIANGLE

I designate as the arithmetic triangle a figure of which the construction is as follows [Fig. 1]. Through an arbitrary point G I draw 2 lines perpendicular to each other, GV and $G\zeta$, on each of which I take as many equal and continuous parts as I like, beginning at G, which I call 1, 2, 3, 4, etc., and these numbers are the indices [*exposans*] of the divisions of the lines.

Then I join the points of the first division, which are on each of the two lines, by another line that forms a triangle of which this line is the *base*.

I also join the two points of the second division by another line that forms a second triangle of which this line is the *base*.

[1] J. Needham, *Science and civilisation in China*, III (Cambridge University Press, New York, 1959), 135.

[2] Russian translation by B. A. Rozenfel'd (Gos. Izdat, Moscow, 1956); see also Selection I.3, footnote 1.

[3] Smith, *History of mathematics*, II, 508–512. See also our Selection II.9 (Girard).

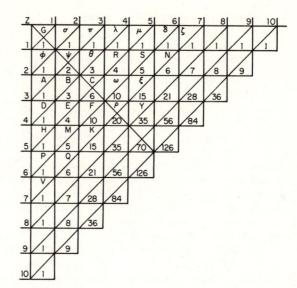

Fig. 1

And joining in this way all the division points which have the same indices I form with them as many *triangles* and *bases*.

I draw through every one of the division points lines parallel to the sides, and these by their intersections form small squares which I call cells [*cellules*].

And the cells that are between two parallels that run from left to right are called *cells of the same parallel rank*, such as the cells G, σ, π, etc., or φ, ψ, θ, etc.

And those that are between two lines that run from the top downward are called *cells of the same perpendicular rank*, such as the cells G, φ, A, D, etc. and these: σ, ψ, B, etc.

And those that the same base traverses diagonally are called *cells of the same base*, such as the following: D, B, θ, λ, and these: A, ψ, π.

The cells of the same base that are equally distant from their extremities are called *reciprocal*, such as these: E, R and B, θ, because the index of the parallel rank of the one is the same as the index of the perpendicular rank of the other, as appears in the example, where E is in the second perpendicular and in the fourth parallel rank, and R is in the second parallel and in the fourth perpendicular rank, reciprocally. It is easy enough to show that those which have their indices reciprocally equal are in the same base and equally distant from its extremities.

It is also quite easy to demonstrate that the index of the perpendicular rank of any cell whatsoever, added to the index of its parallel rank, exceeds the index of its base by unity.

For example, the cell F is in the third perpendicular rank and in the fourth parallel one, and in the sixth base, and its two indices of the ranks $3 + 4$ exceed the index 6 of the base by unity, which results from the fact that the two sides of the triangle are divided into an equal number of parts, but that is rather understood than demonstrated.

The following remark is of the same nature: that every base contains one cell more than the preceding one, and every one contains as many cells as its index has units; the second base $\varphi\sigma$, for instance, has two cells, the third $A\psi\pi$ has three of them, etc.

We now place numbers in each cell and this is done in the following way: the number of the first cell which is in the right angle is arbitrary, but once it has been placed all the other numbers are determined, and for this reason it is called the *generator* of the triangle. And every one of the other numbers is specified by this sole rule:

The number of each cell is equal to that of the cell preceding it in its perpendicular rank plus that of the cell which precedes it in its parallel rank. For instance, the cell F, that is, the number of the cell F, is equal to cell C plus cell E, and so the others.

From this many consequences can be drawn. Here are the most important ones, where I consider the triangles whose generator is unity, but what can be said about them will also apply to the others.

FIRST CONSEQUENCE

In every arithmetic triangle all the cells of the first parallel rank and of the first perpendicular rank are equal to the generator.

Indeed, by the construction of the triangle, every cell is equal to the cell which precedes it in its perpendicular rank plus the cell that precedes it in its parallel rank. Now, the cells of the first parallel rank have no cells which precede them in their perpendicular ranks, nor have those of the first perpendicular rank any in their parallel ranks: hence they are all equal to each other and to the generating first number.

And so φ is equal to G + zero, that is, φ is equal to G.

And so A is equal to φ + zero, that is, φ.

And so σ is equal to G + zero, and π equal to σ + zero.

And so the others.

Using a more modern notation, in which we call P_l^k the cell of parallel rank l and vertical rank k, so that

$$P_l^k = \frac{(k + l - 2)!}{(k - 1)!(l - 1)!},$$

we can write the next "consequences" as follows:

2. $$P_l^k = \sum_{i=1}^{k} P_{l-1}^i; \quad \text{e.g., } \omega = R + \theta + \psi + \varphi;$$

3. $$P_l^k = \sum_{i=1}^{k} P_i^{k-1}; \quad \text{e.g., } C = B + \psi + \sigma;$$

4. $\quad P_l^k - 1 = \sum_{i=1}^{k-1} \sum_{j=1}^{l-1} P_j^i;$ e.g., $\xi - g = R + \theta + \psi + \varphi + \lambda + \pi + \sigma + G,$

where $g = 1$, the generator;

5. $\qquad\qquad P_l^k = P_k^l;$ e.g., $\varphi = \sigma = G, \pi = A = G, D = \lambda = G.$

6. $\qquad$ All $P_l^k = P_k^l$, k fixed; e.g., $\sigma\psi BEM\varphi$ is equal to $\varphi\psi\theta RSN;$

7. $\sum_{l,k=1,\ldots,n} P_l^k = 2 \sum_{i,j=1,\ldots,n-1} P_j^i,$ $\quad k + l = $ fixed number $= a$, $\quad i + j = a - 1;$

e.g., $D + \lambda + B + \theta = 2A + 2\psi + 2\pi;$

8. $\qquad\qquad \sum_{l,k=1,\ldots,n} P_l^k = 2^{n-2}, \quad k + l = n;$

9. $\qquad\qquad 1 + 2 + \cdots + 2^n = 2^{n+1} - 1;$

10. $\sum_{l=n}^{p} P_l^k = 2 \sum_{i=n-1}^{p-1} P_j^i + P_{n-1}^p$ [e.g., $P_4^1 + P_3^2 + P_2^3 = 2(P_3^1 + P_2^2) + P_4^3$],

$k + l = n$, $\quad i + j = n - 1$, $\quad p = n - 2;$ e.g., $D + B + \theta = 2A + 2\varphi + \pi;$

11. $\qquad\qquad P_l^l = 2P_l^{l-1} = 2P_{l-1}^l;$ e.g., $C = \theta + B = 2B.$

TWELFTH CONSEQUENCE

In every arithmetic triangle, if two cells are contiguous in the same base, the upper is to the lower as the number of cells from the upper to the top of the base is to the number of those from the lower to the bottom, inclusive.

Let the two contiguous cells, arbitrarily chosen on the same base, be E, C; then I say that

E	is to	C	as	2	is to	3
lower one		upper one		because there are two cells between E and the first, namely E, H;		because there are three cells between C and the top, namely C, R, μ.

Although this proposition has an infinite number of cases I shall give for it a very short demonstration by supposing two lemmas:

The first one, evident in itself, is that this proportion occurs in the second base; because it is clear enough that φ is to σ as 1 is to 1.

The second one is that if this proposition is true in an arbitrary base, it will necessarily be true in the next base. From which it is clear that it will necessarily be true in all bases, because it is true in the second base because of the first lemma; hence by means of the second lemma it is true in the third base, hence in the fourth base, and so on to infinity.[4]

It is therefore necessary to demonstrate only the second lemma, and this can be done in the following way. Let this proportion be true in an arbitrary base, as in the fourth one D, that is, if D is to B as 1 is to 3, and B to θ as 2 to 2, and θ to λ as 3 to 1, etc., then I say that the same proportion will be true in the next base, $H\mu$, and that, for example, E is to C as 2 is to 3.

Indeed, D is to B as 1 is to 3, by hypothesis.

Hence $\underbrace{D + B}_{E}$ is to B as $\underbrace{1 + 3}_{4}$ is to 3.

In the same way: B is to θ as 2 is to 2, by hypothesis.

Hence $\underbrace{B + \theta}_{C}$ is to B as $\underbrace{2 + 2}_{4}$ is to 2.

But B is to E as 3 is to 4.

Hence, by the double proportion,[5] C is to E as 3 is to 2.   Q.E.D.

The proof can be given in the same way in all the other cases, since this proof is founded only on the fact that this proportion is true in the preceding base, and that every cell is equal to its preceding one plus the one above it, which is true in all cases.[6]

There follow more "consequences," numbered 13–19.[7] The article ends with a "Problem":

Given the indices of the perpendicular and of the parallel rank of a cell, to find the number of the cell, without using the arithmetic triangle.

[4] This seems to be the first satisfactory statement of the principle of complete induction. See H. Freudenthal, "Zur Geschichte der vollständigen Induktion," *Archives Internationales des Sciences 22* (1953), 17–37.

[5] The text has "proportion troublée," probably a misprint for "proportion doublée."

[6] The meaning of this is as follows. Given

$$P_k^l : P_{k-1}^{l+1} = \frac{l}{k-1} \quad \text{(in base } k + l - 1\text{)}.$$

But

$$P_k^l + P_{k+1}^{l+1} = P_{k-1}^{l+1} \quad \text{(rule of formation of the triangle)};$$

hence

$$P_k^{l+1} : P_{k-1}^{l+1} = \frac{l + k - 1}{k - 1},$$

$$P_{k-1}^{l+1} : P_{k-2}^{l+2} = \frac{l + 1}{k - 2},$$

$$P_{k-1}^{l+2} : P_{k-1}^{l+1} = \frac{l + k - 1}{l + 1};$$

hence

$$P_k^{l+1} : P_{k-1}^{l+2} = \frac{l + 1}{k - 1} \quad \text{(in base } k + l\text{)}.$$

[7] For example, consequence 17 states that

$$\sum_{i=1}^{k} P_i^l : \sum_{j=1}^{l} P_k^j = k : l, \quad \text{e.g., } (B + \psi + \sigma) : (B + A) = 3 : 2.$$

These consequences can all be found in the translation of Pascal's paper in Smith, *Source book*, pp. 74–75.

For example, let it be proposed to find the number of the cell ξ of the fifth perpendicular rank and of the third parallel rank.

Having taken all the numbers that precede the index of the perpendicular rank 5, that is, 1, 2, 3, 4, take as many natural numbers beginning with the index of the parallel rank 3, that is, 3, 4, 5, 6.

Now multiply the first numbers into each other, and let the product be 24. Multiply the other numbers into each other, and let the product be 360, which divided by the other product 24, gives 15 as the quotient. This quotient is the desired number.

Indeed, ξ is to the first number of its base V in composed ratio of all the ratios of the cells among themselves, that is,

ξ is to V in composed ratio of $\underbrace{\xi \text{ to } \rho}_{3 \text{ to } 4} + \underbrace{\rho \text{ to } K}_{4 \text{ to } 3} + \underbrace{K \text{ to } Q}_{5 \text{ to } 2} + \underbrace{Q \text{ to } V}_{6 \text{ to } 1}$,

or by the twelfth consequence:

ξ is to V as 3 into 4 into 4 into 5 into 6 into 4 into 3 into 2 into 1,

But V is unity; hence ξ is the quotient of the division of the product of 3 into 4 into 5 into 6 by the product of 4 into 3 into 2 into 1.[8]

Note. If the generator were not unity we should have to multiply the quotient by the generator.

This paper is followed by several others, in which the Pascal triangle is applied.[9] First it is used to sum the arithmetical sequences of different orders 1, 2, 3, 4, etc.; 1, 3, 6, 10, etc., 1, 4, 10, 20, ... (these sequences are called "numbers of the first, second, etc. order" [*ordres numériques*], then to the solution of certain games of chance, to the finding of combinations, to the raising of binomials to different powers, to the summation of the squares, cubes, etc., of the terms of an arithmetical series, etc., and to the proof that (in our present notation) $\int_0^a x^p \, dx = \dfrac{a^{p+1}}{p+1}$, p a positive integer. On this integral see Selection IV.6.

6 FERMAT. TWO FERMAT THEOREMS AND FERMAT NUMBERS

Pierre de Fermat (1601–1665) was a lawyer attached as councilor to the provincial parliament (that is, law court) of Toulouse. Of his contributions to geometry and calculus we speak in Selections III.3 and IV.7, 8. He was the first to take up seriously the challenge offered in number theory by the *Arithmetica* of Diophantus, first made fully available in the original Greek of A.D. *c.* 250 by Claude Bachet in 1621, together with a Latin translation. Fermat communicated his results in letters to his friends or kept them to himself in notes,

---

[8] This means that $P_k^l = \dfrac{l(l+1)\cdots(l+k-2)}{1\cdot2\cdots(k-1)} = C_{k-1}^{l+k-2}$; hence $C_p^n = P_{p+1}^{n-p+1}$, where

$C_p^n = \dfrac{m!}{p!(n-p)!}$, the number of combinations of n elements in groups of p.

[9] Some of this is translated in Smith, *Source book*, pp. 76–79.

many of them as marginal notes to his copy of Bachet. His son Samuel published a second edition of Bachet's Diophantus and added to it his father's marginal notes (Toulouse, 1670).

The extant work of Fermat has been published in the *Oeuvres de Fermat* (4 vols.; Gauthier-Villars, Paris, 1891–1912), in which the Latin texts are accompanied by a French translation (in vol. III, 1896).

We first quote the famous Latin marginal note to Diophantus' Proposition II, 8: "To divide a given square number into two squares," for which Diophantus gives the answer (in our notation) $[a(m^2 + 1)]^2 = (2am)^2 + [a(m^2 - 1)]^2$; for example, $a = \frac{16}{5}$, $m = \frac{1}{2}$; $16 = (\frac{16}{5})^2 + (\frac{12}{5})^2$; see *Oeuvres*, I, 53; French translation, III, 24. Fermat wrote:

In contrast, it is impossible to divide a cube into two cubes, or a fourth power into two fourth powers, or in general any power beyond the square into powers of the same degree; of this I have discovered a very wonderful demonstration [*demonstrationem mirabilem sane detexi*]. This margin is too narrow to contain it.

It is well known that nobody has ever found this *demonstratio sane mirabilis*, but also that nobody has been able to discover a positive integer $n > 2$ for which $x^n + y^n = z^n$ can be solved in terms of positive integers x, y, z. On the enormous literature in this field see P. Bachman, *Das Fermatproblem* (De Gruyter, Berlin-Leipzig, 1919); L. J. Mordell, *Three lectures on Fermat's last theorem* (Cambridge University Press, Cambridge, England, 1921); R. Noguès, *Théorème de Fermat. Son histoire* (Vuibert, Paris, 1932); H. S. Vandiver, "Fermat's last theorem," *American Mathematical Monthly* 53 (1946), 555–578. We shall show (Selection I.9) how Euler proved Fermat's theorem for $n = 3$ and $n = 4$.

Fermat communicated many of his results to the mathematician Bernard Frénicle de Bessy (1605–1675). In a letter of October 18, 1640, written in French, we find, among many observations, the following paragraphs containing another theorem of Fermat, which states that a^{p-1} is divisible by p when p is prime and a, p are relatively prime. Fermat had been interested in Euclid's theorem (*Elements*, Prop. IX, 36) that numbers of the form $2^{n-1}(2^n - 1)$ are perfect, that is, equal to the sum of their divisors including 1 (for example, $6 = 1 + 2 + 3$, $28 = 1 + 2 + 4 + 7 + 14$), if $2^n - 1$ is prime. Such prime numbers $2^n - 1$ Fermat called the *radicals* of the perfect numbers, and he had sent to Father Marin Mersenne some of his conclusions about these radicals in a letter of June 1640.[1] (If n is not prime, $2^n - 1$ cannot be prime; if n is prime, $2^n - 2$ is divisible by n; if n is prime, $2^n - 1$ is divisible only by prime numbers of the form $2kn + 1$; for example, $2047 = 2^{11} - 1 = 23 \times 89$, $2^{11} - 2 = 2046 = 11 \times 186$.) Then, in August 1640, in a letter to Frénicle, Fermat had turned to numbers of the form $2^n + 1$, writing that he was "almost convinced"

---

[1] These radicals $2^n - 1$, when prime, are known as Mersenne numbers M_n. It is clear that n in this case must be prime, but this is not sufficient. For example, $M_{11} = 2047 = 23 \times 89$.

Father Marin Mersenne (1586–1648), a Minorite (Franciscan), was in constant correspondence with the outstanding mathematicians of his day. His *Correspondence* has been published in 8 volumes (ed. C. de Waard; Beauchesne, Édition du Centre National de la Recherche, Paris, 1932–1963).

[*quasi persuadé*] that these numbers are prime when n is a power of 2. We now know that, though this is true for $n = 2, 4, 8, 16$, it stops being true for $n = 32$, which, as Euler showed (*Commentarii Academiae Scientiarum Petropolitanae 1* (1732/33, publ. 1738), 20–48, *Opera omnia*, ser. I, vol. 2, p. 73) is divisible by 641 $(4294967297 = 641 \times 6700417)$.[2] Fermat, on October 10, 1640, after referring to earlier letters, continues:

It seems to me after this that it is important to tell you on what foundation I construct the demonstrations of all that concerns the geometrical progressions, which is as follows:

Every prime number is always a factor [*mesure infailliblement*] of one of the powers of any progression minus 1, and the exponent [*exposant*] of this power is a divisor of the prime number minus 1. After one has found the first power that satisfies the proposition, all those powers of which the exponents are multiples of the exponent of the first power also satisfy the proposition.

Example: Let the given progression be

$$\begin{array}{cccccc} 1 & 2 & 3 & 4 & 5 & 6 \\ 3 & 9 & 27 & 81 & 243 & 729 \quad \text{etc.} \end{array}$$

with its exponents written on top.

Now take, for instance, the prime number 13. It is a factor of the third power minus 1, of which 3 is the exponent and a divisor of 12, which is one less than the number 13, and because the exponent of 729, which is 6, is a multiple of the first exponent, which is 3, it follows that 13 is also a factor of this power 729 minus 1.

And this proposition is generally true for all progressions and for all prime numbers, of which I would send you the proof if I were not afraid to be too long.

But it is true that every prime number is a factor of a power plus 1 in any kind of progression; for, if the first power minus 1 of which the said prime number is a factor has for exponent an odd number, then in this case there exists no power plus 1 in the whole progression of which this prime number is a factor.

Example: Because in the progression of 2 the number 23 is a factor of the power minus 1 which has 11 for exponent, the said number 23 will not be a factor of any power plus 1 of the said progression to infinity.

If the first power minus 1 of which the given prime number is a factor has an even number for exponent, then in this case the power plus 1 which has an exponent equal to half this first exponent will have the given prime as a factor.

The whole difficulty consists in finding the prime numbers which are not factors of any power plus 1 in a given progression, for this, for instance, is useful for finding which of the prime numbers are factors of the radicals of the perfect numbers, and to find a thousand other things as, for example, why it is that the 37th power minus 1 in the progression of 2 has the factor 223. In one word,

---

[2] These numbers $2^n + 1$, $n = 2^t$, when prime, are known as Fermat numbers. See O. Ore, *Number theory and its history* (McGraw-Hill, New York, 1948).

we must determine which are the prime numbers that factor their first power minus 1 in such a way that the exponent of the said power be an odd number—which I think very difficult [*fort malaisé*].

Fermat then continues with other striking properties of powers, also of numbers of the form $2^n + 1$, which, he believed, are all prime if n is a power of 2.[3]

7 FERMAT. THE "PELL" EQUATION

In a letter of February 1657 (*Oeuvres*, II, 333–335; III, 312–313) Fermat challenged all mathematicians (thinking probably in the first place of John Wallis in England) to find an infinity of integer solutions of the equation $x^2 - Ay^2 = 1$, where A is any nonsquare integer. He may have been led to this by his study of Diophantus, who set the problem of finding, for example, a number x such that both $10x + 9$ and $5x + 4$ are squares. If these squares are called u^2 and v^2 respectively, then $u^2 - 2v^2 = 1$, and a solution is $x = 28$. The problem was taken up by De Billy (see below) and later by Euler, who in his "De solutione problematum Diophanteorum per numeros integros," *Commentarii Academiae Scientiarum Petropolitanae 6* (1732/33, publ. 1738), 175–188, *Opera omnia*, ser. I, vol. 2, 6–17, referred to the problem as that of Pell and Fermat. John Pell (1611–1685), an English mathematician, had little to do with the problem, but the problem of Fermat has since been known as that of the Pell equation. It had already been studied by Indian mathematicians, and even in the *Cattle Problem*, attributed to Archimedes, which leads to a "Pell" equation with $A = 4729494 = 2 \cdot 3 \cdot 7 \cdot 11 \cdot 29 \cdot 353$; see T. L. Heath, *A manual of Greek mathematics* (Clarendon Press, Oxford, 1931), 337.

Fermat, after observing that "Arithmetic has a domain of its own, the theory of integral numbers," defines his problem as follows:

Given any number not a square, then there are an infinite number of squares which, when multiplied by the given number, make a square when unity is added.

Example.—Given 3, a nonsquare number; this number multiplied by the square number 1, and 1 being added, produces 4, which is a square.

Moreover, the same 3 multiplied by the square 16, with 1 added makes 49, which is a square.

And instead of 1 and 16, an infinite number of squares may be found showing the same property; I demand, however, a general rule, any number being given which is not a square.

It is sought, for example, to find a square which when multiplied into 149, 109, 433, etc., becomes a square when unity is added.

[3] See note 2.

In the same month (February 1657) Fermat, in a letter to Frénicle, suggests the same problem, and expressly states the condition, implied in the foregoing, that the solution be in integers:

Every nonsquare is of such a nature that one can find an infinite number of squares by which if you multiply the number given and if you add unity to the product, it becomes a square.

Example.—3 is a nonsquare number, which multiplied by 1, which is a square, makes 3, and by adding unity makes 4, which is a square.

The same 3, multiplied by 16, which is a square, makes 48, and with unity added makes 49, which is a square.

There is an infinity of such squares which when multiplied by 3 with unity added likewise make a square number.

I demand a general rule,—given a nonsquare number, find squares which multiplied by the given number, and with unity added, make squares.

What is for example the smallest square which, multiplied by 61 with unity added, makes a square?

Moreover, what is the smallest square which, when multiplied by 109 and with unity added, makes a square?

If you do not give me the general solution, then give the particular solution for these two numbers, which I have chosen small in order not to give too much difficulty.

After I have received your reply, I will propose another matter. It goes without saying that my proposition is to find integers which satisfy the question, for in the case of fractions the lowest type of arithmetician could find the solution.

Connected with this problem are a number of others, assembled by Fermat's friend Jacques de Billy (1602–1669), a Jesuit teacher of mathematics in Dijon, in his *Doctrinae analyticae inventum novum* (ed. S. Fermat; Toulouse 1670), translated in Fermat, *Oeuvres*, III, 325–398. They begin with the Diophantine problem (called a double equation), to make both $2x + 12$ and $2x + 5$ squares (answer $x = 2$). Part III (p. 376) begins (we change to modern notation):

On the procedure for obtaining an infinite number of solutions which give square or cubic values to expressions in which enter more than three terms of different degrees.

1. I shall discuss here in particular expressions which contain the five terms in x^4, x^3, x^2, x, and the constant, but I also wish to discuss expressions with four terms which may be all positive [true], or mixed with negative [false] terms. We wish to give these expressions square values (in the case of five terms), or

cubic ones (in the case of four terms), and this in an infinity of ways. In general we must say that for the square value at least the coefficient of the term in x^4 or the constant term must be a square; as to the cubic values, the coefficient of x^3 or the constant term must be a cube.

Applied to making $x^4 + 4x^3 + 6x^2 + 2x + 7$ a square, De Billy writes $(x^2 + 2x + 1)^2 = x^4 + 4x^3 + 6x^2 + 4x + 1$, which, set equal to the given form, gives $x = 3$.

In the case of $x^4 + 4x^3 + 10x^2 + 20x + 1$ De Billy equates this to $(1 + 10x - 45x^2)^2$, and gets $x = \frac{113}{253}$, then he equates it to $(x^2 + 2x - 1)^2$, and gets $x = -3$, and so on.

Then, by substituting for x the value $x + x_0$, where x is a "primitive" solution, for example $x_0 = -3$, or $x_0 = -4$, and repeating the process, he obtains new solutions. For $x \to x - 3$ he requires that $x^4 - 8x^3 + 28x^2 - 40x + 4$ be a square, which gives $x = \frac{7}{2}$; hence $x = \frac{1}{2}$ is a solution of the original equation. Here he turned a "false" solution into a "true" one. This process can be repeated.

It was from these problems by Fermat that Euler, in the paper of 1732/33, started his research on the "Pell" equation.

8   EULER. POWER RESIDUES

Here follow some contributions of Leonhard Euler (1707–1783) to the theory of numbers. Euler, born in Basel, Switzerland, studied with Johann Bernoulli, was from 1727 to 1741 associated with the Imperial Academy in Saint Petersburg, from 1741 to 1766 with the Royal Academy in Berlin (at the time of Frederick II, "the Great"), and from 1766 to his death again with the Saint Petersburg Academy (at the time of Catherine II, "the Great"). His productivity was enormous, in the writing both of voluminous papers and of huge textbooks, long standard, directly influencing all mathematicians from Lagrange to Riemann. The present extract is from his essay, "Theoremata circa residua ex divisione potestatum relicta" (Theorems on residues obtained by the division of powers), *Novi Commentarii Academiae Scientiarum Petropolitanae* 7 (1758/59, publ. 1761), 49–82, *Opera omnia*, ser. I, vol. 2, 493–518. In this paper Euler lays the foundation of the theory of power residues. We have taken theorems 10–14, in which Euler gives a proof of Fermat's theorem that $a^{p-1} - 1 \equiv 0 \pmod{p}$, p prime and a, p relatively prime.[1] The first nine theorems (in the $\equiv$ notation) are as follows: 1. When p is a prime and a is relatively prime to p, then no term of the geometric progression $1, a, a^2, \ldots, a^n, \ldots$ is divisible by p. 2. When $a^\mu \equiv r$ and $a^\nu \equiv s$, then $a^{\mu+\nu} \equiv rs$. 3. In the geometric progression $1, a, a^2, \ldots, a^n, \ldots$ an infinite set of terms will be $\equiv 1$, and their exponents form an arithmetic progression. 4. When $a^\mu \equiv r$ and $a^{\mu+\nu} \equiv rs$, then $a^\nu \equiv s$. 5. When $a^\lambda \equiv 1$ and $\lambda \neq 0$ is the lowest exponent for which this congruence holds, then the only powers $\equiv 1$ are $1, a^\lambda, a^{2\lambda}, a^{3\lambda}, \ldots$. 6. When $a^{2n} \equiv 1$ then $a^n \equiv \pm 1$ (when $a^{2n} \equiv 1$ and $2n$ is as small as possible, then $a^n \equiv -1$). 7. When $a^\lambda \equiv 1$ and λ is as small as possible, then all residues of the progression $1, a, a^2, \ldots, a^{\lambda-1}$ are different from each other. 8. When $a^\lambda \equiv 1$, and when we divide the powers as follows: $1, a, a^2, \ldots, a^{\lambda-1} |\ a^\lambda, a^{\lambda+1}, \ldots, a^{2\lambda+1} |\ a^{2\lambda}, \ldots, a^{3\lambda-1} |\ a^{3\lambda}, \ldots$, then in each section the

---

[1] See Selection I.6.

residues appear in the same order. 9. There are no more than $p - 1$ different residues, and 1 is always among them. The congruence is always modulo p.

The algorithm of paragraphs 37–46 is the same as that used later to prove that the order of a subgroup is a divisor of the order of the group. Euler's case is that of cyclical groups.

---

37. *Theorem 10. If the number of different residues resulting from the division of the powers* 1, a, a^2, a^3, a^4, a^5, *etc. by the prime number p is smaller than $p - 1$, then there will be at least as many numbers that are nonresidues as there are residues.*

Proof. Let a be the lowest power which, divided by p, has the residue 1, and let $\lambda < p - 1$; then the number of all the different residues will be $= \lambda$ and therefore smaller than $p - 1$. And since the number of all numbers smaller than p is $= p - 1$, there obviously must in our case be numbers that do not appear in the residues. I claim that there are at least λ of them. To prove it, let us express the residues by the terms themselves that produce them, and we get the residues

$$1, a, a^2, a^3, \ldots, a^{\lambda-1},$$

whose number is λ, and, reducing them in the usual way, they all become smaller than p and are all different from each other. As λ is supposed to be $< p - 1$, there exists certainly a number not occurring among those residues. Let this number be k; now I say that, if k is not a residue, then ak and a^2k and a^3k etc. as well as $a^{\lambda-1}k$ do not appear among the residues. Indeed, suppose that $a^\mu k$ is a residue resulting from the power a^μ; then we would have $a^\alpha = np + a^\mu k$ or $a^\alpha - a^\mu k = np$ and then $a^\alpha - a^\mu k = a^\mu(a^{\alpha-\mu} - k)$ would be divisible by p. Now a^μ is not divisible by p, so $a^{\alpha-\mu} - k$ would have to be divisible by p, that is, the power $a^{\alpha-\mu}$ would, if divided by p, give the residue k contrary to the assumption. From this it follows that all the numbers k, ak, a^2k, $\ldots$, $a^{\lambda-1}k$ or numbers derived from them are nonresidues. Moreover, they all are different from each other and their number is $= \lambda$; for if two of them, let us say $a^\mu k$ and $a^\nu k$, divided by p were to give the same residue r, then $a^\mu k = mp + r$ and $a^\nu k = np + r$ and thus $a^\mu k - a^\nu k = (m - n)p$, or $(a^\mu - a^\nu)k = (m - n)p$ would be divisible by p. Now k is not divisible by p, since we have assumed that p is a prime number and $k < p$; then $a^\mu - a^\nu$ would have to be divisible by p; or $a^{\mu-\nu}$ would give, divided by p, the residue 1, which is impossible because $\mu < \lambda - 1$ and $\nu < \lambda - 1$; also $\mu - \nu < \lambda$. Therefore all the numbers k, ak, a^2k, $\ldots$, $a^{\lambda-1}k$, if reduced, will be different and their number is $= \lambda$. Thus there exist at least λ numbers not belonging to the residues so long as $\lambda < p - 1$.

38. *Corollary* 1. Since we have λ different numbers that are residues, and just as many different numbers smaller than p, therefore their total number 2λ cannot be greater than $p - 1$, since there are only $p - 1$ numbers smaller than p.

39. *Corollary* 2. If therefore a^λ is the lowest power which after division by p gives the residue 1, and if $\lambda < p - 1$, then λ is certainly not $> (p - 1)/2$. Hence we have either $\lambda = (p - 1)/2$ or $\lambda < (p - 1)/2$.

40. *Corollary* 3. We have already seen in paragraph 15 that the exponent λ of this lowest power is necessarily smaller than p. Thus either $\lambda = p - 1$ or $\lambda < p - 1$; in which case, when $\lambda < p - 1$, we know now that either $\lambda = (p - 1)/2$ or $\lambda < (p - 1)/2$. Therefore there is no number between $p - 1$ and $(p - 1)/2$ which could ever have the value λ.

41. *Theorem* 11. *Let p be a prime number and a^λ the lowest power of a which, divided by p, gives the residue 1; let $\lambda < (p - 1)/2$; then the exponent λ cannot be greater than $(p - 1)/3$; thus either $\lambda = (p - 1)/3$ or $\lambda < (p - 1)/3$.*

Proof. Since a^λ is the lowest power of a which, divided by p, has the residue 1, there are at most λ different numbers of residues in the series

$$1, a, a^2, \ldots, a^{\lambda - 1}$$

when each term is divided by p. Hence, since $\lambda < p - 1$, there will be exactly $p - 1 - \lambda$ numbers that are nonresidues. Let r be one of them. Then we have seen that all the numbers

$$r, ar, a^2 r, \ldots, a^{\lambda - 1} r,$$

reduced by dividing by p to numbers smaller than p, do not appear as residues. Thus λ numbers are excluded from the residues; but when $\lambda < (p - 1)/2$, then $\lambda < p - 1 - \lambda$ and thus there exist, besides these numbers, some more that are nonresidues. Let s be such a number which is neither a residue nor a nonresidue in the above-mentioned series. Then all the numbers

$$s, as, a^2 s, \ldots, a^{\lambda - 1} s$$

will be nonresidues and these numbers will, as shown in the proof given above, be different from each other. Moreover, none of these numbers, such as $a^\mu s$, occurs in the previous series of nonresidues, that is, we never have $a^\mu s = a^\nu r$. For if $a^\nu r = a^\mu s$, then we would have $s = a^{\nu - \mu} r$, or (for $\mu > \nu$) $s = a^{\lambda + \nu - \mu} r$. This means that s would occur in the first series of nonresidues—contrary to our assumption. Thus, when $\lambda < (p - 1)/2$, there exist so far at least λ numbers that are nonresidues; and thus we have λ residues and 2λ nonresidues and all these numbers are smaller than p. Then it cannot be that their sum 3λ is greater than $p - 1$, or we cannot have $\lambda > (p - 1)/3$. Thus $\lambda = (p - 1)/3$ or $\lambda < (p - 1)/3$ so long as $\lambda < (p - 1)/2$ and p is a prime.

42. *Corollary* 1. If therefore λ is not smaller than $(p - 1)/3$, λ must be $= (p - 1)/3$, if we suppose $\lambda < (p - 1)/2$. If we omit this last restriction and if we know that $\lambda < (p - 1)/3$ does not hold, then λ must necessarily be either $= (p - 1)/3$, or $= (p - 1)/2$, or $= p - 1$.

43. *Corollary* 2. But if $\lambda = (p - 1)/3$ or $\lambda = (p - 1)/2$, then a^{p-1} divided by p gives the residue 1, for if a^λ has the residue 1 then the same holds for $a^{2\lambda}$ and $a^{3\lambda}$.

44. *Theorem* 12. *If a^λ is the lowest power of a which divided by p gives the residue 1, and if $\lambda < (p - 1)/3$, then λ cannot be $> (p - 1)/4$; instead either $\lambda = (p - 1)/4$ or $\lambda < (p - 1)/4$.*

Proof. As the number of all different residues resulting from the division of powers of a by p is $= \lambda$, and as they originate from the following terms,

$$1, a, a^2, \ldots, a^{\lambda-1},$$

then, because of $\lambda < (p-1)/3$, there originate twice that many numbers that are nonresidues from the following two progressions

$$r, ar, a^2r, \ldots, a^{\lambda-1}r$$

and

$$s, as, a^2s, \ldots, a^{\lambda-1}s.$$

The total number of these residues and nonresidues is $= 3\lambda$ and therefore smaller than $p-1$; thus there are still more numbers that are nonresidues. Let t be one such; then, as shown above, all the numbers

$$t, at, a^2t, \ldots, a^{\lambda-1}t,$$

whose number is $= \lambda$, will also be nonresidues. And these numbers not only differ from each other when p is a prime number, but also from all the previous ones, and thus the number of all these residues and nonresidues together $= 4\lambda$. As all of them are smaller than p, we cannot have $4\lambda > p-1$; consequently either λ becomes $= (p-1)/4$ or $\lambda < (p-1)/4$, always supposing that $\lambda < (p-1)/3$ and p is prime.

45. *Corollary* 1. In a similar way it can be shown that, when $\lambda < (p-1)/4$, we never can have $\lambda > (p-1)/5$ and thus we have also here $\lambda = (p-1)/5$ or $\lambda < (p-1)/5$.

46. *Corollary* 2. And in general, if it is known that $\lambda < (p-1)/n$, then one proves in the same way that we cannot have $\lambda > (p-1)/(n+1)$, therefore we must have $\lambda = (p-1)/(n+1)$ or $\lambda < (p-1)(n+1)$.

47. *Corollary* 3. Wherefrom it appears that the number of all numbers that cannot be residues is either $= 0$, or $= \lambda$, or $= 2\lambda$ or any multiple of λ; for if there are more than $n\lambda$ of such numbers, then, if any at all, λ new ones are added to them, so as to make their number $= (n+1)\lambda$; and if this does not yet comprise all the nonresidues, then at once λ new ones are added.

48. *Theorem* 13. *Let p be a prime and a^λ the lowest power of a, giving the residue 1 when divided by p; then the exponent λ is a divisor of the number $p-1$.*

Proof. The number of all different residues is thus $= \lambda$; therefore the number of the remaining numbers smaller than p that cannot be residues will be $= p-1-\lambda$; but this number is (§47) a multiple of λ, say $n\lambda$, so that $p-1-\lambda = n\lambda$, wherefrom results

$$\lambda = \frac{p-1}{n+1}.$$

This makes it clear that when λ is not $= p-1$ then it certainly is equal to a divisor of $p-1$.

49. *Theorem* 14. *Let p be a prime number and a be prime to p; then the power a^{p-1} divided by p has the residue* 1.

Proof. Let a^λ be the lowest power of a giving the residue 1 when divided by p. Then, as we have seen, λ will be $< p$ and we proved above that in this case either $\lambda = p - 1$ or λ is a divisor of the number $p - 1$. In the first case the theorem holds, and a^{p-1} gives, divided by p, the residue 1. In the other case, where λ is a divisor of $p - 1$, we have $p - 1 = n\lambda$; but because the power a^λ gives, divided by p, the residue 1, therefore also all these powers $a^{2\lambda}$, $a^{3\lambda}$, etc. and $a^{n\lambda}$ or a^{p-1} divided by p will give the residue 1. Thus a^{p-1} divided by p will always have the residue 1.

50. *Corollary* 1. Because the power a^{p-1} gives the residue 1 when divided by the prime number p, the formula $a^{p-1} - 1$ is divisible by p, so long as a is a number prime to p, that is, so long as a is not divisible by p.

51. *Corollary* 2. If, therefore, p is a prime, then all powers of exponent $p - 1$, such as n^{p-1}, are divisible by p, or leave 1 as remainder. The latter happens if n is prime to p, the first if this number n is divisible by p.

52. *Corollary* 3. Hence, if p is a prime number, and the numbers a and b are prime to p, then the difference of the powers $a^{p-1} - b^{p-1}$ will be divisible by p. Indeed, since $a^{p-1} - 1$ as well as $b^{p-1} - 1$ are divisible by p, so will also their difference $a^{p-1} - b^{p-1}$ be divisible by p.

. . . .

53. *Scholium.* This is a new proof of the famous theorem that Fermat once stated, and it is completely different from the one I have given in the *Comment. Acad. Petropol., tome VIII*.[2] There I started out from Newton's series expansion of the binomial $(a + b)^n$, using a reasoning seemingly quite remote from the proposition; here, on the other hand, I prove the theorem starting from properties of the powers alone, which makes the proof seem much more natural. Moreover, other important properties of the residues of the powers when they are divided by a prime number come to light. Indeed, it is shown for a prime number p not only that the expression $a^{p-1} - 1$ is divisible by p, but that, under certain conditions, a simpler expression, $a^\lambda - 1$, is divisible by p, and that in that case the exponent λ is always a divisor of $p - 1$.

In the remaining sections Euler proves several theorems on power residues, of which Theorem 19 states that if $a^m \equiv 1 \pmod{p = mn + 1}$, then there always exist numbers x and y such that $ax^n - y^n = 0$. Here p is a prime. Our notation is Euler's, except that we have written a/b where Euler writes $\dfrac{a}{b}$.

[2] "Theorematum quorundam ad numeros primos spectantium demonstratio," *Commentarii Academiae Scientiarum Petropolitanae 8* (1736, publ. 1741), 141–146, *Opera omnia*, ser. I, vol. 2, 35–37. The proof runs as follows. First it is proved by means of the binomial expansion of $(1 + 1)^{p-1}$ that

$$2^{p-1} - 1 = \frac{p(p-1)}{1 \cdot 2} + \frac{p(p-1)(p-2)(p-3)}{1 \cdot 2 \cdot 3 \cdot 4} + \cdots$$

is divisible by p if p is an odd prime. Then by a similar expansion of $(1 + a)^p$ it is shown that $(1 + a)^p - (1 + a) - (a^p - a) \equiv 0 \pmod{p}$ if a is not a multiple of p. Since $2^p - 2 \equiv 0$, the proof follows by complete induction.

9 EULER. FERMAT'S THEOREM FOR $n = 3, 4$

Euler received much of his inspiration for his work on number theory from the study of Fermat (see Selections I.6, 7). Fermat's theorem that $x^n + y^n = z^n$ cannot be solved for positive integers x, y, z, n, $n > 2$, attracted him and he gave proofs for the cases $n = 3$ and $n = 4$. In a paper of 1738 entitled "Theorematum quorundam arithmeticorum demonstrationes," *Commentarii Academiae Scientiarum Petropolitanae 10* (1738, publ. 1747), 125–146, *Opera omnia*, ser. I, vol. 2, 38–58, he gave a proof for the case $n = 4$, adding a number of related theorems. A proof for $n = 4$ had already been given by Fermat's correspondent Bernard Frénicle (c. 1605–1675) in his *Traité des triangles rectangles en nombres* (Paris, 1676). Frénicle here used the so-called method of infinite descent, which Euler also employed. This method is as follows: suppose that a solution of the problem in question is possible in positive integers: then we show how to derive from it a solution in smaller positive integers, and so on. But since this process cannot go on indefinitely, we reach a contradiction and thus show that no solution is possible. For this method, used by Fermat and even before him (G. Eneström, *Bibliotheca Mathematica 14* (1913–14), 347), see, for example, O. Ore, *Number theory and its history* (McGraw-Hill, New York, 1948), 199.

Euler also gave a proof for $n = 4$ in his *Vollständige Anleitung zur Algebra* (Saint Petersburg, 1770), *Opera omnia*, ser. I, vol. 1 (see also ser. I, vol. 5), which has been edited also by J. E. Hofmann (Reclam Verlag, Stuttgart, 2nd ed., 1959). We give the English translation by J. Hewlett, *Elements of algebra* (5th ed., Longman, Orme, London, 1840), 405–413.

Euler's Theorem 1 is preceded by some lemmas and corollaries; lemma 2 states that, if $a^2 + b^2$ is a square and a, b are relative primes, then $a = p^2 - q^2$, $b = 2pq$, where p and q are relative primes, one even, the other odd (hence such a square is always odd).

Theorem 1. The sum of two biquadratic numbers such as $a^4 + b^4$ cannot be a square number unless one of the two biquadratic numbers vanishes.

Proof. I shall change the theorem to be demonstrated in such a way that I shall show that if in one case $a^4 + b^4$ were a square, no matter how large the numbers a and b, then I can progressively find smaller numbers a and b and at the end can reach the smallest integral numbers. Since there are no such smallest numbers of which the sum of the biquadratic numbers is a square, we must conclude that there are no such among the largest numbers.

Let therefore $a^4 + b^4$ be a square and a and b be relative primes, since if they were not relative primes, then I could reduce them by division to primes. Let a be an odd number; then b must be even, since necessarily one number must be even, the other one odd. Let us therefore write

$$a^2 = p^2 - q^2, \quad b^2 = 2pq;$$

here the numbers p and q must be relative primes, the one even, the other odd. But if $a^2 = p^2 - q^2$, then it is necessary that p be odd, because otherwise $p^2 - q^2$ could not be a square. Hence p is an odd number and q an even one. Since $2pq$ must also be a square it is necessary that both p and $2q$ be a square, because p and $2q$ are relative primes. Since $p^2 - q^2$ is a square, it is necessary that

$$p = m^2 + n^2 \quad \text{and} \quad q = 2mn,$$

where again m and n are relative prime numbers, of which one is even, the other odd. But since $2q$ is a square, $4mn$, or mn is a square, hence m and n are squares. If we therefore put

$$m = x^2, \quad n = y^2,$$

then we shall have

$$p = m^2 + n^2 = x^4 + y^4,$$

which must equally be a square. From this it follows that if $a^4 + b^4$ were a square, then also $x^4 + y^4$ would be a square, but it is clear that the numbers x and y would be far smaller than a and b. In the same way we shall from the biquadratic numbers $x^4 + y^4$ again obtain smaller ones, of which the sum is a square, and we progressively reach the smallest biquadratic number among the integers. But since there are no smallest biquadratic numbers of which the sum gives a square, it is clear that there are no very large numbers either. However, if in one pair of the biquadratic numbers one of the terms is zero, then in all remaining pairs the one term vanishes, so that here nothing new results.

Corollary 1. Since therefore the sum of two biquadratic numbers cannot be a square, it is a fortiori impossible that the sum of two biquadratic numbers results in a biquadratic number.

Corollary 2. Although this demonstration pertains only to integers, yet it also shows that we cannot find among fractions two biquadratic numbers of which the sum is a square. Indeed, if $(a^4/m^4) + (b^4/n^4)$ were a square, then $a^4 n^4 + b^4 m^4$, which is a sum of integers, would also be a square, which we have proved to be impossible.

Corollary 3. By means of the same proof we can conclude that no numbers p and q exist such that p, $2q$ and $p^2 - q^2$ are squares; if such numbers existed then there would be values for a and b, which would render $a^4 + b^4$ square; for then $a = \sqrt{p^2 - q^2}$ and $b = \sqrt{2pq}$.

Corollary 4. Suppose therefore $p = x^2$ and $2q = 4y^2$, then $p^2 - q^2 = x^4 - 4y^4$. Then it could not at all happen, that $x^4 - 4y^4$ were a square. Nor could $4x^4 - y^4$ be a square; for then $16x^4 - 4y^4$ would be a square, which reduces it to the former case, because $16x^4$ is a biquadratic number.

Corollary 5. From this it follows that also $ab(a^2 + b^2)$ can never be a square. For the factors $a, b, a^2 + b^2$, all relative primes, would have to be squares, which is impossible.

Corollary 6. In a similar way, there cannot exist relatively prime numbers a and b such as to make $2ab(a^2 - b^2)$ a square. This follows from Corollary 3, where it was proven that no numbers p and q exist such as to make p, $2q$, $p^2 - q^2$ squares. And all this is valid also for numbers that are not relative primes, and the same for fractions according to Corollary 2.

Euler continues with nine more theorems; Theorem 2 states that $a^4 - b^4$ cannot be a square unless $b = 0$ or $b = a$, and Theorem 7 proves Fermat's theorem (a marginal note to Bachet's *Diophantus*, VI, 26) that no triangular number (that is, a number of the form $\frac{1}{2}n(n + 1)$), except 1, can be a biquadratic number.

The following proof of Fermat's theorem on the impossibility of $x^n + y^n = z^n$ $(n > 2)$ for the case $n = 3$ is also taken from Euler's *Vollständige Anleitung zur Algebra* and the English translation by Hewlett, pp. 450–454. The proofs are found in the last section of the book, which does not deal with algebra at all, but with indeterminate equations. Here, in Art. 155, Euler demonstrates that $1 + x^3$ cannot be a cube except for $x = 0$ and $x = -1$, as a special case of an investigation to find out whether the form $a + bx + cx^2 + dx^3$ can be a cube (an investigation started by Fermat; see Selection I.7).

---

243. *Theorem. It is impossible to find any two cubes, whose sum, or difference, is a cube.*

We shall begin by observing, that if this impossibility applies to the sum, it applies also to the difference, of two cubes. In fact, if it be impossible for $x^3 + y^3 = z^3$, it is also impossible for $z^3 - y^3 = x^3$. Now, $z^3 - y^3$ is the difference of two cubes; therefore, if the one be possible, the other is so likewise. This being laid down, it will be sufficient, if we demonstrate the impossibility either in the case of the sum, or difference; which demonstration requires the following chain of reasoning.

I. We may consider the numbers x and y as prime to each other; for if they had a common divisor, the cubes would also be divisible by the cube of that divisor. For example, let $x = 2a$, and $y = 2b$, we shall then have $x^3 + y^3 = 8a^3 + 8b^3$; now if this formula be a cube, $a^3 + b^3$ is a cube also.

II. Since, therefore, x and y have no common factor, these two numbers are either both odd, or the one is even and the other odd. In the first case, z would be even, and in the other that number would be odd. Consequently, of these three numbers, x, y, and z, there is always one that is even, and two that are odd; and it will therefore be sufficient for our demonstration to consider the case in which x and y are both odd: because we may prove the impossibility in question either for the sum, or for the difference; and the sum only happens to become the difference, when one of the roots is negative.

III. If therefore x and y are odd, it is evident that both their sum and their difference will be an even number. Therefore let $\frac{1}{2}(x + y) = p$, and $\frac{1}{2}(x - y) = q$, and we shall have $x = p + q$, and $y = p - q$; whence it follows, that one of the two numbers, p and q, must be even, and the other odd. Now, we have, by adding $(p + q)^3 = x^3$, to $(p - q)^3 = y^3$, $x^3 + y^3 = 2p^3 + 6pq^3 = 2p(p^2 + 3q^2)$; so that it is required to prove that this product $2p(p^2 + 3q^2)$ cannot become a cube; and if the demonstration were applied to the difference, we should have $x^3 - y^3 = 6p^2q + 2q^3 = 2q(q^2 + 3p^2)$, a formula precisely the same as the former, if we substitute p and q for each other. Consequently, it is sufficient for our purpose to demonstrate the impossibility of the formula, $2p(p^2 + 3q^2)$, becoming a cube, since it will necessarily follow, that neither the sum nor the difference of two cubes can become a cube.

IV. If therefore $2p(p^2 + 3q^2)$ were a cube, that cube would be even, and therefore divisible by 8: consequently, the eighth part of our formula, or $\frac{1}{4}p(p^2 + 3q^2)$, would necessarily be a whole number, and also a cube. Now, we know that one of the numbers p and q is even, and the other odd; so that $p^2 + 3q^2$ must be an odd number, which not being divisible by 4, p must be so, or $p/4$ must be a whole number.

V. But in order that the product $\frac{1}{4}p(p^2 + 3q^2)$ may be a cube, each of these factors, unless they have a common divisor, must separately be a cube; for if a product of two factors, that are prime to each other, be a cube, each of itself must necessarily be a cube; and if these factors have a common divisor, the case is different, and requires a particular consideration. So that the question here is, to know if the factors p, and $p^2 + 3q^2$, might not have a common divisor. To determine this, it must be considered, that if these factors have a common divisor, the numbers p^2, and $p^2 + 3q^2$, will have the same divisor; that the difference also of these numbers, which is $3q^2$, will have the same common divisor with p^2. And since p and q are prime to each other, these numbers p^2, and $3q^2$, can have no other common divisor than 3, which is the case when p is divisible by 3.

VI. We have therefore two cases to examine: the one is, that in which the factors p, and $p^2 + 3q^2$, have no common divisor, which happens always, when p is not divisible by 3; the other case is, when these factors have a common divisor, and that is when p may be divided by 3; because then the two numbers are divisible by 3. We must carefully distinguish these two cases from each other, because each requires a particular demonstration.

VII. *Case* 1. Suppose that p is not divisible by 3, and, consequently, that our two factors $p/4$, and $p^2 + 3q^2$, are prime to each other; so that each must separately be a cube. Now, in order that $p^2 + 3q^2$ may become a cube, we have only, as we have seen before, to suppose $p + q\sqrt{-3} = (t + u\sqrt{-3})^3$, and $p - q\sqrt{-3} = (t - u\sqrt{-3})^3$, which gives $p^2 + 3q^2 = (t^2 + 3u^2)^3$, which is a cube. This gives us $p = t^3 - 9tu^2 = t(t^2 - 9u^2)$, and $q = 3t^2u - 3u^3 = 3u(t^2 - u^2)$. Since therefore q is an odd number, u must also be odd; and, consequently, t must be even, because otherwise $t^2 - u^2$ would be even.

VIII. Having transformed $p^2 + 3q^2$ into a cube, and having found $p = t(t^2 - 9u^2) = t(t + 3u)(t - 3u)$, it is also required that $p/4$, and, consequently, $2p$, be a cube; or, which comes to the same, that the formula $2t(t + 3u)(t - 3u)$ be a cube. But here it must be observed that t is an even number, and not divisible by 3; since otherwise p would be divisible by 3, which we have expressly supposed not to be the case: so that the three factors, $2t, t + 3u$, and $t - 3u$, are prime to each other; and each of them must separately be a cube. If, therefore, we make $t + 3u = f^3$, and $t - 3u = g^3$, we shall have $2t = f^3 + g^3$. So that, if $2t$ is a cube, we shall have two cubes f^3, and g^3, whose sum would be a cube, and which would evidently be much less than the cubes x^3 and y^3 assumed at first. Indeed, as we first made $x = p + q$, and $y = p - q$, and have now determined p and q by the letters t and u, the numbers x and y must necessarily be much greater than t and u.

IX. If, therefore, there could be found in large numbers two such cubes as we require, then we should also be able to assign in much smaller numbers two cubes, whose sum would make a cube, and in the same manner we should be led to cubes always less. Now, as it is very certain that there are no such cubes among small numbers, it follows, that there are not any among greater numbers. This conclusion is confirmed by that which the second case furnishes, and which will be seen to be the same.

X. *Case* 2. Let us now suppose, that p is divisible by 3, and that q is not so, and let us make $p = 3r$; our formula will then become $\frac{3}{4}r(9r^2 + 3q^2)$, or $\frac{9}{4}r(3r^2 + q^2)$; and these two factors are prime to each other, since $3r^2 + q^2$ is neither divisible by 2 nor by 3, and r must be even as well as p; therefore each of these two factors must separately be a cube.

XI. Now, by transforming the second factor $3r^2 + q^2$, or $q^2 + 3r^2$, we find, in the same manner as before, $q = t(t^2 - 9u^2)$, and $r = 3u(t^2 - u^2)$; and it must be observed, that since q was odd, t must be here likewise an odd number, and u must be even.

XII. But $\frac{9}{4}r$ must also be a cube; or multiplying by the cube $\frac{8}{27}$ we must have $\frac{2}{3}r$, or $2u(t^2 - u^2) = 2u(t + u)(t - u)$ a cube; and as these three factors are prime to each other, each must of itself be a cube. Suppose therefore $t + u = f^3$, and $t - u = g^3$, we shall have $2u = f^3 - g^3$; that is to say, if $2u$ were a cube, $f^3 - g^3$ would be a cube. We should consequently have two cubes, f^3 and g^3, much smaller than the first, whose difference would be a cube, and that would enable us also to find two cubes whose sum would be a cube; since we should only have to make $f^3 - g^3 = h^3$, in order to have $f^3 = h^3 + g^3$, or a cube equal to the sum of two cubes. Thus, the foregoing conclusion is fully confirmed; for as we cannot assign, in large numbers, two cubes whose sum or difference is a cube, it follows from what has been before observed, that no such cubes are to be found among small numbers.

In Art. 245 Euler shows how to find to two given cubes a^3 and b^3 a third cube x^3 such that $a^3 + b^3 + x^3$ be a cube (for example, when $a = 2, b = 3, x = \frac{124}{19}$ and the sum of their cubes is $(\frac{129}{19})^3$), in Art. 247 he proves that $x^3 \pm y^3 = 2z^3$ is impossible except for $x = y$, and in Art. 248 he shows how to find three cubes whose sum is again a cube (for example, $14^3 + 17^3 + 7^3 = 20^3, 8^3 + 6^3 + 1^3 = 9^3, 3^3 + 4^3 + 5^3 = 6^3$). This last example leads him in Art. 249 to find more sets of three consecutive numbers, which, when cubed and added, give again a cube.

10 EULER. QUADRATIC RESIDUES AND THE RECIPROCITY THEOREM

The present paper, "Observationes circa divisionem quadratorum per numeros primos," *Opuscula analytica* (Saint Petersburg, 1783–1785), I, 64–84, *Opera omnia*, ser. I, vol. 3, 497–512, published in the year of Euler's death, is a new presentation of some of his older work on quadratic residues and also contains new results. It ends with his formulation of what we now call the reciprocity theorem of quadratic residues.

OBSERVATIONS ON THE DIVISION OF SQUARE NUMBERS BY PRIMES[1]

1. *Hypothesis.* If the squares a, b, c, d, etc. of the numbers a^2, b^2, c^2, d^2, etc. are divided by an arbitrary prime number P, then we denote by the Greek letters α, β, γ, δ, etc. the residues left after the division.

2. *Corollary* 1. Since therefore the square aa divided by P leaves the residue α, then, A being the quotient, $aa = AP + \alpha$, and thus $aa - \alpha$ will be divisible by P, and similarly the expressions $bb - \beta$, $cc - \gamma$, $dd - \delta$ will be divisible by the same divisor P.

3. *Corollary* 2. The squares $(a + P)^2$, $(a + 2P)^2$, $(a + 3P)^2$ and in general $(a + nP)^2$ also leave α as residue, if they are divided by the given P. From which it is clear that from the squares of the large numbers the same residues appear upon division by P as from the squares of the smaller numbers.

4. *Corollary* 3. Since furthermore the square $(P - a)^2$, divided by P, gives the same residue as the square a^2, it is clear that (if $a > \frac{1}{2}P$, then $P - a < \frac{1}{2}P$), we obtain all different residues from square numbers which are less than half the divisor P.

5. *Corollary* 4. Hence if we want all residues that are obtained from the division of squares by the given divisor P, then it suffices to consider only those squares of which the roots are not larger than the half of P.

6. *Corollary* 5. Hence if the divisor $P = 2p + 1$, and if we divide all square numbers 1, 4, 9, 16, 25, etc. by it, no more different residues can be produced than there are units in the number p, and those result from the squares of the numbers 1, 2, 3, 4, ..., p. The squares of the following numbers, $p + 1$, $p + 2$, $p + 3$, will reproduce the same residue in reverse order.

7. *Scholium.* It is manifest that the two squares p^2 and $(p + 1)^2$ divided by the number $2p + 1$ give the same residue, since their difference is divisible by $2p + 1$. Hence in general, if the difference of two arbitrary numbers $M - N$ be divisible by $2p + 1$, then it is necessary that each M and N divided by $2p + 1$ give the same remainder. Hence also, since

$$(p + 2)^2 - (p - 1)^2 = 3(2p + 1),$$

and each number $(p + 2)^2$, $(p - 1)^2$ is itself a square, they must give the same residue, and, in general, the square $(p + n + 1)^2$ will give the same residue as the square $(p - n)^2$. Hence it is clear that there are no more residues than there are units in the number p. Whether these residues are either all different or whether some of them are equal is here not decided; and insofar as we admit all kinds of divisors, both cases may occur. However, when the divisor $2p + 1$ is prime, then all these residues are different from each other, as I prove in the following way.

---

[1] The foundations of the theory of quadratic residues, which are laid in this paper, can already be found in Euler's "Demonstrationes theorematis Fermatiani omnem numerum sive integrum sive fractum esse summam quatuor pauciorumve quadratorum" (Demonstration of Fermat's theorem that every number, either integer or fraction, is the sum of four or fewer squares), *Novi Commentarii Academiae Scientiarum Petropolitanae* 5 (1754/55, publ. 1760), 13–58, *Opera omnia*, ser. I, vol. 2, 338–372. This paper also deals with moduli that are not prime. The text has a^2, not aa.

8. *Theorem* 1. *If the divisor $P = 2p + 1$ is a prime number and all squares* 1, 4, 9, 16, ... *up to p^2 are divided by it, then all residues obtained in this way are different from each other, and their number is therefore $= p$.*

Demonstration. Let a and b be two arbitrary numbers less or at any rate not larger than p. We must show that when their squares a^2 and b^2 are divided by the prime number $2p + 1$, they will certainly give different residues. Indeed, if they were to give the same residue, then their difference $aa - bb$ would be divisible by $2p + 1$ and therefore, since $2p + 1$ is prime, in $aa - bb = (a + b)(a - b)$ one of the factors would have to be divisible by $2p$. But since $a < p$ and $b < p$, or at any rate a not $> p$, the sum $a + b$ and a fortiori the difference $a - b$ are less than the divisor $2p + 1$, and hence neither of these two can be divided by $2p + 1$. From this it clearly follows that all squares whose roots are not larger than p certainly give residues different from each other when divided by the prime number $2p + 1$.

The corollaries claim (1) that if 1, 4, 9, 16, etc. are divided by $2p + 1$, the number of residues will be exactly p, (2) that these can be obtained from the numbers $1, 4, 9, 16, \ldots, p$, (3) that there are therefore p residues and p nonresidues.

---

12. *Scholium.* I shall call those numbers less than $2p + 1$, which are excluded from the order of residues, *nonresidues*.[2] Their number is always equal to the number of residues. It is useful to study this difference between residues and nonresidues correctly, and for this purpose I shall present here for some smaller prime divisors the residues and nonresidues,

Divisor 3, $p = 1$		Divisor 5, $p = 2$		Divisor 7, $p = 3$	
Square	1	Squares	1, 4	Squares	1, 4, 9
Residue	1	Residues	1, 4	Residues	1, 4, 2
Nonresidue	2	Nonresidues	2, 3	Nonresidues	3, 5, 6

Divisor 19, $p = 9$	
Squares	1, 4, 9, 16, 25, 36, 49, 64, 81
Residues	1, 4, 9, 16, 6, 17, 11, 7, 5
Nonresidues	2, 3, 8, 10, 12, 13, 14, 15, 18

We shall first find for any prime divisor such memorable properties about these residues and nonresidues that it is worth while to study them properly, so that there seem to be here a considerable number of interesting additions to the theory of numbers.

13. *Theorem* 2. *If in the order of residues derived from the divisor P there are two numbers α and β, then there also occurs their product $\alpha\beta$, if it is less than the divisor P, but if it is larger, then we should take in its place $\alpha\beta - P$ or $\alpha\beta - 2P$, or in general $\alpha\beta - nP$ until a number appears less than P.*

---

[2] This concept, not yet fully developed in the paper of note 1, is here introduced and leads to Theorems 4 and 5.

Demonstration. Let the residues α and β result from the division of the squares aa and bb by the divisor P. Then we have

$$aa = AP + \alpha, \quad \text{and} \quad bb = BP + \beta,$$

so that

$$aabb = ABP^2 + (A\beta + B\alpha)P + \alpha\beta.$$

Hence, if $aabb$ is divided by the divisor P, the residue is $\alpha\beta$, or if $\alpha\beta$ is larger than P, we must take that residue which is obtained when $\alpha\beta$ is divided by P, which will be either $\alpha\beta - P$ or $\alpha\beta - 2P$ or $\alpha\beta - 3P$ or, in general, $\alpha\beta - nP$ such that $\alpha\beta - nP < P$.

The corollaries claim (1) that, if α is a residue, then so also are $\alpha\alpha$, α^3, α^4, etc.; for higher powers we must subtract the necessary multiples of P; (2) that when $P = 2p + 1$ and α is a residue, then, if α^0, α^1, α^2, α^3, α^4, etc. are divided by P, there can be no more than p residues; hence (3) that α^p divided by $2p + 1$ gives $\alpha^0 = 1$, which theorem Euler has proved before.[3] In a scholium he observes that negative residues are also admissible.

---

18. *Theorem 3. If in the sequence of residues obtained from a divisor P there occur two residues α and β, then there also occurs the residue $(\alpha + nP)/\beta$, where n is taken so large that $(\alpha + nP)/\beta$ is an integer. This can always be done.*

Demonstration. Let aa and bb be squares which produce, after division by P, the residues α and β. Then we have

$$aa = AP + \alpha, \quad \text{and} \quad bb = BP + \beta.$$

We now ask for a c such that $c = (a + mP)/b$ is an integer. Then

$$cc = \frac{aa + 2amP + mmPP}{bb} = \frac{\alpha + (A + 2am + mmP)P}{\beta + BP}$$

will be an integer. Since the numerator also has the residue α, and the denominator the residue β, it is clear that, if cc be divided by P, the residue will be reduced to the proposed form. Indeed, let us put, for the sake of brevity, $A + 2am + mmP = D$, so that $cc = (\alpha + DP)/(\beta + BP)$; then if $(\alpha + nP)/\beta = \gamma$, it is necessary to show that $cc = CP + \gamma$, so that the residue from the division of the square cc by the number P may be $= \gamma$. But $\alpha = \beta\gamma - nP$, and hence we can write

$$cc = \frac{\beta\gamma + (D - n)P}{\beta + BP} = CP + \gamma,$$

[3] In this paper, "Theoremata circa divisores numerorum," *Novi Commentarii Academiae Scientiarum Petropolitanae 1* (1747/78, publ. 1750), 20–48, *Opera omnia*, ser. I, vol. 2, 62–85, Theorem 11.

from which it follows that

$$(D - n)P = (\beta C + \gamma B + BCP)P,$$

or

$$D - n = \beta C + \gamma B + BCP,$$

which is exactly the necessary relation between the coefficients of P that will produce integers.

Euler gives another proof, where he takes an n such that $\alpha + nP = \beta\gamma$, then an m such that $a + mP = bc$. Then if $b^2c^2 = EP + \beta\gamma$ and $b^2 = BP + \beta$, $cc = CP + x$, then $b^2c^2 = \beta x + (\beta C + Bx + BCP)P$, or $x = \gamma$. The corollaries state (1) that since 1 is always a residue, $\beta = (1 + nP)/\alpha$ occurs among the residues, so that $\alpha\beta$ is equivalent to 1 among the residues; (2) that to every α there is therefore a β such that $\alpha\beta$ is equivalent to 1, and these two reciprocal residues are different unless both are $+1$ or -1; indeed, from $\alpha^2 = 1 + nP$ it follows that $\alpha = \pm(1 + mP)$; (3) that in this way we can associate every residue in the sequence of residues with its reciprocal, except the solitary 1, or also -1 or $P - 1$ when this appears between the residues. In a scholium Euler points to the importance of this occurrence for the proof of the "most beautiful theorem" (*ad demonstrationem facilem Theorematis pulcerrimi nos manducet*) that every prime of the form $4q + 1$ is the sum of two squares.[4] If α, β, γ are residues and $\mathfrak{A}$, $\mathfrak{B}$, $\mathfrak{C}$ nonresidues, then the $\alpha\beta$ are residues, the $\alpha\mathfrak{A}$ nonresidues, and the $\mathfrak{A}\mathfrak{B}$ residues.

---

23. *Theorem 4. If the divisor P is of the form $4q + 3$, then -1 or $P - 1$ is certainly a nonresidue.*

Demonstration. When we write $P = 2p + 1$, then $p = 2q + 1$, an odd number. Hence the number of all residues will be odd. If -1 were to appear in the sequence of residues, then to every residue α would correspond another residue $-\alpha$, and the sequence of residues could be written as follows:

$$+1, +\alpha, +\beta, +\gamma, +\delta, \text{etc.},$$

$$-1, -\alpha, -\beta, -\gamma, -\delta, \text{etc.},$$

and the number of residues would be even. But since this number is certainly odd, it is impossible that -1 or $P - 1$ should appear in the sequence of residues; hence it belongs in the sequence of nonresidues.

---

[4] Euler published this theorem in his "Demonstratio theorematis Fermatiani omnem numerum primum formae $4n + 1$ esse summam duorum quadratorum," *Novi Commentarii Academiae Scientiarum Petropolitanae* 5 (1754/55, publ. 1760), 3–13, *Opera omnia*, ser. I, vol. 2, 328–337. See also pp. 295–327, and Theorem 5.

The corollaries state (1) that if for prime $P = 4q + 3$ a residue α occurs, then $-\alpha$ or $P - \alpha$ is a nonresidue; if $-\beta$ is a residue, then $+\beta$ is a nonresidue; (2) that if α is the residue of a, then there is no number x with residue $-\alpha$, hence no sum of squares $a^2 + x^2$ exists which is divisible by $4q + 3$; (3) that if α and β are residues belonging to a and b, then $\beta a^2 - \alpha b^2$ is divisible by $4q + 3$; (4) that the form $a^2\beta + \alpha x^2$ is not divisible by $4q + 3$; and (5) that no form $\beta a^2 c^2 + \alpha c^2 x^2$ is divisible by $4q + 3$ unless c^2 is divisible. A scholium gives the sequence of residues for $P = 4q + 3$ in the cases $P = 3, 7, \ldots, 31$, for example,

Divisor 3, Residue 1;
Divisor 7, Residues 1, -3, $+2$;
Divisor 11, Residues 1, $+4$, -2, $+5$, $+3$; $\ldots$
Divisor 31, Residues 1, $+4$, $+9$, -15, -6, $+5$, -13, $+2$, -12, $+7$, -3, -11, $+14$, $+10$, $+8$.

Hence all numbers not larger than half the divisor appear with sign $+$ or $-$, and none appears with both signs. If all the signs are changed, we get the nonresidues. Hence the forms $a^2 + b^2$, $a^2 - 15b^2$, $a^2 - 6b^2$, etc. are not divisible by 31.

---

30. *Theorem 5. If the divisor P is a prime of the form $4q + 1$, then the number -1 or $P - 1$ is certainly a residue.*

The demonstration follows the same indirect reasoning as that of Theorem 4, pairing each residue off against its reciprocal. The corollaries state (1) that if α in this case is a residue, then $-\alpha$ is a residue; (2) that if α belongs to a, $-\alpha$ to b, then $a^2 + b^2$ is divisible by $4q + 1$; (3) that b^2 need not be larger than $4q^2$; (4) and (5) give still other divisible forms. Scholium 1 points to the importance of the reciprocal numbers α, $(1 + nP)/\alpha$ and gives a list of them for all prime divisors from 3 to 29. This fact, writes Euler, gives a simple proof of Fermat's theorem that every prime of the form $4q + 1$ is the sum of two squares. Scholium 2 gives a list of divisors of the form $4q + 1$ with their residues paired against their reciprocals, -1 included in all cases. Then he draws the consequences in:

Scholium 3. Although here the number -1 appears among the residues whenever the divisor is a prime number of the form $4q + 1$, yet, when another prime number s is assigned as the form of prime divisors, it cannot be demonstrated that this number s can be found in the residues. We can mention this theorem:
If the prime divisor is of the form $4ns + (2x + 1)^2$, s being a prime number, then in the residues there occur the numbers $+s$ and $-s$, and another one similar to this:
If the prime divisor is of the form $4ns - (2x + 1)^2$, s being a prime number, then in the residues there occurs the number s, and $-s$ will be among the nonresidues.

When, on the contrary, $-s$ occurs in the residues, and $+s$ in the nonresidues, cannot in general be defined. For special cases, however, the situation can be indicated thus:

When	then the prime divisor must be
$\begin{cases} -2 \text{ is a residue} \\ +2 \text{ is a nonresidue} \end{cases}$	$P = 8n + 3;$
$\begin{cases} -3 \text{ is a residue} \\ +3 \text{ is a nonresidue} \end{cases}$	$P = 12n + 7;$
etc. to ± 23	$P = 92n + 3, 23, 27, \ldots, 87.$

The contemplation of such cases provides this theorem:

If the prime divisor is of the form $4ns - 4z - 1$, excluding all values contained in the form $4ns - (2x + 1)^2$, s being a prime number, then $-s$ occurs among the residues, and $+s$ is a nonresidue.

To these theorems we can still add the following:

If the prime divisor is of the form $4ns + 4z + 1$, excluding all values contained in the form $4ns + (2x + 1)^2$, s being a prime number, then both $+s$ and $-s$ occur among the nonresidues.

I add these theorems so that whoever likes to indulge in such speculation may inquire into their demonstration, since it cannot be doubted that from this the theory of numbers will receive great additions.

Conclusion.[5] These four final theorems, of which the demonstration from now on is desired, can be nicely formulated as follows:

Let s be some prime number, let only the odd squares 1, 9, 25, 49, etc. be divided by the divisor $4s$, and let the residues be noted, which will all be of the form $4q + 1$, of which any may be denoted by the letter α, and the other numbers of the form $4q + 1$, which do not appear among the residues, be denoted by some letter $\mathfrak{A}$, then we shall have

divisor a prime number of the form	then
$4ns + \alpha$	$+s$ is a residue, and $-s$ is a residue;
$4ns - \alpha$	$+s$ is a residue, and $-s$ is a nonresidue;
$4ns + \mathfrak{A}$	$+s$ is a nonresidue, and $-s$ is a nonresidue;
$4ns - \mathfrak{A}$	$+s$ is a nonresidue, and $-s$ is a residue.

---

[5] This is Euler's formulation of the theorem that Legendre and Gauss made known as the "reciprocity theorem of quadratic residues," and of which Legendre's demonstration is given in Selection I.12. The theorem is already contained in undeveloped form in a paper by Euler, "Theoremata circa divisores numerorum in hac forma *paa + qbb* contentorum," *Commentarii Academiae Scientiarum Petropolitanae 14* (1744/46, publ. 1751), 151–181, *Opera omnia*, ser. I, vol. 2, 194–222. It was L. Kronecker who in 1875 pointed out the importance of this theorem of Euler's for the history of the reciprocity theorem: *Werke*, ed. K. Hensel (Teubner, Leipzig), II (1903), 1–10.

11   GOLDBACH. THE GOLDBACH THEOREM

Christian Goldbach (1690–1764) was a German who for many years was in a leading position at the Saint Petersburg Academy of Sciences. He had wide scientific interests, but was especially devoted to number theory and analysis. His correspondence with Euler, which extends from 1729 to 1764, was partly published by Euler's greatgrandson Paul Heinrich Fuss in 1843 and has recently been published in full by A. P. Juškevich and E. Winter, *Leonhard Euler und Christian Goldbach* (Abhandlungen der Deutschen Akademie der Wissenschaften zu Berlin, Klasse für Philosophie, November 1, 1965; Akademie-Verlag, Berlin), with ample commentary. We select from this correspondence part of the letters dealing with the famous Goldbach theorem (or hypothesis). The originals are written in a curious amalgam of German and Latin, the technical expressions being usually given in Latin.

GOLDBACH IN MOSCOW TO EULER IN BERLIN, JUNE 7, 1742
(= May 27 O.S.[1])

I do not believe it useless also to pay attention to those propositions which are very likely, although there does not exist a real demonstration. Even in case they turn out at a later time to be false, yet they may have given occasion for the discovery of a new truth. The idea of Fermat, that every number $2^{2^{n-1}} + 1$ gives a sequence of prime numbers, cannot be correct, as you have already shown,[2] but it would be a remarkable fact if this series were to give only numbers which can be divided into two squares in only one way. Similarly, I also shall hazard a conjecture: that every number which is composed of two prime numbers is an aggregate of as many numbers as we like (including unity), till the combination of all unities [is reached].[3] [Goldbach adds in the margin:] After rereading this I find that the conjecture can be demonstrated in full rigor for the case $n + 1$, if it succeeds in the case for n and if $n + 1$ can be divided into two prime numbers. The demonstration is very easy. It seems at any rate that every number greater than 2 is an aggregate of three prime numbers.[4] [The text of

---

[1] O.S. = old style (the Julian calendar, which in that period differed by eleven days from the Gregorian calendar, N.S., which we use now).

[2] See below, and also Selection I.6

[3] That is, every number n which is a sum of two primes is a sum of as many primes as one wishes up to n. For Euler and Goldbach 1 is a prime number.

[4] This is the first formulation of Goldbach's theorem. When we begin the sequence of primes with 2, this theorem can be formulated as follows: every even number is the sum of two numbers that are either primes or 1. A somewhat more general formulation is that every even number > 2 is the sum of two primes. Then every odd number > 5 is the sum of three primes. For the history of this problem see L. E. Dickson, *History of the theory of numbers* (Carnegie Institution, Washington, D.C., 2nd ed., 1934), I, 421–424 and R. C. Archibald, "Goldbach's theorem," *Scripta Mathematica 3* (1935), 44–50, 153–161, and p. 106 of the above-mentioned edition of the Euler–Goldbach correspondence by Juškevich and Winter.

Goldbach's letter continues:] For example:

$$
4 = \begin{matrix} 1+1+1+1 \\ 1+1+2 \\ 1+3 \end{matrix} \qquad 5 = \begin{matrix} 2+3 \\ 1+1+3 \\ 1+1+1+2 \\ 1+1+1+1+1 \end{matrix}
$$

$$
6 = \begin{matrix} 1+5 \\ 1+2+3 \\ 1+1+1+3 \\ 1+1+1+1+2 \\ 1+1+1+1+1+1 \end{matrix}
$$

EULER IN BERLIN TO GOLDBACH IN MOSCOW, JUNE 30, 1742 (= JUNE 19 O.S.)

When all numbers included in this expression $2^{2^{n-1}} + 1$ can be divided into two squares in only one way, these numbers must also all be prime, which is not the case, for all these numbers are contained in the form $4m + 1$, which, whenever it is prime, can certainly be resolved into two squares in only one way, but when $4m + 1$ is not prime, it is either not resolvable into two squares, or is resolvable in more ways than one. That $2^{32} + 1$, which is not prime, can be divided into two squares in at least two ways I can show in the following way: I. When a and b are resolvable into two squares, then also the product ab will be resolvable into two squares. II. If the product ab and one of the factors a were numbers resolvable into two squares, then also the other factor b would be resolvable into two squares. These theorems can be demonstrated rigorously. Now $2^{32} + 1$, which is divisible into two squares, namely 2^{32} and 1, is divisible by $641 = 25^2 + 4^2$. Hence the other factor, which I will call b for short, must also be a sum of two squares. Let $b = pp + qq$, so that $2^{32} + 1 = (25^2 + 4^2)(pp + qq)$; then

$$
2^{32} + 1 = (25p + 4q)^2 + (25q - 4p)^2
$$

and at the same time

$$
2^{32} + 1 = (25p - 4q)^2 + (25q + 4p)^2;
$$

hence $2^{32} + 1$ is divisible into a sum of two squares in at least two ways. From this, the double reduction can be found a priori, since $p = 2556$ and $q = 409$, hence

$$
2^{32} + 1 = 65536^2 + 1^2 = 622664^2 + 20449^2.
$$

That every number which is resolvable into two prime numbers can be resolved into as many prime numbers as you like, can be illustrated and confirmed by an observation which you have formerly communicated to me, namely that every even number is a sum of two prime numbers. Indeed, let the proposed number n be even; then it is a sum of two primes, and since $n - 2$ is also a sum of two prime numbers, n must be a sum of three, and also four prime numbers, and so on. If, however, n is an odd number, then it is certainly a sum of three prime numbers, since $n - 1$ is a sum of two prime numbers, and can therefore also be resolved into as many primes as you like. However, that every number is a sum of two primes, I consider a theorem which is quite true, although I cannot demonstrate it.[5]

---

12   LEGENDRE. THE RECIPROCITY THEOREM

Euler's studies on quadratic residues and other fields in the theory of numbers were further developed by Adrien Marie Legendre (1752–1833), a geodesist and professor of mathematics at Paris, also known for his work on elliptic integrals and for his *Eléments de géométrie* (Paris, 1794). This last book, published in many editions and translations, did much to reform secondary education in geometry. There also exists an American edition, revised by Charles Davies, a professor at West Point (first ed., Barnes, New York, 1851).

Legendre published, without knowing Euler's paper of 1783, but knowing, of course, much of Euler's earlier works, the law of reciprocity in his lengthy article, "Recherches d'analyse indéterminée," in *Histoire de l'Académie Royale* (Paris, 1785), 465–559, Art. IV. He took up the matter again in his *Essai sur la théorie des nombres* (Paris, An VI = 1797/98), 214–226, from which the text passed without change into the *Théorie des nombres* (Firmin Didot, Paris, 1830), 230–243, from which we translate here a part of the text. In the *Essai* Legendre introduces the symbol $\left(\dfrac{a}{p}\right)$ which is still used to express the law of reciprocity:

$$\left(\frac{a}{p}\right) = +1 \quad \text{if } a \text{ is a quadratic residue (mod } p),$$

$$\left(\frac{a}{p}\right) = -1 \quad \text{if } a \text{ is a quadratic nonresidue (mod } p),$$

a and p being relatively prime. For this symbol the following laws hold (p an odd prime):

$$\left(\frac{a}{p}\right) \equiv a^{\frac{1}{2}(p-1)} \quad (\text{mod } p),$$

$$\left(\frac{a}{p}\right) = \left(\frac{b}{p}\right) \quad \text{if} \quad a \equiv b \quad (\text{mod } p),$$

$$\left(\frac{a}{b}\right)\left(\frac{b}{p}\right) = \left(\frac{ab}{p}\right),$$

$$\left(\frac{-1}{p}\right) = (-1)^{\frac{1}{2}(p-1)}, \quad \left(\frac{2}{p}\right) = (-1)^{\frac{1}{8}(p^2-1)},$$

where we use Gauss's symbol $\equiv$ for a congruence "modulo p."

---

[5] It seems that Euler never tried to prove the theorem, but in a letter to Goldbach of 5/16 December 1752 he stated the additional theorem (which also seems to have been suggested by Goldbach), that every even number of the form $4n + 2$ is equal to the sum of two primes of the form $4m + 1$; for example, $14 = 1 + 13$, $22 = 5 + 17$, $30 = 1 + 29 = 13 + 17$. See the above-mentioned edition of the correspondence, pp. 364, 365.

Goldbach's theorem was first published, independently, by the English mathematician Edward Waring (1734–1798) in his *Meditationes analyticae* (Cambridge, 1776), p. 217; (1782), p. 379, in the form:

Every even number consists of two prime numbers and every odd number either is prime or consists of three prime numbers.

THEOREM CONTAINING A LAW OF RECIPROCITY WHICH EXISTS BETWEEN TWO ARBITRARY PRIME NUMBERS

(166) We have seen that if m and n are two arbitrary prime numbers, odd and different from each other, the concise expressions $\left(\dfrac{m}{n}\right), \left(\dfrac{n}{m}\right)$ represent, the first the remainder of $m^{\frac{1}{2}(n-1)}$ divided by n, the other the remainder of $n^{\frac{1}{2}(m-1)}$ divided by m. We have proved at the same time that these remainders can only be $+1$ or -1. This being so defined, then there exists such a relation between the two remainders $\left(\dfrac{m}{n}\right), \left(\dfrac{n}{m}\right)$ that, if the one is known, the other is immediately determined. Here follows the general theorem which contains this relation.

Whatever may be the prime numbers m and n, if they are not both of the form $4x + 3$, then always $\left(\dfrac{n}{m}\right) = \left(\dfrac{m}{n}\right)$, and if they are both of the form $4x + 3$, then $\left(\dfrac{n}{m}\right) = -\left(\dfrac{m}{n}\right)$. These two general cases are contained in the formula

$$\left(\frac{n}{m}\right) = (-1)^{\frac{1}{2}(m-1)\frac{1}{2}(n-1)}\left(\frac{m}{n}\right).$$

In order to develop the different cases of this theorem we must distinguish, by special letters, the prime numbers of the form $4x + 1$ and those of the form $4x + 3$. In the course of this demonstration we shall denote the first ones by the letters A, a, α and the second ones by the letters B, b, β. Under these assumptions the theorem which we have just announced contains the eight following cases:

I   If $\left(\dfrac{a}{b}\right) = -1$, then $\left(\dfrac{b}{a}\right) = -1$;

II   If $\left(\dfrac{b}{a}\right) = +1$, then $\left(\dfrac{a}{b}\right) = +1$;

III If $\left(\dfrac{B}{b}\right) = +1$, then $\left(\dfrac{b}{B}\right) = -1$;

IV   If $\left(\dfrac{B}{b}\right) = -1$, then $\left(\dfrac{b}{B}\right) = +1$;

V   If $\left(\dfrac{a}{A}\right) = +1$, then $\left(\dfrac{A}{a}\right) = +1$;

VI   If $\left(\dfrac{a}{A}\right) = -1$, then $\left(\dfrac{A}{a}\right) = -1$;

VII If $\left(\dfrac{a}{b}\right) = +1$, then $\left(\dfrac{b}{a}\right) = +1$;

VIII If $\left(\dfrac{b}{a}\right) = -1$, then $\left(\dfrac{a}{b}\right) = +1$.

DEMONSTRATION OF CASES I AND II

(167) I observe first that the equation $x^2 + ay^2 = bz^2$, or more generally the equation $(4f + 1)x^2 + (4g + 1)y^2 = (4n + 3)z^2$, is impossible. Indeed, x and y are supposed to be relative primes, and the first member is therefore always of the form $4k + 1$ or $4k + 2$, while the second member can only be of the form $4k$ or $4k + 3$.

But the equation $x^2 + ay^2 = bz^2$ would be solvable, if two integers λ and μ could be found such that $(\lambda^2 + a)/b$ and $(\mu^2 - b)/a$ were integers.[1] On the other hand, the condition that b is a divisor of $\lambda^2 + a^2$ is $\left(\dfrac{-a}{b}\right) = 1$, or $\left(\dfrac{a}{b}\right) = -1$, and the condition that a is a divisor of $\mu^2 - b$ is $\left(\dfrac{b}{a}\right) = +1$. Hence we could not have at the same time $\left(\dfrac{a}{b}\right) = -1$ and $\left(\dfrac{b}{a}\right) = +1$; moreover, each of these expressions can only be $+1$, or -1, hence

I   If $\left(\dfrac{a}{b}\right) = -1$, then $\left(\dfrac{b}{a}\right) = -1$;

II If $\left(\dfrac{b}{a}\right) = +1$, then $\left(\dfrac{a}{b}\right) = +1$.

We may add that these two propositions are connected in the sense that the one is only a consequence of the other; since if we take the first as $\left(\dfrac{b}{a}\right) = +1$, then $\left(\dfrac{a}{b}\right) = -1$ is impossible, because this would imply $\left(\dfrac{b}{a}\right) = -1$, contrary to the supposition, hence we shall have $\left(\dfrac{a}{b}\right) = +1$.

DEMONSTRATION OF CASES III AND IV

(168) Since B and b are two prime numbers of the form $4n + 3$, we can apply the theorem that it is always possible to satisfy one of the equations

$$Bx^2 - by^2 = +1, \quad Bx^2 - by^2 = -1.[2]$$

Let (1) $\left(\dfrac{B}{b}\right) = +1$. The equation $Bx^2 - by^2 = -1$ cannot exist, for if it were satisfied b would be a divisor of $Bx^2 + 1$, or of $z^2 + B$; but then we would have $\left(\dfrac{-B}{b}\right) = 1$, or $\left(\dfrac{B}{b}\right) = -1$, contrary to the supposition. One of the two equations

---

[1] This was demonstrated in Art. 27, which states: "Given the equation $ax^2 + by^2 = cz^2$, in which the coefficients a, b, c taken individually or in pairs, have neither a square divisor nor a common divisor, then this equation is solvable if three integers λ, μ, ν can be found such that the three quantities $(a\lambda^2 + b)/c$, $(c\mu^2 - b)/a$, $(c\nu^2 - a)/b$ are integers. If these conditions are not fulfilled the equation has no solution."

[2] This was done in Art. 47.

being thus excluded, the other one, $Bx^2 - by^2 = +1$, is necessarily satisfied. From this we see that B is a divisor of $by^2 + 1$ or of $z^2 + b$; hence $\left(\dfrac{-b}{B}\right) = +1$, or $\left(\dfrac{b}{B}\right) = -1$.

Let (2) $\left(\dfrac{B}{b}\right) = -1$, then it is proved in a similar way that the equation $Bx^2 - by^2 = +1$ is impossible, so that the other equation $Bx^2 - by^2 = -1$ is satisfied, hence B is a divisor of $by^2 - 1$ or of $z^2 - b$, which gives $\left(\dfrac{b}{B}\right) = -1$.

Hence

III If $\left(\dfrac{B}{b}\right) = +1$, then $\left(\dfrac{b}{B}\right) = -1$;

IV If $\left(\dfrac{B}{b}\right) = -1$, then $\left(\dfrac{b}{B}\right) = +1$,

from which we see that $\left(\dfrac{B}{b}\right)$ and $\left(\dfrac{b}{B}\right)$ always have opposite sign.

DEMONSTRATION OF CASES V AND VI

(169) Let $\left(\dfrac{a}{A}\right) = +1$, then I say that from this it follows also that $\left(\dfrac{A}{a}\right) = +1$.

Indeed, let β be a prime number of the form $4n + 3$ which is a divisor of the formula $x^2 + a$; then we must have $\left(\dfrac{a}{\beta}\right) = -1$, hence, according to Case I, $\left(\dfrac{\beta}{a}\right) = -1$. Let us consider the impossible equation $x^2 + ay^2 = A\beta z$; this equation would be satisfied (no. 27)[1] if two integers λ and μ could be found such that $(\lambda^2 + a)/A\beta$ and $(\mu^2 - A\beta)/a$ were integers. The first condition is automatically satisfied, since it is necessary that $\left(\dfrac{-a}{A}\right) = -1$, or $\left(\dfrac{a}{A}\right) = +1$, to make $\lambda^2 + a$ divisible by A, and this is true by hypothesis, and to make $\lambda^2 + a$ divisible by β it is necessary that $\left(\dfrac{-a}{\beta}\right) = -1$, or $\left(\dfrac{a}{\beta}\right) = -1$, which also is true.

The second condition demands that $\left(\dfrac{A\beta}{a}\right) = +1$ or $\left(\dfrac{A}{a}\right)\left(\dfrac{\beta}{a}\right) = +1$, but we have already $\left(\dfrac{\beta}{a}\right) = -1$, hence we must have $\left(\dfrac{A}{a}\right) = -1$. This second condition cannot be fulfilled, since the proposed equation is impossible; hence $\left(\dfrac{A}{a}\right) = +1$.

Hence

V If $\left(\dfrac{a}{A}\right) = +1$, then $\left(\dfrac{A}{a}\right) = +1$.

Let now $\left(\dfrac{a}{A}\right) = -1$. We cannot have $\left(\dfrac{A}{a}\right) = +1$, since from this it would result, because of the case that we just demonstrated, that $\left(\dfrac{a}{A}\right) = +1$, contrary to the supposition. Hence we shall have $\left(\dfrac{A}{a}\right) = -1$. Hence

VI If $\left(\dfrac{a}{A}\right) = -1,$ then $\left(\dfrac{A}{a}\right) = -1.$

We have found[3] that when a and A are two prime numbers of the form $4n + 1$ it is always possible to satisfy one of the equations $Ax^2 - ay^2 = \pm 1$, $x^2 - Aay^2 = -1$. The first requires that $\left(\dfrac{A}{a}\right) = +1$ and $\left(\dfrac{a}{A}\right) = +1$; hence if we have $\left(\dfrac{A}{a}\right) = -1$ and $\left(\dfrac{a}{A}\right) = -1$—conditions that can always be derived from each other, as we have shown—then the second equation will be the only one possible, and will necessarily be true, from which there results this theorem:

"A and a being two prime numbers of the form $4n + 1$, if $\left(\dfrac{A}{a}\right) = -1$, or $\left(\dfrac{a}{A}\right) = -1$, then the equation $x^2 - Aay^2 = -1$ will always be possible."

The demonstration of Cases VII and VIII follows similarly; it is shown that of the six possibilities $\pm 1 = ax^2 - b\beta y^2$, $\pm 1 = bx^2 - a\beta y^2$, $\pm 1 = \beta x^2 - aby^2$ (β of the form $4n + 3$ such that $\left(\dfrac{a}{\beta}\right) = -1$) only the pair $+1 = bx^2 - a\beta y^2$, $-1 = bx^2 - a\beta y^2$ are possible.

---

(171) We may remark that the first four cases are demonstrated completely, in a way which leaves nothing to be desired. The other four suppose that when the number a of the form $4n + 1$ is given, then it is also possible to find a number β of the form $4n + 3$ such that β is a divisor of the formula $x^2 + a$, and that hence $\left(\dfrac{a}{\beta}\right) = -1$ holds.

The existence of this auxiliary theorem can immediately be proved when a is of the form $8n + 5$, because if we take $x = 1$ the number $x^2 + a$, now $1 + a$, is of the form $8n + 6$ and therefore divisible by a number of the form $4n + 3$, hence by a prime number of this form, which can be taken as β.

When a is of the form $8n + 1$, then we can observe that this form, considered in connection with multiples of 3, can be divided into two other forms, $24n + 1$ and $24n + 17$. As to the latter form, it suffices to take $x = 1$ and $x^2 + a$, now

---

[3] This was done in Art. 18.

$24n + 18$, is divisible by 3, so that we can take $p = 3$, and the condition $\left(\dfrac{a}{\beta}\right) = 1$ will be satisfied for every prime number a of the form $24n + 17$.

It remains to prove that for every number a of the form $24n + 1$, except unity, we can always find a prime number β of the form $4n + 3$ that will be a divisor of $z^2 + a$, or that will satisfy the condition $\left(\dfrac{a}{\beta}\right) = -1$.

First we can prove easily, by a simple substitution, that every prime number $24n + 1$, which can be expressed in one of the six forms

$$a = 168x + 17, 41, 73, 89, 97, 145,$$

is such that with the corresponding values

$$z = 2, 1, 2, 3, 1, 3$$

the formula $z^2 + a$ will be divisible by 7, so that for all the prime numbers of these forms the value $\beta = 7$ will satisfy the condition $\left(\dfrac{a}{\beta}\right) = -1$.

In the same way it is shown that for $a = 264x + 17, 41, 65, \dots, 241$ the numbers $z = 4, 5, 1, 2, 3, 2, 4, 5, 3, 1$ give $\beta = 11$. Since all prime numbers of the form $4n + 1$ from 73 to 1009, 15 in number, satisfy either the condition $\left(\dfrac{a}{7}\right) = -1$ or $\left(\dfrac{a}{11}\right) = -1$, the hypothesis is verified up to $a = 1009$, and for an infinity of larger primes a. Then Legendre tries to prove the hypothesis in general by using the theorem that, when the prime number a is of the form $8n + 1$, then $2fy^2 + 2gyz + fz^2$ is a quadratic divisor of the form $t^2 + au^2$, showing that prime divisors of $2fy^2 + 2gyz + fz^2$ cannot all be of the form $4n + 1$, hence there must be some of the form $4n + 3$, which will be divisors of $x^2 + a$. Gauss later showed that there are weaknesses in the argument.[4]

Legendre gives many applications, including proofs of theorems that Euler had found by induction in the *Opuscula analytica* (Saint Petersburg, 1783–1785), I; see Selection I.10.

---

[4] The first satisfactory proof of the law of reciprocity was given by Carl Friedrich Gauss, *Disquisitiones arithmeticae* (Fleischer, Leipzig, 1801), Arts. 125–146, now available in an English translation by A. A. Clarke (Yale University Press, New Haven, London, 1966), 82–100. Here, in Art. 135, we find Gauss's formulation of the law: If p is a prime of the form $4n + 1$, $+p$ will be a residue or nonresidue of any prime which is a residue or nonresidue of p. If p is of the form $4n + 3$, then $-p$ will have the same property.

Gauss published five more demonstrations, and another one was found among his papers; see his *Werke*, vols. I, II, also vol. X, part 2, Abh. II, 94–113. German translations of all six proofs by E. Netto appear in Ostwald's *Klassiker*, No. 122 (Engelmann, Leipzig, 1901).

CHAPTER II ALGEBRA

We have seen that mathematical studies in late medieval Latin Europe were stimulated by Latin translations from the Arabic. An important source of information was the treatise on equations written by Mohammed Al-Khwārizmī (*c.* A.D. 825). The Latin translation of its Arabic title, *Liber algebrae et almucabala*, gave the name *algebra* to the theory of equations until the nineteenth century; since that time the term has been used in a much wider sense.

Al-Khwārizmī's book dealt with linear and quadratic equations only. Sixteenth-century Italian mathematicians added the numerical solution of cubic and biquadratic equations. Gradually the study of equations came to involve the study of the character of the roots; the notation changed (with Descartes) to that which we use today, and the importance of algebra grew with its use in coordinate geometry and in the calculus with infinitesimals. One of the most intriguing problems was the question of the number of roots of an equation, which brought in negative and imaginary numbers, and led to the conclusion, in the work of Girard and Descartes, that an equation of degree n can have no more than n roots. The more precise statement, that an equation of degree n always has one root, and hence always has n roots (allowing for multiple roots), became known as the *fundamental theorem of algebra*. After several attempts by D'Alembert, Euler, and others, the proof was finally given by Gauss in 1799. In our selections we have tried to represent this trend.

There is no complete history of algebra, but many data can be found in Cantor, *Geschichte*, in J. Tropfke, *Geschichte der Elementar-Mathematik* (3rd ed.; De Gruyter, Berlin and Leipzig, 1937), III, 20–151, and in Smith, *History of mathematics*, II, 378–521.

1 AL-KHWĀRIZMĪ. QUADRATIC EQUATIONS

Latin Europe, during and after the twelfth century, received much of its information on the decimal position system and on linear and quadratic equations through translations of the work of Muhammad ibn Musa Al-Khwārizmī (or Mohammed the son of Moses from Khorezm, an area on the lower Amu Darya in the present USSR). He flourished in the time of the Baghdad Caliph Al-Ma'mun (813–833) and probably was a member of his "House of Wisdom," a kind of Academy. Al-Khwārizmī's *Arithmetic*, or *Algoritmus de numero Indorum*, was one of the sources by which the Hindu-Arabic decimal position system was introduced into Latin Europe. Here we shall deal with another of his books, his *Algebra*, which, although it contained little that was original, was widely used, in Arabic as well as in Latin, as a source of information on linear and quadratic equations.

The title of this treatise, *Ḥisāb al-jabr w'al-muqābala*, is rendered as *Liber algebrae et almucabola* in the Latin translation by Robert of Chester (*c.* A.D. 1140); we see how the name "algebra" is derived from "al-jabr." The literal meaning of the title is "The calculation of reduction and confrontation,"[1] words denoting, respectively, the transference of negative terms from one side of the equation to the other, and the combination of like terms on the two sides or on the same side. Mohammed's *Algebra* does not use symbolic algebraic notation: he writes out every problem in words, and so does Robert of Chester, except that numbers such as 5, 25, and so on, appear in the Latin manuscript in a notation not very different from ours.

We follow the English translation published by L. C. Karpinski, *Robert of Chester's Latin translation of the Algebra of Khowarizmi* (Macmillan, New York, 1915). The excerpts illustrate the way in which linear and quadratic equations are handled. They are divided into two groups of three types each, namely, in modern notation, (1) $ax^2 = bx$, $ax^2 = b$, $ax = b$; (2) $ax^2 + bx = c$, $ax^2 + b = cx$, $ax^2 = bx + c$. Here a, b, c are positive integers. Mohammed, like most authors before Viète, uses only specific numbers, writing for example, $x^2 + 10x = 39$. He has only what we call positive roots.

Mohammed accompanies his algebraic solution, given in the form of a recipe, by a geometric demonstration, inspired by Euclid's *Elements*. This double way of treating equations can also be followed until the time of Viète and Descartes, and seems to have been based on the feeling that only a geometric demonstration in the tradition of the ancient Greeks had sufficient convincing power. Euclid, in his *Elements*, indeed has a theory of linear and quadratic equations, but (to our feeling) it is hidden behind the façade of geometric theorems and constructions (the so-called application of areas). See Heath, *Euclid's Elements*, I, 383–388; II, 257–267. This same application, although somewhat simplified, appears in Al-Khwārizmī.

An English translation of Al-Khwārizmī's *Algebra* was published by F. Rosen (Oriental Translation Fund, London, 1831). See also S. Gandz, "The sources of al-Khowarizmi's algebra," *Osiris 1* (1936), 263–277.

THE BOOK OF ALGEBRA AND ALMUCABOLA

CONTAINING DEMONSTRATIONS OF THE RULES OF THE EQUATIONS OF ALGEBRA

... Furthermore I discovered that the numbers of restoration and opposition are composed of these three kinds: namely, roots, squares, and numbers.[2] However, number alone is connected neither with roots nor with squares by any ratio. Of these, then, the root is anything composed of units which can be multiplied by itself, or any number greater than unity multiplied by itself: or that which is found to be diminished below unity when multiplied by itself. The square is that which results from the multiplication of a root by itself.

---

[1] *Jabr* is the setting of a bone, hence reduction or restoration; *muqābala* is confrontation, opposition, face-to-face (explanation by Professor E. S. Kennedy).
[2] The term "roots" (*radices*) stands for multiples of the unknown, our x; the term "squares" (*substantiae*) stands for multiples of our x^2; "numbers" (*numeri*) are constants.

Of these three forms, then, two may be equal to each other, as for example:

Squares equal to roots,
Squares equal to numbers, and
Roots equal to numbers.[3]

CHAPTER I. CONCERNING SQUARES EQUAL TO ROOTS[4]

The following is an example of squares equal to roots: a square is equal to 5 roots. The root of the square then is 5, and 25 forms its square which, of course, equals five of its roots.

Another example: the third part of a square equals four roots. Then the root of the square is 12 and 144 designates its square. And similarly, five squares equal 10 roots. Therefore one square equals two roots and the root of the square is 2. Four represents the square.

In the same manner, then, that which involves more than one square, or is less than one, is reduced to one square. Likewise you perform the same operation upon the roots which accompany the squares.

CHAPTER II. CONCERNING SQUARES EQUAL TO NUMBERS

Squares equal to numbers are illustrated in the following manner: a square is equal to nine. Then nine measures the square of which three represents one root.

Whether there are many or few squares, they will have to be reduced in the same manner to the form of one square. That is to say, if there are two or three or four squares, or even more, the equation formed by them with their roots is to be reduced to the form of one square with its root. Further, if there be less than one square, that is, if a third or a fourth or a fifth part of a square or root is proposed, this is treated in the same manner.

For example, five squares equal 80. Therefore one square equals the fifth part of the number 80 which, of course, is 16. Or, to take another example, half of a square equals 18. This square therefore equals 36. In like manner all squares, however many, are reduced to one square, or what is less than one is reduced to one square. The same operation must be performed upon the numbers which accompany the squares.

CHAPTER III. CONCERNING ROOTS EQUAL TO NUMBERS

The following is an example of roots equal to numbers: a root is equal to 3. Therefore nine is the square of this root.

Another example: four roots equal 20. Therefore one root of this square is 5. Still another example: half a root is equal to ten. The whole root therefore equals 20, of which, of course, 400 represents the square.

Therefore roots and squares and pure numbers are, as we have shown, distinguished from one another. Whence also from these three kinds which we have

---

[3] In our notation, $x^2 = ax$, $x^2 = b$, $x = c$.
[4] Latin: *de substantiis numeros coaequantibus*. The examples are $x^2 = 5x$, $\frac{1}{3}x^2 = 4x$, $5x^2 = 10x$.

just explained, three distinct types of equations are formed involving three elements, as

A square and roots equal to numbers,
A square and numbers equal to roots, and
Roots and numbers equal to a square.

CHAPTER IV. CONCERNING SQUARES AND ROOTS EQUAL TO NUMBERS

The following is an example of squares and roots equal to numbers: a square and 10 roots are equal to 39 units.[5] The question therefore in this type of equation is about as follows: what is the square which combined with ten of its roots will give a sum total of 39? The manner of solving this type of equation is to take one-half of the roots just mentioned. Now the roots in the problem before us are 10. Therefore take 5, which multiplied by itself gives 25, an amount which you add to 39, giving 64. Having taken then the square root of this which is 8, subtract from it the half of the roots, 5, leaving 3. The number three therefore represents one root of this square, which itself, of course, is 9. Nine therefore gives that square.

Similarly, however many squares are proposed all are to be reduced to one square. Similarly also you may reduce whatever numbers or roots accompany them in the same way in which you have reduced the squares.

The following is an example of this reduction: two squares and ten roots equal 48 units. The question therefore in this type of equation is something like this: what are the two squares which when combined are such that if ten roots of them are added, the sum total equals 48? First of all it is necessary that the two squares be reduced to one. But since one square is the half of two, it is at once evident that you should divide by two all the given terms in this problem. This gives a square and 5 roots equal to 24 units. The meaning of this is about as follows: what is the square which amounts to 24 when you add to it 5 of its roots? At the outset it is necessary, recalling the rule above given, that you take one-half of the roots. This gives two and one-half which multiplied by itself gives $6\frac{1}{4}$. Add this to 24, giving $30\frac{1}{4}$. Take then of this total the square root, which is, of course, $5\frac{1}{2}$. From this subtract half of the roots, $2\frac{1}{2}$, leaving 3, which expresses one root of the square, which itself is 9.

. . . .

CHAPTER VI. GEOMETRICAL DEMONSTRATIONS[6]

We have said enough, says Al-Khowarizmi, so far as numbers are concerned, about the six types of equations. Now, however, it is necessary that we should

---

[5] This example, $x^2 + 10x = 39$, answer $x = 3$, "runs," as Karpinski notices in his introduction to this translation, "like a thread of gold through the algebras for several centuries, appearing in the algebras of Abu Kamil, Al-Karkhi and Omar al-Khayyami, and frequently in the works of Christian writers," and it still graces our present algebra texts. The solution of this type, $x^2 + ax = b$, is, as we can verify, based on the formula $x = \sqrt{(a/2)^2 + b} - a/2$.

[6] For these geometric demonstrations we must go back, as said, to Euclid's *Elements* (Book VI, Prop. 28, 29; see also Book II, Prop. 5, 6). See also on this subject the introduction to the *Principal works of Simon Stevin*, vol. IIB (Swets-Zeitlinger, Amsterdam, 1958), 464–467.

demonstrate geometrically the truth of the same problems which we have explained in numbers. Therefore our first proposition is this, that a square and 10 roots equal 39 units.

The proof is that we construct [Fig. 1] a square of unknown sides, and let this square figure represent the square (second power of the unknown) which together with its root you wish to find. Let the square, then, be *ab*, of which any side represents one root. When we multiply any side of this by a number (or numbers) it is evident that that which results from the multiplication will be a number of roots equal to the root of the same number (of the square). Since then ten roots were proposed with the square, we take a fourth part of the number ten and apply to each side of the square an area of equidistant sides, of which the length should be the same as the length of the square first described and the breadth $2\frac{1}{2}$, which is a fourth part of 10. Therefore four areas of equidistant sides are applied to the first square, *ab*. Of each of these the length is the length of one root of the square *ab* and also the breadth of each is $2\frac{1}{2}$, as we have just said.

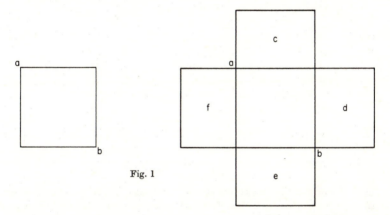

Fig. 1

These now are the areas *c*, *d*, *e*, *f*. Therefore it follows from what we have said that there will be four areas having sides of unequal length, which also are regarded as unknown. The size of the areas in each of the four corners, which is found by multiplying $2\frac{1}{2}$ by $2\frac{1}{2}$, completes that which is lacking in the larger or whole area. Whence it is that we complete the drawing of the larger area by the addition of the four products, each $2\frac{1}{2}$ by $2\frac{1}{2}$; the whole of this multiplication gives 25.

And now it is evident that the first square figure, which represents the square of the unknown [x^2], and the four surrounding areas [$10x$] make 39. When we add 25 to this, that is, the four smaller squares which indeed are placed at the four angles of the square *ab*, the drawing of the larger square, called *GH*, is completed [Fig. 2]. Whence also the sum total of this is 64, of which 8 is the root, and by this is designated one side of the completed figure. Therefore when we subtract from eight twice the fourth part of 10, which is placed at the extremities of the larger square *GH*, there will remain but 3. Five being subtracted from 8, 3 necessarily remains, which is equal to one side of the first square *ab*.

This three then expresses one root of the square figure, that is, one root of the proposed square of the unknown, and 9 the square itself. Hence we take half of ten and multiply this by itself. We then add the whole product of the multiplication to 39, that the drawing of the larger square *GH* may be completed; for the lack of the four corners rendered incomplete the drawing of the whole of this square. Now it is evident that the fourth part of any number multiplied by itself and then multiplied by four gives the same number as half of the number multiplied by itself. Therefore if half of the root is multiplied by itself, the sum total of this multiplication will wipe out, equal, or cancel the multiplication of the fourth part by itself and then by four.

Fig. 2

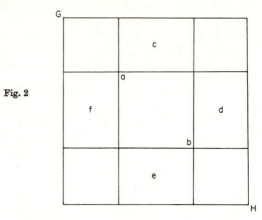

The remainder of the treatise deals with problems that can be reduced to one of the six types, for example, how to divide 10 into two parts in such a way that the sum of the products obtained by multiplying each part by itself is equal to 58: $x^2 + (10 - x)^2 = 58$, $x = 3$, $x = 7$. This is followed by a section on problems of inheritance.

2   CHUQUET. THE TRIPARTY

Nicolas Chuquet of Paris worked in Lyons, where he may have practiced medicine. His extensive work, *Le Triparty en la science des nombres du Maistre Nicolas Chuquet Parisien* (1484), so called because the book is divided into three sections (computation with rational numbers, computation with irrational numbers, and theory of equations), was not printed until 1880, but had considerable influence in manuscript. The book shows that in the mercantile city of Lyons a good deal of arithmetic and algebra was known, comparable to that known in leading cities of Italy and Germany. It can be studied in the *Bollettino di Bibliografia e di Storia delle Scienze Matematiche e Fisiche 13* (1880), 593–814, edited by B. Boncompagni from a manuscript in the Bibliothèque Nationale in Paris.

The third part has become known because of its notation for powers, which has a modern touch, since an exponential notation replaces the common Renaissance, so-called "cossist" notation, which uses special hieroglyphs for what we write (with Descartes) as x, x^2, x^3, x^4, ... (see, for example, the reproduced page of Recorde in Smith, *History of mathematics*, II, 412; also 427–431). In Chuquet we find 12^1 for our $12x$, 12^2 for $12x^2$, 12^3 for $12x^3$, and so on, and this consistently, so that $12^0 = 12$ and $12^{-1} = 12/x$ (-1 is written $1.\tilde{m}$); these exponents are called "denominacions" (this use of negative numbers was quite unusual in those days). Chuquet then shows that $x^m \cdot x^n = x^{m+n}$. How Chuquet does it we see from the following translation of a part of the fourth chapter (pp. 739–740, 746):

How to multiply a difference of number [*une differance de nombre*] by itself or by another similar or dissimilar to it.

Example. He who multiplies $.12^{\underline{0}}$ by $.12^{\underline{0}}$ obtains $.144.$, then he who adds $.0.$ to $.0.$ obtains $.0.$; hence this multiplication gives $.144..$ [1]

This means $12x^0 \cdot 12x^0 = 144x^{0+0} = 144x^0$.

Then he who multiplies $.12^{\underline{0}}$ by $.10^{\underline{2}}$ has first to multiply $.12.$ by $.10.$, which gives $.120.$, and then $.0.$ must be added to $.2..$ Thus the multiplication will give $120^{\underline{2}}$. By the same reasoning he who multiplies $.5^{\underline{1}}$ by $.8^{\underline{1}}$ obtains the multiplication $.40^{\underline{2}}$.

He who also wants to multiply $.12^{\underline{3}}$ by $.10^{\underline{5}}$ must first multiply $.12.$ by $.10.$, obtaining $.120.$, then must add the denominations together, which are $.3.$ and $.5.$, giving $.8..$ Hence the multiplication gives $.120^{\underline{8}}$.

Also he who wants to multiply $.8^{\underline{1}}$ by $.7^{1.\tilde{m}}$ obtains as multiplication $.56.$, then he who adds the denominations together will take $1.\tilde{p}$ with $.1.\tilde{m}$ and obtains $.0..$ Here he obtains the multiplication $.56^{\underline{0}}$. [2]

Similarly, he who would multiply $.8^{\underline{3}}$ by $.7^{1.\tilde{m}}$ will find it convenient first to multiply $.8.$ by $.7..$ He obtains $.56.$, then he must add the denominations, and will take $3.\tilde{p}$ with $.1.\tilde{m}$ and obtain $.2..$ Hence the multiplication gives $.56^{\underline{2}}$ and in this way we must understand other problems.

Then follows a list of powers of 2 with their denominations:

number	1	2	4	8	16	32	64	⋯	1048576
denomination	0	1	2	3	4	5	6	⋯	20

---

[1] Qui multiplie $.12^{\underline{0}}$ par $.12^{\underline{0}}$ montant $.144.$, puis qui adiouste $.0.$ avec $.0.$ monte $.0..$ Ainsi monte ceste multiplicacion $.144..$

[2] $.1.\tilde{p}$ with $.1.\tilde{m}$ means $+1 + (-1)$.

ON EQUIPOLLENCES OF NUMBERS

... And in order better to understand what has been said above about this art and style of abbreviating and equating its terms [*parties*] and of bringing them back to two simple terms as well as one can do it, we shall give here some examples of which the first is as follows—I shall abbreviate:

$$R^2\ \underline{4^2.\tilde{p}.4^1}\ \tilde{p}.2^{\frac{1}{}}.\tilde{p}.1 \quad \text{equal to } .100.^3$$

First I take away $.2^{\frac{1}{}}.\tilde{p}.1$ from both terms and there remains to me $R^2\ \underline{4^2.\tilde{p}.4^1}$ in one term and $99.\tilde{m}.2^{\frac{1}{}}$ in the other. And now that one of the terms is a $\overline{\text{second}}$ root it is convenient to multiply it by itself and we obtain $.4^2.1.\tilde{p}.4^1$ in this term. And similarly we must multiply $.99.\tilde{m}.2^{\frac{1}{}}$ by itself and we obtain $9801.\tilde{m}.396^{\frac{1}{}}\tilde{p}.4^2$ in the other term. Now we still must abbreviate these terms by taking away $.4^2$ from the one and the other term. And then add $.396^{\frac{1}{}}$ to each of them. In this way we shall have $.400^{\frac{1}{}}$ in one term and $.9801.$ in the other term.[4]

After more of this there follows a theory of quadratic equations, in which negative roots are rejected. Chuquet has negative but no fractional exponents. Those we meet, even before Chuquet, in the *Algorismus proportionum* by Nicole Oresme (c. 1323–1382; see Selection

III.1). Here we find a notation $\boxed{\dfrac{1}{2}}$ for $\frac{1}{2}$, $\boxed{\dfrac{p\ \ 1.}{1\ .2.}}$ for $1\frac{1}{2}$, $\boxed{\dfrac{p\ \ 1}{1\ \ 3}}$ for $1\frac{1}{3}$, $\boxed{\dfrac{p\ \ 2}{2\ \ 4}}$ for $2\frac{2}{4}$,

and $\boxed{\dfrac{p\ \ 1.}{1\ \ 2.}}$ 4 for $4^{1\frac{1}{}}$. The p stands for *proportio*. The dots (.) are sometimes present, some-

times absent in the manuscript text reproduced by F. Cajori, *History of mathematical notations* (Open Court, Chicago, 1928), I, 92. A variant $\boxed{1^{p}\frac{1}{2}}$ 4 is found in Cantor, *Geschichte*, II, 121. On the *Algorismus* see *De proportionibus proportionum and Ad pauca respicientes*, ed. E. Grant (University of Wisconsin Press, Madison, 1966), 65–68.

3   CARDAN. ON CUBIC EQUATIONS

The discovery of the numerical solutions of equations of the third degree at the University of Bologna in the early years of the sixteenth century was an important step in the development of algebra. It attracted wide attention, and was discussed in many public disputations. The textbook that laid the whole method open to public inspection was the *Ars magna* (Nuremberg, 1545) by the physician, humanist, mathematician, and scientist-in-general Gerolamo Cardano, or Hieronymus Cardanus, or, in English, Jerome Cardan (1501–1576). Here he stated that Scipio del Ferro at Bologna had discovered the method of solving equations of the type $x^3 + px = q$. Nicolo Tartaglia (c. 1499–1557) had also discovered this,

---

[3] $\sqrt{4x^2 + 4x} + 2x + 1 = 100.$

[4] $\sqrt{4x^2 + 4x} = 99 - 2x; \ 4x^2 + 4x = 9801 - 396x + 4x^2; \ 4x = 9801 - 396x; \ 400x = 9801.$

and then found a method of solving equations of the type $x^3 = px + q$, $x^3 + q = px$. Cardan obtained the solutions from Tartaglia (breaking a pledge of secrecy) and the method of solving cubic equations numerically has ever since been called after Cardan. The *Ars magna* was for many decades the best-known book on algebra, studied by all who were interested, and it lost this position only when Descartes introduced his new methods.

We quote here an English translation of a part of Chapter XI (pp. 29^r-30^r), dealing with the equation $x^3 + px = q$, or in particular $x^3 + 6x = 20$. It is based, as is also the text of Selection II.4, on the translation published in Smith, *Source book*, 204–212. Cardan's notation is quite different from ours, and he expresses the equation by saying: "A cube and unknowns are equal to a number" (Cubus et res aequales numero). For "unknown," our x, he has, like most of his contemporaries, the Latin term *res*, Italian *cosa*, literally, "thing." A cube is conceived as a solid body. By "number" is meant a numerical coefficient, in this case 20.

The book contains solutions for quadratics and for many types of cubes and biquadratics. The coefficients are always positive and specific numbers. Cardan also teaches some properties of equations and their roots. For instance (in Chapter XVII) we read that the equation $x^3 + 10x = 6x^2 + 4$ has three roots, namely 2, $2 + \sqrt{2}$, $2 - \sqrt{2}$, and Cardan sees that their sum adds up to the coefficient of x^2. Cardan is puzzled when imaginaries appear, and keeps them out of the *Ars magna* except in one case (see below), where he meets them in the solution of a quadratic equation. The *casus irreducibilis*, where a real root appears as a sum of the cube roots of two imaginaries (as in $x^3 = 15x + 4$, where $x = 4$, but the Cardan formula gives $x = \sqrt[3]{2 + \sqrt{-121}} + \sqrt[3]{2 - \sqrt{-121}}$) is discussed in the works of Bombelli (1572) and Viète (1591).

On Cardan see O. Ore, *Cardano, the gambling scholar* (Princeton University Press, Princeton, New Jersey, 1953). On the *Ars magna* see J. F. Scott, *A history of mathematics* (Taylor and Francis, London, 1958), 87–92. On Italian mathematicians of the Renaissance, see further E. Bortolotti, *Studi e ricerche sulla storia della matematica in Italia nei secoli XVI e XVII* (Zanichelli, Bologna, 1928).

CONCERNING A CUBE AND UNKNOWNS EQUAL TO A NUMBER

Chapter XI

Scipio del Ferro of Bologna about thirty years ago invented [the method set forth in] this chapter, [and] communicated it to Antonio Maria Florido of Venice, who when he once engaged in a contest with Nicolo Tartaglia of Brescia announced that Nicolo also invented it: and he [Nicolo] communicated it to us when we asked for it, but suppressed the demonstration.[1] With this aid we sought the demonstration, and found it, though with great difficulty, in the manner which we set out in the following.

Demonstration. For example, let the cube of *GH* and six times the side *GH* be

---

[1] Tartaglia and Cardan met in Milan during 1539, after which Tartaglia gave Cardan his method in obscure verses, which he later clarified. Cardan soon mastered the method and knew how to apply it independently. The verses begin as follows:

Quando che'l cubo con le cose appresso	When x^3 together with px
Se agguaglia a qualche numero discreto	Are equal to a q
Trovan dui altri, differenti in esso...	Then take u and v, $u \neq v$...

equal to 20.[2] I take [Fig. 1] two cubes AE and CL whose difference shall be 20, so that the product of the side AC by the side CK shall be 2, i.e., a third of the number of unknowns, and I lay off CB equal to CK; then I say that if it is done thus, the remaining line AB is equal to GH and therefore to the value of the unknown (for it was supposed of GH that it was so). Therefore I complete, after the manner of the first theorem of the 6th chapter of this book,[3] the solids DA, DC, DE, DF, so that we understand by DC the cube of BC, by DF the cube of AB, by DA three times CB times the square of AB, by DE three times AB times the square of BC. Since therefore from AC times CK the result is 2, from 3 times AC times CK will result 6, the number of unknowns, and therefore from AB times 3 AC times CK there results 6 unknowns AB, or 6 times AB, so that 3 times the product of AB, BC, and AC is 6 times AB. But the difference of the cube AC from the cube CK, and likewise from the cube BC, equal to it by hypothesis, is 20; and from the first theorem of the 6th chapter, this is the sum of the solids DA, DE, and DF, so that these three solids make 20. But taking BC minus, the cube of AB is equal to the cube of AC and 3 times AC into the square of CB and minus the cube of BC and minus 3 times BC into the square of AC. By the demonstration, the difference between 3 times CB times the

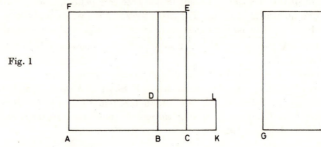

Fig. 1

---

[2] We can follow the reasoning more easily if we take the equation as $x^3 + px = q$, $p = 6$, $q = 20$, and $GH = x$, $AC = u$, $CK = v$, so that the cube $AE = u^3$, $DC = v^3$. Then u and v are selected so that $u^3 - v^3 = q = 20$, $uv = p/3 = 2$. Then we must prove that $AB = u - v = GH = x$. For this purpose we use for $AE = u^3$ the expression for the third power of the binomial ("the first theorem of the 6th chapter," this theorem being stated as a property of solids):

$$u^3 = [(u - v) + v]^3 = (u - v)^3 + 3v(u - v)^2 + 3v^2(u - v) + v^3$$

so that

$$u^3 - v^3 = (u - v)^3 + 3v(u - v)^2 + 3(u - v)v^2 = (u - v)^3 + 3uv(u - v),$$

or

$$u^3 - v^3 = q = (u - v)^3 + p(u - v).$$

Hence, since $q = x^3 + px$ (here Cardan quotes theorems in Euclid's *Elements*, book XI, dealing with the equality of parallelepipeds; the numbering of the propositions differs in ancient editions), we see that $x = u - v$, or $AB = GH$.

[3] Cardan writes in Chapter VI that after Tartaglia had handed over to him his rule he "thought that this would be the royal road to pursue in all cases." And so he established three theorems; in our notation they are:

(a) If $a = u + v$, then $a^3 = u^3 + v^3 + 3(u^2v + uv^2)$;
(b) $u^3 + 3uv^2 > v^3 + 3u^2v$, the difference being $(u - v)^3(u > v)$;
(c) By Euclid's theory of proportions,

$$\frac{u^3 + v^3}{3uv^2 + 3vu^2} = \frac{u^3 - u^2v + uv^2}{3vu^2}$$

square of AC, and 3 times AC times the square of BC, is [3 times] the product of AB, BC, and AC. Therefore since this, as has been shown, is equal to 6 times AB, adding 6 times AB to that which results from AC into 3 times the square of BC there results 3 times BC times the square of AC, since BC is minus. Now it has been shown that the product of CB[4] into 3 times the square of AC is minus; and the remainder which is equal to that is plus, hence 3 times CB into the square of AC and 3 times AC into the square of CB and 6 times AB make nothing. Accordingly, by common sense, the difference between the cubes AC and BC is as much as the totality of the cube of AC, and 3 times AC into the square of CB, and 3 times CB into the square of AC (minus), and the cube of BC (minus), and 6 times AB. This therefore is 20, since the difference of the cubes AC and CB was 20. Moreover, by the second theorem of the 6th chapter, putting BC minus, the cube of AB will be equal to the cube of AC and 3 times AC into the square of BC minus the cube of BC and minus 3 times BC into the square of AC. Therefore the cube of AB, with 6 times AB, by common sense, since it is equal to the cube of AC and 3 times AC into the square of CB, and minus 3 times CB into the square of AC, and minus the cube of CB and 6 times AB, which is now equal to 20, as has been shown, will also be equal to 20. Since therefore the cube of AB and 6 times AB will equal 20, and the cube of GH, together with 6 times GH, will equal 20, by common sense and from what has been said in the 35th and 31st of the 11th Book of the *Elements*, GH will be equal to AB, therefore GH is the difference of AC and CB. But AC and CB, or AC and CK, are numbers or lines containing an area equal to a third part of the number of unknowns whose cubes differ by the number in the equation, wherefore we have the

RULE [5]

Cube the third part of the number of unknowns, to which you add the square of half the number of the equation, and take the root of the whole, that is, the

---

[4] Here begins the text of p. 30ʳ of the *Ars magna*, reproduced in Fig. 2.

[5] This rule is known as Cardan's rule for the case $x^3 + px = q$. In our notation:
Since $u^3 - v^3 = q = 20$, $uv = p/3$, we can find $x = u - v$ by solving a quadratic equation. Since $v = p/3u$, $u^3 - (p/3u)^3 = q$, $u^6 - qu^3 - (p/3)^3 = u^6 - 20u^3 - p = 0$, we find

$$u^3 = \frac{q}{2} \pm \sqrt{\left(\frac{q}{2}\right)^2 + \left(\frac{p}{3}\right)} = 10 \pm \sqrt{100 + 8} = 10 \pm \sqrt{108}$$

Similarly:

$$v^3 = -\frac{q}{2} \pm \sqrt{\left(\frac{q}{2}\right)^2 + \left(\frac{p}{3}\right)^3} = -10 \pm \sqrt{108}.$$

Cardan now states in the "Rule": for u^3 take $10 + \sqrt{108}$ (this is the *binomial*), for v^3 take $-10 + \sqrt{108}$ (this is the *apotome*; both expressions are from Euclid's *Elements*, Book X), so that

$$x = \sqrt[3]{\frac{q}{2} + \sqrt{\left(\frac{q}{2}\right)^2 + \left(\frac{p}{3}\right)^3}} - \sqrt[3]{\frac{q}{2} + \sqrt{\left(\frac{q}{2}\right)^2 + \left(\frac{p}{3}\right)^3}}$$

$$= \sqrt[3]{10 + \sqrt{108}} - \sqrt[3]{-10 + \sqrt{108}}.$$

In Cardan's notation (see Fig. 2, fourth line from the bottom):

℞ V: cub: ℞ 108 *p*: 10 *m*: ℞ V: cubica ℞ 108 *m*: 10;

here *p* stands for "piu," plus, *m* for "meno," minus, and ℞ for "radix." Cardan does not use the signs $+$, $-$, although they were already in use at the time.

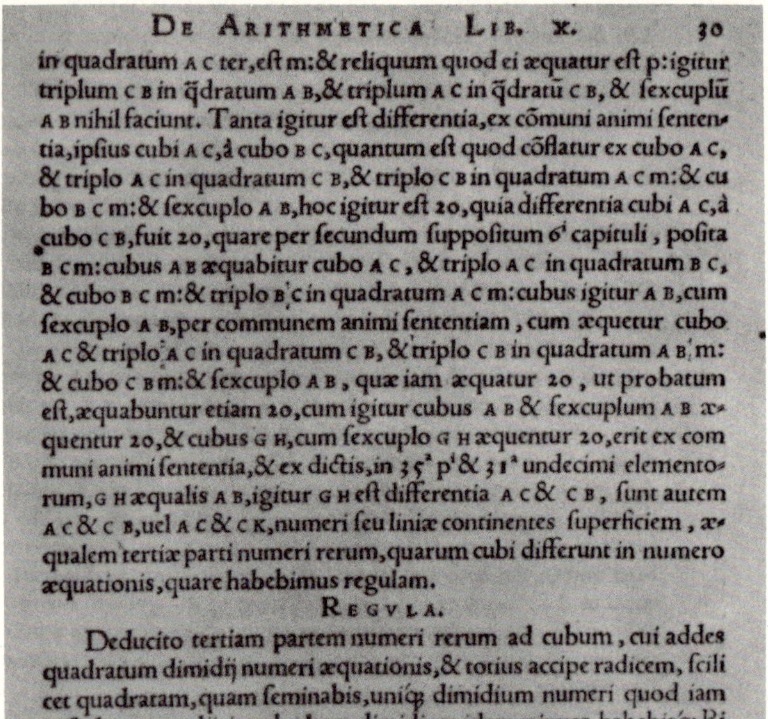

in quadratum a c ter, est m: & reliquum quod ei æquatur est p: igitur
triplum c b in q̃dratum a b, & triplum a c in q̃dratũ c b, & sexcuplũ
a b nihil faciunt. Tanta igitur est differentia, ex cõmuni animi sententia, ipsius cubi a c, â cubo b c, quantum est quod cõflatur ex cubo a c,
& triplo a c in quadratum c b, & triplo c b in quadratum a c m: & cu
bo b c m: & sexcuplo a b, hoc igitur est 20, quia differentia cubi a c, â
cubo c b, fuit 20, quare per secundum suppositum 6ᵗ capituli, posita
b c m: cubus a b æquabitur cubo a c, & triplo a c in quadratum b c,
& cubo b c m: & triplo b c in quadratum a c m: cubus igitur a b, cum
sexcuplo a b, per communem animi sententiam, cum æquetur cubo
a c & triplo a c in quadratum c b, & triplo c b in quadratum a b m:
& cubo c b m: & sexcuplo a b, quæ iam æquatur 20, ut probatum
est, æquabuntur etiam 20, cum igitur cubus a b & sexcuplum a b æ-
quentur 20, & cubus g h, cum sexcuplo g h æquentur 20, erit ex com
muni animi sententia, & ex dictis, in 35ᵃ pᶦ & 31ᵃ undecimi elemento-
rum, g h æqualis a b, igitur g h est differentia a c & c b, sunt autem
a c & c b, uel a c & c k, numeri seu liniæ continentes superficiem, æ-
qualem tertiæ parti numeri rerum, quarum cubi differunt in numero
æquationis, quare habebimus regulam.

Regvla.

Deducito tertiam partem numeri rerum ad cubum, cui addes
quadratum dimidij numeri æquationis, & totius accipe radicem, scili
cet quadratam, quam seminabis, uniç dimidium numeri quod iam
in se duxeras, adijcies, ab altera dimidium idem minues, habebisq̃ Bi
nomium cum sua Apotome, inde detracta ℞ cubica Apotomæ ex ℞
cubica sui Binomij, residuũ quod ex hoc relinquitur, est rei æstimatio.

Exemplum. cubus & 6 positiones, æquantur 20, ducito 2, tertiam partem 6, ad cubum, fit 8, duc 10 dimidium numeri in se,
fit 100, iunge 100 & 8, fit 108, accipe radicem quæ est ℞ 108, & eam geminabis, alte
ri addes 10, dimidium numeri, ab altero mi
nues tantundem, habebis Binomiũ ℞ 108
p: 10, & Apotomen ℞ 108 m: 10, horum
accipe ℞ᶜⁱⁱ cubⁱⁱ & minue illam quæ est Apo

cub⁹ p: 6 reb⁹ æq̃lis 20		
2		20
8 ———— 10		
108		
℞ 108 p: 10		
℞ 108 m: 10		
℞ v: cu. ℞ 108 p: 10		
m: ℞ v: cu. ℞ 108 m: 10		

tomæ, ab ea quæ est Binomij, habebis rei æstimationem, ℞ v: cub: ℞
108 p: 10 m: ℞ v: cubica ℞ 108 m: 10.

Aliud, cubus p: 5 rebus æquetur 10, duc 1, tertiam partem 5, ad
cubum, fit 1, duc 5, dimidium 10, ad quadratum, fit 25, iunge 25 & 1,

H 2　　　　　fiunt

Fig. 2

square root, which you will use, in the one case adding the half of the number which you just multiplied by itself, in the other case subtracting the same half, and you will have a binomial and apotome respectively; then subtract the cube root of the apotome from the cube root of the binomial, and the remainder from this is the value of the unknown. In the example, the cube and 6 unknowns equals 20; raise 2, the 3rd part of 6, to the cube, that makes 8; multiply 10, half the number, by itself, that makes 100; add 100 and 8, that makes 108; take the root, which is $\sqrt{108}$, and use this, in the first place adding 10, half the number, and in the second place subtracting the same amount, and you will have the binomial $\sqrt{108} + 10$, and the apotome $\sqrt{108} - 10$; take the cube root of these and subtract that of the apotome from that of the binomial, and you will have the value of the unknown $\sqrt[3]{\sqrt{108} + 10} - \sqrt[3]{\sqrt{108} - 10}$.

Cardan continues to discuss one case after another. Here are, in our notation, the headings of the different chapters:

11.	$x^3 + ax = b$	20.	$x^3 = ax^2 + bx + c$
12.	$x^3 = ax + b$	21.	$x^3 + a = bx^2 + cx$
13.	$x^3 + a = bx$	22.	$x^3 + ax + b = cx^2$
14.	$x^3 = ax^2 + b$	23.	$x^3 + ax^2 + b = cx$
15.	$x^3 + ax^2 = b$	24.	On the 44 derivative
16.	$x^3 + a = bx^2$		equations (for example,
17.	$x^3 + ax^2 + bx = c$		$x^6 + 6x^4 = 100$)
18.	$x^3 + ax = bx^2 + c$	25.	On imperfect and par-
19.	$x^3 + ax^2 = bx + c$		ticular rules.

Chapter 26 and later chapters also deal with biquadratic equations.

Many examples follow. We occasionally meet negative numbers, which Cardan calls "fictitious" (*fictae*). Another element enters in the following example, taken from Chapter 37, "On the rule of postulating a negative," which involves imaginaries. We substitute modern notation.

I will give as an example:[6] If some one says to you, divide 10 into two parts, one of which multiplied into the other shall produce 30 or 40, it is evident that this case or question is impossible. Nevertheless, we shall solve it in this fashion. Let us divide 10 into equal parts and 5 will be its half. Multiplied by itself, this yields 25. From 25 subtract the product itself, that is 40, which, as I taught you in the chapter on operations in the sixth book, leaves a remainder -15. The root of this added to and then subtracted from 5 gives the parts which multiplied together will produce 40. These, therefore, are $5 + \sqrt{-15}$ and $5 - \sqrt{-15}$.

[6] Here begins the text of p. 66^r of the *Ars magna*, reproduced in Fig. 3.

De Arithmetica Lib. x. 66

exemplum, si quis dicat, diuide 10 in duas partes, ex quarum unius in
reliquam ductu, producatur 30, aut 40, manifestum est, quòd casus
seu quæstio est impossibilis, sic tamē operabimur, diuidemus 10 per
æqualia, & fiet eius medietas 5, duc in se fit 25, auferes ex 25, ipsum
producendum, utpote 40, ut docui te, in capitulo operationum, in sex-
to libro, fiet residuum m: 15, cuius ℞ addita & detracta a 5, ostendit
partes, quæ inuicem ductæ producunt 40, erunt igitur hæ, 5 p: ℞ m:
15, & 5 m: ℞ m: 15.

Demonstratio

Vt igitur regulæ uerus pateat intellectus, sit A B linea, quę dicatur   2
10, diuidenda in duas partes, quarū rectangulum debeat esse 40, est
aūt 40 q̄druplū ad 10, quare nos uolumus
quadruplum totius A B, igitur fiat A D, quadratum A C, dimidij A B, & ex A D auferatur
quadruplum A B, absçʒ numero, ℞ igitur re
sidui, si aliquid maneret, addita & detracta
ex A C, ostenderet partes, at quia tale residu
um est minus, ideo imaginaberis ℞ m: 15, id est differentiæ A D, &
quadrupli A B, quam adde & minue ex A C, & habebis quæsitum, scilicet 5 p: ℞ v: 25 m: 40, & 5 m: ℞ v: 25 m: 40, seu 5 p: ℞ m: 15, & 5
m: ℞ m: 15, duc 5 p: ℞ m: 15 in 5 m: ℞ m: 15, dimissis incruciationi-
bus, fit 25 m: m: 15, quod est p: 15, igitur hoc productum est 40, natu
ra tamē A D, non est eadem cū natura 40, nec A B, quia superficies est

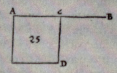

remota à natura numeri, & lineæ, proximius
tamē huic quantitati, quę uere est sophistica,
quoniam per eam, non ut in puro m: nec in
alijs, operationes exercere licet, nec uenari

| 5 p: ℞ m: 15 |
| 5 m: ℞ m: 15 |
| 25 m: m: 15 q̄d. est 40 |

quid sit est, ut addas quadratum medietatis numeri numero produ-
cendo, & à ℞ aggregati minuas ac addas dimidium diuidendi. Exem
plū, in hoc casu, diuide 10 in duas partes, producentes 40, adde 25
quadratū dimidij 10 ad 40, fit 65, ab huius ℞ minue 5, & adde etiam
5, habebis partes secundum similitudinem, ℞ 65 p: 5 & ℞ 65 m: 5. At
hi numeri differunt in 10, non iuncti faciunt 10, sed ℞ 260, & hucusçʒ
progreditur Arithmetica subtilitas, cuius hoc extremum ut dixi, adeo
est subtile, ut sit inutile.

Qvæstio IIII.

Fac de 6 duas partes, quarum quadrata iuncta sint 50, hæc solui
tur per primam, non per secundam regulam, est enim de puro m: ideo
duc 3 dimidium 6 in se, fit 9, minue ex dimidio 50, quod est 25, fit residuum

Fig. 3

Proof. That the true significance of this rule may be made clear, let the line AB [see Fig. 3], which is called 10, be the line which is to be divided into two parts whose rectangle is to be 40. Now since 40 is the quadruple of 10, we wish four times the whole of AB. Therefore, make AD the square on AC, the half of AB. From AD subtract four times AB. If there is a remainder, its root should be added to and subtracted from AC thus showing the parts [into which AB was to be divided]. Even when such a residue is negative, you will nevertheless imagine $\sqrt{-15}$ to be the difference between AD and the quadruple of AB which you should add to and subtract from AC to find what was sought. That is $5 + \sqrt{25-40}$ and $5 - \sqrt{25-40}$, or $5 + \sqrt{-15}$ and $5 - \sqrt{-15}$. Dismissing mental tortures, and multiplying $5 + \sqrt{-15}$ by $5 - \sqrt{-15}$, we obtain $25 - (-15)$ which is $+15$. Therefore the product is 40. However, the nature of AD is not the same as that of 40 or AB because a surface is far from a number or a line. This, however, is closest to this quantity, which is truly puzzling since operations may not be performed with it as with a pure negative number or with other numbers.[7] Nor can we find it by adding the square of half the number to the product number and take away and add from the root of the sum half of the dividend. For example, in the case of dividing 10 into two parts whose product is 40, you add 25, the square of one half of 10, to 40 making 65. From the root of this subtract 5 and then add 5 and according to similar reasoning you will have $\sqrt{65} + 5$ and $\sqrt{65} - 5$. But these numbers differ by 10, and do not make 10 jointly.[8] This subtlety results from arithmetic the final point of which is, as I have said, as subtile as it is useless.

---

4 FERRARI. THE BIQUADRATIC EQUATION

Cardan's *Ars magna* not only presented the numerical solution of cubic equations, but—to the surprise of his contemporaries—also showed how a biquadratic equation can be solved. This was accomplished by a young friend of Cardan's, Ludovico Ferrari (1522–1565), who tried his talents on the equation $x^4 + 6x^2 + 36 = 60x$. The method has since been known as the method of Ferrari.

The text begins with a square AD, of which the side AB is supposed to be itself a square, say $AB = x^2$. Added to AB is a part $BC = p = 3$. Then by means of another addition $CG = y$ the square AH is obtained. Figure 1 shows that the area $LNM = y^2 + 2yp$, where $MD = BC = p$.

---

[7] The sentence is: *quae vere est sophistica, quoniam per eam, non ut in puro m: nec in aliis, operationes exercere licet, nec venari quid sit.* T. R. Witmer (in a translation of the *Ars magna* to be published by the M.I.T. Press) translates this: "This truly is sophisticated, since through it one can (as one cannot in the case of a pure negative) perform operations and pursue a will-o'-the wisp."

[8] Since $x_1 + x_2 = 10$, $x_1 x_2 = 40$, the equation to be solved is $x^2 - 10x + 40 = 0$, hence $x = 5 \pm \sqrt{25 - 40}$. Cardan, puzzled by this "sophistical subtlety," asks whether perhaps $x = \pm 5 + \sqrt{25 + 40}$ will do, but then $x_1 - x_2 = 10$ and not $x_1 + x_2$.

What is shown then is that, if an equation of the form $x^4 + px^2 + qx + r = 0$ is given, it can be written

(A) $(x^2 + p + y)^2 = p^2 + px^2 - qx - r + 2y(x^2 + p) + y^2$

$\qquad\qquad = x^2(p + 2y) - qx + (p^2 - r + 2py + y^2),$

so that the problem is reduced to the finding of a value of y that makes the right-hand member a square in x. This leads to an equation of the third degree in y:

$$4(p + 2y)(p^2 - r + 2py + y^2) - q^2 = 0.$$

Solution of this equation in y leads to the solution of the original biquadratic equation.

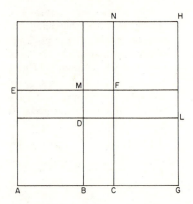

Fig. 1

If therefore AD [Fig. 1] is made 1 fourth power[1] and CD and DE are made 3 squares, and DF is made 9, BA will necessarily be a square and BC will necessarily be 3. Since we wish to add some squares to DC and DE, let these [additions] be [the rectangles] CL and KM. Then in order to complete the square it will be necessary to add the area LNM. This has been shown to consist of the square on GC, which is half the number of [added] squares, since CL is the area [made] from [the product of] GC times AB, where AB is a square, AD having been assumed to be a fourth power. But FL and MN are each equal to GC times CB, by Euclid I, 42,[2] and hence the area LMN, which is the number to be added, is a sum composed of the product of GC into twice CB, that is, into the number of squares which was 6, and GC into itself, which is the number of squares to be added. This is our proof [of the possibility of a solution].

This having been completed, you will always reduce the part containing the fourth power to a root, viz, by adding enough to each side so that the fourth power with the square and number may have a root. This is easy when you take half the number of the squares as the root of the number; and you will at the

---

[1] Cardan writes "square-square," quadratum quadratum ($q^d q^d$), hence x^4.
[2] In Heath's edition of the *Elements* (see Selection II.1) it is I, 43.

same time make the extreme terms on both sides plus, for otherwise the trinomial or binomial changed to a trinomial will necessarily fail to have a root. Having done this, you will add enough squares and a number to the one side, by the third rule,[3] so that the same being added to the other side (in which the unknowns were) will make a trinomial having a square root by assumption; and you will have a number of squares and a number to be added to each side, after which you will extract the square root of each side, which will be, on the one side, 1 square plus a number (or minus a number) and, on the other side, 1 unknown or more, plus a number (or minus a number; or a number minus unknowns), wherefore by the fifth chapter of this book you will have what has been proposed.

QUESTION V [4]

Example. Divide 10 into 3 parts in continued proportion such that the first multiplied by the second gives 6 as product. This problem was proposed by Johannes Colla,[5] who said he could not solve it. I nevertheless said I could solve it, but did not know how until Ferrari found this solution. Put then 1 unknown as the middle number, then the first will be 6/1 unknown, and the third will be $\frac{1}{6}$ of a cube. Hence these together will be equal to 10. Multiplying all by 6 unknowns we shall have 60 unknowns equal to one fourth power plus 6 squares plus 36.[6] Add, according to the 5th rule, 6 squares to each side, and you will have 1 fourth power plus 12 squares plus 36, equal to 6 squares plus 60 unknowns; for if equals are added to equals, the totals are equal. But 1 fourth power plus 12 squares plus 36 has a root, which is 1 square plus 6. If 6 squares plus 60 unknowns also had a root, we should have the job done; but they do not have; hence we must add so many squares and a number to each side, that on the one side there may remain a trinomial having a root, while on the other side it should be made so. Let therefore a number of squares[7] be an unknown and since, as you see in the figure... CL and MK are formed from twice GC into AB, and GC is an unknown, I will always take the number of squares to be added as 2 unknowns, that is, twice GC; and since the number to be added to 36 is LNM it therefore is the square of GC together with the product of twice

---

[3] Rule given earlier in the book.
[4] The problem is to find $y:x = x:z$, $x + y + z = 10$, $xy = 6$, which leads to

$$\frac{6}{x} + x + \frac{1}{6}x^3 = 10.$$

This is written

$$x^4 + 12x^2 + 36 = 6x^2 + 60x, \quad \text{or} \quad (x^2 + 6)^2 = 6x^2 + 60x.$$

This is changed into $(x^2 + 6 + y)^2 = 6x^2 + 60x + 2y(x^2 + 6) + y^2$. The right-hand member is a square if

$$2y^3 + 30y^2 + 72y = 900, \quad y^3 + 15y^2 + 36y = 450,$$

or $y^3 + (12 + \frac{12}{4})y^2 + 36y = \frac{1}{2}(\frac{60}{2})^2$. This is a cubic equation, already discussed by Cardan.

[5] Zuasse de Tonini da Coi, or Johannes Colla, was a mathematician who often conferred with Tartaglia and Cardan.
[6] This means $x^4 + 6x^2 + 36 = 60x$.
[7] Here begins the text of p. 74^{v}, reproduced in Fig. 2.

GC into CB or of GC into twice CB, which is 12, the number of the squares in the original equation. I will therefore always multiply the unknown, half the number of squares to be added, into the number of squares in the original equation and into itself and this will make 1 square plus 12 unknowns to be added on each side, and also 2 unknowns for the number of the squares. We shall therefore have again, by common sense, the quantities written below equal to each other; and each side will have a root, the first, by the third rule,[3] but the second quantity, by an assumption [this is Eq. (A) above]. Therefore the first part of the trinomial multiplied by the third makes the square of half the second part of the trinomial. Thus from half the second part multiplied by itself there results 900, a square, and from the first [multiplied] into the third there result 2 cubes plus 30 squares plus 72 unknowns. Likewise, this may be reduced, since equals divided by equals produce equals, as 2 cubes plus 30 squares plus 72 unknowns equals 900, therefore 1 cube plus 15 squares plus 36 unknowns equals 450.

It is therefore sufficient for the reduction to the rule, if we have always 1 cube plus the number of the former squares, with a fourth of it added to it plus such a multiple of the assumed quantity as the first number of the equation indicates; so that if we had 1 fourth power plus 12 squares plus 36 equals 6 squares plus 60 unknowns we should have 1 cube plus 15 squares plus 36 unknowns equal to 450, half the square of half the number of unknowns. And if we had 1 fourth power plus 16 squares plus 64 equal to 80 unknowns we should have 1 cube plus 20 squares plus 64 unknowns equal to 800. And if we had 1 fourth power plus 20 squares plus 100 equal to 80 unknowns we should have 1 cube plus 25 squares plus 100 unknowns equal to 800. This being understood, in the former example we had 1 cube plus 15 squares plus 36 unknowns equal to 450; therefore the value of the unknown, by the 17th chapter,[8] is

$$\sqrt[3]{287\tfrac{1}{2} + \sqrt{80449\tfrac{1}{4}}} + \sqrt[3]{287\tfrac{1}{2} - \sqrt{80449\tfrac{1}{4}}} - 5.[9]$$

This then is the number of squares which is to be doubled and added to each side (since we assumed 2 unknowns to be added), and the number to be added to each side, by the demonstration, is the square of this, with the product of this by 12, the number of squares.

Cardan continues to analyze this method and gives several more examples, for instance $x^4 + 4x + 8 = 10x^2$, which he reduces to $y^3 + 30 = 2y^2 + 15y$. The book ends, at the conclusion of the 40th chapter, with the exclamation: "Written in five years, may it last as many thousands!"

[8] Cardan here teaches that an equation of the form $x^3 + px^2 + qx = r$ can be reduced to an equation without a term in x^2 by the substitution $y = x + p/3$.

[9] T. R. Witmer has observed that this should be

$$\sqrt[3]{190 + \sqrt{33,903}} + \sqrt[3]{190 - \sqrt{33,903}} - 5$$

(see note 7, previous selection).

torum 1 positio,& quia,ut uides in figura tertiæ regulæ, c L & M R, fiunt ex duplo G c in A B,& G c est 1 positio, ponam numerum qua‑ dratorum addendorum semper 2 positiones,id est duplũ G c,& quia numerus addendus ad 36, est L N M, & ideo quadratum G c cum eo quod fit ex G c duplicato in c B,feu ex G c in duplum c B,& est 1 2,nu merus quadratorum priorum,ducam igitur 1 positionem,dimidium numeri q̃dratorum additorũ,semper in numerum q̃dratorũ priorũ, & in se,& fient 1 q̃dratum p: 1 2 positionibus addenda ex alia parte, & etiam 2 positiones pro numero quadratorum, habemus igitur ite‑ rum ex communi animi sententia , quantitates infrascriptas , inuicem æquales,& utracʒ habent radicem,prima ex regula tertia, sed secun‑

da quantitas ex suppo‑
sito,igitur ducta prima
parte trinomij in ter‑
tiam,fit quadratum di‑
midiæ partis secundæ

1 q̃d˙q̃d.p: 2 pof.p: 1 2. q̃d* p: 1 q̃d.p: 1 2	
pof. additi numeri p: 3 6 æqualia.	
2 pof. 6 q̃dratorũ,p: 60 pof.p: 1 q̃d. p: 1 2	
pof. numeri additi.	

trinomij,quia igitur ex dimidio secundæ in se,fiunt 900,quadrata,& ex prima in tertiam , fiunt 2 cubi p: 30 quadratis p: 7 2 positionibus quadratorum,similiter erit deprimendo per quadrata , quia æqualia per æqualia diuisa,producunt æqualia,ut 2 cu. p: 30 quadratis p: 7 2 positionibus æquantur 900,quare 1 cubus p: 1 5 quadratis p: 3 6 posi tionibus æquantur 450.

Sufficit igitur deducendo ad regulam,habere semper 1 cubum p: numero priorum quadratorum,addita ei quarta parte p:numero po‑ sitionum tali,qualis est numerus equationis primus,ut si habuerimus 1 q̃d˙q̃dratum p: 1 2 quadratis p: 3 6, æqualia 6 quadratis p: 60 posi‑ tionibus,habebimus 1 cubum p: 1 5 quadratis p: 3 6 positionibus æ‑ qualia 450,dimidio quadrati dimidij numeri positionum, & si habe‑ remus 1 q̃d˙q̃dratum p: 1 6 quadratis p:64 æqualia 80 positionibus, haberemus 1 cubum p: 20 quadratis p:64 positionibus æqualia 800, & si haberemus 1 q̃d˙q̃dratum p: 20 quadratis p: 100,æqualia 80 po sitionibus,haberemus 1 cubum p: 2 5 quadratis p: 1 00 positionibus æqualia 800,igitur hoc habito,in priore exemplo habuimus , 1 cub. p: 1 5 quadratis p: 3 6 positionibus æqualia 450, igitur rei æstimatio, per 17ᵐ capitulum,est 1ɛ v:cubica 287 ½ p: ɪ̱ʐ 80449 ¼, p:ɪ̱ʐ v:cubi‑ ca 287 ½, m:ɪ̱ʐ 80449 ¼ m: 5 , hic igitur est numerus quadratorũ,qui duplicatus, est addendus ex utracʒ parte,quia supponuntur 2 res ad‑ dendæ,& numerus addendus ex utracʒ parte,ex demonstratione, est quadratum huius,cum eo quod fit ex hoc in 1 2,numerum quadrato‑

T 2 rum,

Fig. 2

5 VIÈTE. THE NEW ALGEBRA

François Viète (Latinized Franciscus Vieta; 1540–1603) was a French lawyer with important connections at the courts of Henry III and Henry IV. As a mathematician he made contributions to trigonometry and to the solution of higher-degree equations, but he is best remembered as the man who first introduced, in a systematic way, general letters instead of numbers into the theory of equations, something we express by saying that he studied $x^2 + ax + b = 0$ instead of, say, $x^2 + 5x + 6 = 0$. This is, however, not quite the way Viète looked at it, but rather, as we shall see, Descartes's way. Viète saw it as a rediscovery of ancient Greek mathematical methods which would lead to a true mathematics, basic to the search for a universal science, a search later continued by Descartes and Leibniz. For this true mathematics the slogan would hold: *There is no problem that cannot be solved.*

The starting point for Viète was the distinction between analysis and synthesis, as he found it in an ancient commentary to Euclid's *Elements*, Book XIII, Prop. 1–5, which are theorems on the golden section. This commentary—which Viète, in common with most mathematicians of that period, ascribed to Theon of Alexandria (fl. A.D. 370), though it is probably much older (see T. L. Heath, *Euclid's Elements*, III, 442)—runs as follows:

"What is analysis and what is synthesis?

"Analysis is the assumption of that which is sought as if it were admitted and the arrival, by means of its consequences, at something admitted to be true.

"Synthesis is the assumption of that which is admitted and the arrival, by means of its consequences, at something admitted to be true."

Pappus (fl. A.D. 320) uses more elaborate language in the opening words of the seventh book of his *Mathematical collection* (Heath, *Euclid's Elements*, I, 138):

"Analysis then takes that which is sought as if it were admitted and passes from it through its successive consequences to something which is admitted as the result of synthesis. Indeed, in analysis we assume that which is sought as if it were already done, and we inquire what it is from which this results ... until we come upon something already known or belonging to the class of first principles, and such a method we call analysis as being solution backwards.

"But in synthesis, reversing the process, we take as already done that which was last arrived at in the analysis, and, by arranging in their natural order as consequences what were before antecedents, and successively connecting them one with another, we arrive finally at the construction of what was sought; and this we call synthesis."

Most classical Greek mathematics is synthesis. When analysis is used, as for instance in Apollonius' book on conics (Book II, Prop. 44–51), where methods are found to construct diameters and tangents to conics, the analysis is followed by the synthesis, that is, the demonstration. Viète tries to reconstruct this ancient analysis, used, he thinks, by the Greeks, but kept secret. For Viète the most brilliant example of analysis is Diophantus' *Arithmetica*, that is, the algebra of Diophantus, which we have already met in our discussion of Fermat. Viète thought that he could improve on Diophantus by introducing a *general letter algebra*.

Pappus distinguishes between two kinds of analysis, "the one directed to the searching (*zētētikon*) for the truth and called theoretical, and the other directed to the finding (*poristikon*) of what we are told to find and called problematic."[1] In the first, the zetetic,

---

[1] The Greek *zētētikon* is from the verb *zēteō*, to seek; *poristikon* is from *porizō*, to provide. The zetetic method, in Viète's explanation, amounts to the modern analytical method. On the history of the terms "analysis" and "synthesis," see P. Tannery, *Mémoires scientifiques* (Privat, Toulouse, 1926), VI, 425–440 (paper of 1903).

approach, we look for a proof that can be inverted into a synthesis, the demonstration. In the poristic approach we are concerned with the finding of a solution and the inversion is usually a construction (unless the solution is found to be impossible). Viète called this method *Zetetics*.

Viète took these ideas and expanded on them, constructing a calculation with species (*logistica speciosa*) as opposed to the calculation with numerical coefficients (*logistica numerosa*). Typical of the calculation with species was the demand that all terms in the equation be homogeneous. The first exposition of this calculation was in the *In artem analyticem isagoge* (Tours, 1591), of which we translate sections as they appear in the *Opera mathematica*, published in 1646 at Leiden by Franciscus Van Schooten, who added a commentary. A full English translation with commentary has been prepared by J. Winfree Smith (St. John's College, Annapolis, 1955). A French translation by F. Ritter can be found in the *Bollettino di Bibliografia e di Storia delle Scienze Matematiche e Fisiche* (Boncompagni) *1* (1868), 223–244.

It will be seen that Viète, like many other Renaissance mathematicians, paid great attention to Euclid's theory of proportions (*Elements*, Book V) as a cornerstone of the structure of the true mathematics. In this theory Euclid, after having defined the ratio (*logos*) of two magnitudes, and then the equality of two ratios, calls magnitudes that have the same ratio *proportional*. His definition makes it possible to establish proportionality in a geometric way, without regard to the commensurability of the magnitudes, so that it holds for rational as well as irrational (*alogoi*) quantities. See Heath, *Euclid's Elements*, II, 116–119.

It was through Viète that the term "analytics" was introduced into modern mathematics, to replace the term "algebra." As we know, both terms remained, eventually differing in meaning. The term "zetetics" never became popular.

INTRODUCTION TO THE ANALYTIC ART

CHAPTER I. ON THE DEFINITION AND PARTITION OF ANALYSIS, AND ON THOSE THINGS WHICH ARE OF USE TO ZETETICS.

In this chapter Viète refers to Pappus's distinction between analysis and synthesis, and between zetetic and poristic analysis, referring also to Euclid and Theon. There also should be, he writes, a third kind of analysis, the rhetic or exegetic,

---

so that there is a zetetic art by which is found the equation[2] or proportion between the magnitude that is being sought and the given things; a poristic art by which from the equation or proportion the truth of the required theorem is investigated, and an exegetic art by which from the constructed equation or proportion there is produced the magnitude itself that is being sought. And the whole threefold analytical art may be defined as the science of finding the truth

---

[2] Viète writes *aequalitas*, equality, but the term "equation," now used, seems to fit the meaning better. The stress on proportion is due to the respect in which Book V of Euclid's *Elements* was held as a model whereby the contradiction between arithmetic and geometry could be overcome by rigorous mathematical reasoning.

in mathematics. But what truly belongs to the zetetic art is established by the art of logic through syllogisms and enthymemes,[3] of which the foundations are those very symbols[4] by which equations and proportions are obtained ... The zetetic art, however, has its own form of proceeding, since it applies its logic not to numbers—which was the boring habit of the ancient analysis—but through a logistic which in a new way has to do with species.[5] This logistic is much more successful and powerful than the numerical one for comparing magnitudes with one another in equations, once the law of homogeneity has been established and there has been constructed, for that purpose, a traditional series or scale of magnitudes ascending or descending by their own nature from genus to genus, by which scale the degrees and genera of magnitudes in equations may be designated and distinguished.

CHAPTER II. ON THE SYMBOLS FOR EQUATIONS AND PROPORTIONS.

Here Viète takes a number of postulates and propositions from Euclid, such as:

1. The whole is equal to the sum of its parts;
2. Things that are equal to the same thing are equal among themselves;
3. If equals are added to equals, the sums are equal;
8. If like proportionals are added to like proportionals, then the sums are proportional;[6]

15. If there are three or four magnitudes, and the product of the extreme terms is equal to either that of the middle one by itself or that of the middle terms, then these magnitudes are proportional.[7]

CHAPTER III. ON THE LAW OF HOMOGENEOUS QUANTITIES, AND THE DEGREES AND GENERA OF THE MAGNITUDES THAT ARE COMPARED.

The first and supreme law of equations or of proportions, which is called the law of homogeneity, since it is concerned with homogeneous quantities, is as follows:

1. Homogeneous quantities must be compared to homogeneous quantities [*Homogenea homogeneis comparari*].

[3] An enthymeme is a syllogism incompletely stated, perhaps by leaving out the major or the minor premise; for example, in "John is a liar, therefore he is a coward," the premise, "every liar is a coward," is omitted.

[4] Symbols, *symbola*, has here more the meaning of typical rules or stipulations.

[5] Hence the name "logistica speciosa" for Viète's new type of calculation. The term *logistikē* was used by the Greeks for the art of calculation, in contrast to *arithmētikē*, number theory. Viète's term "species" is probably the translation of Diophantus' *eidos*, the term in a particular expression, primarily in reference to the specific power of the unknown it contains. See further the J. Winfree Smith translation of the *Isagoge*, pp. 21–22.

[6] If $a:b = c:d$, then $(a + c):(b + d) = a:b = c:d$.

[7] If a, b, c, d are such that either $ac = b^2$ or $ad = bc$, then either $a:b = b:c$ or $a:b = c:d$.

Indeed, it cannot be known how heterogeneous quantities can be affected among themselves, as Adrastus says.[8] Hence:

If a magnitude is added [*additur*] to a magnitude, it is homogeneous with it.

If a magnitude is subtracted [*subdicitur*] from a magnitude, it is homogeneous with it.

If a magnitude is multiplied [*ducitur*] by a magnitude, the result is heterogeneous with both.

Since they did not, these ancient Analysts, attend to this, the result was much obscurity and blindness.

2. Magnitudes which by their own nature ascend or descend proportionally from genus to genus are called scalars.[9]

The first of the scalar magnitudes is side or root [*latus seu radix*].[10]

The second is square [*quadratum*].

The third is cube.

The fourth is squared square [*quadrati-quadratum*].

The fifth is squared-cube ...

The ninth is cubed-cubed-cube.

And the further ones can from here be named by this series and method ...

The genera of the magnitudes that we have to compare so that they may be named in the order of the scalars are:

(1) Length and breadth [*longitudo, latitudo*],

(2) Plane,

(3) Solid,

(4) Plane-plane,

(5) Plane-solid ...

(9) Solid-solid-solid.

And the further ones can be named from here by this series and method.

.   .   .   .

5. In a series of scalars, the degree in which the magnitude stands compared to the side is called the power [*potestas*]. The other inferior scalars are called parodic[11] grades to this power.

6. The power is pure when it is free from affection. By affection is meant that a homogeneous magnitude is mixed with a magnitude of lower power together with a coefficient.[12]

---

[8] Reference to a reference in Theon: "For Adrastus says that it is impossible to know how heterogeneous magnitudes may be in a ratio to one another." Who Adrastus was does not seem to be known.

[9] *Scalares* means "ladder magnitudes," literally, steps or rungs of a ladder. Viète follows Diophantus in the naming of the powers. The term *scalar*, of vector-analysis fame, is due to W. R. Hamilton (1853).

[10] This is the *cosa*, or *res*, of the cossists, hence x in our notation. The next scalars are x^2 (square), x^3, x^4, and so forth. In Viète these quantities have dimensions.

[11] *Parodic* is from Greek *para, hodas*, on the way, coming up.

[12] x^5 is pure, $x^5 + ax^4$ is affected.

To this Van Schooten adds: "A pure power is a square, cube ... But an affected power is in the second grade: a square together with a plane composed of a side and a length or breadth;[13] in the third grade: a cube together with a solid composed of a square and a length or latitude."

---

7. Adjunct magnitudes which multiply scalars lower in relation to a certain power and thus produce homogeneous magnitudes are called subgradual.

Van Schooten adds: "Subgraduals are length, breadth, plane, solid, plane-plane, etc. Thus if there be a squared square with which is mixed a plane-plane which is the side multiplied with a solid, then the solid will be a subgradual magnitude, and in relation to the squared square the side will be a lower scalar."

CHAPTER IV. ON THE RULES FOR THE CALCULATION BY SPECIES [*logistica speciosa*]

Numerical calculation [*logistica numerosa*] proceeds by means of numbers, reckoning by species by means of species or forms of things, as, for instance, the letters of the alphabet.

Van Schooten adds: "Diophantus operates with numerical calculation in the thirteen books of his *Arithmetica*, of which only the first six are extant, and are now available in Greek and Latin, illustrated by the commentaries of the very erudite Claude Bachet.[14] But the calculation by species has been explained by Viète in the five books of his *Zetetics*,[15] which he has chiefly arranged from selected questions of Diophantus, some of which he solves by his own peculiar method. Wherefore, if you wish to understand with profit the distinction between the two logistics, you must consult Diophantus and Viète together." He then compares specifically certain problems of Diophantus with his and with Viète's solutions.

There are four canonical rules for the calculation by species.

Rule I. *To add a magnitude to a magnitude.*

Take two magnitudes A and B. We wish to add the one to the other. But, since homogeneous magnitudes cannot be affected to heterogeneous ones, those which we wish to add must be homogeneous magnitudes. That one is greater than the other does not constitute diversity of genus. Therefore, they may be

[13] x^2 is pure, $x^2 + ax$ is affected.

[14] Bachet's edition of Diophantus is of 1621, and was the inspiration of Fermat's work on numbers (see selection I.6).

[15] In this work of 1593 Viète gives many examples of his *logistica speciosa*.

fittingly added by means of a coupling or addition; and the aggregate will be *A* plus *B*, if they are simple lengths or breadths. But if they stand higher in the scale, or if they share in genus with those that stand higher, they will be denoted in the appropriate way, say *A* square plus *B* plane, or *A* cube plus *B* solid, and similarly in further cases.

However, the Analysts are accustomed to indicate the affection of summation by the symbol +.

Rule II. *To subtract a magnitude from a magnitude.*

This leads in an analogous way to $A - B$, A square $- B$ square, A is larger than B, also to rules such as $A - (B + D) = A - B - D$; Viète writes = instead of our −.

Rule III. *To multiply a magnitude by a magnitude.*

Take two magnitudes *A* and *B*. We wish to multiply the one by the other.

Since then a magnitude has to be multiplied by a magnitude they will by their multiplication produce a magnitude heterogeneous with respect to each of them; their product will rightly be designated by the word "in" or "under" [*sub*], e.g., *A* in *B*, which will mean that the one has been multiplied by the other, or, as others say, under *A* and *B*, and this simply when *A* and *B* are simple lengths or breadths.[16]

But if the magnitudes stand higher in the scale, or if they share in genus with these magnitudes, then it is convenient to add the names themselves, e.g., *A* square in *B*, or *A* square in *B* plane solid, and similarly in other cases.

If, however, among magnitudes that have to be multiplied, two or more are of different names, then nothing happens in the operation. Since the whole is equal to its parts, the products under the segments of some magnitude are equal to the product under the whole. And when the positive name [*nomen adfirmatum*] of a magnitude is multiplied by a magnitude also of positive name, the product will be positive, and negative [*negatum*] when it is negative.[17]

From which precept it follows that by the multiplication of negative names the product is positive, as when $A - B$ is multiplied by $D - G$; since the product of the positive *A* and the negative *G* is negative, which means that too much is taken away [and similarly negative *B* into positive]. Therefore, in compensation, when the negative *B* is multiplied by the negative *G* the product is positive.[18]

The denominations of the factors that ascend proportionally from genus to genus in magnitude behave, therefore, in the following way:

A side multiplied by a side produces a square,

A side multiplied by a square produces a cube . . .

---

[16] In arithmetic the custom was to use "in": *ducta in*; in geometry, "under": a rectangle is "under" its sides.

[17] + in + is +; + in − is −.

[18] $(A - B)(D - G) = AD - AG - BD + BG$.

And conversely, a square multiplied by a side produces a cube . . .
A solid multiplied by a solid-solid produces a solid-solid-solid,
And conversely, and so on in that order.

Rule IV. *To divide a magnitude by a magnitude.*

This leads in an analogous way to such expressions as $\dfrac{B \text{ plane}}{A}$, $\dfrac{B \text{ cube}}{A \text{ plane}}$, and

so forth. Furthermore to add $\dfrac{Z \text{ plane}}{G}$ to $\dfrac{A \text{ plane}}{B}$; the sum will be

$$\frac{G \text{ in } A \text{ plane} + B \text{ in } Z \text{ plane}}{B \text{ in } G}.$$

To multiply $\dfrac{A \text{ plane}}{B}$ by Z; the result will be $\dfrac{A \text{ plane in } Z}{B}$.

CHAPTER V. CONCERNING THE LAWS OF ZETETICS.

The way to do Zetetics is, in general, directed by the following laws:

1. If we ask for a length, but the equation or proportion is hidden under the cover of the data of the problem, let the unknown to be found be a side.
2. If we ask for a plane . . . let the unknown to be found be a square.

. . . .

9. If the element that is homogeneous under a given measure happens to be combined with the element that is homogeneous in conjunction, there will be antithesis.

These laws amount to introducing (1) x, (2) x^2, (3) x^3, (4) the law of homogeneity, as in $x = ab$; and to (5) denoting the unknown by vowels $A, E, \ldots$ and the given magnitudes by consonants, $B, G, D, \ldots$, (6) constructing $x^2 = ab + cd$, or, as Viète writes it: A square equal to B in $C + D$ in F; (7) forming $ax \pm bx$ ("homogeneous in conjunction"); (8) forming $x^3 + ax^2 - bx^2 = c^2d + e^3$; (9) passing from $x^3 + ax^2 + bx^2 - c^2d + e^2f = g^3$ to $x^3 + ax^2 - bx^2 = c^2d - e^2f + g^3$ ("antithesis"). Then Viète continues with Propositions marked (10) to (12), which state that an equation is not changed by antithesis, by hypobibasm, and by parabolism. Hypobibasm means dividing by the unknown, as passing from $x^3 + ax^2 = b^2x$ to $x^2 + ax = b^2$, parabolism is dividing by a known magnitude. Nos. (13) and (14) deal with the relation of equations to proportions.

These are the titles of the next chapters:

VI.   Concerning the examination of theorems by means of the poristic art.
VII. Concerning the function of the rhetic art.
VIII. The notation of equations and the epilogue to the art.[19]

[19] Chapter VI mentions the retracing of the zetetic process by synthesis; Chapter VII the special application of the analytic art, after solution, to special arithmetic and geometric problems. Here Viète speaks of the "exegetic art." Chapter VIII is the discussion of different possible expressions and equations, stressing homogeneity. There are 29 rules.

This chapter ends as follows:

---

29. Finally, the analytic art, now having been cast into the threefold form of zetetic, poristic, and exegetic, appropriates to itself by right the proud problem of problems, which is

THERE IS NO PROBLEM THAT CANNOT BE SOLVED.[20]

This paper is followed (in the 1646 edition) by *Ad logisticam speciosam notae priorae*. Originally it was followed by *Ad logisticam speciosam notae posteriores*, which was already lost at the time of Van Schooten. A French translation of the *Notae priorae* follows the Ritter translation of the *Isagoge* in the *Bollettino de Bibliographie e di Storia* (Boncompagni) *1* (1868), 245–275. It contains 56 propositions concerning algebraic identities and geometric problems. After this comes the *Zeteticorum libri quinque* (1593), inspired by Diophantus and solving more problems by means of the *logica speciosa*. We give an example, taken from Viète's *Zetetics*, comparing Diophantus' method and Viète's:

I. To divide a given number into two numbers with a given difference (Diophantus, *Arithmetic*, I, Prob. 1).

Diophantus: Let the given number be 100, and the difference be 40; let the smaller number be x, then the larger will be $x + 40$. Then $x + (x + 40) = 2x + 40 = 100$, hence $2x = 60, x = 30, x + 40 = 70$.

Viète: Let the given number be D, and the difference be B; let the smaller side be A, then the larger will be $A + B$. Then $A + (A + B) = A2 + B = D$, hence $A2 = D - B$, $A = D\frac{1}{2} - B\frac{1}{2}, A + B = D\frac{1}{2} + B\frac{1}{2}$.

In modern notation: $D = a, B = b, A = x$, then $x + (x + b) = a, 2x = a - b, x = \frac{1}{2}(a - b), x + b = \frac{1}{2}(a + b)$. The transition from ancient algebra to our modern one is very clear here.

6 GIRARD. THE FUNDAMENTAL THEOREM OF ALGEBRA

Albert Girard (1595–1632), a native of Lorraine who worked in Holland, was the editor of the works of Simon Stevin. His *L'invention nouvelle en l'algèbre* (Amsterdam, 1629; edited by D. Bierens de Haan, Muré, Leiden, 1884) was based on Stevin's *Arithmétique* of 1585, but where Stevin's voluminous treatise was, in the main, only a well-written presentation of known results, leading up to the solution of third- and fourth-degree equations, Girard's much shorter book contains many new results. Its fame is based on its formulation of the fundamental theorem of algebra, so that he takes complex roots seriously (contrary to Stevin). In presenting the section of the book in which Girard formulates this theorem (pp. E2^v–F2^r), we have maintained the Stevin terminology, but not his notation. We have therefore kept the words "quantity" for term, "denominator" for exponent, and "number" for coefficient. A discussion of Girard's book can be found in H. Bosmans, "La théorie des équations dans 'L'invention nouvelle en l'algèbre' d'Albert Girard," *Mathésis 41* (1926), 59–67, 100–109, 145–155.

---

[20] Quod est, Nullum non problema solvere.

Since the theorem which will follow requires some new expressions, we shall begin with definitions.

Definition I. A simple equation is one that has only one term[1] equal to a number; otherwise it is called composed or mixed.

Explication. As when x^2 is equal to 49; or $12x$ equal to 24; hence when one term is equal to the other, the equation is simple. But when there are more terms than two, it is composed or mixed, as when x^2 is equal to $6x + 40$, or similar equations.[2]

Definition II. When one term is compared to another, the first is called the subject, or the antecedent, the other one is called the predicate [*parangon*[3]] or consequent.

For example in $3x^2 - 4x = 70$, $3x^2 - 4x$ is the subject, 70 the predicate.

Definition III. A complete equation is one that has all the terms without leaving one out.

Definition IV. An incomplete equation is a mixed equation that does not have all its terms.

Thus, $x^6 = 11x^5 + 13x^4 - 7x^3 + 6x^2 + 9x - 31$ is complete, but $x^4 = 5x^2 + 36$ or $x^3 = 12x - 16$ are incomplete.

Definition V. A mixed equation that has only one default[4] is almost complete; one that has two defaults is complete but for two [complette à deux pres]; and in a similar way we have equations complete but for three, etc.

Thus, $x^3 = 7x - 6$ is almost complete.

[1] We translate "grandeur" by "term."

[2] Girard writes "when 1② is equal to 49: or 12① equal to 24," "when 1② is equal to 6① + 40." This is Stevin's notation. From now on we paraphrase the examples used to illustrate the Definitions.

[3] French *paragon*, from Spanish *parangón*, model, type.

[4] A *default* is a missing term.

Definition VI. A primitive equation is one in which the denominators of the terms are relative primes.[5]

Thus, $x^4 = 6x^3 - 13x + 16$ is primitive.

Definition VII. A derivative equation is one in which the denominators of the terms have common divisors [*sont entr'eux composez*].

Thus, $x^6 = 7x^4 - 9x^2 + 12$, where x^3, x^2, x, x^0 are the primitives or $x^3 = 17$, where x^3 and x^0 are the primitives, as Stevin says in his *Arithmétique*, Def. 27°.

Definition VIII. In mixed equations the highest term[6] is called the maximum or high extremity [*maxime, ou haute extrémité*]; the one that is one degree lower is called the first mixed; the one that is one degree lower still is called the second mixed; and so on, so that x^0 is the closure or lowest extremity [*la fermeture ou basse extrémité*].

For example, if $x^9 = 3x^8 - 10x^6 + 4x + 12$, then x^9 is the maximum, $3x^8$ the first mixed, $10x^6$ the third mixed, $4x$ the eighth mixed, and 12 is the closure.

Definition IX. In mixed equations there are three orders; the first is called the prior order, when the numbers of algebra are the subject (as partly unknown) and the closure or common number is the predicate or parangon (as the only otherwise known). The second order is the alternative, in which the even quantities are separated from the odd ones in such a way that the high extremity is + not −. The third order is the posterior one, in which the high extremity has the sign +, with the number 1.[7]

Definition X. The alternate order of equations is that in which the maximum or high extremity has no other number than unity, with the sign +, and all odd denominators or characters are on one side, and the even ones on the

---

[5] Def. 27 of Stevin's *Arithmétique* (1585) makes a distinction between polynomials such as $ax + b, px^2 + qx + r, bx^3 + mx + n$, which are called primitive, and those such as $ax^2 + b$, $cx^3 + d, px^4 + qx^2 + q, bx^6 + snx^2 + n$, which can be obtained from primitive polynomials by replacing x by a power of x. These polynomials are called derivative.

[6] The term with the highest exponent.

[7] Number = coefficient.

other side, to wit the ones as the subject, the other ones as the predicate. Which serves to find the original signs again, when the equation in question is re-ordered.

For example, let the equation be $x^7 = 4x^6 + 14x^5 - 56x^4 - 49x^3 + 196x^2 + 36x - 144$; put in alternate order it is $x^7 - 14x^5 + 49x^3 - 36x = 4x^6 - 56x^4 + 196x^2 - 144$.

Definition XI. When several numbers are given, let the total sum be called the first faction; the sum of all their products two by two be called the second faction; the sum of all their products three by three be called the third faction; and always so on until the end, but the product of all the numbers is the last faction: and so there are as many factions as there are numbers given.

Let the numbers be 2, 4, 5; then the first faction is 11, the second faction is $8 + 10 + 20 = 38$, the third faction is 40. If the numbers are 2, -3, 1, 3, then we get in succession 3, -7, -27, -18; when they are 1, 2, 3, 4, -1, -1, -2 the factions are 6, -14, -56, 49, 196, -36, -144.[8]

$$
\begin{array}{ccccccc}
& & & 1 & & & \\
& & 1 & & 1 & & \\
& & 1 & 2 & 1 & & \\
& 1 & 3 & & 3 & 1 & \\
1 & 4 & 6 & & 4 & 1 &
\end{array}
$$

Definition XII. When several unities are placed at the sides, and other numbers in the middle, find by means of addition the figure which may be called the triangle of extraction: and let the unity above signify simple arithmetic, and the others stand for algebra; then let 1, 1 be called the rank of the x, and 1, 2, 1 the rank of the x^2, furthermore let 1, 3, 3, 1 be called the rank of the x^3, and so always on till infinity.[9]

Theorem I. If a set of numbers are present, then the number of products of every faction can be found by the triangle of extraction: and by its rank in accordance with the multitude of the numbers.

Thus, if four numbers are given, take rank 4 in the triangle, which gives 1, 4, 6, 4, 1. The first 1 is the unity of the maximum, the 4 gives the number of terms of the sum, the 6 that of the double products, and so forth.

[8] There is an error here: the factions are 6, 0, -42, -21, 84, 20, -48.
[9] On this triangle of extraction see Selection I.5 on Pascal's *Triangle arithmétique*. Girard finds 2 as $1 + 1$, 3 as $1 + 2$, 6 as $3 + 3$, etc. "Simple arithmetic" means ordinary number (x^0), the other numbers come from algebra: x, x^2, etc.

Theorem II. All equations of algebra receive as many solutions as the denomination of the highest term shows, except the incomplete,[10] and the first faction of the solutions is equal to the number of the first mixed, their second faction is equal to the number of the second mixed; their third to the third mixed, and so on, so that the last faction is equal to the closure, and this according to the signs that can be observed in the alternate order.

Explication. Let a complete equation $x^4 = 4x^3 + 7x^2 - 34x - 24$ be given, then the denominator of the highest term is 4, which means that there are four certain solutions, no more and no less, namely 1, 2, -3, 4: here 4 is the number of the first mixed, 7 of the second mixed, and so on. But to see the thing in its perfection we must take the signs which we can see in the alternate order, as $x^4 - 7x^2 - 24 = 4x^3 - 34x$. Then the numbers with their signs (according to the order of the quantities) will be 4, -7, -34, -24, which are the four factions of the four solutions.

Otherwise let $x^4 = 4x^3 - 6x^2 + 4x - 1$, and in alternate order $x^4 + 6x^2 + 1 = 4x^3 + 4x$, of which the numbers with the signs, according to the order of the terms, are 4, 6, 4, 1, which are factions of the four solutions 1, 1, 1, 1, and so the others (observe here that when the solutions are simple unities, then the factions are the numbers of the triangle of extraction of the rank of the highest quantity); similarly in the equation of the 10th definition, which is

$$x^7 = 4x^6 + 14x^5 - 56x^4 - 49x^3 + 196x^2 + 36x - 144,$$

there are 7 solutions, namely 1, 2, 3, 4, -1, -2, -3, which are discussed in the 10th and 11th definitions.

As to the incomplete equations, they have not always so many solutions, nevertheless we can well explain the solutions whose existence is impossible, and show wherein lies the impossibility because of the defectiveness and incompleteness of the equation. Such as $x^3 = 7x - 6$, here we still have the three solutions, namely 1, 2, -3, and all incomplete equations such as this one can be put in the form of complete ones: $x^3 = 0x^2 + 7x - 6$ in order to find all the solutions as it has been done before. For instance, $x^3 = 167x - 26$ will complete as $x^3 = 0x^2 + 167x - 26$, and in alternate order $x^3 - 167x = 0x^2 - 26$, the numbers with their signs (according to the order of the quantities) will be 0, -167, -26. Hence we must find three numbers which have such factions, namely, their sum must be 0, the sum of their double products -167, and the product of all three -26. Now, if we have found one of the three, say -13, as

---

[10] This phrase "except the incomplete" seems to imply some restrictions, which Girard tries to remove in his "Explication," However, many authors seem to be willing to give Girard priority in the formulation of the fundamental theorem of algebra. There exists a formulation of the fundamental theorem by Peter Rothe, a Nuremberg mathematician, in his *Arithmetica philosophica* (Nuremberg, 1600), where he states that equations have at most as many roots as their degree indicates. Descartes, in his *Géométrie* (1637), admits that an equation can be said to have as many roots as the degree indicates, if imaginary roots are taken into account (Selection II.7). Proofs of the theorem are attempted in the eighteenth century; see below, Selections II.10, 11, 12.

we did above, then since the product of the three was -26, the product of the two others will be 2, and as the sum of the three numbers is 0, and one of them is -13, hence the sum of the two others will be 13. The question is now reduced to this one: to find two numbers of which the sum is 13 and the product 2. Observe that we say two numbers, it will therefore be an equation of which the major term is $x^2 \cdots$ and hence $x^2 + 2 = 13x$, in alternate form. If we restore this to the ordinary order we have $x^2 = 13x - 2$, hence the numbers of the solution will be $6\frac{1}{2} + \sqrt{40\frac{1}{4}}$ and also $6\frac{1}{2} - \sqrt{40\frac{1}{4}}$, which together with -13 will be the three required solutions.

In the same way, if $x^4 = 4x - 3$, then the four factions will be 0, 0, 4, 3, and consequently the four solutions will be

$$1$$
$$1$$
$$-1 + \sqrt{-2}$$
$$-1 - \sqrt{-2}$$

(observe that the product of the last two is 3). We must therefore always remember to keep this in mind: if someone were to ask what is the purpose of the solutions that are impossible, then I answer in three ways: for the certitude of the general rule, and the fact that there are no other solutions, and for its use. The use is easy to see, since it serves for the invention of solutions of similar equations as we can see in Stevin's Arithmetic, in the 5th difference of the 71st problem ... [11]

By this means you will find that nobody before has solved the equations with all the solutions.

One of Stevin's problems is $x^3 = 6x^2 - 10x + 3$, for which he finds only the solution 3. Girard finds also $1\frac{1}{2} + \sqrt{\frac{5}{4}}$ and $1\frac{1}{2} - \sqrt{\frac{5}{4}}$; there are some other examples as well.

Girard, continuing, studies algebraic equations in more detail. Among his results are expressions for the sums s_k of the kth powers of the roots. In our modern notation his results can be written as follows: if the equation is $x^n = a_1 x^{n-1} + a_2 x^{n-2} + \cdots + a_n$, so that a_1 is the "first mixed," then

$$s_1 = a_1, \quad s_2 = a_1^2 - 2a_2, \quad s_3 = a_1^3 - 3a_1a_2 + 3a_3,$$
$$s_4 = a_1^4 - 4a_1^2 a_2 + 4a_1 a_3 + 2a_2^2 - 4a_4.$$

Girard writes

$$\text{``}A, Aq - B2, Acub - AB3 + C3$$
$$Aqq - AqB4 + AC4 + Bq2 - D4$$

sera la somme des solutions, quarez, cubes, quaré-quarez."

[11] See *The principal works of Simon Stevin*, IIB (Swets-Zeitlinger, Amsterdam, 1958), 648. Stevin, in this problem LXXI, discusses the general theory of cubic equations. See also H. Bosmans, "La résolution des équations du 3ᵉ degré d'après Stevin," *Mathésis 37* (1923), 246–254, 304–311, 341–347.

The last part of the book is entitled, "On the measure of the area of spherical triangles and polygons, newly invented," and contains the theorem:

Every spherical polygon bounded by arcs of great circles has as many surface degrees as the sum of all its interior angles exceeds the sum of the interior angles of a rectilinear polygon of the same name when the area of the sphere is taken as 720 surface degrees.

The theorem is accompanied by a rather cumbersome proof. The theorem was discovered earlier by Thomas Harriot (1603), but remained in manuscript. See J. A. Lohne, "Thomas Harriot als Mathematiker," *Centaurus 11* (1965), 19–45. The theorem is sometimes called after Legendre, who explained it in a very simple fashion, in *Eléments de géométrie* (Paris, 1794), Livre VII, Proposition 23.

7 DESCARTES. THE NEW METHOD

The search for a universal mathematics leading to a universal science, on which Viète had been meditating, appears again, in much stronger philosophical form, in the work of René Descartes (Latinized Renatus Cartesius; 1596–1650). Descartes, a French gentleman of independent means, was educated by the Jesuits; after a term as a soldier, he lived in Holland during the most productive part of his life. He used to connect his search for a general method with a mystical experience, on November 10, 1619 or 1620, of which he wrote that, "full of enthusiasm, I discovered the foundations of a wonderful science" (*mirabilis scientiae fundamenta*). What was on his mind was first laid down in his 21 *Regulae ad directionem ingenii* (Rules for the guidance of our mental powers), written prior to or in 1629, first published in 1692 in Dutch and in 1701 in Latin; see Descartes, *Oeuvres*, X, 359–469. Here we find some of the ideas of Viète again expressed, but then developed in Descartes's own way. We follow the translation in N. K. Smith, *Descartes: Philosophical writings* (St. Martin's Press, New York, 1953).

In Rule IV, "In the search for the truth of things a method is indispensable," we find:

> For the human mind has in it a something divine, wherein are scattered the first seeds of useful modes of knowledge. Consequently it often happens that, however neglected and however stifled by distracting studies, they spontaneously bear fruit. Arithmetic and geometry, the simplest of the sciences, are instances of it. We have evidence that the ancient geometers made use of a certain analysis which they applied to the solution of all problems, although, as we find, they invidiously withheld knowledge of this method from posterity. There is now flourishing a certain kind of arithmetic, called algebra, which endeavors to accomplish in regard to numbers what the ancients achieved in respect to geometrical figures. These two sciences are no other than spontaneous fruits originating from the innate principles of the method in question.

But when Descartes first studied mathematics, he was disappointed in his search for his method of true understanding:

> For truly there is nothing more futile than to occupy ourselves so much with mere numbers and imaginary figures that it seems that we could be content to

rest in the knowledge of such trifles ... When, however, I afterwards bethought myself how it could be that the first discoverers of philosophy refused to admit to the study of wisdom anyone not versed in mathematics, as if they viewed mathematics as being the simplest of all disciplines, and as altogether indispensable for training and preparing our human powers for the understanding of other more important sciences, I could not but suspect that they were acquainted with a mathematics very differerent from that which is commonly cultivated in our day. Not that I imagined that they had a complete knowledge of it. Their extravagant exultations, and the sacrifices they offered for the simplest discoveries, show quite clearly how rudimentary their knowledge must have been. I am convinced that certain primary seeds of truth implanted by nature in the human mind ... had such vitality in that rude and unsophisticated ancient world, that the mental light by which they discerned virtue to be preferable to pleasure ... likewise enabled them to recognize true ideas in philosophy and mathematics, even though they were not yet able to obtain complete mastery of them. Certain vestiges of this true mathematics I seem to find in Pappus and Diophantus, who, though not belonging to that first age, yet lived many centuries before our time.[1] These writers, I am inclined to believe, by a certain baneful craftiness, kept the secrets of this mathematics to themselves ... Instead they have chosen to propound ... a number of sterile truths, deductively demonstrated with great show of logical subtlety, with a view to winning an amazed admiration, thus dwelling indeed on the results obtained by way of their method, but without disclosing the method itself—a disclosure which would have completely undermined that amazement. Lastly, in the present age there have been certain very able men who have attempted to revive this mathematics. For it seems to be no other than this very science which has been given the barbarous name, algebra—provided, that is to say, that it can be extricated from the tortuous array of numbers and from the complicated geometrical shapes by which it is overwhelmed, and that it be no longer lacking in the transparency and unsurpassable clarity which, in our view, are proper to a rightly ordered mathematics.

Then Descartes asks what is meant by mathematics:

What, on more attentive consideration, I at length came to see is that those things only were referred to mathematics in which order or measure is examined, and that with respect to measure it makes no difference whether it be in numbers, shapes, stars, sounds or any other object that such measure is sought, and that there must therefore be some general science which explains all that can be

[1] Like Viète's, Descartes's starting points are Pappus and Diophantus. With Viète, who speaks of improving on or rescuing this art of analysis (which he dates up to Plato), Descartes believes that this art was well developed in ancient times and kept a semisecret. The dependence of Descartes on Viète is not clear; Descartes claimed not to have seen the *logistica speciosa* until he himself found his own method.

inquired into respecting order and measure, without application to any other special subject-matter, and that this is what is called *mathematica universalis*, no specially devised designation, but one already of long standing, and of current use as covering everything on account of which the other sciences are called parts of mathematics.

Carrying out this program, Descartes states several propositions which he later works out in more detail in his *Discours de la méthode* and its appendix, the *Géométrie* (1637; see Selection II.8), as, for example, in Rule XVI, on the use of letters instead of numbers:

Thus, for instance, if we seek the base of a right-angled triangle with the given sides 9 and 12, the arithmetician will say that it is $\sqrt{225}$ or 15. But we shall substitute a and b for 9 and 12, and shall find the base to be $\sqrt{a^2 + b^2}$. In this way the two parts a and b, which in the number notation were confused, are kept distinct. Also, the realization that terms like "root," "square," "cube," "biquadratic" for proportions which follow by continuous order, are misleading.

For though a magnitude may be entitled a cube or a biquadratic, it should never be presented to the imagination otherwise than as a line or a surface ... What above all requires to be noted is that the root, the square, the cube, etc., are merely magnitudes in continued proportion, which always implies the freely chosen unit of which we have spoken above [in Rule XIV].[2] The first proportional is related to this unit immediately or by one single relation, the second by the mediation of the first and the second, and so by three relations, etc. We therefore entitle the magnitude, which in algebra is called the root, the first proportional; that called the square we shall speak of as being the second proportional, and similarly in the case of the other.

In Rule XVIII Descartes shows how he envisages addition, subtraction, multiplication, and division of line segments, in which he represents the product of two line segments a and b not only as a rectangle, but also as a line. All these ideas were later carried out in his *Géométrie*.

8 DESCARTES. THEORY OF EQUATIONS

Descartes carried out his ideas on the algebraic representation of geometrical quantities in his *Géométrie*, published as appendix I to his *Discours de la méthode pour bien conduire sa raison et chercher la vérité dans les sciences* (Discourse of the method for conducting reason

---

[2] This is the place where Descartes's new algebra is born, the algebra that can be used for coordinate geometry. When 1 is a unit length and x an arbitrary line segment, the proportion $1 : x = x : x^2$ allows us to express x^2 as a line segment. It is here that Descartes breaks with Viète's condition of homogeneity. See further Selections III.3, 4.

well and seeking the truth in the sciences; Leiden, 1637); *Oeuvres de Descartes*, ed. C. Adam and P. Tannery (12 vols.; Paris, 1897–1918), VI, 1–78, 367–485. In this *Géométrie* he showed how the "Algebra of the Moderns," that is, the Renaissance algebra of Cardan and his successors, can be applied to the "Geometry of the Ancients," that is, to the Greek geometry of Apollonius and Pappus. How he did it we show in Selection III.4. But at the same time, in Book Three of the *Géométrie*, he applied it to the theory of equations, adding new discoveries of his own in a presentation of the subject that begins to have a modern look, since his authority made it acceptable to a growing number of mathematicians. He presents some important theorems that are valid for the case in which the right-hand member of an equation is zero, such as the proposition that $x - a$ is a factor of the left-hand member if $x = a$ is a root of the equation. His text also contains a formulation of the fundamental theorem of algebra, as well as the so-called "rule of signs" called after him. We present the section of Book Three that contains these results, in a translation based on *The Geometry of René Descartes*, translated from the French and Latin by D. E. Smith and M. L. Latham (Open Court, Chicago, London, 1925; Dover, New York, 1954), 159–163, 175; this edition also has the original French text.

In the translation we have left the term "dimension" for our degree, "true" for positive roots and "false" for negative ones. The terms "real" and "imaginary" are also in Descartes's text; the use of these terms in our modern sense begins here.

Note that Descartes's coefficients have special numerical values, so that he ignores Viète's *logistica speciosa*[1] as well as his own way of denoting constants by letters $a, b, c, \ldots$ But he introduces x, y for the unknown and the variables and this has become standard practice.

It is necessary that I make some general statements concerning the nature of equations, that is, of sums composed of several terms, in part known, in part unknown, of which some are equal to the others, or rather, all of which considered together are equal to zero, because this is often the best way to consider them.

Know then that in every equation there are as many distinct roots, that is, values of the unknown quantity, as is the number of dimensions of the unknown quantity.[2]

---

[1] As we have said, Descartes's relation to Viète is not clear. The same holds for his relation to the English mathematician and astronomer Thomas Harriot (1560–1621), also known for his description of newly named Virginia (1588). Harriot wrote (probably about 1610) the influential *Artis analyticae praxis* (London, 1631), with a theory of equations which in several aspects anticipates Descartes's treatment. He wrote $aaa - 3bba \mathrel{=\!=\!=} + 2.ccc$ for our $x^3 - 3b^2x = 2c^3$ (see Selection I.2) and used $>$ and $<$ in our present sense: *Signum majoritatis ut $a > b$ significat a majorem quam b*. John Wallis (1616–1703), in his *Treatise of algebra* (Oxford, 1685), claimed that Descartes borrowed heavily from Harriot: "Hariot hath laid the foundations on which Des Cartes (though without naming him) hath built the greatest part (if not the whole) of his Algebra or Geometry." Wallis missed the point, but his statement foreshadows the tension between English and Continental mathematicians typified in the later Newton–Leibniz priority struggle. See J. F. Scott, *The mathematical work of John Wallis* (Taylor and Francis, London, 1938), chap. IX.

[2] This is Descartes's formulation of the fundamental theorem of algebra. For its further, more exact, formulation and proof, see Selections II.6 on Girard, II.10 on Euler, and II.12 on Gauss.

Suppose, for example, $x = 2$ or $x - 2 = 0$,[3] and again, $x = 3$, or $x - 3 = 0$. Multiplying together the two equations $x - 2 = 0$ and $x - 3 = 0$, we have $xx - 5x + 6 = 0$, or $xx = 5x - 6$. This is an equation in which x has the value 2 and at the same time x has the value 3. If we next make $x - 4 = 0$ and multiply this by $xx - 5x + 6 = 0$, we have $x^3 - 9xx + 26x - 24 = 0$, another equation, in which x, having three dimensions, has also three values, namely, 2, 3, and 4.

It often happens, however, that some of the roots are false, or less than nothing. Thus, if we suppose x to represent the defect[4] of a quantity 5, we have $x + 5 = 0$ which, multiplied by $x^3 - 9xx + 26x - 24 = 0$, gives $x^4 - 4x^3 - 19xx + 106x - 120 = 0$, as an equation having four roots, namely three true roots, 2, 3, and 4, and one false root, 5.

It is evident from this that the sum of an equation containing several roots is always divisible by a binomial consisting of the unknown quantity diminished by the value of one of the true roots, or plus the value of one of the false roots. In this way, the dimension of an equation can be lowered.

On the other hand, if the sum of an equation is not divisible by a binomial consisting of the unknown quantity plus or minus some other quantity, then this latter quantity is not a root of the equation. Thus the last equation $x^4 - 4x^3 - 19xx + 106x - 120 = 0$ is divisible by $x - 2$, $x - 3$, $x - 4$, and $x + 5$, but is not divisible by x plus or minus any other quantity, which shows that the equation can have only the four roots, 2, 3, 4, and 5.

We can determine from this also the number of true and false roots that any equation can have, as follows: An equation can have as many true roots as it contains changes of sign, from $+$ to $-$ or from $-$ to $+$; and as many false roots as the number of times two $+$ signs or two $-$ signs are found in succession.[5] Thus, in the last equation, since $+x^4$ is followed by $-4x^3$, giving a change of sign from $+$ to $-$, and $-19xx$ is followed by $+106x$ and $+106x$ by -120, giving two more changes, we know there are three true roots; and since $-4x^3$ is followed by $-19xx$ there is one false root.

It is also easy to transform an equation so that all the roots that were false shall become true roots, and all those that were true shall become false. This is done by changing the sign of the second, fourth, sixth, and all even terms, leaving unchanged the signs of the first, third, fifth, and other odd terms. Thus, if instead of

$$+x^4 - 4x^3 - 19xx + 106x - 120 = 0$$

[3] Descartes writes: "x equal to 2 or $x - 2$ equal to nothing, and again $x \propto 3$, or $x - 3 \propto 0$." He does not use the sign $=$, already introduced by Recorde and Harriot; see note 1.

[4] A defect [*défaut*] is the negative of a positive number; thus -5 is the defect of 5, that is, the remainder when 5 is subtracted from zero.

[5] This is the sign rule as stated by Descartes. It was formulated in a more precise manner by Isaac Newton in his *Arithmetica universalis* (Cambridge, 1707) and by C. F. Gauss in "Beweis eines algebraischen Lehrsatzes," *Crelle's Journal für die reine und angewandte Mathematik 3* (1828), 1–4; *Werke*, III, 65–70. We now can express it in the following way: If $f(x) = 0$ is an equation of degree n with real coefficients, where $f(x) = a_0 x^n + a_1 x^{n-1} + \cdots + a_{n-1} x + a_n$, then the number of positive roots is equal to or an even number less than the number of variations in the signs of successive terms. Multiple roots have to be counted in accordance with their multiplicity, and zero is not a positive root.

we write

$$+x^4 + 4x^3 - 19xx - 106x - 120 = 0,$$

we get an equation having only one true root, 5, and three false roots, 2, 3, and 4.

If the roots of an equation are unknown and it be desired to increase or diminish each of these roots by some known number, we must substitute for the unknown quantity another quantity greater or less by the given number.[6] Thus, if it be desired to increase by 3 the value of each root of the equation

$$x^4 - 4x^3 - 19xx + 106x - 120 = 0,$$

put y in the place of x, and let y exceed x by 3, so that $y - 3 = x$. Then for xx put the square of $y - 3$, or $yy - 6y + 9$; for x^3 put its cube, $y^3 - 9yy + 27y - 27$; and for x^4 put its fourth power [*quarré de quarré*], or $y^4 - 12y^3 + 54yy - 108y + 81$. Substituting these values in the above equation, and combining, we have

$$
\begin{array}{r}
y^4 - 12y^3 + 54yy - 108y + 81 \\
+ 4y^3 - 36yy + 108y - 108 \\
- 19yy + 114y - 171 \\
- 106y + 318 \\
- 120 \\
\hline
\end{array}
$$

or
$$
\begin{array}{l}
y^4 - 8y^3 - 1yy + 8y \quad = 0, \\
y^3 - 8yy - 1y + 8 \quad = 0,
\end{array}
$$

whose true root is now 8 instead of 5, since it has been increased by 3.

If, on the other hand, it is desired to diminish by 3 the roots of the same equation, we must make $y + 3 = x$ and $yy + by + 9 = xx$, and so on, so that instead of

$$x^4 + 4x^3 - 19xx - 106x - 120 = 0$$

we have

$$
\begin{array}{r}
y^4 + 12y^3 + 54yy + 108y + 81 \\
+ 4y^3 + 36yy + 108y + 108 \\
- 19yy - 114y - 171 \\
- 106y - 318 \\
- 120 \\
\hline
y^4 + 16y^3 + 71yy - 4y - 420 = 0.
\end{array}
$$

It should be observed that increasing the true roots of an equation diminishes the false roots by the same amount; and on the contrary diminishing the true roots increases the false roots; while diminishing either a true or a false root by a

[6] This change of variable of an equation by means of substitution of the type $y = x + a$ is not new, and is, as we observed in Selection II.4, note 8, one of the substitutions used by Cardan in his *Ars magna* (1545). Descartes here shows its use in the search for positive and negative roots. The notation, as elsewhere in Descartes's writings, strikes us as quite modern.

quantity equal to it makes the root zero; and diminishing it by a quantity greater than the root renders a true root false or a false root true. Thus by increasing the true root 5 by 3, we diminish each of the false roots, so that the root previously 4 is now only 1, the root previously 3 is zero, and the root previously 2 is now a true root, equal to 1, since $-2 + 3 = +1$. This explains why the equation $y^3 - 8yy - y + 8 = 0$ has only three roots, two of them, 1 and 8, being true roots, and the third, also 1, being false; while the other equation $y^4 - 16y^3 + 71yy - 4y - 420 = 0$ has only one true root, 2, since $+5 - 3 = +2$, and three false roots, 5, 6, and 7.

By going from Cardan via Viète to Descartes we have bypassed some important writers on algebra, among them Rafael Bombelli (c. 1530–after 1572) and Simon Stevin (1546–1620). Bombelli introduced operations with imaginary numbers in connection with the solution of the cubic equation. Stevin's algebra, contained in his *Arithmétique* (Leiden, 1585), can be studied in an English paraphrase in the *Principal works*, IIB (Swets-Zeitlinger, Amsterdam, 1958).

Bombelli and Stevin replaced the old notation for the powers of the unknown, by which each power was expressed by its own symbol (a method that had been used by Diophantus), by a notation which used numbers to express exponents, so that we find in Stevin ① for x, or a, etc., ② for x^2, or a^2, etc., ⓪ for x^0, or a^0, etc. Another algebraist, William Oughtred (1574–1660), a Cambridge-educated English minister, author of *Clavis mathematicae* (London, 1631), has been credited with the introduction or improvement of mathematical symbols, such as the use of $\times$ for multiplication, and the introduction of $::$ for proportion. See F. Cajori, *William Oughtred* (Open Court, Chicago, 1916), who also gives a sample of his work on equations in English translation. Those interested in the history of mathematical notations should consult F. Cajori, *History of mathematical notations* (2 vols.; Open Court, Chicago, 1928, 1929). On Oughtred's older contemporary Harriot see note [1], and J. A. Lohne, "Thomas Harriot als Mathematiker," *Centaurus 11* (1965), 19–45.

9 NEWTON. THE ROOTS OF AN EQUATION

Isaac Newton (1642–1727) was born on the day then called December 25, 1642, according to the old (Julian) calendar. The present (Gregorian) calendar was introduced (not without great opposition) into England and her American colonies in 1752, when it was decreed that September 3 should be renamed September 14, and the year, which used to begin on March 25, should henceforth begin on January 1. Those who count retroactively, therefore, call Newton's birthday January 5, 1643, and this is the reason why the year of Newton's birth is taken as 1642 in some accounts and 1643 in others.

Newton entered Trinity College, Cambridge, in 1661, succeeded Isaac Barrow in the Lucasian professorship in Cambridge in 1669, moved to London in 1696 as Warden of the Mint, and became Master of the Mint in 1699. From 1703 till his death he was president of the Royal Society. The period from 1664 to 1669, during part of which he lived at his father's place in Lincolnshire to escape the plague, can be referred to as his "golden period," in which he laid the foundation of his discoveries in the calculus, in mechanics, and in optics. His most famous work is the *Philosophiae naturalis principia mathematica* (London, 1687), which contains his celestial mechanics based on his law of universal attraction. Most of his other works were published many years after they were written. Among them is his algebra, the *Arithmetica universalis*, edited by William Whiston (Cambridge, 1707), written as the text of lectures between *c*. 1673 and *c*. 1683. We use here the English translation, *Universal arithmetick* (London, 1720). On Newton's work see further H. W. Turnbull, *The mathematical discoveries of Newton* (Blackie, Glasgow, 1945).

The *Arithmetica* contains methods of solving equations in 77 problems, investigations on the roots, including the search for divisors of polynomials, common divisors of two polynomials, and a formulation of Descartes's sign rule. We present here a section (pp. 202–207) in which Newton deals with symmetric functions of the roots (see Selection II.6 on Girard) and then shows how to find upper limits of the roots. He uses the term "affirmative" for what we call positive, as opposed to "negative," "dimension" for our degree, "quantity" of a term for its coefficient, "rectangle" for the product of two terms, and "content" for that of more than two terms.

Newton has first shown that in the equation (in our notation)

$$x^n + a_1 x^{n-1} + a_2 x^{n-2} + \cdots = 0$$

$-a_1$ is the sum of the roots, a_2 "the aggregate of the rectangles of each two of the roots," and so forth. Then he continues (p. 202), writing $p = -a_1$, $q = -a_2$, and so on (we have somewhat modernized the spelling):

From the generation of equations it is evident, that the known quantity of the second term of the equation, if its sign be changed, is equal to the aggregate of all the roots under their proper signs; and that of the third term, equal to the aggregate of the rectangles of each two of the roots; that of the fourth, if its sign be changed, is equal to the aggregate of the contents under each three of the roots; that of the fifth is equal to the aggregate of the contents under each four, and so on ad infinitum.

Let us assume $x = a$, $x = b$, $x = -c$, $x = d$, etc. or $x - a = 0$, $x - b = 0$, $x + c = 0$, $x - d = 0$, and by the continual multiplication of these we may generate equations, as above. Now, by multiplying $x - a$ by $x - b$ there will be produced the equation $xx \genfrac{}{}{0pt}{}{-a}{-b} x + ab = 0$; where the known quantity of the second term, if its signs are changed, viz. $a + b$, is the sum of the two roots a and b, and the known quantity of the third term is the only rectangle contained under both. Again, by multiplying this equation by $x + c$, there will be produced the cubic equation $x^3 \genfrac{}{}{0pt}{}{-a}{+c}{-b} xx \genfrac{}{}{0pt}{}{+ab}{-ac}{-bc} x + abc = 0$, where the known quantity of the second term having its signs changed, viz. $a + b - c$, is the sum of the roots a, and b, and $-c$; the known quantity of the third term $ab - ac - bc$ is the sum of the rectangles under each two of the roots a and b, a and $-c$, b and $-c$; and the known quantity of the fourth term under its sign changed, $-abc$, is the only content generated by the continual multiplication of all the roots, a by b into $-c$.

Moreover, by multiplying that cubic equation by $x - d$, there will be produced this biquadratic one;

$$
\begin{array}{ccccc}
 & & +ab & & \\
 & -a & -ac & +abc & \\
 & -b & -bc & -abd & \\
x^4 & x^3 & xx & x & -abcd = 0. \\
 & +c & +ad & +bcd & \\
 & -d & +bd & +acd & \\
 & & -cd & &
\end{array}
$$

Where the known quantity of the second term under its signs changed, viz. $a + b - c + d$, is the sum of all the roots; that of the third, $ab - ac - bc + ad - bd - cd$, is the sum of the rectangles under every two roots; that of the fourth, its signs being changed, $-abc + abd - bcd - acd$, is the sum of the contents under each ternary; that of the fifth, $-abcd$, is the only content under them all. And hence we first infer, that of any equation whose terms involve neither surds nor fractions all the rational roots, and the rectangles of any two of the roots, or the contents of any three or more of them, are some of the integral divisors of the last term;[1] and therefore when it is evident that no divisor of the last term is either a root of the equation, or rectangle, or content of two or more roots, it will also be evident that there is no root, or rectangle, or content of roots, except what is surd.

Let us suppose now, that the known quantities of the terms of any equation under their signs changed, are p, q, r, s, t, v, etc. viz. that of the second p, that of the third q, of the fourth r, of the fifth s, and so on. And the signs of the terms being rightly observed, make $p = a$, $pa + 2q = b$, $pb + qa + 3r = c$, $pc + qb + ra + 4s = d$, $pd + qc + rb + sa + 5t = e$, $pe + qd + rc + sb + ta + 6v = f$,[2] and so on in infinitum, observing the series of the progression. And a will be the sum of the roots, b the sum of the squares of each of the roots, c the sum of the cubes, d the sum of the biquadrates, e the sum of the quadrato-cubes, f the sum of the cubo-cubes, and so on. As in the equation $x^4 - x^3 - 19xx + 49x - 30 = 0$, where the known quantity of the second term is -1, of the third -19, of the fourth $+49$, of the fifth -30; you must make $1 = p$, $19 = q$, $-49 = r$, $30 = s$. And there will thence arise $a = (p =) 1$, $b = (pa + 2q = 1 + 38 =) 39$, $c = (pb + qa + 3r = 39 + 19 - 147 =) -89$,

[1] This sentence has been translated from the original Latin text, since the English text is garbled. The term *surd* is usually used for incommensurable roots of a commensurable number, as $\sqrt{2}$, but also means any nonrational number; "quantities partly rationall and partly surde," writes Recorde in his *Pathewaie to knowledge* (London, 1551). The term, meaning deaf or mute, is a Latin translation of an Arabic translation of the Greek term *alogos*, meaning *irrational*.

[2] These sums are known as sums of Newton, although they were known to Girard (Selection II.6). The general formulas for the equation $a_0 x^n + a_1 x^{n-1} + a_2 x^{n-2} + \cdots + a_{n-1} x + a_n = 0$ are $a_0 s_1 + a_1 = 0$, $a_0 s_2 + a_1 s_1 + 2a_2 = 0$, $a_0 s_3 + a_1 s_2 + a_2 s_1 + 3a_3 = 0, \ldots, a_0 s_n + a_1 s_{n-1} + a_2 s_{n-2} + \cdots + na_n = 0$; $a_0 s_{n+k} + a_1 s_{n+k-1} + a_2 s_{n+k-2} + \cdots + a_n s_k = 0$ for $k = 0, 1, 2, \ldots$.
Here s_i is the sum of the ith powers of the roots. They go in essential back to C. Maclaurin, *A treatise of algebra* (London, 1748), Part II, 141–143.

$d = (pc + qb + ra + 4s = -89 + 741 - 49 + 120 =) 723$. Wherefore the sum of the roots will be 1, the sum of the squares of the roots 39, the sum of the cubes -89, and the sum of the biquadrates 723, viz. the roots of that equation are $1, 2, 3$, and -5, and the sum of these $1 + 2 + 3 - 5$ is 1; the sum of the squares, $1 + 4 + 9 + 25$, is 39; the sum of the cubes, $1 + 8 + 27 - 125$, is -89; and the sum of the biquadrates, $1 + 16 + 81 + 625$, is 723.

OF THE LIMITS OF EQUATIONS.

And hence are collected the limits between which the roots of the equation shall consist, if none of them is impossible. For when the squares of all the roots are affirmative, the sum of the squares will be affirmative, and by the same argument, the sum of the biquadrates of all the roots will be greater than the biquadrate of the greatest root, and the sum of the cubo-cubes greater than the cubo-cube of the greatest root.

Wherefore, if you desire the limit which no roots can pass, seek the sum of the squares of the roots, and extract its square root. For this root will be greater than the greatest root of the equation. But you will come nearer the greatest root if you seek the sum of the biquadrates, and extract its biquadratic root; and yet nearer, if you seek the sum of the cubo-cubes, and extract its cubo-cubical root; and so on in infinitum.[3]

Thus, in the precedent equation, the square root of the sum of the squares of the roots, or $\sqrt{39}$, is $6\frac{1}{2}$ nearly, and $6\frac{1}{2}$ is farther distant from 0 than any of the roots $1, 2, 3, -5$. But the biquadratic root of the sum of the biquadrates of the roots, viz. $\sqrt[4]{723}$, which is $5\frac{1}{4}$ nearly, comes nearer to the root that is most remote from nothing, viz. -5.

Rule II. If, between the sum of the squares and the sum of the biquadrates of the roots you find a mean proportional, that will be a little greater than the sum of the cubes of the roots connected under affirmative signs. And hence, the half sum of this mean proportional, and of the sum of the cubes collected under their proper signs, found as before, will be greater than the sum of the cubes of the affirmative roots, and the half difference greater than the sum of the cubes of the negative roots.[4]

And consequently, the greatest of the affirmative roots will be less than the cube root of that half sum, and the greatest of the negative roots less than the cube root of that semi-difference.

Thus, in the precedent equation, a mean proportional between the sum of the squares of the roots 39, and the sum of the biquadrates 723, is nearly 168. The sum of the cubes under their proper signs was, as above, -89, the half sum of

---

[3] If x_μ is the largest root in absolute value, then for any positive integer m $|x_\mu| < \sqrt[2m]{s_{2m}}$, and $\sqrt[2m]{s_{2m}} - |x_\mu|$ will tend to zero with increasing m. The proof follows from the fact that s_{2m} is always positive.

[4] Let $s_3 = \check{s}_3 - \bar{s}_3$, where $\check{s}_3$ is the sum of the third powers of the positive roots, and $-\bar{s}_3$ that of the negative roots. Then $\sqrt{s_2 s_4} > \check{s}_3 + \bar{s}_3$, $\frac{1}{2}(\sqrt{s_2 s_4} + \check{s}_3 - \bar{s}_3) > \check{s}_3$, $\frac{1}{2}(\sqrt{s_2 s_4} - \check{s}_3 + \bar{s}_3) > \bar{s}_3$.

this and 168 is $39\frac{1}{2}$, the semi-difference $128\frac{1}{2}$. The cube root of the former, which is about $3\frac{1}{2}$, is greater than the greatest of the affirmative roots 3. The cube root of the latter, which is $5\frac{1}{21}$ nearly, is greater than the negative root -5. By which example it may be seen how near you may come this way to the root, where there is only one negative root or one affirmative one.

Rule III. *And yet you might come nearer still*, if you found a mean proportional between the sum of the biquadrates of the roots and the sum of the cubo-cubes, and if from the semi-sum and semi-difference of this, and of the sum of the quadrato-cube of the roots, you extracted the quadrato-cubical roots. For the quadrato-cubical root of the semi-sum would be greater than the greatest affirmative root, and the quadrato-cubic root of the semi-difference would be greater than the greatest negative root, but by a less excess than before. Since therefore any root, by augmenting or diminishing all the roots, may be made the least, and then the least converted into the greatest, and afterwards all besides the greatest be made negative, it is manifest how any root desired may be found nearly.

Rule IV. *If all the roots except two are negative, those two may be both together found this way.*

The sum of the cubes of those two roots being found according to the precedent method, as also the sum of the quadrato-cubes, and the sum of the quadrato-quadrato-cubes of all the roots: between the two latter sums seek a mean proportional, and that will be the difference between the sum of the cubo-cubes of the affirmative roots, and the sum of the cubo-cubes of the negative roots nearly; and consequently, the half sum of this mean proportional, and of the sum of the cubo-cubes of all the roots, will be the sum of the cubo-cubes of the affirmative roots, and the semi-difference will be the sum of the cubo-cubes of the negative roots. Having therefore both the sum of the cubes, and also the sum of the cubo-cubes of the two affirmative roots, from the double of the latter sum subtract the square of the former sum, and the square root of the remainder will be the difference of the cubes of the two roots. And having both the sum and difference of the cubes, you will have the cubes themselves. Extract their cube roots, and you will nearly have the two affirmative roots of the equation. And if in higher powers you should do the like, you will have the roots yet more nearly. But these limitations, by reason of the difficulty of the calculus, are of less use, and extend only to those equations that have no imaginary roots. Wherefore I will now shew how to find the limits another way, which is more easy, and extends to all equations.

Rule V. *Multiply every term of the equation by the number of its dimensions, and divide the product by the root of the equation. Then again multiply every one of the terms that come out by a number less by unity than before, and divide the product by the root of the equation. And so go on, by always multiplying by numbers less by unity than before, and dividing the product by the root, till at length all the terms are destroyed, whose signs are different from the sign of the first or highest term, except the last. And that number will be greater than any affirmative root; which being writ in the terms that come out for the root, makes the aggregate of those which were each time produced by multiplication to have always the same sign with the first or highest term of the equation.*[5]

As if there was proposed the equation $x^5 - 2x^4 - 10x^3 + 30xx + 63x - 120 = 0$. I first multiply this thus; $\overset{5}{x^5} - \overset{4}{2x^4} - \overset{3}{10x^3} + \overset{2}{30xx} + \overset{1}{63x} - \overset{0}{120}$. Then again multiply the terms that come out divided by x, thus; $\overset{4}{5x^4} - \overset{3}{8x^3} - \overset{2}{30xx} + \overset{1}{60x} + \overset{0}{63}$, and dividing the terms that come out again by x, there comes out $20x^3 - 24xx - 60x + 60$; which, to lessen them, I divide by the greatest common divisor 4, and you have $5x^3 - 6xx - 15x + 15$. These being again multiplied by the progression $3, 2, 1, 0$, and divided by x, become $15xx - 12x - 15$, and again divided by 3 become $5xx - 4x - 5$. And these multiplied by the progression $2, 1, 0$, and divided by $2x$ become $5x - 2$. Now, since the highest term of the equation x^5 is affirmative, I try what number writ in these products for x will cause them all to be affirmative. And by trying 1, you have $5x - 2 = 3$ affirmative; but $5xx - 4x - 5$, you have -4 negative. Wherefore the limit will be greater than 1. I therefore try some greater number, as 2. And substituting 2 in each for x, they become

$$5x \quad - 2 = 8$$
$$5xx - 4x - 5 = 7$$
$$5x^3 - 6xx - 15x \quad + 15 = 1$$
$$5x^4 - 8x^3 - 30xx + 60x \quad + 63 = 79$$
$$x^5 - 2x^4 - 10x^3 + 30xx + 63x - 120 = 46$$

Wherefore, since the numbers that come out $8.7.1.79.46$. are all affirmative, the number 2 will be greater than the greatest of the affirmative roots. In like manner, if I would find the limit of the negative roots, I try negative numbers. Or that which is all one, I change the signs of every other term, and try affirmative ones. But having changed the signs of every other term, the quantities in which the numbers are to be substituted, will become

$$5x \quad + 2$$
$$5xx + 4x \quad - 5$$
$$5x^3 + 6xx - 15x \quad - 15$$
$$5x^4 + 8x^3 - 30xx - 60x \quad + 63$$
$$x^5 + 2x^4 - 10x^3 - 30xx + 63x + 120$$

Out of these I choose some quantity wherein the negative terms seem most prevalent; suppose $5x^4 + 8x^3 - 30xx - 60x + 63$, and here substituting for x the numbers 1 and 2, there come out the negative numbers -14 and -33. Whence the limit will be greater than -2. But substituting the number 3, there

---

[5] This rule is known as Newton's rule, and can be expressed in modern language as follows: If $f(z) = a_0 z^n + a_{n-1} z^{n-1} + a_2 z^{n-2} + \cdots + a_n = 0$, $a_0 > 0$, then every number $z = L$ for which $f(z)$ and its derivatives $f'(z), f''(z), \ldots, f^{n-1}(z)$ are positive is an upper limit for the positive roots of $f(z) = 0$. The rule can be proved as a corollary to the theorem of Fourier (1831; also called after Budan); see, for instance W. S. Burnside and A. W. Panton, *The theory of equations* (3rd ed.; Hodges, Figgis and Co., Dublin, 1892), 170, 182.

comes out the affirmative number 234. And in like manner in the other quantities, by substituting the number 3 for x, there comes out always an affirmative number, which may be seen by bare inspection. Wherefore the number -3 is greater than all the negative roots. And so you have the limits 2 and -3, between which are all the roots.

---

10   EULER. THE FUNDAMENTAL THEOREM OF ALGEBRA

We have seen how Girard, in 1629, formulated the principle that an algebraic equation of degree n has n roots. The first attempt to prove it was made by Jean Le Rond D'Alembert (1717–1783), after 1759 permanent secretary of the French Academy. It can be found in his "Recherches sur le calcul intégral," *Histoire de l'Académie Royale, Berlin, 1746* (1748), 102–224. It is from this attempt that the theorem is still known as the theorem of D'Alembert. The theorem had received a new importance in those days because of its application to the integration of rational fractions by means of partial functions. In D'Alembert's words:

"In order to reduce in general a differential rational function to the quadrature of the hyperbola or to that of the circle, it is necessary, according to the method of M. Bernoulli (Acad. Paris 1702),[1] to show that every rational polynomial, without a divisor composed of a variable x and of constants, can always be divided, when it is of even degree, into trinomial factors $xx + fx + g$, $xx + hx + i$, etc., of which all coefficients f, g, h, i, are real. It is clear that this difficulty affects only the polynomial that cannot be divided by any binomial, $x + a$, $x + b$, etc., because we can always by division reduce to zero all the real binomials, if there are any, and it can easily be seen that the products of these binomials will give real factors $xx + fx + g$."

D'Alembert's proof, which used geometrical arguments, was not very convincing. As a matter of fact, D'Alembert did not even prove the existence of a root, but only showed the form the root takes, and for this adds to his proof a dissertation on complex numbers and complex functions, something fairly new at that time. The proof given at the same time by D'Alembert's friend Euler was different, and, despite certain weaknesses,[2] more appealing. The proof can be found under the title "Recherches sur les racines imaginaires des équations," *Histoire de l'Académie Royale, Berlin, 1747* (1749), 222–228; *Opera omnia*, ser. I, vol. 6, 78–147. It was a purely algebraic proof. First Euler shows that when a root $x + y\sqrt{-1}$ exists there is also one of the form $x - y\sqrt{-1}$, so that there must be a factor of the form $xx + px + q$. After an example of how to decompose an equation of the fourth degree into two quadratic factors, Euler proves three theorems: (1) that an equation of odd

---

[1] Johann Bernoulli, "Solution d'un problème concernant le calcul intégral," *Histoire de l'Académie Royale des Sciences, Paris, 1702*, 399–410; *Opera omnia*, I, 393–400. Here Bernoulli shows how to integrate pdx/q, when p and q are polynomials in x. He first reduces by division the degree of p to that below the degree of q, and then reduces the differential to a sum of terms of the form $adx/(x + f)$. This gives, in his words, a sum of logarithmic terms, or quadratures of the hyperbola, which are either real or imaginary, also expressible as circular arcs or sectors. Leibniz, in the *Acta Eruditorum* of 1702, had indicated a similar procedure, but Bernoulli's work was more complete; see Leibniz, *Mathematische Schriften*, part 2, vol. 1 (1858), 350–361.

[2] These weaknesses were brought out by Gauss in his dissertation (1799); see Selection II.12.

degree has at least one root, (2) that one of even degree has either no real roots or pairs of such roots, and (3) that an equation of even degree with negative absolute term has at least one positive and one negative root. All three theorems are demonstrated geometrically. Then follows:

Theorem 4. Every equation of the fourth degree, as

$$x^4 + Ax^3 + Bx^2 + Cx + D = 0,$$

can always be decomposed into two real factors of the second degree.

Demonstration. It is known that by setting $x = y - \frac{1}{4}A$ we can change this equation into another one of the same degree without the second term, and, since this transformation can always be performed, let us suppose that in the proposed equation the second term is already missing, and that we have to resolve this equation:

$$x^4 + Bx^2 + Cx + D = 0,$$

into two real factors of the second degree. It is clear that these factors will be of the form

$$(xx + ux + \alpha)(xx - ux + \beta) = 0.$$

If now we compare this product with the proposed equation, we shall find

$$B = \alpha + \beta - uu, \quad C = (\beta - \alpha)u, \quad D = \alpha\beta,$$

from which we derive

$$\alpha + \beta = B + uu, \quad \beta - \alpha = \frac{C}{u},$$

hence

$$2\beta = uu + B + \frac{C}{u} \quad \text{and} \quad 2\alpha = uu + B - \frac{C}{u},$$

and since we have $4\alpha\beta = 4D$ we obtain the equation

$$u^4 + 2Buu + BB - \frac{CC}{uu} = 4D$$

or

$$u^6 + 2Bu^4 + (BB - 4D)uu - CC = 0,$$

from which the value of u must be found. And since the absolute term $-CC$ is essentially negative, we have shown that this equation has at least two real roots. When we take one of them as u, then the values of α and β will also be real, and hence the two supposed factors of the second degree $xx + ux + \alpha$ and $xx - u\alpha + \beta$ will be real.   Q.E.D.

Among the corollaries to Theorem 4 is the statement that the resolution into real factors is now also proved for the fifth degree, and Scholium II points out that, if the roots of the given fourth-degree equation are a, b, c, d, then the sixth-degree equation in u, u being the sum of two roots of the given equation, will have the six roots $a + b$, $a + c$, $a + d$, $c + d$, $b + d$, $b + c$. Since $a + b + c + d = 0$, we can write for u the values $p, q, r, -p, -q, -r$, and the equation in u becomes

$$(uu - pp)(uu - qq)(uu - rr) = 0.$$

Theorem 5. Every equation of degree 8 can always be resolved into two real factors of the fourth degree.

The proof follows the same reasoning as before. First the term in x^7 is eliminated, so that the two supposed factors can be written $x^4 - ux^3 + \alpha x^2 + \beta x + \gamma = 0$ and $x^4 + ux^3 + \delta x^2 + \epsilon x + \zeta = 0$. Since u expresses the sum of four roots of the eighth-degree equation, it can have $\dfrac{8 \cdot 7 \cdot 6 \cdot 5}{1 \cdot 2 \cdot 3 \cdot 4} = 70$ values, and it will satisfy an equation of the form

$$0 = (uu - pp)(uu - qq)(uu - rr)(uu - ss)\cdots,$$

with 35 factors. The absolute term is negative, and the reasoning continues as before.

Corollary 2 states that the theorem is also proved for degree 9, and Corollary 3 that it is also proved for degrees 6 and 7, since we have only to multiply such an equation by xx or x to obtain an equation of degree 8. Scholium II tries to make certain that not only u, but also the other coefficients $\alpha, \beta, \gamma, \ldots$ are real, a reasoning to which Lagrange and Gauss later objected. Euler continues in Theorem 6 with similar proofs for equations of degree 16, and then, in general:

Theorem 7. Every equation of which the degree is a power of the binary, as 2^n (n being a number greater than 1), can be resolved into two factors of degree 2^{n-1}.

The proof, similar to the previous ones, leads to an equation in u of degree

$$N = \frac{2^n}{2^{n-1}} \frac{2^n - 1}{2^{n-1} - 1} \frac{2^n - 2}{2^{n-1} - 2} \frac{2^n - 3}{2^{n-1} - 3} \cdots \frac{2^{n-1} + 1}{1},$$

which is "oddly even" [*impairement pair*], that is, $\frac{1}{2}N$ is odd, so that the absolute term of the equation in u is again negative.

Scholium. We thus have here a complete demonstration of the proposition, which is usually presupposed in analysis, especially in the integral calculus, and which claims that every rational function of a variable x, as

$$x^m + Ax^{m-1} + Bx^{m-2} + \text{etc.}$$

can always be resolved into real factors, either simple ones of the form $x + p$, or double ones of the form $xx + px + q$.

Euler believes that the proof is solid ("je crois qu'on n'y trouvera rien à redire"), but to strengthen the argument he gives extra proofs for degrees 6, $4n + 2$, $8n + 4, \ldots, 2^n p$, p an odd number. In the first and second cases he shows that there exists at least one real factor of the second degree, in the case $8n + 4$ at least one factor of the fourth degree, and in the case $2^n p$ a factor of degree 2^{n-1}.

The second part of the paper proves that all nonreal roots are of the form $M + N\sqrt{-1}$, M and N being real. This implies extensive discussion of all operations with complex numbers, including the raising to imaginary powers, the taking of logarithms, and the formation of trigonometric functions of complex angles. This section could enter straight into any modern elementary text on complex numbers.

11 LAGRANGE. ON THE GENERAL THEORY OF EQUATIONS

Joseph Louis Lagrange (1736–1813), born in Turin of partly French ancestry, became professor of mathematics at the Royal Artillery School of Turin before he was twenty; from 1766 to 1787 he was Academician in Berlin (as successor to Euler), and from 1787 to his death Academician and professor of mathematics in Paris. His fundamental work on algebra dates from his Berlin period. Indeed, his paper, "Réflexions sur la résolution algébrique des équations," *Nouveaux Mémoires de l'Académie Royale, Berlin, 1770* (1772), 134–215; *1771* (1773), 138–254; *Oeuvres* (Gauthier-Villars, Paris), III (1869), 205–421, opens a new period in the study of the theory of equations. By relating the general nature of the roots of an equation to the theory of permutations Lagrange introduces the point of view which in the next century would lead to the theory of Galois. After a methodical investigation of the existing methods of solution of quadratic, cubic, and biquadratic equations, such as those of Cardan, Ferrari, and Euler, and of equations of higher degree, such as that of

Tschirnhaus,[1] Lagrange came to the conclusion that the general structure of the solution of an equation of degree μ is determined by certain equations called resolvents (*réduites*), of which the degree is either $1 \cdot 2 \cdot 3 \cdots \mu$ or a divisor of this number. Because of the length of the paper we can give only the sections that present the core of Lagrange's theory.

Lagrange takes $f(x, y, z, \ldots)$ as a rational function of $x, y, z, \ldots,$[2] and $x', x'', x''', \ldots$ as the roots of the equation $x^\mu + mx^{\mu-1} + nx^{\mu-2} + px^{\mu-3} + \cdots = 0$.

---

90. To begin with the simplest cases, let us suppose that the given equation is only of the second degree, and that we ask for the equation by which such a function $f(x', x'')$ will be determined.[2] I write $t = f(x', x'')$ so that t will be the unknown of the required equation, and, as x' and x'' are both determined by the same equation,

$$x^2 + mx + n = 0,$$

I write, for greater generality, x instead of x' and y instead of x'', and so I obtain the equation

$$t - f(x, y) = 0,$$

and our task is to eliminate x and y by means of the two equations

$$x^2 + mx + n = 0, \quad y^2 + my + n = 0.$$

Let $t - f(x, y) = X$; then we shall first eliminate x from the equation $X = 0$ by means of the equation $x^2 + mx + n = 0$, which will give an equation that I shall denote by $Y = a$, and in which Y will be a rational function of the quantities t, m, n, and y. Then we shall eliminate y from this last equation by means of the other equation $y^2 + my + n = 0$, and we shall have the final equation $T = c$, where T will be a rational function of t, m, n.

I observe now that, since the roots of the equation $x^2 + mx + n = 0$ are x' and x'', if we denote by X' and X'' the values of X that result from the substitution of these roots for x, then we shall have . . .

$$Y = X'X'',$$

---

[1] Walter von Tschirnhaus (1651–1708), "Methodus auferendi omnes terminas intermedios ex datu aequatio meo," *Acta Eruditorum* (1683), 204–207, tried to reduce an equation of degree n: $x^n + a_1 x^{n-1} + \cdots + a_n = 0$ with roots $x_1, x_2, \ldots, x_n$ by a substitution $y = \varphi(x) = \alpha_0 + \alpha_1 x + \cdots + \alpha_{n-1} x^{n-1}$ to the form $y^n - k^n = 0$. He formed the equation

$$\varphi(y) = (y - y_1)(y - y_2) \cdots (y - y_n) = y^n + b_1 y^{n-1} + \cdots + b_n,$$

where $y_i = \varphi(x_i)$, and tried to make as many as possible of the b_i zero by selecting the α_i in an appropriate way. See W. S. Burnside and A. W. Panton, *The theory of equations* (3rd ed.; Hodges, Figgis and Co., Dublin, 1892), 424–429.

[2] Lagrange writes $f[(x)(y)(z) \cdots], f[(x')(x'')]$.

and similarly, since x' and x'' are also the roots of the equation $y^2 + my + n = 0$, if we denote by Y' and Y'' the values of Y that result from the substitution of x' and x'' for y, we shall have

$$T = Y'Y''.$$

Now we have $X' = t - f(x', y)$, $X'' = t - (f(x'', y)$; hence

$$Y = [t - f(x', y)] \times [t - f(x'', y)],$$

and therefore

$$Y' = [t - f(x', x')] \times [t - f(x'', x')],$$
$$Y'' = [t - f(x', x'')] \times [t - f(x'', x'')],$$

so that we obtain

$$T = [t - f(x', x'')] \times [t - f(x'', x')] \times [t - f(x', x')] \times [t - f(x'', x'')].$$

If we now consider the function[3] $f(x^2)$ write $t - f(x^2) = \xi$, and eliminate x from the equation $\xi = 0$ by means of the equation $x^2 + mx + n = 0$, then we shall have the equation $\theta = 0$, where θ will be a rational function of t and m, n. Denoting by ξ' and ξ'' the values of ξ that result from the substitution of x', x'' for x, we shall have

$$\theta = \xi'\xi''.$$

But we have

$$\xi' = t - f(x', x'), \quad \xi'' = t - f(x'', x''),$$

hence

$$\theta = [t - f(x', x')] \times [t - f(x'', x'')].$$

Write

$$\Theta = [t - f(x', x'')] \times [t - f(x'', x')],$$

and we shall have $T = \Theta\theta$, or $\Theta = T/\theta$, so that, since T and θ are rational functions of t, m, and n, it is clear that Θ will also be a rational function of t, m, n.

The equation $T = 0$ can therefore be decomposed into two, $\Theta = 0$ and $\theta = 0$, and, since the first is the one that gives the value of $f(x^2)$, we see that the determination of the proposed function $f(x', x'')$ will depend uniquely on the other equation, $\Theta = 0$.

[3] Lagrange writes $f[(x)^2]$ for $f[(x), (x)]$.

Hence, in order to find this equation $\Theta = 0$ that solves the problem, we have only to eliminate from the equations

$$t - f(x, y) = 0, \quad t = f(x^2) = 0$$

the unknowns x and y by means of the equations $x^2 + mx + n = 0$, $y^2 + my + n = 0$, and, denoting by $T = 0$ and $\theta = 0$ the resulting equations, we shall have immediately $\Theta = T/\theta$.[4]

91. We see from the expression for Θ that the equation $\Theta = 0$, which serves to determine the value of the function $f(x', x'')$, is of the second degree, and that its two roots are $f(x', x'')$ and $f(x'', x')$. Indeed, as the roots x' and x'' are determined in the same way by the equation $x^2 + mx + n = 0$, it is clear that the two functions $f(x', x'')$ and $f(x'', x')$, which differ from each other only by the interchange of the roots x' and x'', must also be determined by the same equation.

If the function $f(x', x'')$ were of the form $f(x'', x')$, so that

$$f(x', x'') = f(x'', x'),$$

then we would have

$$\Theta = [t - f(x', x'')]^2 ;$$

hence the equation $\Theta = 0$ would be simply

$$t - f(x', x'') = 0,$$

from which we see that the function in question would in this case be determined by a linear equation, and hence will be given by a rational expression in m and n.

92–98. This same method is now applied to a cubic equation, $x^3 + mx^2 + nx + p = 0$, where the function $f(x', x'', x''')$ has to be determined. There are six permutations of x', x'', x'''. The corresponding equation $T = 0$ will be of the form $T = \Theta\theta\theta_1\theta_2\theta_3$. T will be a rational function in t, m, n, p. $\Theta = [t - f(x', x'', x''')] \times [t - f(x'', x', x''')] \times [t - f(x'', x''', x')] \times [t - f(x', x''', x'')] \times [t - f(x''', x', x'')] \times [t - f(x''', x'', x')]; \theta = [t - f(x'^3)] \times [t - f(x''^3)] \times [t - f(x'''^3)]; \theta_1 = [t - f(x'^2, x'')] \times \cdots \times [t - f(x''^2, x'')]$. To find $\Theta = 0$ we must

---

[4] As a simple example of Lagrange's method, take $x^2 + mx + n = 0$ and $f(x', x'') = x' + 2x''$. Then

$$X = t - x - 2y,$$
$$T = (t^2 + mt - 3n)^2 + 2m(t^2 + mt - 3n)(2t + 3n) + 4n(2t + 3n)^2,$$
$$Y = (t - 2y)^2 + m(t - 2y) + n,$$
$$\xi = t - 3x,$$
$$\theta = t^2 + 3mt + 9n,$$
$$\Theta = (t - x'' - 2x')(t - x' - 2x'') = t^2 + 3mt + (2m^2 + n).$$

eliminate the unknowns x, y, z from the five equations $t - f(x, y, z) = 0$, $t - f(x^2, y) = 0$, $t - f(x, y^2) = 0$, $t - f(x, y, z) = 0$, $t - f(x^3) = 0$ with the aid of the three equations $x^3 + mx^2 + nx + p = 0$ and the similar ones for y and z. Then writing the resulting equations as $T = 0$, $T_1 = 0$, $T_2 = 0$, $T_3 = 0$, $\theta = 0$, we find $T_1 = \theta\theta_1$, $T_2 = \theta\theta_2$, $T_3 = \theta\theta_3$, $T = \Theta\theta\theta_1\theta_2\theta_3$, hence $\Theta = T\theta^2/T_1T_2T_3$.

This is a long and painful method, and for equations of higher degree it becomes even more so. But the method shows the nature of the resolvent $\Theta = 0$. The degree of Θ in the case of an equation of degree μ will be that of the number of permutations of μ elements, hence $1 \cdot 2 \cdot 3 \cdots \mu = \pi$ will always be a rational function of t and the coefficients $n, m, p, \ldots$ of the proposed equations, hence also the coefficients of t in Θ, which can therefore also be found directly. Here Cramer's book and Waring's *Meditationes algebraicae* are quoted.[5] When $f(x', x'', \ldots)$ is unchanged under certain permutations of the $x', x'', \ldots$, the degree of $\Theta = 0$ will go down, from π to $\pi/2$, $\pi/3$, etc. When f contains only λ of the μ roots, the degree of $\Theta = 0$ will, in general, be $\dfrac{1 \cdot 2 \cdot 3 \cdots \mu}{1 \cdot 2 \cdot 3 \cdots (\mu - \lambda)}$.

---

99. From all that we have shown it follows therefore, in general, (1) that all functions of the roots x' x'', x''', $\ldots$ of the same equation that are *alike* [*fonctions semblables*, that is, invariant under the same and only under the same permutations of the roots] are necessarily of the same degree, (2) that this degree will be equal to the number $1 \cdot 2 \cdot 3 \cdots \mu$ (μ being the degree of the given equation), or of a factor [*sous-multiple*] of this number; (3) that in order to find directly the simplest equation $\theta = 0$ by which an arbitrary given function of $x', x'', x''', \ldots$ can be determined we have only to look for the different values that this function can receive under the permutations of the quantities $x', x'', x''', \ldots$ among themselves. Then, taking these values for the roots of the required equations, we can determine by known methods the coefficients of this equation.

100. If we now have found, either by the solution of the equation $\theta = 0$ or otherwise, the value of a given function of the roots $x', x'', x''', \ldots$, then I say that the value of any other function of these same roots can be found, and this, in general, by means of a simple linear equation, with the exception of some particular cases which require an equation of second degree, or of the third degree, etc. This problem seems to me one of the most important of the theory of equations, and the general Solution that we shall give of it will serve to throw a new light on this part of Algebra.

Expressions for y in t when the two functions denoted by t and y are alike are discussed in Arts. 100–102. An example is given for $x^4 + mx^3 + nx^2 + px + q = 0$, which is supposed

[5] For Cramer's book see Selection III.10. As to Edward Waring (1734–1798), the Lucasian professor of mathematics in Cambridge, his *Meditationes algebricae* (Cambridge, 1770) contains many theorems on the roots, especially also imaginary roots, of algebraic equations. Here he published (without proof) the theorem that carries his name: "Every integer is either a cube or the sum of 2, 3, 4, 5, 6, 7, 8 or 9 cubes; either a biquadrate, or the sum of 2, 3, etc. or 19 biquadrates." See on Waring also Selection I.11, note 5.

to be divisible by $x^2 + fx + g = 0$. Then $g = (p - nf + mf^2 - f^3)/(m - 2f)$ gives g expressed rationally in f. But if $f = m/2$, $p = mn/2 - m^3/8$, then this value of f gives $g = 0/0$, and g must be found from

$$g^2 - (n - m^2/4)g + q = 0.$$

This is exactly the case where the value $f = m/2$ is a double root of the equation for f, which is of the sixth degree. In 103 the case is discussed in which the functions t and y are no longer alike. The result is summarized in the following theorems.

---

104. Therefore:

1. If we have two arbitrary functions t and y of the roots x', x'', x''', ... of the equation

$$x^\mu + mx^{\mu - 1} + nx^{\mu - 2} + \cdots = 0,$$

and if these functions are such that all the permutations of the roots x' x'', x''', ... that make the function y vary at the same time make the function t vary also, then we can, generally speaking, express the value of y in t and in $m, n, p, \ldots$ by a rational expression, such that when a value of t is known the corresponding value of y is also known; we say *generally* speaking since if it happens that the known value of t is a double, or triple, etc. root of the equation in t, in that case the corresponding value of y will depend on a quadratic, or cubic, etc. equation, of which all the coefficients will be rational functions of t and of $m, n, p, \ldots$.

2. If the functions t and y are such that the function t maintains the same value under permutations that make the function y vary, then we cannot find the value of y in t and in m, n, p except by means of an equation of the second degree, if to one value of t there correspond two different values of y, or of the third degree, if to one value of t there correspond three different values of y, and so on. The coefficients of these equations in y will be, generally speaking, rational functions of t and of $m, n, p, \ldots$ such that when a value of t is given we shall have y by means of a simple solution of an equation of the second or of the third degree, etc. But it may happen that the known value of t is a double or triple, etc. root of the equation in t; in that case the coefficients of the equations in question will still depend themselves on an equation of the second or of the third degree, etc.

From this we can derive the necessary conditions for the determination of the values of the roots x', x'', x''', ... themselves with the aid of those of an arbitrary function of these roots, since we have for this purpose only to take the simple root x in the place of the function y, and to apply to this case the preceding conclusions.

In Arts. 105–108 applications are given to equations of the third and fourth degree. Then:

109. These observations give, if I mistake not, the true principles of the solution of equations and the most suitable analysis to guide us to it: all is reduced, as we see, to a kind of calculus of combinations, by which one finds a priori the results that one might expect. It would be fitting to apply these principles to equations of the fifth and higher degrees, of which the solution is at present unknown; but this application requires too large a number of investigations and combinations. Moreover, success remains very doubtful, so that we shall not for the present pursue this work. However, we hope to return to it at some other time, and we shall here be satisfied in the exposition of the foundations of a theory which seems to us new and general . . .[6]

. . . .

115. We add an example, taken from geometry. We take the well-known problem in which one draws through the corner D of a square $ACDB$ [Fig. 1] a straight line MN in such a way that the part MN of it contained between the two adjacent sides AC, AB of the square, continued to M, N, has a given length.

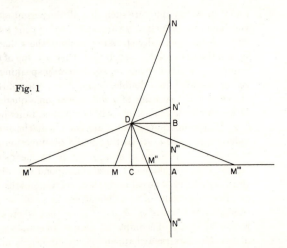

Fig. 1

Calling a the side of the square and b the given length of the line MN, we take, in order to find the position of that line, the unknown $CM = x$; then $MD = \sqrt{x^2 + a^2}$, and the two similar triangles MCD, MAN will give immediately

$$x : \sqrt{a^2 + x^2} = (a + x) : MN = (a + x) : b;$$

[6] For further understanding of Lagrange's ideas in the framework of modern algebra, consult A. Speiser, *Theorie der Gruppen endlicher Ordnung* (Springer, Berlin, 1927); J. Vuillemin, *La philosophie de l'algèbre de Lagrange*, Conférences du Palais de la Découverte D 71 (Paris, 1960); I. Bachmacova, "Le théorème fondamental de l'algèbre et la construction des corps algébriques," *Archives Internationales d'histoire des Sciences 13* (1960), 211–222.

from which we derive the equation

$$bx = (a + x)\sqrt{a^2 + x^2},$$

which, after removal of the radical and arrangement in order of x, becomes

$$x^4 + 2ax^3 + (2a^2 - b^2)x^2 + 2a^3x + a^4 = 0,$$

which is, as we see, of the fourth degree.

Let us now see if, according to the vary nature of the problem, we cannot find some relation between the roots of this equation that will make it possible to decompose it into equations of lower degree.

To achieve this I observe that one can actually draw through the point D four lines that fulfill the conditions of the problem; they are the lines MN, $M'N'$, $M''N''$, $M'''N'''$, such that the roots of the preceding equation will be the lines CM, CM', CM'', CM''', of which the last two, as we see, are negative.

Let us call the lines x', x'', x''', x^{iv}. Then because of the similarity of triangles MDC, DNB we have:

$$MC : CD = DB : BN;$$

but, since $M'N'$ must be equal to MN and $CD = DB$, it is easy to see that also $M'C = BN$. Thus we have the proportion

$$x' : a = a : x'',$$

or

$$x'x'' - a^2 = 0.$$

We could conclude, by the principle of sufficient reason, that such a relation must also exist between the two other roots x''', x^{iv}, but, if we wish to convince ourselves of this relation a posteriori, we have only to consider that because of $M''N'' = N''M'''$ we also must have $CM''' = BN''$, so that then, because of the similarity of the triangles DCM'', DBN'', we have $CM'' : CD = DB : BN'' = DB : CM'''$, that is,

$$x''' : a = a : x^{iv},$$

hence

$$x'''x^{iv} - a^2 = 0.$$

Since, therefore we have two similar equations, one between x''', x^{iv}, and since these equations remain the same on changing x' into x'' and x''' into x^{iv}, it follows from principles established before that the equation of the fourth degree, derived above, will necessarily be decomposable into two equations of the second degree, such as

$$x^2 - f'x + g' = 0,$$
$$x^2 - f''x + g'' = 0,$$

where f' and f'' will be roots of an equation of the second degree, just like g' and g'', but, since $g' = x'x''$ and $g'' = x'''x^{\mathrm{iv}}$, we will have $g' = g'' = a^2$. Hence the two factors of the equation in question will be

$$x^2 - f'x + a^2 = 0,$$
$$x^2 - f''x + a^2 = 0.$$

Multiplying the two we shall have

$$x^4 - (f' + f'')x^3 + (f'f'' + 2a^2)x^2 - a^2(f' + f'')x + a^4 = 0,$$

hence

$$f' + f'' = -2a, \quad f'f'' + 2a^2 = 2a^2 - b^2;$$

therefore

$$f'f'' = -b^2,$$

so that the equation which has f' and f'' as its roots will be

$$f^2 + 2af - b^2 = 0.$$

Finally, it is clear that if in the equation of the fourth degree in x we put $x = az$ we shall have an equation in x of the type called *reciprocal*, in which we can therefore eliminate all odd powers of the unknown by making $z = (1 - y)/(1 + y)$, so that the proper substitution for this purpose is $x = a(1 - y)/(1 + y)$.

If we take the value of y from this equation, we have

$$y = \frac{a - x}{a + x} = 1 - \frac{2x}{a + x},$$

but, since

$$\frac{x}{a + x} = \frac{\sqrt{a^2 + x^2}}{b} = \frac{MD}{MN},$$

we shall have

$$y = 1 - \frac{2MD}{MN} = \frac{MN - 2MD}{MN} = \frac{2DR}{MN} = \frac{2DR}{b},$$

where R is the midpoint of the line MN. From this we see that we would have arrived at a fourth-degree equation without odd powers of the unknown if we had taken the line DR as the unknown. This is what Newton did in the solution of the problem in the *Universal arithmetic*,[7] but we must confess, it seems to me, that such a choice of the unknown is not very natural, and that it can only be

---

[7] On this book see Selection II.9. It contains a number of geometrical problems; Lagrange seems to refer to Problem 29.

done, so to say, *post festum*. At any rate it seems to me that the principle on which Newton lets it depend has not all the evidence that we may rightly demand in a matter of this kind.

---

12 LAGRANGE. CONTINUED FRACTIONS

We have seen in Selection I.1 that Leonardo of Pisa introduced a type of continued fraction. We meet an algorithm equivalent to our modern continued fractions in Raffaele Bombelli's *L'Algebra* (Bologna, 1572), a book best known for its treatment of imaginaries, and in the *Trattato del modo brevissimo di trovare la radice quadra delle numeri* (Bologna, 1613) by Pietro Antonio Cataldi (1552–1626). Both men were professors at the University of Bologna. Their algorithm is equivalent to what we now write

$$\sqrt{M} = \sqrt{a^2 + r} = a + \frac{r}{2a} + \frac{r}{2a} + \cdots,$$

where M is a positive integer and a^2 the square closest to M and less than M. Bombelli computed several partial fractions of $\sqrt{13} = \sqrt{9 + 4}$, Cataldi of $\sqrt{18} = \sqrt{16 + 2}$. Translations of part of their computations can be found in Smith, *Source book*, 80–84; the texts, in the original Italian, can be found in A. Favaro, "Notizie storiche sulle funzioni continue dal secolo decimoterzo al decimosettimo," *Bullettino di Bibliografia e di Storia delle Scienze Matematiche e Fisiche* (Boncompagni) 7 (1874), 451–589. Here we also find the text in which John Wallis, in his *Arithmetica infinitorum* (Oxford, 1656), after having obtained his infinite product for $\square = 4/\pi$ (Selection IV.13), announces that Lord William Brouncker (1620?–1684), "that Most Noble Man, after having considered this matter, saw fit to bring this quantity by a method of infinitesimals peculiar to him, to a form which can thus be conveniently written

$$\square = 1\tfrac{1}{2} \frac{9}{2} \frac{25}{2} \frac{49}{2} \frac{81}{2} \text{ etc.,"}$$

and pointed out how the partial fractions are successively larger and smaller than $\square$, the process converging to it [*ad numerum justum acceditur*].

The theory was further developed by Euler in several papers; see O. Perron, *Die Lehre von den Kettenbrüchen* (3rd ed.; Teubner, Stuttgart, 1954), I, 190. In "De fractionis continuis dissertatio," *Commentarii Academiae scientiarum Petropolitanae 9, 1737* (1744), 98–137, *Opera omnia*, ser. I, vol. 14, 187–215, Euler proved that a periodic continued fraction is the root of a quadratic equation. The converse theorem was proved by Lagrange in his "Addition au Mémoire sur la résolution des équations numériques," *Histoire de l'Académie Royale, Berlin, 24, 1768* (1770), 111–180, *Oeuvres*, II (1868), 581–652. The "Mémoire" to which this was an addition appeared in the same *Histoire 23, 1767* (1769), 311–352, *Oeuvres*, II, 539–578. In this paper Lagrange had shown how, given a numerical equation of any degree with a positive root, this root can be expressed in a continued fraction. We present here, from the "Addition," Lagrange's proof of the converse theorem.

In order to give this proof, Lagrange writes the quadratic equation

$$E_1 x^2 - 2\epsilon x - E = 0,$$

where E, E_1 are integers such that $\epsilon^2 + EE_1 > 0$. Hence the roots are real. Let

$$x = \frac{\epsilon + \sqrt{B}}{E_1}, \quad B = \epsilon^2 + EE_1,$$

and let λ be the integer closest to and less than x; then

$$x = \lambda + \frac{1}{x_1},$$

and x satisfies an equation of the form

$$E_2 x_1^2 - 2\epsilon_1 x_1 - E_1 = 0.$$

Lagrange proves that

$$\epsilon^2 + E_1 E_2 = \epsilon^2 + EE_1 = B.$$

Writing

$$x_1 = \lambda_1 + \frac{1}{x_2},$$

where λ_1 is the integer closest to and less than x_1, and carrying on in the same way, he obtains a series of "transformed equations":

(A)

$$E_{\gamma+1} x_\gamma^2 - 2\epsilon_\gamma x_\gamma - E_\gamma = 0, \quad \gamma = 1, 2, 3, \ldots,$$

where

$$x_\gamma = \frac{\epsilon_\gamma + \sqrt{B}}{E_{\gamma+1}} = \lambda_{\gamma+1} + \frac{1}{x_{\gamma+1}},$$

and for x the continued fraction

$$x = \lambda_1 + \frac{1}{\lambda_2} + \frac{1}{\lambda_3} + \cdots.$$

---

34. Now I claim that the continued fraction which expresses the value of x will always be necessarily periodic.

To prove this theorem we shall begin by proving generally that, whatever be the proposed equation, we shall necessarily always arrive at transformed equations in which the first and last terms have different signs.

We omit this proof, which is based on a theorem in Lagrange's *Mémoire* of 1769 that in constructing the series of transformed equations we must arrive at one of which only one root is > 1. Then in sec. 35 Lagrange takes (A) to be this equation. Then in (A) and in the following transformed equations

---

the first and the last terms will be of different signs, so that the numbers

$$E_\gamma, E_{\gamma+1}, E_{\gamma+2}, \ldots$$

all have the same sign. Now we have

$$B = \epsilon_\gamma^2 + E_\gamma E_{\gamma+1} = \epsilon_{\gamma+1}^2 + E_{\gamma+1}E_{\gamma+2},$$

hence, since $E_\gamma, E_{\gamma+1}, E_{\gamma+2}, \ldots$ have the same sign, the products $E_\gamma E_{\gamma+1}$, $E_{\gamma+1}E_{\gamma+2}, \ldots$ will necessarily be positive, from which it follows:

1°) that we shall have

$$\epsilon_\gamma^2 < B, \quad \epsilon_{\gamma+1}^2 < B,$$

that is (if we disregard the sign),

$$\epsilon_\gamma < B, \quad \epsilon_{\gamma+1} < B,$$

and so on to infinity;

2°) that, since the numbers $E, E_1, E_2, \ldots$ are all integers,

$$E_\gamma < B, \quad E_{\gamma+1} < B, \quad E_{\gamma+2} < B.$$

Hence, since B is given, it is clear that there are only a certain number of integers that are $< B$ or $< \sqrt{B}$. The numbers

$$E_\gamma, E_{\gamma+1}, E_{\gamma+2}, \ldots, \quad \epsilon_\gamma, \epsilon_{\gamma+1}, \epsilon_{\gamma+2}, \ldots,$$

can therefore have only a certain number of different values. Thus, in the one as well as in the other of these sequences it is necessary that, when we go to infinity, the same terms return an infinite number of times. By the same token, it is also necessary that the same combination of corresponding terms in the two sequences return an infinite number of times, from which it follows that we shall necessarily have, for example,

$$E_{\gamma+\delta+\nu} = E_{\gamma+\delta}, \quad \epsilon_{\gamma+\delta+\nu} = \epsilon_{\gamma+\delta},$$

or, taking $\gamma + \delta = \mu$,

$$E_{\mu+\nu} = E_\mu, \quad \epsilon_{\mu+\nu} = \epsilon_\mu.$$

Hence, since

$$B = \epsilon_\mu^2 + E_\mu E_{\mu+1} = \epsilon_{\mu+\nu}^2 + E_{\mu+\nu} E_{\mu+\nu+1},$$

we have

$$E_{\mu+\nu+1} = E_{\mu+1}.$$

But, since

$$x_\mu = \frac{\epsilon_\mu + \sqrt{B}}{E_{\mu+1}} \quad \text{and} \quad x_{\mu+\nu} = \frac{\epsilon_{\mu+\nu} + \sqrt{B}}{E_{\mu+\nu+1}},$$

$\chi_{\mu+\nu} = \chi_\mu$, so that the continued fraction will necessarily be periodic.

36. Indeed, we see, by means of the previous formulas, that, if

$$E_{\mu+\nu} = E_\mu \quad \text{and} \quad \epsilon_{\mu+\nu} = \epsilon_\mu,$$

we shall have

$$E_{\mu+\nu+1} = E_{\mu+1}, \quad \lambda_{\mu+\nu+1} = \lambda_{\mu+1}, \quad \epsilon_{\mu+\nu+1} = \epsilon_{\mu+1},$$

and so on, so that in general the terms of the three sequences

$$E, E_1, E_2, \ldots, \quad \epsilon, \epsilon_1, \epsilon_2, \ldots, \quad \lambda_1, \lambda_2, \ldots,$$

which will have for index [*exponant*] $\mu + n\nu + \pi$, are the same as the preceding terms whose indices are $\mu + \pi$, if we take for n some positive integer.

Every one of these three sequences thus will be periodic, beginning with the terms E_μ, ϵ_μ, $\lambda_{\mu+1}$, and their periods will have ν terms, after which the same terms will recur in the same order, to infinity.

Then Lagrange goes on to prove that, as soon as in the sequence $E, E_1, E_2, \ldots$ we arrive at two consecutive terms, such as E_γ, $E_{\gamma+1}$, that have the same sign, we are certain that one of these two terms will already be one of the periodic terms. Also that if two corresponding terms, say $E_{\gamma+3}$, $\epsilon_{\gamma+3}$, of the sequences $E_\gamma, E_{\gamma+1}, \ldots, \epsilon_\gamma, \epsilon_{\gamma+1}, \ldots$ are given, all the preceding terms in these sequences are given. And finally, if we have arrived at these terms E_γ, $E_{\gamma+1}$ of the same sign, then the one $< \sqrt{B}$ will be the periodic one.

---

42. If we have $\epsilon = 0$, so that $x = \sqrt{E/E_1}$,[1] then $B = EE_1$, from which we see that of the two numbers E, E_1 the smallest will be $< \sqrt{B}$, and the largest will be $> \sqrt{B}$, hence in this case, if the number E/E_1 of which the square root has to be taken, is < 1, then the sequence will be periodic from the first term E

[1] The use of the solidus / for the fraction bar is the only change in notation we have made. Lagrange's notation, like Euler's, is not, or hardly, different from ours.

on, and if this number is >1, then the period cannot start lower than the second term.

Lagrange continues by giving a method to simplify the procedure, and then studies the case in which both $+$ and $-$ appear in the continued fraction.

13 GAUSS. THE FUNDAMENTAL THEOREM OF ALGEBRA

The first satisfactory proof of this theorem was presented by Carl Friedrich Gauss (1777–1855) in his Helmstädt doctoral dissertation of 1799 under the title *Demonstratio nova theorematis omnem functionem algebraicam rationalem integram unius variabilis in factores reales primi vel secundi gradus resolvi posse* (New proof of the theorem that every integral rational algebraic function of one variable can be decomposed into real factors of the first or second degree), *Werke*, III, 3–56. After a criticism in Secs. 1–12 of the previous demonstrations by D'Alembert (1746), Euler (1749), De Foncenet (1759), and Lagrange (1772), Gauss proceeds as follows. It will be seen that in the geometrical language he uses he transfers geometrical continuity to arithmetical quantities without proof, but at least at one place expresses his conviction that he can make this aspect of his proof also strictly rigorous. This can indeed be accomplished by the methods of Bolzano and Weierstrass.

13. *Lemma.* If m is an arbitrary positive integer, then the function $\sin x^m - \sin m\varphi \cdot r^{m-1}x + \sin (m-1)\varphi \cdot r^m$ is divisible by $x^2 - 2 \cos \varphi \cdot rx + r^2$.

The proof is given by direct division.

14. *Lemma.* If the quantity r and the angle φ are so determined that the equations

$$r^m \cos m\varphi + Ar^{m-1} \cos (m-1)\varphi + Br^{m-2} \cos (m-2)\varphi + \text{etc.}$$
$$+ Krr \cos 2\varphi + Lr \cos \varphi + M = 0, \quad (1)$$

$$r^m \sin m\varphi + Ar^{m-1} \sin (m-1)\varphi + Br^{m-2} \sin (m-2)\varphi + \text{etc.}$$
$$+ Krr \sin 2\varphi + Lr \sin \varphi = 0 \quad (2)$$

exist, then the function

$$x^m + Ax^{m-1} + Bx^{m-2} + \text{etc.} + Kx^2 + Lx + M = X$$

will be divisible by the quadratic factor $x^2 - 2 \cos rx + r^2$, unless $r \sin \varphi = 0$. If $r \sin \varphi = 0$, then the same function is divisible by $x - r \cos \varphi$.

The proof is given by taking the functions

$$\sin \varphi \cdot rx^m \quad - \quad \sin m\varphi \cdot r^m x \quad + \quad \sin (m-1)\varphi \cdot r^{m+1},$$
$$A \sin \varphi \cdot rx^{m-1} - A \sin (m-1)\varphi \cdot r^{m-1}x + A \sin (m-2)\varphi \cdot r^m,$$
$$B \sin \varphi \cdot rx^{m-3} - B \sin (m-2)\varphi \cdot r^{m-2}x + B \sin (m-3)\varphi \cdot r^{m-1},$$
$$\dots\text{etc}\dots\dots\dots\dots\dots\dots\dots\dots\dots\dots\dots\dots\dots\dots\dots\text{etc}\dots\dots\dots$$
$$K \sin \varphi \cdot rx^2 \quad - \quad K \sin 2\varphi \cdot r^2 x \quad + \quad K \sin \varphi \cdot r^3,$$
$$L \sin \varphi \cdot rx \quad - \quad L \sin \varphi \cdot rx \quad \quad * \quad \quad ,$$
$$M \sin \varphi \cdot r \quad \quad * \quad \quad + \quad M \sin (-\varphi) \cdot r,$$

which are each divisible by $x^2 - 2 \cos \varphi \cdot xr + r^2$ (according to the first lemma) and which, added up, give $\sin \varphi \cdot rX + 0 + 0$. When $r = 0$, X is divisible by $x - r \cos \varphi$; when $\sin \varphi = 0$, then $\cos \varphi = \pm 1$, $\cos 2\varphi = \pm 1$, $\cos 3\varphi = \pm 1$, etc. and X becomes zero for $x = r \cos \varphi$.

15. The previous theorem is usually given with the aid of imaginaries, cf. Euler, *Introductio in analysin infinitorum*, I, p. 110;[1] I found it worth while to show that it can be demonstrated in the same easy way without their aid. Hence it is clear that, in order to prove our theorem, we only have to show: *If some function X of the form $x^m + Ax^{m-1} + Bx^{m-2} + $ etc. $+ Lx + M$ is given, then r and φ can be determined in such a way that the equations (1) and (2) are valid.* Indeed, from this it follows that X possesses a real factor of the first or second degree; division by it necessarily gives a real quotient of lower degree... We shall now prove this theorem.

16. We consider a fixed infinite plane (the plane of our Fig. 1) and in it a fixed infinite straight line GG passing through the fixed point C. In order to

Fig. 1

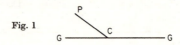

express all line segments by numbers we take an arbitrary segment as unit, and erect at an arbitrary point P of the plane, with distance r from center C and with angle $GCP = \mu$, a perpendicular equal to the value of the expression

$$r^m \sin m\varphi + Ar^{m-1} \sin (m-1)\varphi + \text{etc.} + Lr \sin \varphi.$$

I shall denote this expression by T. I consider the distance r always as positive, and for points on the other side of the axis the angle φ must either be taken as larger than two right angles, or (what amounts to the same thing) as negative. The end points of this perpendicular (which have to be taken as above the plane for positive T, as below the plane for negative T, and for vanishing T as in the plane) form a continuous, curved surface, infinite in all directions, which I shall call for the sake of brevity the *first surface*. In exactly the same way we can refer to the same plane, the same center, and the same axis another surface, with altitude above every point of the plane equal to

$$r^m \cos m\varphi + Ar^{m-1} \cos (m-1)\varphi + \cdots + Lr \cos \varphi + M;$$

[1] This is the book from which Selection V.15 is taken. See also note 8 below.

this expression I shall always denote by U. This surface, also continuous and infinite in all directions, will be distinguished from the other one by the name of *second surface*. From this it is clear that our entire task is then to prove that there exists at least one point that lies at the same time in the plane, in the first surface, and in the second surface.

17. It can easily be understood that the first surface lies partly above, and partly below the plane, since we can take the distance r from the center so large that the first term $r^m \sin m\varphi$ in T surpasses all following terms; if then the angle φ is conveniently chosen, this term can become positive as well as negative. The fixed plane must therefore be intersected by the first surface. I shall call this intersection the *first curve*, and it will be determined by $T = 0$. The same reasoning shows that the plane is intersected by the second surface; this intersection will be called the *second curve*, and its equation will be $U = 0$. Both curves will, properly speaking, consist of several branches, which may be entirely separated from each other, but each by itself forms a continuous curve. Indeed, the first curve will always be a so-called reducible curve, since the axis GC must be considered a part of this curve, because $T = 0$ for $\varphi = 0$ or $\varphi = 180°$ for any value of r. We prefer, however, to consider the totality of all branches, which pass through all points for which $T = 0$, as one single curve (as is customary in higher geometry). The same happens for all branches passing through the points for which $U = 0$. Now our problem has been reduced to the task of proving that there exists in the plane at least one point at which one of the branches of the first curve is intersected by one of the branches of the second curve. This makes it necessary to study more closely the behavior of these curves.

18. First I observe that each curve is algebraic, and, referred to orthogonal coordinates, of order m. Indeed, if the origin of the abscissae is taken at C and the direction of the abscissa x is measured toward G and that of the ordinate y toward P, then $x = r \cos \varphi$, $y = r \sin \varphi$, and generally, for arbitrary n:

$$r^n \sin n\varphi = nx^{n-1}y - \frac{n(n-1)(n-2)}{1 \cdot 2 \cdot 3} x^{n-3}y^3 + \frac{n \cdots (n-4)}{1 \cdots 5} x^{n-5}y^5 - \text{etc,}$$

$$r^n \cos n\varphi = x^n - \frac{n(n-1)}{1 \cdot 2} x^{n-2}y^2 + \frac{n(n-1)(n-2)(n-3)}{1 \cdot 2 \cdot 3 \cdot 4} x^{n-4}y^4 - \text{etc.}$$

T and U consist therefore of several terms of the form $ax^\alpha y^\beta$, where α, β are positive integers, whose sum has m as its maximum value. Moreover, it is easy to see that all terms of T contain the factor y, so that the first curve, to express it exactly, consists of the line with equation $y = 0$ and a curve of order $m - 1$. However, we do not need to take this difference into consideration.

It is of more importance to investigate whether the first and second curves have infinite branches, and what their number and character will be. At an infinite distance from the point C the first curve, with equation

$$\sin m\varphi + \frac{A}{r} \sin (m-1)\varphi + \frac{B}{rr} \sin (m-2)\varphi \text{ etc } = 0$$

coincides with that curve whose equation is $\sin m\varphi = 0$. This consists only of m straight lines intersecting at C; the first of these is the axis GCG', the other ones make with this axis the angles $(1/m)180°, (2/m)180°, (3/m)180°, \ldots$. The first curve therefore has $2m$ infinite branches, which divide the circumference of a circle described with infinite radius into $2m$ equal parts, such that its circumference is intersected by the first branch in the intersection of the circle with the axis, by the second branch at distance $(2/m)180°$, by the third one at distance $(3/m)180°$, etc. It follows similarly that the second curve at infinite distance from the center has the curve represented by the equation $\cos m\varphi = 0$ as its asymptote. This curve consists of the totality of m straight lines which also intersect in C at equal angles, but in such a way that the first one forms with the axis CG the angle $(1/m)90°$, the second one the angle $(3/m)90°$, the third one the angle $(5/m)90°$, etc. The second curve therefore also has $2m$ infinite branches, which each form the middle between two neighboring branches of the first curve, so that they intersect the circumference of the circle of infinite radius in points which are $(1/m)90°, (3/m)90°, (5/m)90°, \ldots$ away from the axis. It is also clear that the axis itself always forms two infinite branches of the first curve, namely, the first and the $(m + 1)$th. This situation of the branches is well illustrated by Fig. 2, constructed for the case $m = 4$; the branches of the second curve are here dotted to distinguish them from those of the first curve. This also occurs in Fig. 4.[2] Since these results are of the utmost importance, and some readers might be offended by infinitely large quantities, I shall show in the next section how these results can also be obtained without the help of infinite quantities.

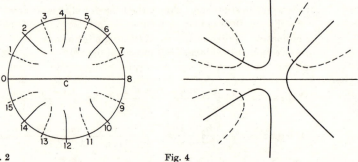

Fig. 2   Fig. 4

19. Theorem. *Under the conditions mentioned before we can construct a circle with center C, on the circumference of which there exist $2m$ points at which $T = 0$, and as many points at which $U = 0$; they are situated in such a way that each point of the second kind lies between two of the first kind.*

Let the sum of all coefficients (taken positive) $A, B, \ldots, K, L, M$ be $= S$; let furthermore R be at the same time $> S\sqrt{2}$ and > 1.[3] I then say that on the

---

[2] [Footnote by Gauss] Fig. 4 is constructed assuming $X = x^4 - 2xx + 3x + 10$; so that readers less familiar with general and abstract investigations can study in a concrete example how both curves are situated. The length of line $CG = 10$ $(CN = 1.26255)$.

[3] [Footnote by Gauss] For $S > \sqrt{\frac{1}{2}}$ the second condition is contained in the first one, for $S < \sqrt{\frac{1}{2}}$ the first condition in the second one.

circle described with radius R the conditions exist indicated in the theorem. I denote for short by 1 that point of its circumference which is distant $(1/m)45°$ from the point of intersection of the circle with the left-hand side of the axis. Hence for (1) $\varphi = (1/m)45°$. Similarly I denote by (3) that point which is distant $(3/m)45°$ from the point of intersection and for which therefore $\varphi = (3/m)45°, \ldots$ up to the point $(8m - 1)$ which is distant $[(8m - 1)/m]45°$ from that point of intersection if we always proceed in the same direction, or $(1/m)45°$ if we move in the opposite way. Thus there are in total $4m$ points in the circumference at equal distances from each other. Then there exists between $(8m - 1)$ and (1) a point for which $T = 0$, a similar point lies between (3) and (5), between (7) and (9), $\ldots$; their number is $2m$. In the same way we see that the single points for which $U = 0$ lie between (1) and (3), between (5) and (7), $\ldots$; their number is therefore also $2m$. Apart from these $4m$ points there are no other points on the circumference for which T or $U = 0$.

Proof. I. At point (1) $m\varphi = 45°$, and therefore

$$T = R^{m-1}\left(R\sqrt{\tfrac{1}{2}} + A \sin (m - 1)\varphi + \frac{B}{R} \sin (m - 2)\varphi + \text{etc.} + \frac{L}{R^{m-2}} \sin \varphi\right).$$

The sum $A \sin (m - 1)\varphi + (B/R) \sin (m - 2)\varphi$ etc certainly cannot be larger than S, and hence must be smaller than $R\sqrt{\tfrac{1}{2}}$; the value of T at this point is therefore certainly positive. Hence, a fortiori, T is positive when m lies between $45°$ and $135°$, that is, T has always a positive value from point (1) to point (3). The same reasoning shows that T is everywhere positive from point (9) to point (11), and, generally speaking, from some point $(8k + 1)$ to point $(8k + 3)$, where k means any integer. In a similar way we see that T is negative everywhere between (5) and (7), between (13) and (15), etc., and, generally speaking, between $(8k + 5)$ and $(8k + 7)$, so that in all these intervals it can nowhere be $= 0$. But since at (3) the value is positive and at (5) negative, it must be $= 0$ somewhere between (3) and (5)[4] and in the same way between (11) and (13), etc. up to the interval between $(8m - 1)$ and (1) inclusive, so that together $T = 0$ at $2m$ points.

II. That there are, apart from these $2m$ points, no others of the same property can be seen in the following way. Since there are none between (1) and (3), between (5) and (7), etc., other such points would exist only if in one of the intervals from (3) to (5) or from (7) to (9), etc. there were at least two of them. In that case, however, T would have to be in the same interval at some point either a *maximum* or a *minimum*,[5] hence $dT/d\varphi = 0$. But

$$\frac{dT}{d\varphi} = mR^{m-2}\left(R \cos m\varphi + \frac{m - 1}{m} A \cos (m - 1)\varphi + \text{etc.}\right)$$

---

[4] This is one of the places where Gauss accepts on visual evidence a theorem that now requires proof.

[5] This theorem is named after Michel Rolle (1652–1719), in whose *Méthode pour résoudre les égalitez* (Paris, 1691) it can be found without proof and without special emphasis. It appeared in other eighteenth-century works, as in Euler's *Institutiones calculi differentialis* (Saint Petersburg, 1755), sec. 298 (*Opera omnia*, ser. I, vol. 18, 503). See F. Cajori, *Bibliotheca Mathematica* (3d ser.) *11* (1910–11), 300–313.

and $\cos m\varphi$ is always negative between (3) and (5) and [in value] $> \sqrt{\frac{1}{2}}$. From this we can easily see that $dT/d\varphi$ is a negative quantity in this whole interval; and in the same way we see that it is everywhere positive between (7) and (9), everywhere negative between (11) and (13), etc. In none of these intervals therefore can it be $= 0$, so that our assumption was wrong. Hence, etc.

III. Here Gauss shows in the same way that $dU/d\varphi$ cannot be 0 in the intervals (1) and (3), (5) and (7), etc., so that there are on the circumference of the circle no more than $2m$ points where $U = 0$.

That part of the theorem which teaches that there are no more than $2m$ points at which $T = 0$, and no more than $2m$ points at which $U = 0$, can also be demonstrated by representing $T = 0$, $U = 0$ as curves of order n, which are intersected by a circle, being a curve of the second order, in no more than $2m$ points, as is stated in higher geometry.[6]

20. If another circle with radius larger than R is described around the same center and is divided in the same way, then here also there exists between the points (3) and (5) a single point at which $T = 0$, and similarly between (7) and (9), etc., and it can easily be seen that such points between (3) and (5) on both circumferences are the closer the less the radius of the larger circle differs from the radius R. The same also happens if the circle is described with a radius somewhat smaller than R, but still larger than $S\sqrt{2}$ and 1. From this we see that the circumference of the circle described with radius R is actually *intersected* by a branch of the first curve at that point between (3) and (5) where $T = 0$; the same holds for the other points where $T = 0$. It is also clear that the circumference of this circle is intersected by a branch of the second curve at all $2m$ points for which $U = 0$. These conclusions can also be expressed in the following way: If a circle of sufficient size is described around the center C, then $2m$ branches of the first curve and as many branches of the second curve enter into it, and in such a way that every two neighboring branches of the first curve are separated from each other by a branch of the second curve. See Fig. 2, where the circle is now no longer of infinite, but of finite magnitude; the numbers added to the separate branches should not be confused with the numbers by which I have denoted for short, in the previous and in this paragraph, certain limiting points on the circumference.

21. It is now possible to deduce from the relative position of the branches which enter into the circle that inside the circle there must be an intersection of

---

[6] Gauss refers to the theorem named after Etienne Bézout (1730–1783), but also announced by other authors of his time. It was only insufficiently proved in Gauss's day.

a branch of the first curve with a branch of the second curve, and this can be done in so many ways that I hardly know which method is to be preferred to another. The following method seems to be the clearest: We indicate by O (Fig. 2) that point of the circumference of the circle in which it is intersected by the left-hand side of the axis (which itself is one of the $2m$ branches of the first curve); the next point, at which a branch of the second curve enters, by 1; the point next to this, at which a branch of the first curve enters, by 2, etc., up to $4m - 1$. At every point indicated by an even number, therefore, a branch of the second curve enters into the circle, but a branch of the first curve at every point indicated by an odd number. Now it is known from higher geometry that every algebraic curve (or the single parts of an algebraic curve when it happens to consist of several parts) either runs into itself or runs out to infinity in both directions and that therefore, if a branch of an algebraic curve enters into a limited space, it necessarily has to leave it again.[7] From this we can easily conclude that every point indicated by an even number (or, for short, *every even point*) must be connected with another even point by a branch of the first curve inside the circle, and that in a similar way every point indicated by an odd number is connected with another similar point by a branch of the second curve. Although this connection of two points may be quite different because of the nature of the function X, so that it cannot in general be determined, yet it

[7] [Footnote by Gauss] It seems to be sufficiently well demonstrated that an algebraic curve can neither be suddenly interrupted (as e.g., occurs with the transcendental curve with equation $y = 1/\log x$), nor lose itself after an infinite number of terms (like the logarithmic spiral), and nobody, to my knowledge, has ever doubted it. But if anybody desires it, then on another occasion I intend to give a demonstration which will leave no doubt. Moreover, it is clear in the present case that if a branch, for instance 2, were nowhere to leave the circle (Fig. 3), one could enter the circle between O and 2, then go around this

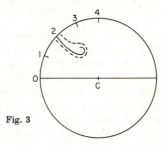

Fig. 3

whole branch (which has to lose itself in the space of the circle), and at last leave the circle again between 2 and 4, without meeting the first curve anywhere on the way. But this is patently absurd, since at the point at which you enter the circle you have the first surface above you, but where you leave the circle it is below you. Hence you would necessarily meet the first surface somewhere, and this at a point of the first curve.—From this reasoning, based on the principles of the geometry of position [*geometria situs*], which are no less valid than the principles of the geometry of magnitudes, it also follows that, if you enter the circle on a branch of the first curve, you can leave it at another point by always staying on the first curve, but it does not follow that the path is continuous in the sense accepted in higher geometry [see note 8]. It is here sufficient that the path be a continuous line in a general sense, that is, nowhere interrupted, but everywhere coherent.

can easily be shown that, *whatever this connection may be, there will always be an intersection of the first with the second curve.*

22. The proof of this necessity can best be given in an indirect way [*apagogice*].[8] We shall assume that the connection of pairs of all even points and of pairs of all odd points can be arranged in such a way that no intersection results of a branch of the first curve with a branch of the second curve. Since the axis is a part of the first curve, point O will clearly be connected with point $2m$. The point 1 therefore cannot be connected with a point situated outside of the axis, that is, with no point indicated by a number larger than $2m$, since otherwise the connecting curve would necessarily intersect the axis. If therefore we suppose that 1 is connected with the point n, then n will be $< 2m$. By a similar reasoning we find that when 2 is connected with n', $n' < n$, since otherwise the branch $2 \cdots n'$ must necessarily intersect the branch $1 \cdots n$. Point 3, for the same reason, must be connected with a point situated between 4 and n', and it is clear that, if we suppose 3, 4, 5, ... to be connected with n'', n''', n'''', ..., n''' is situated between 5 and n'', n'''' between 6 and n''', etc. From this it follows that at last we come to a point h which is connected with the point $h + 2$. The branch which at point $h + 1$ enters into the circle must in this case intersect the branch connecting the points h and $h + 2$. But since the one of these two branches belongs to the first, the other to the second curve, it is clear that our assumption is contradictory, and that therefore there exists necessarily somewhere an intersection of the first with the second curve.

If we combine this result with the previous one, then we arrive from all the investigations explained above at the rigorous proof of the theorem that *every integral rational algebraic function of one variable can be decomposed into real factors of the first and second degree.*

In the last two sections Gauss (1) observes that the same reasoning could have led to the conclusion that there exist at least m intersections of the first and second curve, (2) notes that the proof, here based on geometrical principles, could also have been presented in a purely analytical form, and (3) gives a short sketch of a different proof.

Gauss, during his lifetime, returned to the theorem more than once, and gave three more proofs. The last one, of 1849, took up again the ideas of the first one, but now using imaginaries. Gauss added that he avoided using them in 1799, but in 1849 it seemed to him no longer necessary. All the proofs can be found in a German translation by E. Netto in Ostwald's *Klassiker*, No. 14 (Engelmann, Leipzig, 1890).

[8] Gauss may refer here to Euler's definition of a continuous curve in his *Introductio in analysin infinitorum* (Lausanne, 1748), vol. II, cap. 19; *Opera omnia*, ser. I, vol. 9, p. 11: A line is called a continuous curve if its nature is expressed by one definite function, where (vol. I, cap. 14; *Opera omnia*, ser. I, vol. 8, p. 18) a function is a variable quantity z, defined as an analytical expression composed of this variable quantity and constant numbers or quantities, such as $a + 3z$, $az + b\sqrt{(aa - zz)}$, c^z, etc.

On Euler's proof and Gauss's critique, see also A. Speiser's note in Euler, *Opera omnia*, ser. I, vol. 29 (1956), Einleitung, pp. VIII–X.

14 LEIBNIZ. MATHEMATICAL LOGIC

Gottfried Wilhelm von Leibniz (1646–1716), German philosopher, mathematician, historian, and linguist, must be credited with one of the attempts to create an algebra of logic. Many scholars see in the work of the Spanish scholastic philosopher Raimundo Lullio (1235–1315) an earlier attempt, and indeed, in Lullio's *Ars magna* we find a description of a kind of logical machine for combining certain classes of concepts occurring in theology, which leads, among many other consequences, to tables of permutations and combinations. This early attempt at a *scientia universalis* had some influence on Descartes (who did not care for it: "the art of Lullio serves rather to speak without judgment on things that we do not know than to understand them," in the second section of the *Discours de la méthode*) and other mathematicians, including Leibniz. We have seen that Cardan, who called his mathematics *Ars magna*, and Viète, who thought that with his *speciosa* he could solve all mathematical questions, played with similar ideas of a science beyond the existing science. Leibniz, from his early days when he studied permutations and combinations until the end of his life, attempted such generalizations. In a letter to Christiaan Huygens, dated September 8, 1679, he wrote (in French):

"But after all the progress that I have made in these matters [he speaks of his calculus, and of systems of transcendental equations such as $x^z + z^x = b$, $x^x + z^z = c$, b, c constants], I am still not satisfied with Algebra in that it gives neither the shortest ways nor the most beautiful constructions of Geometry. This is the reason that with respect to this I believe that we need still another Analysis which is purely geometric or linear, which expresses for us directly position [*situs*] as Algebra expresses magnitude. And I believe that I see the means thereto, and that one could represent figures and even machines and motions in characters, as Algebra represents numbers or quantities, and I send you an essay that seems to me of a certain importance"; Huygens, *Oeuvres complètes*, VIII (1899), 216.

This essay is also printed in the same *Oeuvres*, 219–224, and reminds one a little of Grassmann's point calculus. With it Leibniz claims to prove, for instance, that a plane and a sphere intersect in a circle; but his technical apparatus is quite different. A related manuscript, dated 1676 and entitled *Characteristica geometrica*, is printed in Leibniz, *Mathematische Schriften*, part 2, vol. 1 (1858), 141–178. The next paper, also related, is called *De analysi situs* (178–211).

See on this calculus H. Freudenthal, "Leibniz und die Analysis situs," *Consejo superior de investigaciones científicas. Homenaje a Millás-Vallicrosa (Barcelona)* *1* (1954), 612–621.

Leibniz's search for a general science of all sciences, to be carried on by a method that he called *characteristica generalis* or *lingua generalis*, also appears in a letter, written in French toward the end of his life, to Pierre Rémond de Montmort (1678–1719) in Paris, known in the theory of probability. Written March 14, 1714, it mentions Leibniz's idea of a *speciosa* more general than that which Viète had tried to embody in his algebra:

"I would like to give a method of Speciosa Generalis [*Spécieuse Générale*], in which all truths of reason would be reduced to a kind of calculus. This could at the same time be a kind of language or universal script, but very different from all that have been projected hitherto, because the characters and even the words would guide reason [*y dirigeroient la raison*], and the errors (except those of fact) would only be errors of computation. It would be very difficult to form or to invent this Language or Characteristic, but very easy to learn it without any Dictionaries. It would also serve to estimate the degree of likelihood (because we have not sufficient data to arrive at truths that are certain), and to see what is necessary to obtain them. And this estimate would be very important for its application to life and for the deliberations of practice, where by estimating the probabilities one miscalculates most often in more than half the cases [*on se mécompte le plus souvent de plus de la moitié*]." This can be found in C. I. Gerhardt, *Die philosophischen Schriften von G. W. Leibniz* (7 vols.; Weidman, Berlin, 1875–1890), III, 611–615. In vol. VII, 3–247, twenty pages of Leibniz on this subject are collected, with an introduction. See also L. Couturat, *La logique de Leibniz* (Alcan, Paris, 1901), and the section on Leibniz in C. I. Lewis, *A survey of symbolic logic* (University of California Press, Berkeley, 1918). Couturat has added other Leibniz notes on the same subject in *Opuscules et fragments inédits de Leibniz* (Alcan, Paris, 1903). English translations of some of Leibniz's essays in this field can be found in L. E. Loemker, *G. W. Leibniz, Philosophical papers and letters* (University of Chicago Press, Chicago, 1956), Vol. 1.

The following selection is a translation of paper XX of the Gerhardt collection (VII, 236–247), which, together with a translation of paper XIX, is based on an appendix in the book by C. I. Lewis. Both essays represent one of the later forms in which Leibniz expressed his *calculus universalis*, the calculus for displaying the most universal relations of scientific concepts, in which we recognize a first attempt at mathematical logic. The essay is of uncertain date, but was written after 1685, and was never published until Gerhardt brought it to light. Added to it is a segment of Gerhardt's paper XIX (vol. VII, 228–235), which is similar to XX, but also contains the concept of "subtraction." Here Leibniz writes $A + B$, where in XX he writes $A \oplus B$, for "both A and B," the class made up of the two classes A and B in extension.

A paraphrase of the Lewis translation of paper XX with comment can be found in W. and M. Kneale, *The development of logic* (Clarendon Press, Oxford, 1962), 340–343.

For further information on the history of formal logic see H. Scholtz, *Concise history of logic* (Philosophical Library, New York, 1961). On the development of mathematical logic see also Jean van Heijenoort, ed., *From Frege to Gödel: A source book in mathematical logic, 1879–1931* (Harvard University Press, Cambridge, Massachusetts, 1967).

Definition 1. Terms which can be substituted for one another wherever we please without altering the truth of any statement [*salva veritate*], are the *same* [*eadem*] or *coincident* [*coincidentia*]. For example, "triangle" and "trilateral," for in every proposition demonstrated by Euclid concerning "triangle" "trilateral" can be substituted without loss of truth.

$A = B$[1] *signifies* that A and B are the same, or as we say of the straight line XY and the straight line YX [Fig. 1], $XY = YX$, or the shortest path of a [point] moving from X to Y coincides with that from Y to X.

Definition 2. Terms which are not the same, that is, terms which cannot always be substituted for one another, are *different* [*diversa*]. Such are "circle" and "triangle," or "square" (supposed perfect, as it always is in Geometry) and "equilateral quadrangle," for we can predicate this last of a rhombus, of which "square" cannot be predicated.

$A \neq B$ *signifies* that A and B are different, as, for example, R̲ Y̲ S̲ X̲ the straight lines XY and RS.

[1] Leibniz writes $A \infty B$, the symbol ∞ or ∞ being at that time a common symbol for equality (see, for instance, Descartes, Selection II.8).

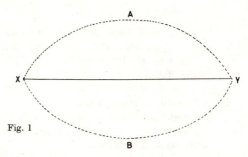

Fig. 1

Proposition 1. *If* $A = B$, *then also* $B = A$. *If anything be the same with another, then that other will be the same with it.* For since $A = B$ (by hyp.), it follows (by def. 1) that in the statement $A = B$ (true by hyp.) B can be sub-stituted for A and A for B; hence we have $B = A$.

Proposition 2. *If* $A \neq B$, *then also* $B \neq A$. *If any term be different from another, then that other will be different from it.* Otherwise we should have $B = A$, and in consequence (by the preceding prop.) $A = B$, which is contrary to hypothesis.

Proposition 3. *If* $A = B$ *and* $B = C$, *then* $A = C$. *Terms which coincide with a third term coincide with each other.* For if in the statement $A = B$ (true by hyp.) C be substituted for B (by def. 1, since $C = B$), the resulting proposition will be true.

Corollary. If $A = B$ and $B = C$ and $C = D$, then $A = D$; and so on. For $A = B = C$, hence $A = C$ (by the above prop.). Again, $A = C = D$; hence (by the above prop.) $A = D$.

Thus since equal things are the same in magnitude, the consequence is that things equal to a third are equal to each other. The Euclidean construction of an equilateral triangle makes each side equal to the base, whence it results that they are equal to each other. If anything be moved in a circle, it is sufficient to show that the paths of any two successive periods, or returns to the same point, coincide, from which it is concluded that the paths of any two periods whatever coincide.

Proposition 4. *If* $A = B$ *and* $B \neq C$, *then* $A \neq C$. *If of two things which are the same with each other, one differ from a third, then the other also will differ from that third.* For if in the proposition $B \neq C$ (true by hyp.) A be substituted for B, we have (by def. 1, since $A = B$) the true proposition $A \neq C$.

Definition 3. *A is in L*, or *L contains A*, is the same as to say that L can be made to coincide with a plurality of terms, taken together, of which A is one.[2]

Definition 4. Moreover, all those terms such that whatever is in them is in L, are together called *components* (*componentia*) with respect to the L thus com-posed or constituted.

$B \oplus N = L$ *signifies* that B is in L; and that B and N together compose or constitute L. The same thing holds for a larger number of terms.[3]

[2] Definition 3 of paper XIX says: If a plurality of terms taken together coincide with one, then any one of the plurality is said to *be in* (*inesse*) or to be *contained in* (*conteneri*) that one with which they coincide, and that one is called the *container* (*continens*).

[3] Leibniz here introduces the logical addition of terms, distinguishing its symbol from that for arithmetical addition, +, by a circle around it. He did not make this distinction for the symbol for equality; see note 1.

Definition 5. I call terms one of which is in the other *subalternates* (*subalternantia*), as A and B if either A is in B or B is in A.

Definition 6. Terms neither of which is in the other [I call] *disparate* (*disparata*).

Axiom 1. $B \oplus N = N \oplus B$, or transposition here alters nothing.[4]

Postulate 1. If an arbitrary term is given, then it is possible to find some term different from it, and even disparate, or one that is not in the other.[5]

Postulate 2. Any plurality of terms, as A and B, can be added to compose a single term, $A \oplus B$ or L.

Axiom 2. $A \oplus A = A$. If nothing new be added, then nothing new results, or repetition here alters nothing. (For 4 coins and 4 coins are 8 coins, but not 4 coins and the same 4 coins already counted.)

Proposition 5. *If A is in B and $A = C$, then C is in B. That which coincides with the inexistent, is inexistent.*[6] For in the proposition, A is in B (true by hyp.), the substitution of C for A (by def. 1 of coincident terms, since, by hyp., $A = C$) gives, C is in B.

Proposition 6. *If C is in B and $A = B$, then C is in A. Whatever is in one of two coincident terms, is in the other also.* For in the proposition, C is in B, the substitution of A for B (since $A = B$) gives, C is in A.[7] (This is the converse of the preceding.)

Proposition 7. *A is in A. Any term whatever is contained in itself.* For A is in $A \oplus A$ (by def. of "inexistent," that is, by def. 3) and $A \oplus A = A$ (by ax. 2). Therefore (by prop. 6), A is in A.

Proposition 8. *If $A = B$, then A is in B. Of terms which coincide, the one is in the other.* This is obvious from the preceding. For (by the preceding) A is in A —that is (by hyp.), in B.

Proposition 9. *If $A = B$, then $A \oplus C = B \oplus C$. If terms which coincide be added to the same term, the results will coincide.* For if [Fig. 2] in the proposition, $A \oplus C = A \oplus C$ (true per se), for A in one place be substituted B which coincides with it (by def. 1), we have $A \oplus C = B \oplus C$.

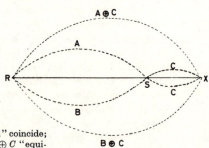

Fig. 2. A "triangle" and B "trilateral" coincide; $A \oplus C$ "equilateral triangle" and $B \oplus C$ "equilateral trilateral" coincide.

---

[4] This is the commutative law for logical addition.

[5] Couturat here remarks that the second part of this postulate is not valid if we take as term the whole universe of discourse, since then there exists no disparate term.

[6] Here "inexistent" means "existent in"; see note 2.

[7] Gerhardt (and Lewis) have "A is in B."

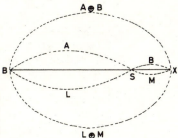

Fig. 3. A "triangle" and L "trilateral" coincide; B "regular" coincides with M "most capacious of equally-many-sided figures with equal perimeters"; "regular triangle" coincides with "most capacious of trilaterals making equal peripheries out of three sides."

Scholium. This proposition cannot be converted—much less, the two which follow. A method for finding an illustration of this fact will be exhibited below, in the problem which is prop. 23.

Proposition 10. *If $A = L$ and $B = M$, then $A \oplus B = L \oplus M$. If terms which coincide be added to terms which coincide, the results will coincide.* For [Fig. 3] since $B = M$, $A \oplus B = A \oplus M$ (by the preceding), and putting L for the second A (since, by hyp., $A = L$) we have $A \oplus B = L \oplus M$.

Scholium. This proposition cannot be converted, for if $A \oplus B = L \oplus M$ and $A = L$, still it does not follow that $B = M$—and much less can the following be converted.

Proposition 11. *If $A = L$ and $B = M$ and $C = N$, then $A \oplus B \oplus C = L \oplus M \oplus N$. And so on. If there be any number of terms under consideration, and an equal number of them coincide with an equal number of others, term for term, then that which is composed of the former coincides with that which is composed of the latter.* For (by the preceding, since $A = L$ and $B = M$) we have $A \oplus B = L \oplus M$. Hence, since $C = N$, we have (again by the preceding) $A \oplus B \oplus C = L \oplus M \oplus N$.

Proposition 12. *If B is in L, then $A \oplus B$ will be in $A \oplus L$. If the same term be added to what is contained and to what contains it, the former result is contained in the latter.* For [Fig. 4] $L = B \oplus N$ (by def. of "inexistent"), and $A \oplus B$ is in $B \oplus N \oplus A$ (by the same), that is, $A \oplus B$ is in $L \oplus A$.[8]

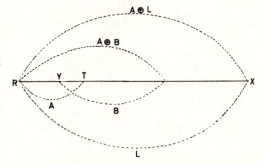

Fig. 4. B "equilateral," L "regular," A "quadrilateral." "Equilateral" is in or is an attribute of "regular"; hence "equilateral quadrilateral" is in "regular quadrilateral" or "perfect square." YS is in RX; hence $RT \oplus YS$, or RS, is in $RT \oplus RX$, or in RX.

[8] Expressed in a more modern symbolism, the reasoning is: If $B < L$, then there exists an N such that $L = B + N$. But $A + B < (A + B) + N$, and since $(A + B) + N = A + (B + N) = A + L = L + A$, $A + B < A + L$. As Couturat remarks, Leibniz uses here, implicitly, the associative law for logical addition.

Scholium. This proposition cannot be converted; for if $A \oplus B$ is in $A \oplus L$, it does not follow that B is in L.

Proposition 13. If $L \oplus B = L$, then B is in L. If the addition of any term to another does not alter that other, then the term added is in the other. For [Fig. 5] B is in $L \oplus B$ (by def. of "inexistent") and $L \oplus B = L$ (by hyp.), hence (by prop. 6) B is in L.

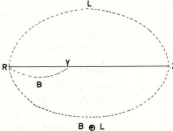

Fig. 5. $RY \oplus RX = RX$; hence RY is in RX.
RY is in RX; hence $RY \oplus RX = RX$.

Let L be "parallelogram" (every side of which is parallel to some side), B be "quadrilateral." "Quadrilateral parallelogram" is the same as "parallelogram." Therefore to be quadrilateral is in [the intension of] "parallelogram." Reversing the reasoning, to be quadrilateral is in "parallelogram." Therefore, "quadrilateral parallelogram" is the same as "parallelogram."

Proposition 14. If B is in L, then $L \oplus B = L$. Subalternates compose nothing new; or if any term which is in another be added to it, it will produce nothing different from that other. (Converse of the preceding.) If B is in L, then (by def. of "inexistent") $L = B \oplus P$. Hence (by prop. 9) $L \oplus B = B \oplus P \oplus B$, which (by ax. 2) is $= B \oplus P$, which (by hyp.) is $= L$.

Proposition 15. If A is in B and B is in C, then also A is in C. What is contained in the contained, is contained in the container. For [Fig. 6] A is in B (by hyp.),

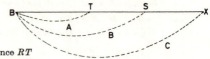

Fig. 6. RT is in RS, and RS is in RX; hence RT is in RX. A "quadrilateral," B "parallelogram," C "rectangle."

hence $A \oplus L = B$ (by def. of "inexistent"). Similarly, since B is in C, $B \oplus M = C$, and putting $A \oplus L$ for B in this statement (since we have shown that these coincide), we have $A \oplus L \oplus M = C$. Therefore (by def. of "inexistent") A is in C.

To be quadrilateral is in [the intension of] "parallelogram," and to be parallelogram is in "rectangle" (that is, a figure every angle of which is a right angle). If instead of concepts per se we consider individual things comprehended by the concept, and put A for "rectangle," B for "parallelogram," C for "quadrilateral," the relations of these can be inverted. For all rectangles are comprehended in the number of the parallelograms, and all parallelograms in the number of the quadrilaterals. Hence also, all rectangles are contained amongst (in) the quadrilaterals. In the same way, all men are contained amongst (in) all

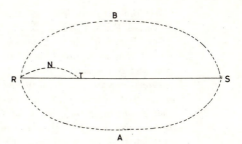

Fig. 7. $RT, N; RS, A; SR \oplus RT, B$.
To be trilateral is in [the intension
of] "triangle," and to be triangle is
in "trilateral." Hence "triangle"
and "trilateral" coincide. Similarly,
to be omniscient is to be omnipotent.

the animals, and all animals amongst all the material substances, hence all men
are contained amongst the material substances. And conversely, the concept of
material substance is in the concept of animal, and the concept of animal is in
the concept of man. For to be man contains [or implies] being animal.

Scholium. This proposition cannot be converted, and much less can the
following.

Corollary. If $A \oplus N$ is in B, N also is in B. For N is in $A \oplus N$ (by def. of
"inexistent").

Proposition 16. *If A is in B and B is in C and C is in D, then also A is in D.*
And so on. *That which is contained in what is contained by the contained, is in the
container.* For if A is in B and B is in C, A also is in C (by the preceding). Whence
if C is in D, then also (again by the preceding) A is in D.

Proposition 17. *If A is in B and B is in A, then A = B. Terms which contain
each other coincide.* For [Fig. 7] if A is in B, then $A \oplus N = B$ (by def. of "inexistent"). But B is in A (by hyp.), hence $A \oplus N$ is in A (by prop. 5). Hence
(by coroll. prop. 15) N also is in A. Hence (by prop. 14) $A = A \oplus N$, that is,
$A = B$.

Proposition 18. *If A is in L and B is in L, then also A $\oplus$ B is in L. What is
composed of two, each contained in a third, is itself contained in that third.* For
[Fig. 8] since A is in L (by hyp.), it can be seen that $A \oplus M = L$ (by def. of
"inexistent"). Similarly, since B is in L, it can be seen that $B \oplus N = L$. Putting these together, we have (by prop. 10) $A \oplus M \oplus B \oplus N = L \oplus L$. Hence
(by ax. 2) $A \oplus M \oplus B \oplus N = L$. Hence (by def. of "inexistent") $A \oplus B$ is
in L.

Proposition 19. *If A is in L and B is in L and C is in L, then A $\oplus$ B $\oplus$ C is
in L.* And so on. *Or in general, whatever contains terms individually, contains also
what is composed of them.* For $A \oplus B$ is in L (by the preceding). But also C is in
L (by hyp.), hence (once more by the preceding) $A \oplus B \oplus C$ is in L.

Fig. 8. RYS is in RX; YST
is in RX; hence RT is in RX. A
"equiangular," B "equilateral,"
$A \oplus B$ "equiangular equilateral"
or "regular," L "square." "Equiangular" is in [the intension of]
"square," and "equilateral" is in
"square." Hence "regular" is in
"square."

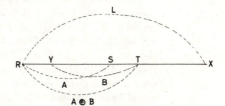

Scholium. It is obvious that these two propositions and similar ones can be converted. For if $A \oplus B = L$, it is clear from the definition of "inexistent" that A is in L, and B is in L. Likewise, if $A \oplus B \oplus C = L$, it is clear that A is in L, and B is in L, and C is in L. Also that $A \oplus B$ is in L, and $A \oplus C$ is in L, and $B \oplus C$ is in L. And so on.

Proposition 20. If A is in M and B is in N, then A $\oplus$ B is in M $\oplus$ N. If the former of one pair be in the latter and the former of another pair be in the latter, then what is composed of the former in the two cases is in what is composed of the latter in the two cases. For [Fig. 9] A is in M (by hyp.) and M is in $M \oplus N$ (by def. of "inexistent"). Hence (by prop. 15) A is in $M \oplus N$. Similarly, since B is in N and N is in $M \oplus N$, then also (by prop. 18) $A \oplus B$ is in $M \oplus N$.

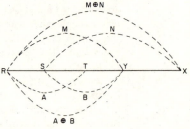

Fig. 9. RT is in RY and ST is in SX; hence $RT \oplus ST$, or RY, is in $RY \oplus SX$, or in RX. If A be "quadrilateral" and B "equiangular," $A \oplus B$ will be "rectangle." If M be "parallelogram" and N "regular," $M \oplus N$ will be "square." Now "quadrilateral" is in [the intension of] "parallelogram," and "equiangular" is in "regular," hence "rectangle" (or equiangular quadrilateral") is in "regular parallelogram or square."

Scholium. This proposition cannot be converted. Suppose that A is in M and $A \oplus B$ is in $M \oplus N$, still it does not follow that B is in N; for it might happen that B as well as A is in M, and whatever is in B is in M, and something different in N. Much less, therefore, can the following similar proposition be converted.

Proposition 21. If A is in M and B is in N and C is in P, then A $\oplus$ B $\oplus$ C is in M $\oplus$ N $\oplus$ P. And so on. Whatever is composed of terms which are contained, is in what is composed of the containers. For since A is in M and B is in N (by the preceding), $A \oplus B$ is in $M \oplus N$. But C is in P, hence (again by the preceding) $A \oplus B \oplus C$ is in $M \oplus N \oplus P$.

Proposition 22. Two disparate terms, A and B, being given, to find a third term, C, different from them and such that with them it composes subalternates A $\oplus$ C and B $\oplus$ C—that is, such that although A and B are neither of them contained in the other, still $A \oplus C$ and $B \oplus C$ shall one of them be contained in the other.

Solution. If we wish that $A \oplus C$ be contained in $B \oplus C$, but A be not contained in B, this can be accomplished in the following manner: Assume (by post. 1) some term, D, such that it is not contained in A, and (by post. 2) let $A \oplus D = C$, and the requirements are satisfied.

For [Fig. 10] $A \oplus C = A \oplus A \oplus D$ (by construction) $= A \oplus D$ (by ax. 2). Similarly, $B \oplus C = B \oplus A \oplus D$ (by construction). But $A \oplus D$ is in $B \oplus A \oplus D$ (by def. 3). Hence $A \oplus C$ is in $B \oplus C$. Which was to be done. SY and YX are disparate. If $RS \oplus SY = YR$, then $SY \oplus YR$ will be in $XY \oplus YR$.

Let A be "equilateral," B "parallelogram," D "equiangular," and C "equiangular equilateral" or "regular," where it is obvious that although "equilateral" and "parallelogram" are disparate so that neither is in the other, yet

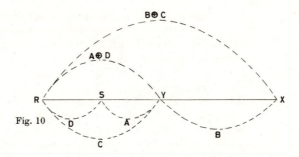

Fig. 10

"regular equilateral" is in "regular parallelogram" or "square." But, you ask, will this construction prescribed in the problem succeed in all cases? For example, let A be "trilateral," and B "quadrilateral"; is it not then impossible to find a concept which shall contain A and B both, and hence to find $B \oplus C$ such that it shall contain $A \oplus C$, since A and B are incompatible? I reply that our general construction depends upon the second postulate, in which is contained the assumption that any term and any other term can be put together as components. Thus God, soul, body, point, and heat compose an aggregate of these five things. And in this fashion also quadrilateral and trilateral can be put together as components. For assume D to be anything you please which is not contained in "trilateral," as "circle." Then $A \oplus D$ is "trilateral and circle," which may be called C. But $C \oplus A$ is nothing but "trilateral and circle" again. Consequently, whatever is in $C \oplus B$ is also in "trilateral," in "circle," and in "quadrilateral." But if anyone wish to apply this general calculus of compositions of whatever sort to a special mode of composition, for example, if one wish to unite "trilateral" and "circle" and "quadrilateral" not only to compose an aggregate but so that each of these concepts shall belong to the same subject, then it is necessary to observe whether they are compatible. Thus immovable straight lines at a distance from one another can be added to compose an aggregate but not to compose a continuum.

Proposition 23. *Two disparate terms, A and B, being given, to find a third, C, different from them* [and such that $A \oplus B = A \oplus C$].

Solution. Assume (by post. 2) $C = A \oplus B$, and this satisfies the requirements. For since A and B are disparate (by hyp.)—that is (by def. 6), neither is in the other—therefore (by prop. 13) it is impossible that $C = A$ or $C = B$. Hence these three are different, as the problem requires. Thus $A \oplus C = A \oplus A \oplus B$ (by construction), which (by ax. 2) is $= A \oplus B$. Therefore $A \oplus C = A \oplus B$. Which was to be done.

Proposition 24. *To find a set of terms, of any desired number, which differ each from each and are so related that from them nothing can be composed which is new, or different from every one of them* [i.e., such that they form a group with respect to the operation $\oplus$].

Solution. Assume (by post. 1) any terms, of any desired number, which shall be different from each other, A, B, C, and D, and from these let $A \oplus B = M$, $M \oplus C = N$, and $N \oplus D = P$. Then A, B, M, N, and P are the terms required. For (by construction) M is made from A and B, hence A, or B, is in M, and

M in N, and N in P. Hence (by prop. 16) any term which here precedes is in any which follows. But if two such are united as components, nothing new arises; for if a term be united with itself, nothing new arises; $L \oplus L = L$ (by ax. 2). If one term be united with another as components, a term which precedes will be united with one which follows, hence a term which is contained with one which contains it, as $L \oplus N$, but $L \oplus N = N$ (by prop. 14). And if three are united, as $L \oplus N \oplus P$, then a couple, $L \oplus N$, will be joined with one, P. But the couple, $L \oplus N$, by themselves will not compose anything new, but one of themselves, namely the latter, N, as we have shown; hence to unite a couple, $L \oplus N$, with one, P, is the same as to unite one, N, with one, P, which we have just demonstrated to compose nothing new. And so on, for any larger number of terms. Q.E.D.

Here follows a segment of paper XIX in Gerhardt's collection which contains the concept of "subtraction."

Theorem 8. If terms which coincide be subtracted from terms which coincide, the remainders will coincide.

If $A = L$ and $B = M$, then $A - B = L - M$. For $A - B = A - B$ (true per se), and the substitution, on one or the other side, of L for A and M for B, gives $A - B = L$. Q.E.D.

[*Note in the margin of the manuscript.*] In dealing with concepts, subtraction [*detractio*] is one thing, negation another. For example, "nonrational man" is absurd or impossible. But we may say: An ape is a man except that it is not rational, or [They are] men except in those respects in which man differs from the beasts, as in the case of Grotius' iambus [*Homines nisi qua Bestiis differt homo, ut in iambo Grotii*].[9] "Man" minus "rational" is something different from "nonrational man." For "man" minus "rational" = "brute." But "nonrational man" is impossible. "Man" − "animal" − "rational" is Nothing. Thus subtractions can give Nothing or simple nonexistence—even less than nothing—but negations can give the impossible.

[9] To Professor L. J. Rogier in Nijmegen, Netherlands, I owe the full quotation:

EUCHARISTIA

Procul profani, qui, quod os et quod manus
Oculique monstrant, nec quid ultra, creditis,
Queis una mens exsensa; ventri obnoxia,
Humoque prona gens, et, ut summam loquar,
Homines, nisi qua bestiis differt homo.

—*Hug. Grotii Poemata omnia* (ed. quarta; Lugdunum Batavorum, 1645), p. 21

[Away, ye profane, who believe what the mouth and what the hands and eyes show and not anything beyond, to whom the mind alone is apart from the senses; obedient to the stomach, and a race prone on the ground, and, to summarize, men, except insofar as man differs from the beasts.]

CHAPTER III GEOMETRY

Medieval geometry was mostly taken from Euclid, whose work was partially known through the widely circulating works of Boethius (sixth century) and which became available through translations by Gerhard of Cremona, Johannes Campanus, and others in the twelfth and thirteenth centuries. The first printed Euclid, in the Latin version of Campanus, was that of Erhard Ratdolt in Venice (1482), a beautiful piece of work with many figures, and from that time on full or partial editions, in the original Greek and in translations, appeared in several countries. During the sixteenth century published editions of the works of Archimedes, Apollonius, and Pappus increased geometrical knowledge and curiosity.

Goniometry and trigonometry, developed by Arabic-writing authors influenced by Ptolemy as well as by Indian and Chinese authorities, became part of the Latin inheritance through Regiomontanus' *De triangulis omnimodis* (c. 1464; first printed in Nuremberg, 1533); see Selection III.2. Because of the absence of a special notation, it kept the appearance of a particular section of computational geometry. This computational aspect was strengthened during the sixteenth century by the publication of trigonometric tables, culminating in the *Opus Palatinum* with all six trigonometric functions tabulated to ten decimal places for every 10″, compiled by Georg Joachim Rhaeticus (Neustadt, 1596).

With Fermat and Descartes a new period opens in the history of geometry. However, since some scholars see in Oresme's work an early attempt at coordinate geometry, we begin with a sample of his writing. At the same time, we should not forget that Ptolemy, in his *Geography* (c. A.D. 125), had already explained the measurement of points on the earth with the aid of latitude and longitude (terms taken over by Oresme). In Oresme's time (fourteenth century) Ptolemy's manuscripts were little known in Europe and not yet used by cartographers, but Oresme may have consulted a manuscript copy of the *Geography*.

At the time of Descartes and Fermat, cartography had become a well-developed mathematical science, and cartographers were accustomed to use many types of map projection and thus many types of coordinate system. However, there is no evidence that Descartes and Fermat were influenced by cartography; their starting point was the possibility of applying the new algebra of the sixteenth century as developed by Cardan and Viète to the geometry of the ancient Greeks.

1 ORESME. THE LATITUDE OF FORMS

We have taken from medieval authors only those selections (I.1, II.1) that establish a direct link with the Arabic world. A special Source Book will be necessary to deal with the medieval scholastic contribution to science. However, we make an exception for the work of Oresme, since there are indications that this forms a direct link with Renaissance science.

The study of the Aristotelian categories of quality and quantity focused attention on the difference between the *intensio* and *remissio* of a quality and the increase and decrease of a quantity. A quantity, as a solid, increases by adding another body to it, but a quality, as wisdom, is intensified in quite another way, and so is heat, considered a quality before, in the seventeenth century, thermometry was invented. Another quality was motion, and change in velocity was widely discussed, notably in the thirteenth and fourteenth centuries. This gave rise to the application of quantitative ideas to qualities.

One form in which this application occurred was that of *calculatio*, which applied the Euclidean theory of proportions to theological concepts, and to qualities including motion. There was indeed in the work of some of the *Calculatores* (such as Thomas Bradwardine, archbishop of Canterbury, *c.* 1290–1349) a beginning of kinematics. There was also some work on infinite series in connection with Zeno's paradoxes, for instance, by Richard Suiseth, of Merton College, Oxford, *c.* 1345, called *the* Calculator; see C. B. Boyer, *The history of the calculus* (Dover, New York, 1959), 74–79.

Another form was the geometrical representation of the variability of the intensity of a quality, and with this is connected the name of Nicole Oresme (*c.* 1323–1382), M.A. of the University of Paris (1349), who was dean of Rouen Cathedral (1361–1377) and afterward Bishop of Lisieux. He represented, as one of the first to do so, such a variable value, and especially that of a velocity, for any point of a body or for any instant of time by a line segment plotted in a given direction, and thus drawing the first graph.

We find this idea in several works written by him or ascribed to him, of which the *Tractatus de latitudinibus formarum*, probably written by a pupil, was published in Padua (1482). There were later editions, which show that Oresme's ideas may have had circulation among Renaissance mathematicians. He uses the concept of *uniformity* and *difformity* (combined in *uniformiter difformis*, uniformly difform), when the change of intensity upon displacement along the base, the *longitudo*, is proportional to the amount of this displacement. To every point of the *longitudo* corresponds a line segment called *latitudo*, indicating the intensity. The plotted latitudes form a plane surface, and so we obtain the notion of a "rectangular" heat, when the intensity is uniform, or a "trapezoidal" velocity when the velocity increases uniformly. The end points of the latitudes form the graph, the *linea summitatis* (summit line).

We take our selection from a manuscript by Oresme recently published, *Quaestiones super geometriam Euclides*, ed. H. L. L. Busard (Brill, Leiden, 1961), with an English translation, from which we take the vital Question 10. It is preceded in Questions 1–9 by a number of different topics current in fourteenth-century scholastic mathematics, such as the discussion of the existence of an infinite circle and the commensurability or incommensurability of the diagonal and the side of a square.

In Question 10 and in the seven following questions Oresme expounds his theory of *latitudines*, which was already referred to in the preceding question. The first question he puts in this connection is: can a quadrilateral be uniformly difform in altitude? After an objection has first been advanced, Oresme gives his reply. Next he discusses how the different qualities can be represented. He concludes the question with a spirited and highly personal defense, in which he appeals, among others, to Aristotle.

Oresme considers a place or body (*subjectum*) in which he takes a straight line segment, the *longitudo*, and at each point of it, perpendicular to the segment in a plane through it, a line segment representing the value of a certain intensity.

Further information on the *Calculatores* and Oresme can be found in E. J. Dijksterhuis, *The mechanization of the world picture* (Clarendon Press, Oxford, 1961), 185–200. On the *Quaestiones* see also J. Murdock, "Oresme's Commentary to Euclid," *Scripta Mathematica 27* (1964), 67–91, and Selection V.9. For other mathematical works by Oresme see his *De proportionibus proportionum and Ad pauca respicientes*, ed. and trans. E. Grant (University of Wisconsin Press, Madison, 1966), and Marshall Clagett, *The science of mechanics in the Middle Ages* (University of Wisconsin Press, Madison, 1959), who brings to our attention that a certain Giovanni di Casali may have used a graph in 1346, hence before Oresme, whose work dates from between 1348 and 1362 (pp. 332–333, 414).

(*a*) The altitude of a surface is judged by the line drawn perpendicular to the base, as might appear from a figure.

(*b*) A surface is called uniformly or equally high if all the lines by which the altitude is judged are equal; a surface is called difformly high if these lines are unequal and extend up to a line not parallel to the base (the summit line).

(*c*) The altitude is called uniformly difform if every three or more equidistant altitudes exceed one another in an arithmetical proportion [Fig. 1], i.e., the

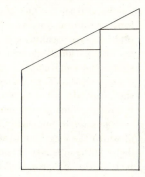

Fig. 1

first extends as much above the second as the second above the third, from which it appears that the summit line is a straight line which is not parallel to the base.

(*d*) The altitude is called difformly difform if the altitudes do not exceed one another in this way. In that case their summit line is not straight and the difformity in altitude varies with the variation of this summit line.

Furthermore:

(*a*) Of a quality two things are represented, *viz.* the *intensio per gradum* and the *extensio per subjectum*, and consequently such a quality is imagined to have two dimensions.

H. L. L. Busard remarks that this means that the extensity of an extended quality of any kind can be designated by a line or a plane (called the longitude) described in the subject. The intensity of the quality from point to point in the subject has to be represented by lines (called latitudes) erected perpendicular to the longitude of the same quality. The latitude thus acts as a variable ordinate in a system of coordinates, while the longitude is not to be identified with the variable abscissa; there is only one longitude with an infinite number of latitudes.[1]

For this reason it is sometimes said that a quality has a latitude and a longitude instead of an intensity and an extensity.

(*b*) A quality may be imagined as belonging to a point or an indivisible subject, such as the soul, but also to a line and even to a surface and a body.

Conclusion 1: The quality of a point or an indivisible subject can be represented by a line, because it has only one dimension, *viz.* intensity. From this it follows that such a quality, such as knowledge or virtue, cannot be called uniform or difform, just as a line is not called uniform or difform. It also follows that properly speaking one cannot refer to the latitude of knowledge and virtue, because no longitude can be associated with it, whereas every latitude presupposes a longitude.

Conclusion 2: The quality of a line can be represented by a surface, of which the longitude is the rectilineal extensity of the subject and the latitude the intensity, which is represented by perpendiculars erected on the subject-line.

Conclusion 3: Similarly the quality of a surface can be represented by a body of which the length and the breadth form the extensity of the surface and the depth is the intensity of the quality. For the same reason the quality of a whole body might be represented by a body of which the length and the breadth would be the extensity of the whole body and the depth the intensity of the quality.

However, someone might doubt: If the quality of a line is here represented by a surface and the quality of a surface by a body having three dimensions, the quality of a body will no doubt be represented by something having four dimensions in a different kind of quantity.

I say that it is not necessary to give a fourth dimension. In fact, if one imagines that a *punctus fluens* causes a line, a line a surface, a surface a body, it

---

[1] The extensity corresponds to an infinity of intensities.

is not necessary for a *corpus fluens* to cause a fourth kind of quantity but only a body, and because of this Aristotle, in *De Caelo* I,[2] says that according to this method of representation no passage from a body to a different kind of quantity is possible. In the case under consideration one should reason in the same way.

It is therefore necessary to speak of the quality of a line, and analogously it is considered what has to be said of the quality of a surface or a body.

Conclusion 4: A uniform linear quality can be represented by a rectangle that is uniformly high, in such a way that the base represents the extensity and the summit line is parallel to the base.

A uniformly difform quality can be represented by a surface that is uniformly difformly high, in such a way that the summit line is not parallel to the base. This can be proved: the intensities of the points of the quality are proportional to the altitudes of the perpendiculars erected in the corresponding points of the base.

A quality can be uniformly difform in two ways, just as a surface can be uniformly difformly high in two ways:

(*a*) Such a quality may terminate at zero degree and is then represented by a surface which is uniformly difformly high down to zero, i.e., by a right-angled triangle;

(*b*) Such a quality may terminate at both ends at some degree and is then represented by a quadrilateral, the summit line of which is a straight line not parallel to the base, i.e., by a right-angled trapezium.

Conclusion 5: By means of the above it can be proved that a uniformly difform quality is equal to the medium degree [Fig. 2], i.e., just as great as a uniform quality would be at the degree of the middle point, and this can be proved in the same way as for a surface.[3]

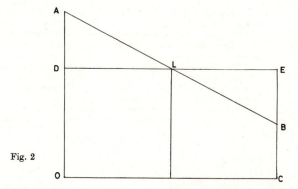

Fig. 2

---

[2] This rejection of a fourth dimension is based on Aristotle's *De Caelo*, I, 1; 268a31–268b2. It remains the prevailing attitude in Renaissance days even where, as in Cardan or Viète, quadratic equations are related to planes and cubic equations to solids (see Selections II.3, 5). Despite an occasional remark by Pascal, Wallis, and Lagrange, only the nineteenth century took the geometry of four dimensions seriously. See Selections III.3, note 3, and IV.12.

[3] Conclusion 5 is often referred to as the Mertonian rule. See E. J. Dijksterhuis, *The mechanization of the world picture* (Clarendon Press, Oxford, 1961), 197. See also Selection IV.4.

Last conclusion: A difformly difform quality is represented by a surface of which the line representing the subject is the base, while the summit line is a nonstraight line, not parallel to the base. Such a difformity may be imagined in an infinite number of different ways, for the summit line may vary in a great many ways.

However, someone might say: It is not necessary to represent a quality in this way. I say that the representation is good, as also appears in Aristotle,[4] for he represents time by a line. In the same way in *Perspectiva* the *virtus activa* is represented by a triangle.[5] Moreover according to this representation one can understand more easily what is said about uniformly difform qualities, and consequently the representation is good.

This means that, since qualities are represented by surfaces, the equality of two surfaces may also be transferred to the qualities which they represent. In this case, therefore, one has to prove that surface $OCBA$ = surface $OCED$, and from this equality it then follows that the uniformly difform quality that is represented by $OCBA$ is equal to the uniform quality that is represented by $OCED$.

2 REGIOMONTANUS. TRIGONOMETRY

Trigonometry was developed into a independent branch of mathematics by Islamic writers, notably by Nasir ed-dīn at-Tūsī (or Nasir Eddin, 1201–1274). The first publication in Latin Europe to achieve the same goal was Regiomontanus' *De triangulis omnimodis* (On triangles of all kinds; Nuremberg, 1533).

[4] Aristotle, *Physica*, IV, 11; 220a4–20. In lines 219b1–2 Aristotle defines time as "numerus motus secundum prius et posterius." Here he tries to explain that the "now-moment," on the one hand, divides time into two parts (past–future), but, on the other hand, makes it continuous. He compares time to a line on which a point makes a division but also constitutes continuity on the line.
[5] The *virtus activa* is the light diffused from the source of light (*lumen*). Later, in Question 17, Oresme concludes: "Such a force or such a light extends uniformly difformly, or in other words: it is a uniformly difform quality. This appears plausible because—since the force does not extend uniformly—it seems to diminish as the distance increases; this diminution has to take place proportionally, i.e., uniformly difformly" [Fig. 3]. The *Perspectiva* mentioned is the one written by Witelo (Vitellio), a Polish mathematician of the thirteenth

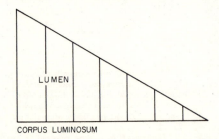

Fig. 3

LUMEN

CORPUS LUMINOSUM

century, first printed in Nuremberg (1535), a book that was widely read, and on which Kepler wrote a book, *Ad Vitellionem paralipomena* (Frankfurt a. M., 1604).

XX.

In omni triangulo rectangulo, si super uertice acuti anguli, secundũ
quantitatem lateris maximi circulum descripseris, erit latus ipsum acu
tum, subtendens angulum sinus rectus conterminalis sibi arcus dictũ
angulum respicientis: lateri autem tertio sinus complementi arcus di:
cti æqualis habebitur.

Sit triangulus rectanglus a b c, angulum c rectum habens, & a acutum, su
per cuius uertice a secundum quãtitatem lateris maximi a b, maximo scilicet an
gulo oppositi describatur circulus b e d, cuius circũferentiæ occurrat latus a c

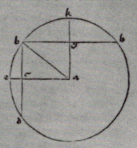

quoad satis est prolongatũ in e puncto. Di:
co quòd latus b c angulo b a c oppositum
est sinus arcus b e dictum angulum subten
dentis. Latus autem tertium, scilicet a c, æ:
quale est sinui recto complementi arcus b e.
Extendatur enim latus b c occurrendo cir:
cumferentiæ circuli in puncto d . à punctis
autem a quidem centro circuli exeat semidi
ameter a k æquedistans lateri b c. & à pun
cto b corda b h æquedistans lateri a c . se:
cabunt autem se necessario duæ lineæ b h &
a k, angulis a b h & b a k acutis existenti
bus, quod fiat in puncto g. Quia itaq; semidi
ameter a e cordam b d secat orthogonali:
ter propter angulum a c b rectum, secabit & ipsam per æqualia ex tertia tertij ele
mentoꝶ. & arcum b d per æqualia ex 29.eiusdem. quemadmodum igitur tota li
nea b d per diffinitionem corda est arcus b d, ita medietas eius, línea scilicet b c
est sinus dimídij arcus b e respicientis angulũ b a e siue b a c . quod asseruit pri
ma pars theorematis nostri. Secundam deinceps partem ueram cõfiteberis, si pri
us per 34.primi angulum a g b rectum esse didiceris. semidiameter enim a k, &
cordam b h, & arcũ eius ex supra memoratis medijs per æqua scindet . quare per
diffinitionem línea recta b g sinus erit arcus b k. Est autem línea b g æqualis la

C 3   teri tri

Fig. 1

Johannes Müller (1436–1476), called Regiomontanus from his birthplace, Königsberg in Franconia, was an instrument maker, mathematician, astronomer, and humanist who settled at Nuremberg and died in Rome as adviser to the pope on calendar reform. His trigonometry, finished in 1464, remained in manuscript until 1533. The book, reprinted at Basel in 1561, was widely studied during the sixteenth century. It deals with both plane and spherical trigonometry without using formulas: all theorems and demonstrations are verbal, with frequent references to Euclid's *Elements*. The trigonometric concepts used are the sine (*sinus* or *sinus rectus*) and versed sine (*sinus versus*), conceived as line segments and expressed as parts of a given radius (*sinus totus*) $R = 60.000$.[1]

The English translation we give of parts of the book is based on that by B. Hughes in *Regiomontanus on triangles* (University of Wisconsin Press, Madison, 1967), 59. First our text leads up to the sine law of plane triangles; then we present Regiomontanus' way of stating the cosine law for spherical triangles. In the figures capital letters A, B, ... are used instead of Regiomontanus' a, b, ...; see Fig. 1.

Book I. *Theorem* 20. In every right triangle, if we describe a circle with center a vertex of an acute angle and radius the length of the longest side,[2] then the side subtending this acute angle is the right sine [*sinus rectus*] of the arc adjacent to that side and opposite the given angle; the third side is equal to the sine of the complement of the arc.

If a right triangle ABC [Fig. 2] is given with C the right angle and A an acute angle, around the vertex of which a circle BED is described with the longest

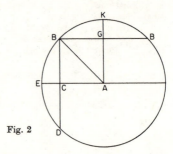

Fig. 2

side—that is, the side opposite the largest angle—as radius, and if side AC is extended sufficiently to meet the circumference of the circle at point E, then side BC opposite angle BAC is the sine of arc BE subtending the given angle, and furthermore the third side AC is equal to the right sine of the complement of arc BE.

---

[1] Hence Regiomontanus' *sinus* or *sinus rectus* of an angle α is our $R \sin \alpha$, and his *sinus versus* of an angle α is our $R \operatorname{versin} \alpha = R(1 - \cos \alpha)$. See Selection I.4, note 2.

[2] Regiomontanus does not use the term "hypotenuse," herein following Euclid. Neither does he use the term "trigonometry," which appears first in the title of the book of Bartolomeus Pitiscus, *Trigonometria* (Heidelberg, 1595).

Then, extending BC to CD, just as by definition the entire line BD is the chord of arc BD, so also its half, namely line BC, is the sine of the half-arc BE opposite angle BAE or BAC.

Book I. *Theorem* 28. When the ratio of two sides of a right triangle is given, its angles can be ascertained.

One of the two sides is opposite the right angle or else none is. First, if side AB, whose ratio to side AC is known, is opposite right angle ACB, then the angles of this triangle become known.

For instance, if in triangle ABC (Fig. 1), $AB:BC = 9:7$, then multiply 7 into $R = 60.000$ (the whole right sine, *sinus rectus totus*) and divide by 9. The quotient, 46667, corresponds to about 51°3′, the value of angle ABC.[3]

Book II opens with the sine law for plane triangles.

Book II. *Theorem* 1. In every rectilinear triangle the ratio of one side to another side is as that of the right sine of the angle opposite one of the sides to the right sine of the angle opposite the other side.

As we said elsewhere, the sine of an angle is the sine of the arc subtending that angle. Moreoever, these sines must be related through one and the same radius of the circle or through several equal radii. Thus, if triangle ABG [Fig. 3] is a

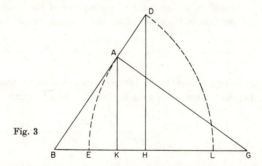

Fig. 3

rectilinear triangle, then the ratio of side AB to side AG is as that of the sine of angle AGB to the sine of angle ABG; similarly, that of side AB to BG is as that of the sine of angle AGB to the sine of angle BAG.

[3] Regiomontanus had sine tables for several values of R, which he may have computed himself, or taken from other, perhaps Arabic, sources. The table for $R = 60.000$ was first published in the *Tabulae directionum projectionumque* (Augsburg, 1490). The Basel edition of *De triangulis* also has a table.

If triangle ABG is a right triangle, we will provide the proof directly from Theorem I.28 above. However, if it is not a right triangle yet the two sides AB and AG are equal, the two angles opposite the sides will also be equal and hence their sines will be equal. Thus from the two sides themselves it is established that our proposition is verified. But if one of the two sides is longer than the other—for example, if AG is longer—then BA is drawn all the way to D, until the whole line BD is equal to side AG. Then around the two points B and G as centers, two equal circles are understood to be drawn with the lengths of lines BD and GA as radii respectively. The circumferences of these circles intersect the base of the triangle at points L and E, so that arc DL subtends angle DBL, or ABG, and arc AE subtends angle AGE, or AGB. Finally two perpendiculars AK and DH, from the two points A and D, fall upon the base. Now it is evident that DH is the right sine of angle ABG and AK is the right sine of angle AGB. Moreover, by VI.4 of Euclid,[4] the ratio of AB to BD, and therefore to AG, is as that of AK to DH. Hence what the proposition asserts is certain.

Then follow many applications; for instance, Theorem 2 shows how to find the sides of a triangle if their sum is known together with the angles opposite them.

Book V. *Theorem* 2. In every spherical triangle that is constructed from the arcs of great circles, the ratio of the versed sine of any angle to the difference of two versed sines, of which one is the versed sine of the side subtending this angle while the other is the versed sine of the difference of the two arcs including this angle, is as the ratio of the square of the whole right sine to the rectangular product of the sines of the arcs placed around the mentioned angle.

In this theorem we recognize, in geometric and hence homogeneous form, the cosine law for a spherical triangle. We omit the proof, which is quite complicated. In our notation:

$$\frac{R \text{ versin } \alpha}{R \text{ versin } a - R \text{ versin } (b - c)} = \frac{R^2}{R \sin b \cdot R \sin c},$$

where a, b, c are the sides and α is the angle opposite a in the spherical triangle on a sphere of radius R. The expression can be reduced to

$$\cos a = \cos b \cos c + \sin b \sin c \cos \alpha.$$

---

[4] Theorem 4 of Book VI of Euclid's *Elements* states that in similar triangles corresponding sides are proportional.

3 FERMAT. COORDINATE GEOMETRY

Analytic geometry (the term itself, in its present meaning, appears first in the beginning of the nineteenth century) can be dated back to the works on coordinate geometry by Descartes (1637; Selections II.7, 8) and Fermat. Fermat's papers, probably written about the same time as Descartes's work, were posthumously published by his son in *Varia opera mathematica* (Toulouse, 1679), and thus had less influence than the work of his rival. Both authors were moved by the same spirit: they wanted to show how the Renaissance algebra of Cardan and his successors could be applied to the geometry of the Greeks, notably to Apollonius' theory of loci as preserved by Pappus. In carrying out their program they differed in their methods. Fermat used the sixteenth-century notation of Viète, in which, as we have seen, our $Dx = By$ is written "D in A aequatur B in E," and in which the homogeneity of the formulas is preserved: when D and A represent line segments, then "A in D aeq. Z pl" stands for "A times D is equal to the area Z (Z plane)." Descartes introduced the notation still in use in which known constant quantities are indicated by the letters $a, b, \ldots$, unknown or variable quantities by $x, y, \ldots$, their squares, cubes, and so on by $aa = a^2, a^3, \ldots, xx = x^2, x^3$, and so on. Descartes also rejected the homogeneity of the formulas (see Selections III.4, 5).

Descartes's discussion consists in giving examples of his method. Fermat, starting with loci expressed by straight lines and following these with loci expressed by conic sections, has a method that shows some similarity with our way of introducing analytic geometry.

Both Descartes and Fermat used as an important test case for their methods the so-called problem of Pappus, found in Book VII of Pappus' *Collection* (*Synagōgē*), written at about the end of the third century A.D. On this problem see M. R. Cohen and I. E. Drabkin, *A source book in Greek science* (Harvard University Press, Cambridge, 1948), 79–80, and T. L. Heath, *History of Greek mathematics* (Clarendon Press, Oxford, 1921), II, 400–401.[1] Here follows Pappus' text, which is preceded by a remark that Apollonius, in the third book of his *Conics* (*c.* 220 B.C.), mentions "the locus for three and four lines." Then Pappus continues:

"But this locus of three and four lines, of which Apollonius says, in his third book, that Euclid has not treated it completely, he himself has also not been able to achieve it, and he has not even been able to add anything to what Euclid has written about it . . .

"Here we shall show what is that locus of three and four lines . . . Let three straight lines be given in position. Let there pass through the same point to these three straight lines three others at given angles, and let the ratio of the rectangle taken on two of these lines to the square of the third be given. Then the point will be on a solid line given in position, that is, on one of the three conics.[2] And if one passes straight lines at given angles to four straight lines given in position, and if the ratio of the rectangle taken on two of them to that taken on the other two is given, then the point will also be on a conic section given in position. On the other hand, if there are only two straight lines, then it is known that the locus is plane, but if there are more than four lines, then the locus of the point is no longer

---

[1] Pappus' *Collection* can be consulted in the French translation by P. Ver Eecke: *Pappus d'Alexandrie. La collection mathématique* (2 vols.; Declès de Brouwer, Paris, Bruges, 1933; Paris, 1959). The quoted text is on pp. 507–510.

[2] Pappus distinguishes between plane, solid, and linear problems. The plane problems require only circles and straight lines for their construction, the solid ones require general conic sections, and the linear ones require more complex curves. This distinction is taken over by Fermat as well as by Descartes (see Selection III.4).

one of those that are known; it belongs to those that we simply call linear loci (without knowing anything more about their nature or their properties). Nobody has made the synthesis of these loci, nor has anyone shown how they serve as such loci, not even for the one that would seem to be the first and most obvious. These loci appear in the following proposition.

"If there are six given lines, and the ratio of the solid with three of the drawn lines as sides to the solid with the other three lines as sides is given, then the point will also be on a line given in position.

"If there are more than six lines, then we can no longer say that we give the ratio between some object based on four lines and another object based on the other lines, because there is no figure that can be based on more than three dimensions.[3] And yet certain recent writers have permitted themselves to interpret such things, but when they referred to the product of a rectangle by a square or a rectangle they ceased being intelligible. Yet they might have expressed and indicated their meaning generally by means of compound ratios, both in the case of the previous propositions and in the case of those now under discussion, in the following way.

"Through a point we draw two lines given in position to other lines at given angles, and the product consisting of the ratio of one of those drawn lines to another, and of the ratio of another couple of these drawn lines, and that of a third couple, and finally that of the last drawn line to a (specially) given line—if there are seven lines in all—or of that of the two last ones, if there are eight of them, is given. Then the point will be on a line given in position.

"The same can be said for any number of lines, even or odd. But, as I have said, for each of these loci which follow that for four lines, there has not been made any synthesis which permits us to know the line."

The problem of Pappus can be stated in modern terms as follows [Fig. 1] (it is understood that it is a problem in the plane). If to a line $L(x, y) = ax + by + c = 0$ a line M is drawn through a point $P(x_0, y_0)$ at angle α intersecting the line $L = 0$ in Q, then $PQ = \pm L(x_0, y_0) \times$ cosec $\alpha / \sqrt{a^2 + b^2} = $ const. $L(x_0, y_0)$. Hence if the $2n$ lines are given by the equations $L_i = 0, M_i = 0, i = 1, 2, \ldots, n$, then the locus of P is given by the equation $L_1 L_2 \cdots L_n \pm \lambda M_1 M_2 \cdots M_n = 0$, the value of λ depending on the given α, and to a solution with positive λ there always corresponds another one with negative λ. When $2n - 1$ lines are given: $L_i = 0, i = 1, \ldots, n$, $M_a = 0, a = 1, \ldots, n - 1$, then the locus is given by some

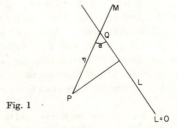

Fig. 1

[3] Acceptance of a geometry of four dimensions had to wait until the nineteenth century. See F. Cajori, *A history of mathematics* (Macmillan, New York, 2d ed., 1938), 184, 256, and Selection III.1, note 2. The first to build a systematic geometry of more than three dimensions was H. Grassmann, in his *Ausdehnungslehre* of 1844.

equation of the form $L_1 L_2 \cdots L_n \pm \mu M_1 M_2 \cdots M_{k-1} M_k^2 M_{k+1} \cdots M_{n-1} = 0$. For $n = 2$ the locus is a conic section.

It should be noted that in the title of Fermat's paper selected here: *Ad locos planos et solidos isagoge* (Introduction to plane and solid loci; *Oeuvres*, I, 92–103, French translation, III, 85–96), the term "plane locus" refers to a locus that can be constructed with the aid of straightedge and compass only, the term "solid locus" to one in which a conic section different from a circle or a straight line appears. When curves of degree higher than two appeared, the problem, or locus, was called linear. These terms appear in Pappus, and not only Fermat but also Descartes and others used them (see Selection III.4). A modern translation of the title would be: *Introduction to loci consisting of straight lines and curves of the second degree.*

There is no doubt that several ancient authors have written on loci, witness Pappus, who, at the beginning of his seventh book, states that Apollonius had written on plane loci and Aristaeus on solid loci.[4] But it seems that to them the study of loci did not come quite easily; this we gather from the fact that for several loci they did not give a sufficiently general account, as will be clarified by what follows here.

We shall therefore submit this science to an appropriate and particular analysis, so that from now on a general way to the study of loci shall be opened.

As soon as in a final equation [*aequalitas*] two unknown quantities appear, there exists a locus, and the end point of one of the two quantities describes a straight or a curved line. The straight line is the only one of its kind, but the types of curves are infinite: a circle, a parabola, a hyperbola, an ellipse, etc.

Whenever the end point of the unknown quantity describes a straight line or a circle, we have a plane locus; when it describes a parabola, hyperbola, or ellipse, then we have a solid locus; if other curves appear, then we say that the locus is a linear locus [*locus linearis*]. We shall not add anything to this last case, since the study of the linear locus can easily be derived from that of plane and solid ones by means of reductions.

The equations can be easily visualized [*institui*], when the two unknown quantities are made to form a given angle, which we usually take to be a right one, with the position and the end point of one of them given. Then, if no one of the unknown quantities exceeds a square, the locus will be plane or solid, as will be clear from what we shall say.

Let NZM be a straight line given in position, N a fixed point [Fig. 2] on it. Let NZ be one unknown quantity A, and the segment ZI, applied to it at given angle NZI, be equal to the other unknown quantity E. When D times A is equal to B times E, the point I will describe a straight line given in position, since B is to D as A is to E.[5] Hence the ratio of A to E is given, and, since the angle at Z is given, the form of the triangle NIZ, and with it the angle INZ, is given. But the point N is given and the straight line NZ is given in position: hence NI is given in position and it is easy to make the synthesis [*compositio*].

[4] Aristaeus flourished about the end of the the the fourth century B.C.

[5] D in A aequetur B in E, ut B ad D, ita A ad E. Let $A = x = NZ$, $E = y = ZI$; then $Dx = By$, or $B : D = x : y$, which is the equation of line NI.

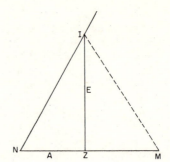

Fig. 2

To this equation all equations can be reduced of which the terms [*homogenea*] are partly given, partly mixed with the unknowns A and E, either multiplied with the given quantities, or appearing simply. Let Z pl $- D$ times A equal B times E. Let D times R be Z pl. Then we will find that B is to D as $R - A$ is to E. Let us take MN equal to R, then point M is given, hence MZ is equal to $R - A$. Hence the ratio of MZ to ZI is known, but the angle at Z is given, hence also the form of the triangle IZM. We conclude that the straight line MI is given in position. Thus point I will be on a straight line given in position.[6] We reach the same result without difficulty for any equation containing the terms A and E.

This is the first and simplest equation of a locus, by means of which all the loci dealing with a straight line can be found; see, for example, the seventh proposition of Book I of Apollonius *On plane loci*, which has since found a more general formulation and construction. This equation also leads to the following elegant proposition, which we discovered with its help:

Let any number of straight lines be given in position. From some point draw to them straight lines at given angles. If the sum of the products of the lines thus drawn with the given lines is equal to a given area, then the point will be on a straight line given in position.[7]

[6] When $Z - Dx = By$ (Z, Dx, By are rectangles), then if $Z = DR$ (R a line), $D(R - x) = By$, or $B : D = (R - x) : y$; Z pl, we have seen, means Z is a plane (area).

[7] Fermat was one of the mathematicians who tried to reconstruct Apollonius' book *On plane loci* with the aid of the detailed accounts of it preserved by Pappus. Fermat's reconstruction is in the *Oeuvres*, I, 3–51. Proposition 7 is: "If through two given points at a given angle two lines are led in given ratio, and the endpoint of one line stays on a plane locus [hence a straight line or circle] given in position, then this will be the same for the endpoint of the other."

Hence, if A and B (Fig. 3) are the given points, AH and BD are drawn at the given angle AKB, and AH/BD is in the given ratio, then, if H moves on line HG, D moves on the straight line DE. Here $AG \perp HG$, $BE \perp ED$. This follows from $AH/BD = AG/BE$.

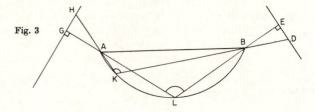

Fig. 3

We omit a great number of other propositions, which could be considered as corollaries to those of Apollonius.

The second species of equations of this kind are of the form

$$A \text{ times } E \text{ is } Z \text{ pl,}^{[8]}$$

in which case point I traces a *hyperbola*. Draw NR parallel to ZI; through any point, such as M, on the line NZ, draw MO parallel to ZI. Construct the rectangle NMO equal in area to Z pl. Through the point O, between the asymptotes NR, NM, describe a hyperbola; its position is determined and it will pass through point I, since we have assumed, as it were, AE—that is to say, the rectangle NZI—equivalent to the rectangle NMO [Fig. 4].

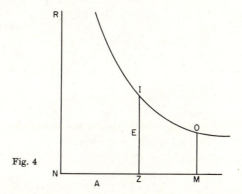

Fig. 4

To this equation we may reduce all those whose terms are in part constant, or in part contain A or E or AE.

If we let

$$D \text{ pl} + A \text{ times } E \text{ equal } R \text{ times } A + S \text{ times } E,$$

then we obtain, by well-known methods,

$$R \text{ times } A + S \text{ times } E - A \text{ times } E \text{ equal } D \text{ pl.}$$

Let us construct a rectangle on two sides such that the terms R times $A + S$ times $E - A$ times E are contained in it; then the two sides will be $A - S$ and $R - E$ and the rectangle on them will be equal to R times $A + S$ times $E -$ A times $E - R$ times E.

If now we subtract R times S from D pl, then the rectangle on $A - S$ and $R - E$ will be equal to D pl $- R$ times S.

Take NO equal to S and ND, parallel to ZI, equal to R. Through point D, draw DP parallel to NZ; through point O, draw OV parallel to ND; prolong ZI to P [Fig. 5].

[8] $xy = Z$.

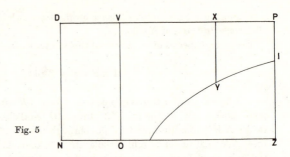

Fig. 5

Since $NO = S$ and $NZ = A$, we have $A - S = OZ = VP$. Similarly, since $ND = ZP = R$ and $ZI = E$, we have $R - E = PI$. The rectangle on PV and PI is therefore equal to the given area D pl $- R$ times E. The point I is therefore on a hyperbola having PV, VO as asymptotes.[9]

If we take any point X, the parallel XY, and construct the rectangle VXY, and through point Y we describe a hyperbola between the asymptotes PV, VO, it will pass through point I. The analysis and construction are easy in every case.

The next species of equations for loci arises if we have A^2 equal to E^2, or in given ratio to E^2, or, again if $A^2 + A$ times E is in given ratio to E^2. Finally this type includes all the equations whose terms are of the second degree containing either A^2, E^2, or the rectangle on A and E. In all these cases the point I traces a *straight line*, as can easily be demonstrated.[10]

Other cases leading to a straight line are $(x^2 + xy) : y^2 = a^2 : b^2$ (Fermat considers only positive values). Then it is shown that the cases which we write in the form $x^2 = ay$, $y^2 = ax$, $b^2 - x^2 = ay$, $b^2 + x^2 = ay$ all lead to parabolas. Then follows $b^2 - x^2 = y^2$, which leads to a circle, as well as $b^2 - 2ax - x^2 = y^2 + 2cy$.

But if $Bq - Aq$ is to Eq in a given ratio, then the point I will be on an *ellipse*.

Let MN be equal to B, and let an ellipse be described with M as vertex, NM as diameter, and N as center, of which the ordinates [*applicatae*] are parallel to the straight line ZI. The squares of the ordinates must have a given ratio to the rectangle formed by the segments of the diameter. Then the point I will be on this ellipse. Indeed, the square on NM — the square on NZ is equal to the rectangle formed by the segments of the diameter [Fig. 6].[11]

---

[9] Let $D + xy = Rx + Sy$, then $Rx + Sy - xy = D$, or $(x - S)(R - y) = D - RS$, reducible to the case of note 6.

[10] $x^2 = y^2$, $x^2 : y^2 = a^2 : b^2$, $(x^2 + xy) : y^2 = a^2 : b^2$ all lead to straight lines.

[11] This is the case $(b^2 - x^2) : y^2 = p^2 : q^2$. If $MN = b = M'N$, $NZ = x$, then $(b - x)(b + x)/y = $ const. $= (MZ : M'Z)/IZ$. This way of defining an ellipse is found, for instance, in Archimedes.

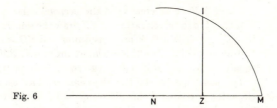

Fig. 6

To this equation can be reduced all those in which Aq is on one side of the equation and Eq with an opposite sign and a different coefficient on the other side. If the coefficients are the same and the angle a right angle, the locus will be a circle, as we have said. If the coefficients are the same, but the angle is not a right angle, the locus will be an ellipse.

Moreover, though the equations include terms which are products of A or E by given magnitudes, the reduction may nevertheless be made by the artifice which we have already employed.

When $(b^2 + x^2):y^2$ is a given ratio, I lies on a hyperbola. Then follows "the most difficult of all equations," which contains not only x^2 and y^2, but also xy. Fermat analyzes the case $b^2 - 2x^2 = 2xy + y^2$, which, as he shows, represents an ellipse.

Finally Fermat returns to Apollonius' book on plane loci, and at the end solves one more problem on loci:

A single example will suffice to indicate the general method of construction. Given two points N and M, required the locus of the points such that the sum of the squares of IN, IM shall be in a given ratio to the triangle INM [Fig. 7].

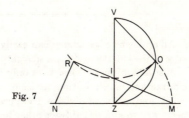

Fig. 7

Let $NM = B$, let E be the line ZI drawn at right angles to NM, and let A be the distance NZ. According to well-known methods we find that A^2 bis $+ B^2 - B$ times A bis $+ E^2$ bis is to rectangle B times E in a given ratio.[12] Following in treatment the procedures previously explained we have the suggested construction.

[12] $NM = B$, $ZI = E = y$, $NZ = A = x$. Then $(2x^2 + B^2 - 2Bx + 2y):By = $ const.

Bisect NM at Z; erect at Z the perpendicular ZV; make the ratio $4ZV$ to NM equal to the given ratio. On VZ draw the semicircle VOZ, inscribe ZO equal to ZM, and draw VO. With V as center and VO as radius draw the circle OIR. If from any point R on this circle, we draw RN, RM, I say that the sum of the squares of RN and RM is in the given ratio to the triangle RNM.

The constructions of the theorems on loci could have been much more elegantly presented if this discovery had preceded our previous revision of the two books *On plane loci*. Yet, we do not regret this work, however precocious or insufficiently ripe it may be. In fact, there is for science a certain fascination in not denying to posterity works that are as yet spiritually incomplete; the labor of the work, at first simple and clumsy, gains strength as well as stature through new inventions. It is quite important that the student should be able to discern clearly the progress which appears veiled as well as the spontaneous development of the science.

---

4 DESCARTES. THE PRINCIPLE OF NONHOMOGENEITY

Descartes, as we have seen (Selections II.7, 8), presented his application of algebra to geometry in the *Géométrie*, published in 1637 as Appendix I to his *Discours de la méthode*. We present here the beginning of Book I of this *Géométrie*, where Descartes explains his principle of nonhomogeneity, based on the proportions $1 : a = a : a^2$, $a : a^2 = a^2 : a^3$, and so on, which leads him to a notation that is substantially the same as we use in our modern theory of equations, in which we have no hesitation in writing, say, $x^3 + ax^2 + bx + c = 0$ instead of $x^3 + ax^2 + b^2x + c^3 = 0$, and use x, y, z for the unknowns, a, b, c for the given quantities. Descartes then applied his reformed algebra to the geometry of the Ancients, which led to coordinate geometry. Our translation is based on the same Smith-Latham text on which Selection II.8 is based. The title of Book I is *Problems the construction of which requires only straight lines and circles*.

All problems in geometry can easily be reduced to such terms that a knowledge of the lengths of certain straight lines is sufficient for their construction.

Just as arithmetic consists of only four or five operations, namely, addition, subtraction, multiplication, division, and the extraction of roots, which may be considered a kind of division, so in geometry, to find required lines it is merely necessary to add or subtract other lines; or else, taking one line which I shall call the unit in order to relate it as closely as possible to numbers, and which can in general be chosen arbitrarily, and having given two other lines, to find a fourth line which shall be to one of the given lines as the other is to the unit (which is the same as multiplication); or, again, to find a fourth line which is to one of the given lines as the unit is to the other (which is equivalent to division); or, finally, to find one, two, or several mean proportionals between the unit and some other line (which is the same as extracting the square root, cube root, etc., of the given line). And I shall not fear to introduce these arithmetical terms into geometry, for the sake of greater clarity.

For example, let AB [Fig. 1] be taken as the unit, and let it be required to multiply BD by BC. I have only to join the points A and C, and draw DE parallel to CA; then BE is the product of BD and BC.

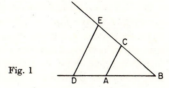

Fig. 1

If it be required to divide BE by BD, I join E and D, and draw AC parallel to DE; then BC is the result of the division.

Or, if the square root of GH [Fig. 2] is desired, I add, along the same straight line, FG equal to the unit; then, bisecting FH at K, I describe the circle FIH about K as a center, and draw from G a perpendicular and extend it to I, and GI is the required root. I do not speak here of cube root, or other roots, since I shall speak more conveniently of them later.

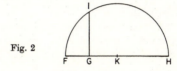

Fig. 2

Often it is not necessary thus to draw the lines on paper, but it is sufficient to designate each by a single letter. Thus, to add the lines BD and GH, I call one a and the other b, and write $a + b$. Then $a - b$ will indicate that b is subtracted from a; ab that a is multiplied by b; a/b^1 that a is divided by b; aa or a^2 that a is multiplied by itself; a^3 that this result is multiplied by a, and so on, indefinitely. Again, if I wish to extract the square root of $a^2 + b^2$, I write $\sqrt{a^2 + b^2}$; if I wish to extract the cube root of $a^3 - b^3 + abb$, I write $\sqrt[3]{a^3 - b^3 + abb},^2$ and similarly for other roots. Here it must be observed that by a^2, b^3, and similar expressions, I ordinarily mean only simple lines, which, however, I name squares, cubes, etc., so that I make use of the terms employed in algebra.

It should also be noted that all parts of a single line should as a rule be expressed by the same number of dimensions, when the unit is not determined in the problem. Thus, a^3 contains as many dimensions as abb or b^3, these being the component parts of the line which I have called $\sqrt[3]{a^3 - b^3 + abb}$. It is not, however, the same thing when the unit is determined, because it can always be understood, even where there are too many or too few dimensions; thus, if it be

---

[1] Descartes writes $\dfrac{a}{b}$.

[2] Descartes writes $\sqrt{C.a^3 - b^3 + abb}$.

required to extract the cube root of $a^2b^2 - b$, we must consider the quantity a^2b^2 divided once by the unit, and the quantity b multiplied twice by the unit.

Finally, so that we may be sure to remember the names of these lines, a separate list should always be made as often as names are assigned or changed. For example, we may write $AB = 1$, that is AB is equal to 1; $GH = a$, $BD = b$, and so on.[3]

If, then, we wish to solve any problem, we first suppose the solution already effected,[4] and give names to all the lines that seem needful for its construction—to those that are unknown as well as to those that are known. Then, making no distinction between known and unknown lines, we must unravel the difficulty in any way that shows most naturally the relations between these lines, until we find it possible to express a single quantity in two ways. This will constitute what we call an equation, since the terms of one of these two expressions are equal to those of the other. And we find as many such equations as there are supposed to be unknown lines; but if, after considering everything involved, so many cannot be found, it is evident that the question is not entirely determined. In such a case we may choose arbitrarily lines of known length for each unknown line to which there corresponds no equation.

If there are several equations, we must use each in order, either considering it alone or comparing it with the others, so as to obtain a value for each of the unknown lines; and so we must combine them until there remains a single unknown line which is equal to some known line, or whose square, cube, fourth power, fifth power, sixth power, etc., is equal to the sum or difference of two or more quantities, one of which is known, while the others consist of mean proportionals between unity and this square, or cube, or fourth power, etc., multiplied by other known lines. I express this as follows:

$$z = b,$$

or

$$z^2 = -az + bb,$$

or

$$z^3 = az^2 + bbz - c^3,$$

or

$$z^4 = az^3 - c^3z + d^4, \text{ etc.}$$

That is, z, which I take for the unknown quantity, is equal to b; or, the square of z is equal to the square of b diminished by the square of z, plus the square of b multiplied by z, diminished by the cube of c; and similarly for the others.

Thus, all the unknown quantities can be expressed in terms of a single quantity, whenever the problem can be constructed by means of circles and straight lines, or also by conic sections, or even by some other curve only one or two degrees more composed.

But I shall not stop to explain this in more detail, because I should deprive you of the pleasure of mastering it yourself, as well as of the advantage of training your mind by working over it, which is in my opinion the principal benefit to be derived from this science. Because I find nothing here so difficult

---

[3] Descartes writes $AB \backsim 1$, etc.; see Selection II.8, note 3.

[4] This is the "analysis" of Pappus, see Selection II.5.

that it cannot be worked out by any one at all familiar with ordinary geometry and with algebra, who will consider carefully all that is set forth in this treatise.

That is why I shall content myself with the statement that if the student, in solving these equations, does not fail to make use of division wherever possible, he will surely reach the simplest terms to which the problem can be reduced.

And if it can be solved by ordinary geometry, that is, by the use of straight lines and circles traced on a plane surface, when the last equation shall have been entirely solved there will remain at most only the square of an unknown quantity, equal to the product of its root by some known quantity, increased or diminished by some other quantity also known.[5] Then this root or unknown line can easily be found. For example, if I have $z^2 = az + bb$, I construct [Fig. 3] a right triangle NLM with one side LM, equal to b, the square root of the

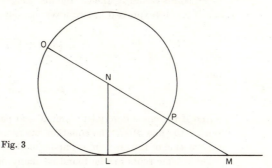

Fig. 3

known quantity bb, and the other side, LN, equal to $\frac{1}{2}a$, that is to half the other known quantity which was multiplied by z, which I suppose to be the unknown line. Then prolonging MN, the hypotenuse [*la baze*] of this triangle, to O, so that NO is equal to NL, the whole line OM is the required line z. It is expressed in the following way:

$$z = \tfrac{1}{2}a + \sqrt{\tfrac{1}{4}aa + bb}.$$

But if I have $yy = -ay + bb$, where y is the quantity whose value is desired, I construct the same right triangle NLM, and on the hypotenuse MN lay off NP equal to NL, and the remainder PM is y, the desired root. Thus I have

$$y = -\tfrac{1}{2}a + \sqrt{\tfrac{1}{4}aa + bb}.$$

In the same way if I had

$$x^4 = -ax^2 + b^2,$$

PM would be x^2 and I should have

$$x = \sqrt{-\tfrac{1}{2}a + \sqrt{\tfrac{1}{4}aa + bb}},$$

and so for other cases.

[5] Hence $z^2 = az \pm b$.

Finally, if I have $z^2 = az - bb$, I make NL equal to $\frac{1}{2}a$ and LM equal to b as before; then [Fig. 4], instead of joining the points M and N, I draw MGR parallel to LN, and with N as a center describe a circle through L cutting MGR

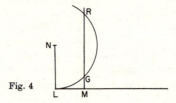

Fig. 4

in the points G and R; then z, the line sought, is either MG or MR, for in this case it can be expressed in two ways, namely,

$$z = \tfrac{1}{2}a + \sqrt{\tfrac{1}{4}aa - bb}$$

and

$$z = \tfrac{1}{2}a - \sqrt{\tfrac{1}{4}aa - bb}.$$

And if the circle described about N and passing through L neither cuts nor touches the line MGR, the equation has no root, so that we may say that the construction of the problem is impossible.[6]

These same roots can be found by many other methods. I have given these very simple ones to show that it is possible to construct all the problems of ordinary geometry by doing no more than the little covered in the four figures that I have explained. This is one thing which I believe the ancient mathematicians did not observe, for otherwise they would not have put so much labor into writing so many books in which the very sequence of the propositions shows that they did not have a sure method of finding them all, but rather gathered those propositions on which they had happened by accident.

Then Descartes takes up Pappus' problem (see Selection III.3) for the case of four lines. Here he introduces (Fig. 5) two segments AB and BC which he calls $AB = x$, $BC = y$. At this point he introduces what later would be called (oblique) coordinates. The four lines are AB, AD, EF, GH, to which at given angles are drawn from C the lines CB, CD, CF, and CH. He expresses the locus as an equation of the second degree. After some remarks on the case of more than four lines, he goes on to Book II (see Selection III.5).

---

[6] Descartes here follows the common practice of his day, which considered only the types $z^2 + az - b^2 = 0$, $z^2 - az - b^2 = 0$, and $z^2 - az + b^2 = 0$ of quadratic equations, ignoring the type $z^2 + az + b^2 = 0$, since it has no positive roots (a is a segment, hence positive). Only much later (Newton) did mathematicians begin to associate coordinates with negative numbers. All coordinates in Descartes are positive. The name "coordinate" does not appear in Descartes; this term is due to Leibniz.

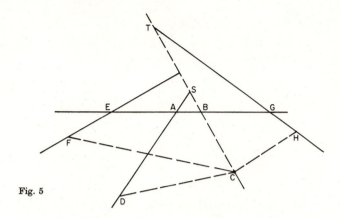

Fig. 5

5 DESCARTES. THE EQUATION OF A CURVE

In his *Géométrie* of 1637, Descartes applied his reformed algebra (see Selection II.7) to the geometry of the Ancients. In Book I he applies his coordinate method to Pappus' problem (see the previous Selection). The required locus can then be expressed by a relation between two variables which he denotes by x and y and in which we recognize oblique "Cartesian" corodinates.

"Since there is only one condition to be expressed . . . we may give any value we please to either the one or the other of the unknown quantities x or y, and find the value of the other from this equation. It is evident that when no more than five lines are given, the quantity x, which is not used to express the first of the lines, can never be of degree [*dimension*] higher than the second. Assigning thus a given value to y, we have only $x^2 = \pm\, ax \pm b^2$ [*il ne restera que* $xx \infty + ou - ax + ou - bb$], and therefore the quantity x can be found with ruler and compasses, by a method already explained" (Smith and Latham, *The geometry of René Descartes*, p. 34; see Selection II.8).

Then, in Book II, after a classification of the problems of geometry into plane, solid, and linear ones (according to Pappus; see Selection III.2). Descartes suggests that a further classification of these "linear" curves is desirable, but that the classical distinction between geometrical and mechanical curves does not seem justified, since circles and straight lines can also be considered instruments [*machines*]. He then discusses some of these mechanical ways of describing a curve, and gives ((pp. 49–55 of the Smith-Latham translation) the following example of his coordinate method:

I wish to know the *genre*[1] of the curve *EC* [Fig. 1], which I imagine to be described by the intersection of the ruler *GL* and the rectilinear plane figure *CNKL*, whose side *KN* is produced indefinitely in the direction of *C*, and which,

[1] Earlier in Book II, Descartes has defined the *genre* of a curve. In our terms: If an algebraic curve has degree $2n - 1$ or $2n$, its *genre* is n. This terminology may have been inspired by the problem of Pappus. Newton (see Selection III.8) translates *genre* by *genus*.

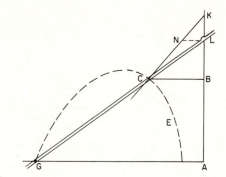

Fig. 1

being moved in the same plane in such a way that its side KL always coincides with some part of the line BA (produced in both directions), imparts to the ruler GL a rotary motion about G (the ruler being so connected to the figure $CNKL$ that it always passes through L).[2] If I wish to find out to what *genre* this curve belongs, I choose a straight line, as AB, to which to refer all its points, and in AB I choose a point like A at which to begin the calculation. I say that I choose the one and the other, because we are free to choose them as we like, for while it is necessary to use care in the choice in order to make the equation as short and simple as possible, yet no matter what line I should take instead of AB the curve would always prove to be of the same *genre*, a fact easily demonstrated.

Then I take on the curve an arbitrary point, as C, at which I will suppose that the instrument to describe the curve is applied. Then I draw through C the line CB parallel to GA. Since CB and BA are unknown and indeterminate quantities, I shall call one of them y and the other x. But in order to find the relation between these quantities I consider also the known quantities which determine the description of the curve, as GA, which I shall call a; KL, which I shall call b; and NL, parallel to GA, which I shall call c. Then I say that as NL is to LK, or as c is to b, so CB, or y, is to BK, which is therefore equal to $\frac{b}{c}y$. Then BL is equal to $\frac{b}{c}y - b$, and AL is equal to $x + \frac{b}{c}y - b$. Moreover, as CB is to LB, that is, as y is to $\frac{b}{c}y - b$, so AG or a is to LA or $x + \frac{b}{c}y - b$. Multiplying the second by the third, we get $\frac{ab}{c}y - ab$ equal to

$$xy + \frac{b}{c}yy - by,$$

[2] The instrument thus consists of three parts: (1) a ruler AK of indefinite length, fixed in the plane; (2) a ruler GL, also of indefinite length, passing through a pivot G in this plane (but not on AK); and (3) a triangle LNK, KN indefinitely extended toward KC, to which the ruler GL is connected at L so as to make the triangle slide with its side KL along AB.

which is obtained by multiplying the first by the last. Therefore, the required equation is

$$yy = cy - \frac{cx}{b}\,y + ay - ac.$$

From this equation we see that the curve EC belongs to the first *genre*, it being, in fact, a hyperbola.

If in the instrument used to describe the curve we substitute for the straight line CNK this hyperbola or some other curve of the first *genre* lying in the plane $CNKL$, the intersection of this curve with the ruler GL will describe, instead of the hyperbola EC, another curve, which will be of the second *genre*.

Thus, if CNK be a circle having its center at L, then we shall describe the first Conchoid of the Ancients,[3] while if we use a parabola having KB as diameter we shall describe the curve which, as I have already said, is the first and simplest of the curves required in the problem of Pappus, that is, the one which furnishes the solution when five lines are given in position.[4]

Then Descartes continues with his solution of the problem of Pappus, which leads him to the consideration of conic sections and other curves with several types of equations, such as

$$y^2 = 2y - xy + rx - x^2,$$

$$y^3 - 2ay^2 - a^2y + 2a^3 = axy,$$

$$x^2 = ry - \frac{r}{q}\,y^3,$$

$$y^2 - by^2 - cdy + bcd + dxy = 0.$$

Here also is Descartes's method of finding the equation of a normal to a curve. This method was in a sense opposed to that of Fermat, whose method was based on finding first the equation of a tangent to a curve (see Selection IV.8) and thus came close to the idea of a derivative.

Book III of the *Géométrie* contains algebra; see Selection II.7.

6   DESARGUES. INVOLUTION AND PERSPECTIVE TRIANGLES

Girard Desargues (1593–1662) was an architect and military engineer, who lived at Lyons, and for some time also at Paris, where he met other mathematicians, including Descartes. His work in the field of perspective, in which he derived many sweeping generalizations, was

---

[3] Pappus mentions four types of conchoid (shell curves); the first is the one we still call a conchoid, in polar coordinates $r = a + b \sec \theta$. It is a curve of the third degree, therefore of the second *genre* of Descartes.

[4] This is also a curve of the second *genre*.

little appreciated in his day, the more so since he used a peculiar technical language largely borrowed from botany. This was the time when under Descartes's and Fermat's influence algebra and infinitesimal methods were applied to geometry, so that the more general appreciation of purely geometric methods had to wait until in the nineteenth century projective geometry was developed. Many of these projective concepts are found in the *Brouillon proiect d'une atteinte aux evenemens des rencontres du Cone avec un Plan, par L. S. G. D. L.*, which "Le Sieur Girard Desargues Lyonnais" published in 1639 in 50 copies. All were lost until 1845 when Professor Michel Chasles at Paris discovered a transcript made in 1679 by the mathematician Philippe de La Hire, a pupil of Desargues. This copy was published in the *Oeuvres de Desargues*, ed. N. G. Poudra (2 vols.; Leiber, Paris, 1864), I, 103–230. Around 1950 an original copy was found in the Bibliothèque Nationale in Paris, which has been utilized in the edition of the *Brouillon proiect* in R. Taton, *L'Oeuvre mathématique de G. Desargues* (Presses Universitaires, Paris, 1951). The central concept of the *Brouillon proiect* is that of involution, and of all the technical terms introduced by Desargues this is the only one that has survived.

Three pairs of points on a line in involution had already been discussed in Pappus' *Mathematical collection*, in his commentary on a lost book by Euclid, *On porisms*. Whatever Pappus' influence on Desargues may have been, Desargues's exposition remains highly original. Here follow some parts leading up to involution and some applications. It may help to translate some of Desargues's terms into modern language.

Ordinance of straight lines (Ordonnance de lignes droites)	Pencil of lines
Top of an ordinance (But d'une ordonnance)	Top of pencil
Trunk (Tronc)	Straight line with points on it
Knots (Noeuds)	Points on a line (also: ends of a segment) through which pass other lines
Branch (Rameau)	Each of these lines, hence ray
Twig (Brin)	Segment on one of the lines
Engaged or disengaged common point (Point commun engagé ou dégagé)	Point on or outside the segment
Point pair mixed with another pair (Points d'un couple mêlés aux points d'un autre couple)	Point pairs on a line forming overlapping segments
Unmixed (Démêlés)	When the segments do not overlap
Rectangle of segments (Rectangle de pièces)	Product of segments
Rectangles relative to each other (Rectangles relatifs entre eux)	When there are 4 points $AA'BB'$ on a line: $A'B\,A'B'$ and $AB\,AB'$
Twin rectangles (Rectangles gémeaux entr'eux)	When there are 6 points, $AA'BB'CC'$ on a line: Products as $A'B\,A'B'$ and $A'C\,A'C'$

Some definitions follow in Desargues's language. Two pairs of points CG and DF are given on a line.

Points of a pair mixed with those of another pair. To indicate the way the points of one of these couples are situated with respect to the points of the other couple [Fig. 1]: When one of the points of one pair C is between and its mate

Fig. 1
```
F    C      D           G
F      C    G           D
F      D    C           G
```

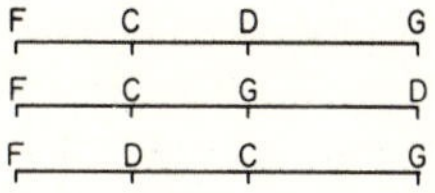

(accouplé) G is outside the points of the other pair, then we say here that the points of one of the pairs are *mixed* with the points [*meslez aux poincts*] of the other pair . . .

Points of a pair unmixed with the points of another pair. When the points of one pair lie both similarly either between or outside the points of the other pair, then we say here that the points of one pair are *unmixed* with the points of the other pair.

Tree. When there exists on a straight line AH [Fig. 2] a point A common to and similarly engaged or disengaged to the two segments of each of the three

Fig. 2
```
B   D   G   A   H F         C
```

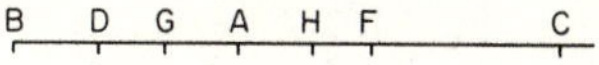

pairs AB, AH; AC, AG; AD, AF, of which the rectangles are all equal, then such a condition on a straight line is here called a *Tree* [*Arbre*], of which the line itself is the *Trunk* [*Tronc*].[1]

Stump. The point, such as A, common to each of the six segments AB, AH, AC, AG, AD, AF is called the *Stump* [*Souche*].

Branch. Each of these same six segments AB, AH, AC, AG, AD, AF is called a *Branch* [*Branche*].

Involution. And when on a straight line AH [Fig. 3] there exist three pairs of points BH, CG, DF situated in such a way that the two points of each of the

Fig. 3
```
A G   D   B H      F          C
```

pairs are either both mixed or unmixed with the two points of the other pairs, and such that the rectangles thus relative to the segments between these points are in the same ratio to each other as their twins, taken in the same order, are in ratio to each other; then such a disposition of the three pairs of points on a straight line is here called *Involution*.[2]

---

[1] This is the first definition of involution, with the aid of the stump (central point, *souche*) A, as follows: $AB \times AH = AC \times AG = AD \times AF$.

[2] Two pairs as GD, GF, or GB, GH, where D and F, B and H, each belong to the same pair and G to another, are called *twins* [*couples de brins gemelles entre elles*]. Hence this definition of involution is that $\dfrac{CB \times CH}{GB \times GH} = \dfrac{CD \times CF}{GD \times GF} = \dfrac{AC}{AG}$, where A is the central point. This second definition, independent of the central point A, is equivalent to the modern way of saying that the cross ratios (GC, AB) and (GC, FH) are equal. Desargues gives a rather complicated proof of the identity of the two definitions.

Desargues now studies the elliptic and hyperbolic cases and then introduces the harmonic set, which he calls four points in involution.

Four points in involution. We may conceive the words *four points in involution* to express two cases [*événements*] of the same sort, since one or the other of these two events results: the first where four points on a line each at a finite distance yield three consecutive segments of which either end segment is to the middle one as the sum of the three is to the other end segment, and the second in which three points are at finite distances on a line with a fourth point at an infinite distance, in which case the points likewise yield three segments of which one end segment is to the middle segment as the sum of the three is to the other end segment. This is incomprehensible and seems at first to imply that the three points at a finite distance yield in this case two segments that are equal to each other with the midpoint as both stump and endpoint coupled at an infinite distance.

Thus we should take careful note that a line which is bisected and then produced to infinity is one of the cases of the involution of four points.

Now follow many theorems on harmonic point sets, in rather complicated form. Then, after having introduced pencils of rays in involution, and having demonstrated what we call the property of projectivity of an involution, Desargues continues with metric properties of such pencils.

Mutually corresponding rays [*Rameaux correspondans entre eux*]. In the case of but four points *B, D, G, F* [Fig. 4] in involution on a line through which there

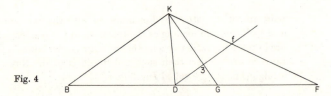

Fig. 4

pass four rays *BK, DK, GK, FK* radiating in a pencil from the point *K*, the pairs of rays *DK, FK* or *GK, BK* that pass through the corresponding points *D, F* or *B, G* are here called corresponding rays.

In this case, when the two corresponding rays *BK, GK* are perpendicular to each other, they bisect each of the angles between the other two corresponding rays *DK, FK*.

Since the line *Df* was drawn parallel to any one of the rays *BK* which is perpendicular to its corresponding ray GK, the line Df is also perpendicular to the ray GK.

Furthermore, because of the parallelism of BK and Df, the ray GK bisects Df at the point 3.

Thus the two triangles $K3D$, $K3f$ each have a right angle at the point 3, and they have the sides $3K$, $3D$, and $3K$, $3f$ which include the equal angles $K3D$, $K3f$ equal to each other.

Since the two triangles K3D, K3f are equal and similar, the ray GK bisects one of the angles DKF between the corresponding rays DK, FK, and the ray BK clearly bisects the other of the angles made by the same corresponding rays DK, FK.

When any one of the rays GK bisects one of the angles DKF between the two other mutually corresponding rays DK, FK, this ray GK is perpendicular to its corresponding ray BK which also bisects the other angle included by the corresponding rays DK, FK.

This is true since when the line Df was drawn perpendicular to any ray GK, the two triangles $K3D$ and $K3f$ each have a right angle at the point 3 and furthermore each has an equal angle at the point K and also a common side $K3$, consequently they are similar and equal and the ray GK bisects Df at the point 3.

Therefore, the ray BK is parallel to the line Df and it also is perpendicular to the corresponding ray GK.

When in a plane, there is a pencil of four lines BK, DK, GK, FK from the vertex K, and when two of these lines as BK and GK are mutually perpendicular and bisect each of the angles which the two others FK, DK make, it follows that these four lines cut any other line $BDGF$ lying in their plane in four points B, D, G, F which are arranged in involution.

When a line FK in a plane bisects one of the sides Gb of the triangle BGb at f, and when through the point K which is thus determined on one of the other two sides Bb there passes another line KD parallel to the bisected side Gb, the four points B, D, G, F determined by this construction on the third side BG of the triangle are in involution.[3]

When from the angle B which subtends the bisected side Gb there passes another line Bp parallel to the bisected side Gb, the four points F, f, K, p determined on the line FK by the three sides BG, Gb, Bb of the triangle BGb and the line BP are themselves in involution.

And in the second case, by drawing the line Bf in a way similar to the line GB, the three points G, f, b at a finite distance and the point at infinite distance are in involution, since to these pass four branches of a pencil whose vertex is at B and which consequently determines on the line FK four points F, f, K, p in involution.

Then as the line FGB cuts on the line bf a segment Gf equal to the segment bf, the side of a triangle such as bfK, this is equivalent to saying that this line FGB is double the side bf of the triangle bfK.

When in a plane a line FGB is double one of the sides bf of a triangle bfK, and when from the point B which it determines on either side bK of the two other

---

[3] Construction of the fourth ray of a harmonic pencil.

sides of the same triangle, there passes a line Bp parallel to the double side bf, then this construction yields on the third side Kf of the triangle bfK, four points F, f, K, p which are in involution.

This is evident when the line BF is drawn.

When from the angle K subtended by the double side bf there passes a line KD parallel to the double side bf, this construction gives on the line FB four points F, B, D, G which are in involution, as is evident when the line KG is drawn.

This theory swarms with similar means of deciding when four or three pairs of points are in involution on a line, but these are sufficient to open the mine of that which follows.

The last part of the *Brouillon* deals mainly with the polar properties of conics, obtained by the general intersection of a cone or cylinder (combined in the term *rouleau*, in accordance with Desargues's unifying principle, so that conics are *coupes de rouleau*).

Another outstanding contribution of Desargues's is found in an appendix of three propositions to a book published by his pupil, the master craftsman Abraham Bosse, called *Manière universelle de M. Desargues pour pratiquer la perspective* (Paris, 1648), 340–343.[4] The first contains the theorem on perspective triangles called after Desargues. The text can be found in the *Oeuvres de Desargues*, I, 430–433, or in Taton's book, pp. 206–212. The theorem is stated in the following words:

When the lines HDa, HEb, EDc, lag, flb, HKl, Dg, EKf [Fig. 5] either in different planes or in the same plane, meet in any order or direction [*biais*] whatsoever, and in similar points,[5] then the points c, f, g lie in one line cfg. For whatever form the figure takes, and in all cases, the lines being in different planes, [say] *bac, lag, clb* in one plane and EDc, KDg, EKf in another, then the points c, f, g are in each of the two planes, hence they are in one straight line cfg. And when the lines are in the same plane

Fig. 5

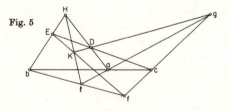

---

[4] The title of this book in the *Oeuvres* of 1864 is incorrect; see Taton, p. 67.

[5] Such as Da, Eb, Kl meeting at H.

the theorem is proved by an argument involving repeated application of the theorem of Menelaus concerning a transversal in a triangle. Then the converse theorem is maintained, both for different planes and for the same plane. The argument, which is not very clear, can be studied in English translation in Smith, *Source book*, 307–311, if it is understood that Desargues's notation

$$cD - cE \begin{cases} gD - gK \\ \\ fK - fE \end{cases}$$

means $\dfrac{cD}{cE} = \dfrac{gD}{gK} \times \dfrac{fK}{fE}$ and expresses the application of Menelaus' theorem to transversal *fcg* of triangle *DEK*.

In a final remark Desargues observes that the figure formed by two perspective triangles in space is transformed by oblique parallel projection on the plane of one of the triangles into two perspective triangles of the plane. The two figures correspond "line to line, point to point, and reasoning to reasoning . . . and the properties of the figures can be discussed from either figure."

7 PASCAL. THEOREM ON CONICS

Blaise Pascal (1623–1662) was the son of the mathematically gifted official Etienne Pascal, and was introduced as a boy by his father to the circle of savants around Father Marin Mersenne, to which Descartes and Desargues belonged. After talking to Desargues, and probably reading his *Brouillon proiect* the 16-year-old Blaise published a "petit placard en forme d'affiche" (handbill) containing his *Essay pour les coniques* (Paris, 1640), which announces the "theorem of Pascal" concerning the so-called "hexagrammum mysticum." Like Desargues's *Brouillon*, it had only a small circulation; at present only two copies are known. Pascal uses some of the original terms of Desargues, such as *ordonnance de lignes* for *pencil of lines*.

The text is published in *Oeuvres de Pascal*, ed. L. Brunschwicg and P. Boutroux (Paris, 1908), I, 243–260, with a reproduction of the "handbill"; also in R. Taton, *L'Oeuvre mathématique de G. Desargues* (Presses Universitaires, Paris, 1951), 190–194. See also R. Taton, "L' 'Essay pour les coniques' de Pascal," *Revue d'Histoire des Sciences et de leurs Applications 8* (1955), 1–18. We base our translation on that in Smith, *Source book*, 326–330.

ESSAY ON CONICS

First Definition. When several straight lines meet at the same point, or are parallel to each other, all these lines are said to be of the same order or of the same pencil [*ordonnance*], and the totality of these lines is termed an order of lines or a *pencil* of lines.

Definition II. By the expression "conic section," we mean the circumference of the circle, the ellipse, the hyperbola, the parabola, and the rectilinear angle;

since a cone cut parallel to its base, or through its vertex, or in the three other directions which produce respectively an ellipse, a hyperbola, and a parabola, produces in the conic surface, either the circumference of a circle, or an angle, or an ellipse, a hyperbola, or a parabola.

Definition III. By the word line used alone, we mean a straight line.[1]

Lemma I. If in the plane M, S, Q [Fig. 1] two straight lines MK, MV are drawn from point M and two lines SK, SV from point S; and if K is the point

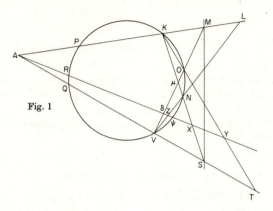

Fig. 1

of intersection of the lines MK, SK; V the point of intersection of the lines MV, SV; A the point of intersection of the lines MA, SA; and μ the point of intersection of the lines MV, SK; and if through two of the four points A, K, μ, V, which do not lie in the same line with points M, S, such as the points K, V, a circle passes cutting the lines MV, MK, SV, SK at points O, P, Q, N, then I say that the lines MS, NO, PQ are of the same order.[2]

Lemma II. If through the same line several planes are passed, and are cut by another plane, all lines of intersection of these planes are of the same order as the line through which these planes pass.

On the basis of these two lemmas and several easy deductions from them, we can demonstrate that if the same things are granted as for the first lemma, that is, through points K, V, any conic section whatever passes cutting the lines

---

[1] "Par le mot de droite mis seul, nous entendons ligne droite." The first definition establishes the projective equivalence of intersection and parallel lines, and is thus based on Desargues's concept of points at infinity. Similarly, the second definition establishes that any plane intersection of an (oblique) cone is a conic section. Pascal, again following Desargues, breaks with the Apollonian tradition of the "triangle par l'axe," in which the only sections of an (oblique) cone considered are in planes perpendicular to the triangle formed by a plane through the "axis."

[2] Remembering that "are of the same order" means "pass through the same finite or infinite point," we recognize "Pascal's theorem" for a circle. Notice that for Pascal his theorem is only a lemma, which indeed it was; he used it to find many other properties. We can only conjecture how Pascal proved his lemma, but it is likely that, like Desargues, for his theorem, he used the theorem of Menelaus.

MK, MV, SK, SV in points P, O, N, Q, then the lines MS, NO, PQ will be of the same order. This constitutes a third lemma.[3]

By means of these three lemmas and certain deductions therefrom, we propose to give a complete text on the elements of conics,[4] that is to say, all the properties of diameters and other straight lines, of tangents, etc., the construction of the cone from substantially these data, the description of conic sections by points, etc.

Having done this, we shall state the properties which follow, doing this in a more general manner than usual. Take, for example, the following: If in the plane MSQ, in the conic PKV, there are drawn the lines AK, AV, cutting the conic in points P, K, Q, V, and if from two of these four points that do not lie in the same line with point A—say the points K, V—and through two points N, O, taken on the conic, there are produced four lines KN, KO, VN, VO, cutting the lines AK, AV, at points L, M, T, S, then I maintain that the proportion composed of the ratios of the line PM to the line MA, and of the line AS to the line SQ, is the same as the proportion composed of the ratio of the line PL to the line LA, and of the line AT to the line TQ.[5]

We shall also demonstrate [Fig. 2] that if there are three lines DE, DG, DH that are cut by the lines AP, AR at points F, G, H, C, γ, B and if the point E be

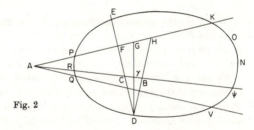

Fig. 2

fixed in the line DC, [a] the proportion composed of the ratios of the rectangle $EF.FG$ to the rectangle $EC.C\gamma$, and of the line $A\gamma$ to the line AG, is the same as the ratio of the rectangle $EF.FH$ to the rectangle $EC.CB$, and of the line AB to the line AH.[6] [b] The same is also equal to the ratio of the rectangle $FE.FD$ to the rectangle $CE.CD$. Consequently, if a conic section passes through

---

[3] This is the extension of Pascal's theorem to any conic, including the degenerate case (hence "Pascal's theorem" in modern axiomatics, the special case already known to Pappus).

[4] We know that Pascal (*Oeuvres*, II, 220) worked on this treatise, and Leibniz in a letter of 1676 reported on his study of the manuscript. After that time it disappeared.

[5] This theorem is equivalent to the statement that the cross ratios ($ALMP$) and ($ASTQ$) are equal.

[6] [a] $\dfrac{EF \times FG}{EC \times C\gamma} \times \dfrac{A\gamma}{AG} = \dfrac{EF \times FH}{EC \times CB} \times \dfrac{AB}{AH}.$

This is Pappus' theorem; in modern terms: the four rays of the pencil $D(A, C, \gamma, B)$ cut out from two intersecting lines point ranges with equal cross ratios (the term "cross ratio" is due to W. K. Clifford, 1869).

[b] This ratio $\dfrac{EF \times FH}{EC \times CB} \times \dfrac{AB}{AH}$ is also equal to $\dfrac{FE \times FD}{CE \times CD}.$

(footnote continued)

the points E, D, cutting the lines AH, AB in points P, K, R, ψ, [c] the proportion composed of the ratio of the rectangle of these lines EF, FG, to the rectangle of the lines EC, $C\gamma$, and of the line γA to the line AG, will be the same as the ratio of the rectangle of the lines FK, FP to the rectangle of the lines CR, $C\psi$, and of the rectangle of the lines AR, $A\psi$ to the rectangle of the lines AK, AP.

We can also show that if four lines AC, AF, EH, EL [Fig. 3] intersect in points N, P, M, O, and if a conic section cuts these lines in points C, B, F, D,

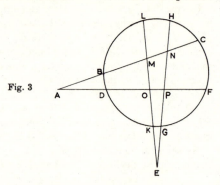

Fig. 3

H, G, L, K, the proportion consisting of the ratios of rectangle $MC.MB$ to rectangle $PF.PD$, and of rectangle $AD.AF$ to rectangle $AB.AC$, is the same as the proportion composed of the ratios of rectangle $ML.MK$ to the rectangle $PH.PG$, and of rectangle $EH.EG$ to rectangle $EK.EL$.[7]

We can also demonstrate a property stated below, due to M. Desargues of Lyons, one of the great minds of this time and one of the best versed in mathematics, particularly in conics, whose writings on this subject, although few in number, give abundant proof of his knowledge to those who seek for information. I should like to say that I owe the little that I have found on this subject to his writings, and that I have tried to imitate his method, as far as possible, in which he has treated the subject without making use of the triangle through the axis.[8]

Giving a general treatment of conic sections, the following is the remarkable property under discussion: If in the plane MSQ [Fig. 1] there is a conic section

---

This can be reduced to Menelaus' theorem for the triangle:

$$FH \times AB \times CE = AH \times CB \times DF,$$

also used by Desargues.

[c] $\dfrac{EF \times FG}{EC \times C\gamma} \times \dfrac{\gamma A}{AG} = \dfrac{FK \times FP}{Cr \times C\psi} = \dfrac{Ar \times A\psi}{AK \times AP}$.

The last identity expresses a relation between the segments that the conic intercepts on the sides of the triangle AFC.

[7] $\dfrac{MC \times MB}{PF \times PD} \times \dfrac{AD \times AF}{AB \times AC} = \dfrac{ML \times MK}{PH \times PG} \times \dfrac{EH \times EG}{EK \times EL}$.

This formula establishes the relation between the segments a conic section cuts out on the sides of quadrilateral $MNOP$.

[8] See note 1.

PQN, on which are taken four points K, N, O, V from which are drawn the lines KN, KO, VN, VO, in such a way that through the same four points only two lines may pass, and if another line cuts the conic at points R, ψ, and the lines KN, KO, VN, VO in points X, Y, Z, δ, then as the rectangle $ZR \cdot Z\psi$ is to the rectangle $\gamma R \cdot \gamma \psi$, so the rectangle $\delta R \cdot \delta \psi$ is to the rectangle $XR \cdot X\psi$.[9]

We can also prove that if in the plane of the hyperbola, the ellipse, or the circle AGE [Fig. 4] of which the center is C the line AB is drawn touching the

Fig. 4

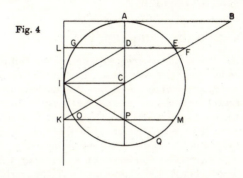

section at A, and if having drawn the diameter we take line AB such that its square shall be equal to one-fourth of the rectangle of the figure and if CB is drawn, then for any line such as DE, parallel to line AB and cutting the section in E and the lines AC, CB in points D, F, the property holds that if the section AGE is an ellipse or a circle, the sum of the squares of the lines DE, DF will be equal to the square of the line AB;[10] and in the hyperbola, the difference between the same squares of the lines DE, DF will be equal to the square of the line AB.

We can also deduce from this several problems; for example:

From a given point to draw a tangent to a given conic section;

To find two diameters that meet at a given angle;

To find two diameters cutting at a given angle and having a given ratio.

There are many other problems and theorems, and many deductions which can be made from what has been stated above, but the distrust which I have owing to my little experience and capacity, does not allow me to go further into the subject until it has passed the examination of able men who may be willing to take this trouble. After that if someone thinks the subject worth continuing, we shall endeavor to extend it as far as God gives us the strength.

At Paris, M.DC.XL.

---

[9] This is Desargues's theorem on the involution cut out by a pencil of conics on a transversal.

[10] $AB^2 = DE^2 + DF^2$, equal to the square of the minor axis, determines the ellipse. This equality is equivalent to the equation of the ellipse in its modern form, $x^2/a^2 + y^2/b^2 = 1$, and is related to propositions of Apollonius and Desargues. In the case of a circle, "the fourth of the rectangle of the figure" is the square of the radius.

The "petit placard" of Pascal suffered the same fate as the "brouillon" of Desargues: it soon got lost. Huygens wrote in 1656 to P. de Carcavy that he had never seen these documents. Leibniz had a look at Pascal's many notes, and referred in his own notes to the *Essay*, but the notes of Pascal are now lost. The first to rediscover Pascal's theorem was Colin Maclaurin (1727); the first to republish it, now with a demonstration, was William Braikenridge, like Maclaurin a resident of Edinburgh, in his *Exercitatio geometrica de descriptione linearum curvarum* (London, 1733), using what we now call projective pencils. On Maclaurin see Selection V.13.

8 NEWTON. CUBIC CURVES

Newton's *Enumeratio linearum tertii ordinis* was first published together with his *Opticks* in London in 1704, but it was written much earlier, perhaps in or after 1676. It represents a definite step ahead in the development of analytic, or algebraic, geometry. Where algebraic curves of higher degree than two had so far been studied only in certain special forms, such as the cissoid, or the folium of Descartes, Newton undertook the classification of all curves of the third degree; moreover, he freely used (a novelty at the time) positive and negative values of the coordinates. He was the first to study the character of these curves by means of their points of intersection with a straight line. The exposition is brief, and is hard to follow; James Stirling, as early as 1717, published a commentary. There exists an English translation by C. R. M. Talbot, *Sir Isaac Newton's "Enumeration of lines of the third order"* (Bohn, London, 1860), from which we here reproduce some sections. Talbot adds a very detailed commentary.

The general idea of Newton's work can be summed up as follows. First, certain affine properties of conics, such as asymptotes and diameters, are extended to third-order curves. Then the curve is studied with the aid of a set of oblique coordinate axes. If we write the equation of the curve in the form

$$a_0y^3 + (b_0 + b_1x)y^2 + (c_0 + c_1x + c_2x^2)y + (d_0 + d_1x + d_2x^2 + d_3x^3) = 0,$$

then, if $b_0 = b_1 = 0$, the x-axis is a "diameter," that is, the locus of the centers of gravity of the three points of intersection on each ordinate.

We further observe that there exists at least one (real) asymptote. If the x-axis is taken parallel to this asymptote, then $a_0 = 0$. The locus of the midpoints of chords parallel to this asymptote is the hyperbola with the equation (taking $b_1 \neq 0$)

$$2y(b_0 + b_1x) + c_0 + c_1 + c_2x^2 = 0.$$

This curve has two asymptotes, which we can select as x- and y-axes. Then the equation of the hyperbola will be of the form $Axy + B = 0$, so that $b_0 = c_1 = c_2 = 0$, and the cubic curve will have the equation

$$b_1xy^2 + c_0y + d_0 + d_1x + d_2x^2 + d_3 = 0,$$

in which we recognize Case I of Newton. There are three other cases, which Newton enumerates, but he does not prove that four exhausts the number of cases. This was done by François Nicole, *Histoire et mémoires de l'Académie de Paris* (1729), 194. Further classification of cubic equations leads Newton to 72 species, of which the last is the cubic parabola $y = ax^3 + bx^2 + dx + d$. The principle of classification based on the study of the roots of equations obtained by taking one of the axes parallel to an asymptote is rather arbitrary, and many other classifications have been given; for instance, Julius Plücker, *System der analytischen Geometrie* (Duncker, Berlin, 1835), 220–241, found 219 types.

ENUMERATION OF LINES OF THE THIRD ORDER

SECTION I. THE ORDER OF LINES

Geometrical lines[1] are best divided into orders, according to the dimensions of the equation expressing the relation between absciss and ordinate, or, which is the same thing, according to the number of points in which they can be cut by a straight line. So that a line of the first order will be a straight line; those of the second or quadratic order will be conic sections and the circle; and those of the third or cubic order will be the cubic Parabola, the Neilian Parabola, the Cissoid of the ancients, and others we are about to describe. A curve of the first genus (since straight lines are not to be reckoned among curves) is the same as a line of the second order, and a curve of the second genus is the same as a line of the third order.[2] And a line of the *infinite order* is one which a straight line may cut in an infinite number of points, such as the spiral, cycloid, quadratrix, and every line generated by the infinitely continued rotations of a radius.

SECTION II. THE PROPERTIES OF CONIC SECTIONS ARE ANALOGOUS TO THOSE OF CURVES OF HIGHER ORDERS

The chief properties of conic sections have been much treated of by geometers, and the properties of curves of the second and higher genera are very similar to them, as will be shown in the following enumeration of their principal properties:

1. *Of Curves of the Second Genus, Their Ordinates, Diameters, Vertices, Centres, and Axes.* If parallel straight lines terminated by the curve be drawn in a conic section, the straight line bisecting two of them will bisect all the others, and is called a diameter, and the bisected lines are called ordinates to the diameter, and the intersection of all the diameters is the centre; the intersection of the diameter with the curve is called the vertex; that diameter, whose ordinates are rectangular to it, being called the axis. In like manner, in curves of the second genus, if any two parallel straight lines are drawn, meeting the curve in three points; the straight line which cuts these parallel lines so that the sum of the two segments meeting the curve on one side of the secant, will cut in the same ratio all lines parallel to these, provided they also meet the curve in three

---

[1] Descartes, in his *Géométrie*, Book II, calls curves that can be expressed by a single equation (which in his case is always an algebraic equation), "courbes géométriques"; see D. E. Smith and M. L. Latham, trans., *The Geometry of René Descartes* (Open Court, Chicago, London, 1925; Dover, New York, 1954), 48.

[2] This division of geometrical curves into curves of genus [*genre*] 1 (conic sections), genus 2 (third- and fourth-degree curves), genus 3 (curves of degree 5 and 6) can also be found in Descartes, on the same page as referred to in note 2; see Selection III.4, note 1.

points; that is, so that the sum of the two parts on one side of the secant, shall equal the third part on the other side. These three parts, thus equal, may be called ordinates, and the secant or cutting line to which the ordinates are applied, the diameter; the intersection of diameter and curve, the vertex; and the intersection of two diameters, the centre. The diameter having rectangular ordinates, if any exist, may also be called an axis; and where all the diameters meet in a point, that point will be the general centre.

2. *Of Asymptotes and Their Properties.* A hyperbola of the first genus will have two asymptotes; that of the second genus will have three; of the third, four, and no more; and so on for the rest. And as the segments of any straight line intercepted between the conic hyperbola and its asymptotes on each side are equal, so in hyperbolas of the curve and its three asymptotes in three points, the sum of those two segments of the secant, which are drawn from any two asymptotes on the same side to two points of the curve, will be equal to the third part, which is drawn from the third asymptote on the contrary side, to the third point in the curve.[3]

3. *Of Latera Recta and Transversa.* And in like manner, as in the nonparabolic conic sections, the square of the ordinate—i.e., the rectangle contained by the ordinates on each side of the diameter[4]—is to the rectangle of the segments of the diameter of the ellipse and hyperbola, terminated at the vertices; as a certain given line called the latus rectum, to the part of the diameter which lies between the vertices, and is called the latus transversum; so in nonparabolic curves of the second genus, the product of three ordinates, is to the product of the abscisses of the diameter between the ordinates and the three vertices of the curve, in a given ratio; in which ratio, if three lines are taken to three segments of the diameter between the vertices of the curve, each to each, then these three lines may be considered as the latera recta of the curve, and the three segments of the diameter between the vertices as its latera transversa. And as in the conic parabola, which has but one vertex to a diameter, the rectangle under the ordinates, is equal to the rectangle under the absciss between the vertex and ordinates, and a given straight line called the latus rectum: so in curves of the second genus, which have only two vertices to the same diameter, the product of the three ordinates is equal to the product of the two parts of the diameter cut off between the ordinates and the two vertices, and a certain given straight line, which may be called the latus rectum.[5]

Paragraph 4 deals with the ratio of the products of the segments of parallel lines.

---

5. *Of Hyperbolic and Parabolic Branches, and Their Directions.* All infinite branches of curves of the second and higher genus, like those of the first, are

---

[3] This can be proved with the aid of a diameter. For a full explanation see Talbot, *Newton's "Enumeration,"* p. 37.

[4] It will be remembered that "the rectangle contained by two lines x and y" means the product xy.

[5] See Talbot, *Newton's "Enumeration,"* pp. 38–40 for full documentation.

either of the hyperbolic or parabolic sort. I define a hyperbolic branch as one which constantly approaches some asymptote, a parabolic branch to be that which, although infinite, has no asymptote. These branches are easily distinguished by their tangents; for supposing the point of contact to be infinitely distant, the tangent of the hyperbolic branch will coincide with the asymptote, but the tangent of the parabolic branch being at an infinite distance, vanishes, and is not to be found. The asymptote to any branch is, therefore, found by seeking for the tangent to a point in that branch at an infinite distance. The direction of the branch may be found, by determining the position of a straight line parallel to the tangent referred to a point in the curve infinitely distant; for such straight line will have the same direction as the infinite branch itself.

SECTION III. THE REDUCTION OF ALL CURVES OF THE SECOND GENUS TO FOUR CASES OF EQUATIONS

All lines of the first, third, fifth, seventh, or odd orders, have at least two infinite branches extending in opposite directions; and all lines of the third order have two branches of the same kind, proceeding in opposite directions, towards which no other of their infinite branches proceed (except only the Cartesian parabola).

Case I (Fig. 1). If the branches be hyperbolic, let GAS be their asymptote, and let CBc be any line drawn parallel to it, meeting the curve on each side (if

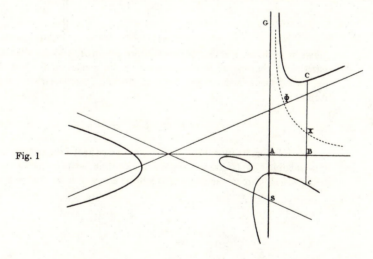

Fig. 1

possible); let this line be bisected in X, which will be the locus of a hyperbola, say $X\phi$, one of whose asymptotes is AG. Let its other asymptote be AB, and the equation defining the relation between absciss AB and ordinate BC, if $AB = x$, $BC = y$, will always be of the form

$$xy^2 + ey = ax^3 + bx^2 + cx + d,$$

where the terms b, c, d, a, e, designate given quantities, affected by their proper signs $+$ or $-$, of which any may be deficient, so that the figure, by reason of their absence, be not changed to a conic section. It may be, however, that this conic hyperbola coincides with its asymptotes; that is, the point X may fall in the straight line AB; and in that case the term $+ey$ is deficient.

Case II (Fig. 2). But if the straight line CBc is not bounded by the curve at each end, but only meets the curve in one point, draw AB any straight line

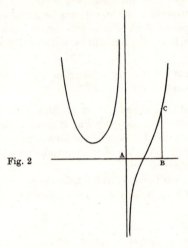

Fig. 2

given in position, meeting the asymptote AS in A, and draw another line BC parallel to that asymptote, and meeting the curve in C; then the equation expressing the relation of the ordinate BC and absciss AB always assumes the form

$$xy = ax^3 + bx^2 + cx + d.$$

Case III (Fig. 3). But if the opposite branches are of the parabolic sort, let the straight line CBc be drawn, if possible, meeting each branch of the curve, and being bisected in B, the locus of B will be a straight line. Let this straight line

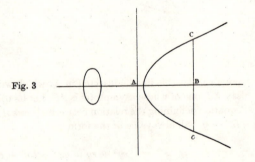

Fig. 3

be AB, terminating at any point A, then the equation expressing the relation of ordinate BC and absciss AB always assumes the form

$$y^2 = ax^3 + bx^2 + cx + d.$$

Case IV (Fig. 4). But when CBc only meets the curve in one point, and, therefore, cannot be bounded at both ends, let that one point be C; and at the point

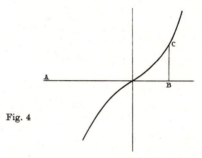

Fig. 4

B let CBc meet another straight line given in position, AB, and terminating at any point A, then the equation expressing the relation between ordinate BC and absciss AB always assumes the form

$$y = ax^3 + bx^2 + cx + d.$$

The Names of the Curves. In the enumeration of these cases of curves, we shall call that which is included within the angle of the asymptotes in like manner as the hyperbola of the cone, the *inscribed* hyperbola; that which cuts the asymptotes, and includes within its branches the parts of the asymptotes so cut off, the *circumscribed* hyperbola; that which, as to one branch, is inscribed, and, as to the other, circumscribed, we shall call the *ambigenous* hyperbola; that which has branches concave to each other and proceeding towards the same direction, the *converging* hyperbola; that which has branches convex to each other, and proceeding towards contrary directions, the *diverging* hyperbola; that which has branches convex to contrary parts and infinite towards contrary sides, the *contrary branched* hyperbola; that which, with reference to its asymptote, is concave at the vertex, and has diverging branches, the *conchoidal* hyperbola; that which cuts the asymptote in contrary flexures, having on both sides contrary branches, the *serpentine* hyperbola; that which intersects its conjugate, the *cruciform* hyperbola; that which intersects and returns in a loop upon itself, the *nodate* hyperbola; that which has two branches meeting at an angle of contact, and there stopping, the *cusped* hyperbola; that which has an infinitely small conjugate oval, i.e., a conjugate point, the *punctate* hyperbola; that which, from the impossibility of two roots, has neither oval, node, cusp, nor conjugate point, the *pure* hyperbola.

In the same way we shall speak of parabolas, as *converging, diverging, contrary branched, cruciform, nodate, cusped, punctate,* and *pure.*

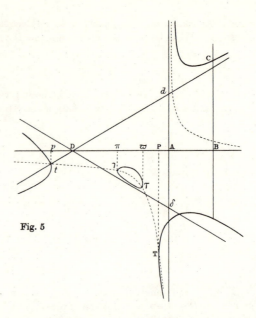

Fig. 5

In the case of the first-mentioned equation, if ax^3 is positive (Fig. 5), the figure will be a triple hyperbola with six hyperbolic branches progressing to infinity alongside of three asymptotes, no one of which is parallel to another, two alongside of each hyperbola, on contrary sides. And these asymptotes, if the term bx^2 is not deficient, will cut one another at three points, making a triangle ($Dd\delta$); but if the term bx^2 is deficient, all the points will converge to one

Fig. 6

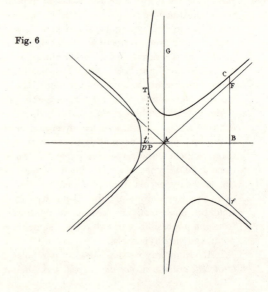

point. In the former case, take $AD = -b/2a$ and $Ad = A\delta = b/2\sqrt{a}$, join Dd, $D\delta$; and Ad, Dd, $D\delta$ will be the three asymptotes. In the latter case (Fig. 6) draw any ordinate BC parallel to the principal ordinate AG, and in it produced in each direction take BF, Bf, equal to each other, and in the ratio to AB of $\sqrt{a}:1$, join AF, Af; and AG, AF, Af will be the three asymptotes. This hyperbola we call *redundant*, because it exceeds the conic hyperbola in the number of its hyperbolic branches.

Then follows a paragraph on diameters.

SECTION IV. THE ENUMERATION OF CURVES. OF THE NINE REDUNDANT HYPER-
BOLAS, HAVING NO DIAMETER, AND THREE ASYMPTOTES, MAKING A TRIANGLE

When a redundant hyperbola has no diameter (Fig. 5), let the four roots be found of the equation $ax^4 + bx^3 + cx^2 + dx + e^2/4 = 0$. Erect the ordinates PT, $\ddot{\omega}\tau$, πl, pt which are tangent to the curve at the points T, τ, l, t, and by touching give the limits of the curve by which this species is determined. Indeed, if all the roots AP, $A\ddot{\omega}$, $A\pi$, AP (Figs. 5, 7), are real and unequal, and of

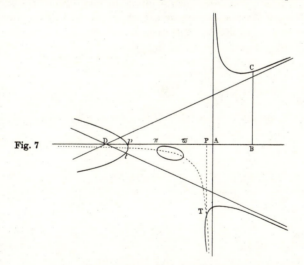

Fig. 7

the same sign, the curve consists of three hyperbolas (inscribed, circumscribed, and ambigenous), and of an oval. One hyperbola lies towards D, another towards d, and the third towards δ; and the oval always lies within the triangle $Dd\delta$ within the limits τl, in which it is touched by the ordinates πl and $\ddot{\omega}\tau$.

This is the 1st species.

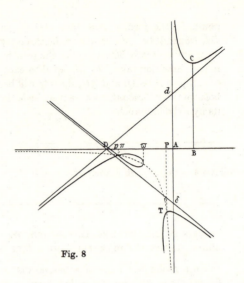

Fig. 8

If the two greatest roots $A\pi$, AP (Fig. 8), or the two least AP, $A\bar{\omega}$ (Fig. 9), equal to each other, and all of the same sign, the oval and circumscribed hyperbola will coalesce, their points of contact l and t, or T and τ, coming together, and the branches of the hyperbola, intersecting one another, run on into the oval, making the figure nodate.

This is the 2d species.

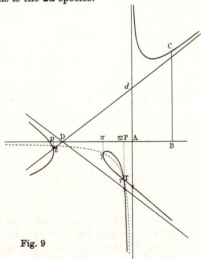

Fig. 9

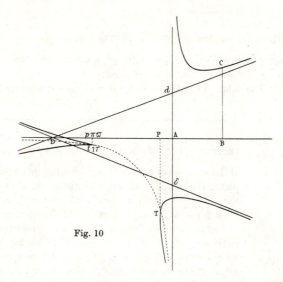

Fig. 10

If the three greatest roots AP, $A\pi$, $A\omega$ (Fig. 10), or the three least roots $A\pi$, $A\tilde{\omega}$, AP (Fig. 11), are equal to each other, the node becomes a sharp cusp; because the two branches of the circumscribed hyperbola meet at an angle of contact, and extend no farther.

This is the 3d species.

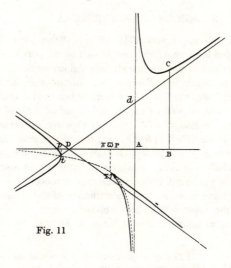

Fig. 11

The total number of species is 72. The last six are

(a) The five divergent parabolas $y^2 = ax^3 + bx^2 + ca + d$, classified according to the roots of $ax^3 + bx^2 + ax + d = 0$, and

(b) The cubic parabola $y = ax^3 + bx^2 + cx + d$ (species 72).

SECTION V. THE GENERATION OF CURVES BY SHADOWS[7]

If the shadows of curves caused by a luminous point, be projected on an infinite plane, the shadows of conic sections will always be conic sections; those of curves of the second genus will always be curves of the second genus; those of the third genus will always be curves of the third genus; and so on *ad infinitum*.

And in the same manner as the circle, projecting its shadow, generates all the conic sections, so the five divergent parabolas, by their shadows, generate all other curves of the second genus. And thus some of the more simple curves of other genera might be found, which would form all curves of the same genus by the projection of their shadows on a plane.

There follow a paragraph on double points, Section VI on "The organic description of curves," and Section VII on "The construction of equations by the description of curves."

9 AGNESI. THE VERSIERA

Maria Gaetana Agnesi (1718–1799) of Milan, sister of the composer Maria Teresa Agnesi, occupied for a time (with the consent of the pope) a chair at the university of Bologna until she retired to devote herself to religious work. Her *Instituzioni analitiche* (Milan, 1748; French translation, Paris, 1775; English translation, London, 1801), in two beautiful volumes, was the first comprehensive textbook on the calculus after L'Hôpital's early book. This book, as well as the *Traité du calcul intégral* (2 vols.; Paris, 1754, 1756) by Louis Antoine de Bougainville, the later explorer, was soon superseded by Euler's great texts on the calculus (1755–1770). Among the many curves discussed by Agnesi we find the *versiera*, $xy^2 = a^2(a - x)$. The word is derived from Latin *vertere*, to turn, but is also an abbreviation of Italian *avversiera*, female devil. Some wit in England once translated it "witch," and the silly pun is still lovingly preserved in most of our textbooks in the English language.

We have taken Agnesi's introduction of the versiera as one of our selections to honor the first important woman mathematician since Hypatia (fifth century A.D.). The curve had already appeared in the writings of Fermat (*Oeuvres*, I, 279–280; III, 233–234) and of

---

[7] This short paragraph on the *genesis curvarum per umbras* contains Newton's theorem that every curve of the third order can be obtained from a divergent parabola by central projection from one plane on another. For this theorem see also the twenty-second lemma of Newton's *Principia* and its first book, Props. 25, 26.

others; the name *versiera* is from Guido Grandi.[1] We translate the text of Book I, pp. 380–382, where it appears as Problem III of a set which in Problem I introduces the cissoid, in Problem II the curve $y^2(2ax - x^2) = ax - x^2$, and in Problem IV the conchoid. Cissoid and conchoid are curves known from antiquity.

Problem III. Given the semicircle ADC with AC as diameter [Fig. 1], to find outside of it the point M such that MB, the line orthogonal to the diameter AC, intersecting the circle in D, makes the ratio $AB:BD$ equal to that of $AC:BM$, and as an infinite number of such points M can be found, satisfying that condition, we inquire after their locus.

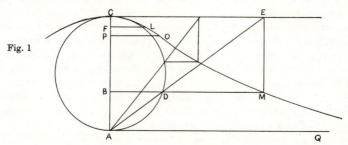

Fig. 1

Let M be one of these points, and let $AC = a$, $AB = x$, $BM = y$, then, because of the properties of the circle $BD = \sqrt{ax - xx}$, and because of the above condition, we shall have $AB:BD = AC:BM$; i.e., $x:\sqrt{ax - xx} = a:y$; and therefore $y = \dfrac{a\sqrt{ax - xx}}{x}$, or $y = a\sqrt{\dfrac{a - x}{x}}$ is the equation of the required curve, which is called the *Versiera*.

Since $AB = x$, $BM = y$, AC is the x-axis and AQ, parallel to BM, is the axis of the ordinate y. If we set first $x = 0$, then $y = \infty$, and therefore AQ is the asymptote of the curve. If $y = 0$, then $a\sqrt{a - x} = 0$, hence $x = a$; when therefore $x = a$, the curve will touch the axis AC, and will pass through the point C, which is its vertex. If $x = \frac{1}{2}a$, then $y = a$; when $x = AP = \frac{3}{4}a$, then $y = a\sqrt{\frac{1}{3}}$; when $x = AF = \frac{4}{5}a$, then $y = a\sqrt{\frac{1}{4}}$. Suppose x is greater than a, then the expression under the radical sign becomes negative, and the curve imaginary. In order to see if the curve is concave or convex to the axis AC, we form the proportion: As $CP = \frac{1}{4}a$ (which corresponds to $x = \frac{3}{4}a$) is to $y = a\sqrt{\frac{1}{3}}$, so is $CF = \frac{1}{5}a$ (which corresponds to $x = \frac{4}{5}a$) to the fourth, which will be $\frac{4}{5}a\sqrt{\frac{1}{3}}$; but $x = \frac{4}{5}a$ gives us $y = a\sqrt{\frac{1}{4}}$, and $\frac{4}{5}a\sqrt{\frac{1}{3}}$ is smaller than $a\sqrt{\frac{1}{4}}$; therefore the curve will be concave to the AC axis;[2] but toward the asymptote AQ it must also be convex, so it will be partly concave and partly convex and therefore will

[1] G. Grandi, *Quadratura circuli et hyperbolae* (Pisa, 1703). The curve is type 63 in Newton's classification (Selection III.8). The first to use the term "witch" in this sense may have been B. Williamson, *Integral calculus*, 7 (1875), 173; see *Oxford English Dictionary*.

[2] This reasoning is not very convincing, but Agnesi promises to find the point of inflection later; see the next footnote.

have a point of inflection [*flesso contrario*], which can be found by a method to be given in due time.[3] Since, if we take x negative, the term under the radical sign in the denominator becomes negative, y must be imaginary and the curve will have the shape shown in the figure, which indicates that it has a branch similar and equal to the branch CLM in the part of negative y.[4]

10 CRAMER AND EULER. CRAMER'S PARADOX

Gabriel Cramer (1704–1752) was a Swiss mathematician who gave lessons at Geneva. He was the editor of Jakob Bernoulli's *Opera* (Geneva, 1744) and Johann Bernoulli's *Opera omnia* (Geneva, 1742), and the author of the *Introduction à l'analyse des lignes courbes algébriques* (Geneva, 1750), a large volume containing the most complete exposition of algebraic curves existing at that time, going far beyond Newton's *Enumeratio*. It discusses many properties of asymptotes and the singularities of algebraic curves, deals with special curves, and also contains "Cramer's paradox," which follows here. In an appendix Cramer gives the general rule for the solution of n linear equations with n unknowns (in his case $n = 3$), which can be regarded as a contribution to the theory of determinants, although Cramer has no special notation; he only describes the composition of the numerator and the denominator of the solution. "Cramer's paradox" states that an algebraic curve C_n of degree n is not always uniquely determined by $\frac{1}{2}n(n + 3)$ points—$\frac{1}{2}n(n + 3)$ being the number of the coefficients of its general equation minus 1—since for $n > 2$ this number is not larger than the number n^2 of the intersections of the C_n with another C_n. For instance, a C_3 is not always uniquely determined by 9 points. The "paradox" had already appeared in Colin Maclaurin's *Geometria organica* (London, 1720), and in Euler (see below). The general rule for the solution of n linear equations with n unknowns was already known to Leibniz (and to the Japanese mathematician Seki Kōwa, or Seki Takakusu, 1603).

In Chapter III Cramer argues that an equation of order v has in general $\frac{1}{2}vv + \frac{3}{2}v$ coefficients (one of which can always be made 1); he then reasons as follows:

38. From this it follows that it is possible, in general [*régulièrement*], to pass a curve of order v through $\frac{1}{2}vv + \frac{3}{2}v$ given points, or that a curve of order v is determined, and its equation given, when $\frac{1}{2}vv + \frac{3}{2}v$ points are given through which it must pass.

A line of the first order is thus determined by two given points, a curve of the second order by 5, one of the third by 9, one of the fourth by 14, one of the fifth by 20, etc.

---

[3] On p. 561, Book II, Agnesi finds by taking $ddy = 0$ that the point $x = \frac{3}{4}a$, $y = a\sqrt{\frac{1}{3}}$ is a point of inflection. The supposition $ddy = 0$ gives only $x = 0$ and $x = a$. There is no attempt to test whether higher derivatives are zero or different from zero. (On this see Selection V.13, Maclaurin).

[4] On pp. 392–393, Book I, Agnesi proves that we also obtain the versiera when we draw an arbitrary line BDM parallel to CE, draw ADE, and draw EM parallel to CA; then M is a point of the versiera. On the versiera see further G. Loria, *Bibliotheca mathematica 11* (1897), 7–12; (ser. 3) *3* (1902), 127–130; G. Eneström, *ibid. 12* (1911–1912), 175–176.

Then follows an example to prove this statement by means of five points through which a conic has to pass. This leads to five linear equations for the five coefficients of $A + By + Cx + Dyy + Exy + xx = 0$. To solve these equations Cramer says, in a footnote, that he believes he has "found a rather easy and general rule, when there are an arbitrary number of equations and of unknowns of which none occurs to a degree higher than the first. It can be found in Appendix 1." He carries out his computation by selecting the coordinate system in a convenient way. Then follows the study of the points of intersection of a curve with a straight line, and of two curves, with examples. This leads up to Art. 46:

---

46. It has been demonstrated[1] that if there are two variables and two un-determined equations expressing the relation between these variables with constants, one of them of order m and the other of order n, then if, with the aid of these two equations, one eliminates one of these variables, the remaining equation, in its ultimate form, has at most mn dimensions. In this equation there can therefore be only mn roots at most. Hence, two algebraic curves described in the same place can meet each other at most in as many points as there are unities in the product of the numbers which are the exponents of their order.[2] For instance, a curve of the third order can meet a curve of the fourth order in at most 12 points, and a curve of the 5th order can only meet a curve of the 12th order in at most 60 points.

47. This principle seems at first to be in contradiction to that of art. 38. It is always possible to describe a curve of the second order through five points, whatever be the position of these five points. If three of them are on a straight line, then this line will cut in three points the curve of the second order that passes through the five given points. But we have seen (art. 39 or the prec.) that a line can intersect a curve of the second order in only two points. How to recon-cile these two opposite consequences? There is only one way. That is to say that, in this case, the curve of the second order that passes through the five given points is not a curve, but the system of two lines of which one is the same that passes through the three points that were given as lying in a straight line, and the other passes through the two remaining points. Computation will confirm the truth of this reconciliation.

Here follows an example in coordinates and a general conclusion, quoting Maclaurin, *Geometria organica*, p. 137.[3]

[1] [Footnote by Cramer] This principle, a purely algebraic one, can be demonstrated in Algebra. As I know of nobody who has given the demonstration, I have found it necessary to present one in Appendix 2.

[2] [Footnote by Cramer] Mr. Maclaurin has demonstrated the same thing, but I do not believe that his demonstration has been published. See Philos. Trans. 39, p. 143 (1732).

[3] On this book see C. Tweedie, "The 'Geometria organica' of Colin Maclaurin," *Proceedings of the Royal Society, Edinburgh, 36* (1915–16), 87–150, with a paraphrase of the contents. The "paradox" is §64, Corollary 2 of Section V. See also C. Tweedie, "A study of the life and writings of Colin Maclaurin," *Mathematical Gazette 8* (1915–16), 132–151; *9* (1919), 303–305.

If, in the number $\frac{1}{2}vv + \frac{3}{2}v$ of points through which one can and wants to pass a curve of order v, there are more than tv of these points that are on a curve of order t inferior to v, then the required curve of order v is not a unique curve, but the system of two or more curves, one of which is this same curve of order t on which there are tv given points. For otherwise two curves, one of order t, the other of order v, would intersect in more than tv points, which is impossible (art. 48).

48. Another contradiction between art. 46 and art. 38 is the following. Since a curve of order m can meet a curve of order n in only mn points, a curve of order v will meet a curve of the same order in only vv points. If therefore vv is equal to or larger than the number $\frac{1}{2}vv + \frac{3}{2}v$—which is that of the points that determine a curve of order v—then it is possible to pass more than one curve of order v through $\frac{1}{2}vv + \frac{3}{2}v$ points, which seems contrary to art. 38. Thus two curves of the third order can intersect each other in 9 points, if one takes these points so as to let a curve of third order pass through them; then it is clear that the two curves intersecting in these 9 points equally satisfy the requirement. The equation of the curve passing through these 9 points is therefore not determined. In the same way two curves of the fourth order can intersect one another in 16 points. And we have established (art. 38) that 14 points are taken among the 16 in which these two curves intersect one another, each curve satisfies the problem, which therefore is undetermined.

This contradiction is resolved by the remark at the end of art. 38: if we have as many equations as we need—speaking generally—to determine all the coefficients of the equation which is taken to represent the curve that has to pass through a certain number of given points, then it may yet happen that these coefficients remain undetermined. Then the assumed equation remains undetermined and represents an infinity of curves of the same order. From which it follows that if the 9 points through which we wish to pass a curve of the third order are such that two curves of this order can pass through them, then it will be possible to pass through these same 9 points an infinity of curves of the third order . . . This is a real paradox [*Ce qui est un véritable paradoxe*].

Euler, at about the same time, had taken up the same problem that puzzled Cramer in his paper "Sur une contradiction apparente dans la doctrine des lignes courbes," *Mémoires de l'Académie des Sciences de Berlin* (1748; publ. 1750), 219–233; *Opera omnia*, ser. I, vol. 26, 33–45. Euler discusses the same contradictions, giving some examples of k systems of k equations that do not uniquely determine their roots. The example in the case of a cubic curve is the following.

I shall consider only one case, that in which the nine points are placed in a square

$$
\begin{matrix}
a & b & c \\
d & e & f \\
g & h & i
\end{matrix}
$$

and the axis is drawn through the points d, e, f, the point e being the origin of the abscissas. Denoting the interval between two points by a, we have for the abscissa $x = 0$ three values of the ordinate [*appliquée*] y, which are 0, $+a$, $-a$, and the same three values correspond also to the abscissa $x = a$ and $x = -a$. To these values corresponds the equation

$$my(yy - aa) = nx(xx - aa),$$

where the ratio of the coefficients m and n may be anything, so that there exists an infinity of curves of the third order all passing through the given points.

It is true that this equation also contains straight lines and conic sections, since for $n = 0$ there will be three straight lines ac, df, and gi; if $m = 0$ there will be three straight lines ag, bh, and ci; if $m = n$ there will be a straight line aei and an ellipse passing through the points b, c, f, d, g, h, and if $m = -n$ the curve of the third order will be composed of a straight line ceg and an ellipse passing through the points a, b, d, f, h, i. But for all other ratios that one can impose on m and n there will always be a true curve of the third degree.

Euler mentions the theorem that a curve of degree m and one of degree n can intersect, in general, in mn points, in the sense that the number of intersections cannot be greater than mn. He gives a proof in the paper directly following the paper on the "contradiction," pp. 234–248; *Opera omnia*, ser. I, vol. 26, 46–59). He had already discussed this question in pp. 474–482 of his *Introductio in analysin infinitorum* of 1748; the opening paragraphs of the two articles are almost identical. In the paper of 1748 Euler introduces his elimination method based on the formation of the products of the differences of the roots of the two equations of degree m and n; see W. S. Burnside and A. W. Panton, *The theory of equations* (Hodges, Dublin, 1892), 348.

Cramer and Euler had corresponded on the "paradox" in letters of 1744 and 1745 (*Opera omnia*, ser. I, vol. 26, XI–XII). At that time Cramer did not fully understand the nature of the problem, but Euler explained it to him.

11 EULER. THE BRIDGES OF KÖNIGSBERG

Leibniz, in a letter to Christiaan Huygens of September 8, 1679, expressed the need he felt for a type of calculation different from the ordinary algebra (see Selection II.14). Here he mentioned a possible *analysis situs*, or *geometria situs*.

Huygens was skeptical, and Leibniz did not pursue this aspect of his philosophy very far. Nothing was published on this *geometria situs* until Euler, in 1736, used Leibniz' term to denote a topological problem, namely the "Königsberg bridge problem." This has little to do with Leibniz' mathematics of *situs*, and it is possible that Euler had only heard of Leibniz' use of the term through an oral tradition, perhaps through one of the Bernoullis. Whatever may have happened, Euler's interpretation of the term *geometria situs*, or, as we also say, *analysis situs*, has won out. It is a field that started (before Euler) as a set of disconnected theorems, puzzles, and brain teasers, until during the nineteenth century a more systematic approach emerged. Now we see it as a branch of topology.

Euler's problem of the seven bridges of Königsberg[1] was published in the *Commentarii Academiae Scientiarum Petropolitanae 8* (1736), 128–140, and can be found in the *Opera omnia*, ser. I, vol. 17, 1–10. We translate the first 13 sections, after which Euler continues with other combinations of bridges.

A German translation of the whole paper appears in A. Speiser, *Klassische Stücke der Mathematik* (Zürich and Leipzig, 1925), 127–138, and a French paraphrase in E. Lucas, *Récréations mathématiques* (2nd ed.; Paris, 1891), 21–38.

THE SOLUTION OF A PROBLEM BELONGING TO THE *GEOMETRIA SITUS*

1. Besides that part of geometry which treats of quantities and has been studied eagerly at all times, there is another, so far almost unknown, first mentioned by Leibniz, which he named *Geometria situs*. This part concerns itself with that which can be determined by position [*situs*] alone, and with the analysis of the properties of position; here quantities will be ignored and the calculus of quantities not used. But what kind of problems belong to this geometry, and what method has to be utilized for their solution, is not yet certain enough. Thus, when recently a problem was mentioned, seemingly belonging to geometry, but such that it did not call for the determination of a quantity, now admitted of a solution by the calculus of quantities, I did not hesitate to refer it to the Geometria situs, especially since in its solution position only came into consideration, whereas calculus was of no use. Hence I shall set forth the method that I discovered for the solution of such problems, to serve here as a sample of Geometria situs.

2. The problem, supposedly quite well known, was as follows: At Königsberg in Prussia there is an island *A*, called "der Kneiphof," and the river surrounding it is divided into two branches as can be seen in Fig. 1. Over the branches of this river lead seven bridges, *a*, *b*, *c*, *d*, *e*, *f*, and *g*. Now the question is whether one can plan a walk so as to cross each bridge once and not more than once. I was told that some deny this possibility, others are doubtful, but that nobody affirms it. Wherefrom I formulated the following problem, framed in a very

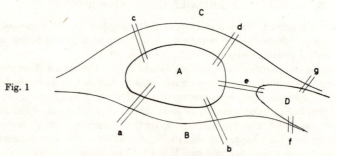

Fig. 1

---

[1] Former seaport of East Prussia, on the Pregel, now Kaliningrad.

general way for myself: Whatever the shape of the river and its division into branches may be, and whatever the number of bridges, to find out whether it is possible or not to cross each bridge exactly once.

3. As concerns the Königsberg problem of the seven bridges, it could be solved by a complete enumeration of all walks possible; then we would know if one of them fulfills the condition or none. This method, however, is, because of the great number of combinations, too difficult and cumbersome. Moreover, it could not be applied to other questions where still more bridges exist. If the investigation were to be conducted in this way, then there would be found much that was not called for at all; that is the reason, no doubt, why this way would be so arduous. That is why I dropped this method and looked for another, leading only so far that it shows whether such a walk can be found or not; for I suspected that such a method would be much simpler.

4. My whole method is based on the proper designation of the bridges, using the capitals A, B, C, D to indicate the single regions separated from each other by the river. If one thus reaches region B from region A, crossing bridge a or b, then I denote this transition by the letters AB, where the first gives the region from which the traveler comes, whereas the second gives the region where he arrives after crossing the bridge. If the traveler then goes from the region B over the bridge f into the region D, this transition is denoted by the letters BD. These two transitions AB and BD, carried out in succession, I denote by ABD only, because the middle one B indicates the region into which the first transition leads, as well as the region out of which the second transition leads.

5. In the same way, if the traveler goes from the region D over the bridge g to the region C, I denote these three successive transitions by the four letters $ABDC$. For these four letters $ABDC$ indicate that the traveler, finding himself initially in the region A, has passed into the region B, whence he proceeded into region D, and finally from there arrived at C; but as these regions are separated from each other by rivers, the traveler must necessarily cross three bridges. A crossing of four bridges is thus indicated by five letters, and then the traveler crosses an arbitrary number of bridges, then his path will be denoted by a number of letters greater by one than the number of bridges. The crossing of seven bridges requires therefore eight letters for its description.

6. By this method of description I do not pay any attention to what bridges are crossed, that is, when the transition from one region to another can be accomplished on different bridges, then it is irrelevant which one is used, as long as it leads to the region indicated. If therefore the path over the seven bridges could be planned so that it crosses each once and only once, then it could be represented by eight letters, where these letters would have to succeed one another in such a way that the immediate succession of the letters A and B appears twice; since there are two bridges a and b connecting the regions A and B; for the same reason the succession of letters A and C should also appear twice in this series of eight letters; moreoever, the succession AD as well as BD and CD must each appear once.

7. Our question is now reduced to another, namely, whether from the four letters A, B, C, and D a series of eight letters can be formed in which all these successions appear as often as is prescribed. However, before one sets out to find

such an arrangement one should better attempt to show whether such a one exists or not. For if one could show that no such arrangement is at all possible, then all the effort to find it will be useless. That is why I invented a rule, which permits one to decide in this and all similar questions without difficulty whether such an arrangement of letters is possible.

8. To find such a rule, I observe a single region A, to which an arbitrary number of bridges a, b, c, d, etc. lead. Of those bridges I first pay attention only to a. When the traveler crosses this bridge, he must either have been in A before he crossed or arrive at A after the crossing; according to our method of notation the letter A will appear twice, no matter where the path started, in A or not. And if five bridges lead to A, then in our notation crossing them all will make the letter A appear three times. And when the number of bridges is an arbitrary odd number then, by increasing the number by one and dividing by two, we obtain the number of times the letter A must appear.

9. Hence, in the case of the Königsberg bridges (Fig. 1) we have five bridges leading to the island A, namely a, b, c, d, e. The letter A must therefore appear three times in the symbol of the path. As three bridges lead to B, B must appear twice, and the same way D and C will occur twice. In the series of eight letters, which indicates the crossing of the seven bridges, A must appear three times but B, C, and D each twice; and this is in no way possible in a series of eight letters. This shows that the desired crossing of the seven Königsberg bridges cannot be carried out.

10. In a similar way one can always decide whether a path exists that leads over each bridge just once, if only the number of bridges leading to each region is odd. For such a path always exists when the number of bridges, increased by one, equals the sum of all numbers indicating how often each letter must appear. In case this sum, as in our example, is greater than the number of bridges increased by one, then such a path can in no way be laid out. The rule I gave, to determine from the number of bridges leading to A how often in the symbol of the path the letter A appears, is independent of whether all bridges, as in Fig. 2, come from a single region B or from different regions; for I consider only the region A and inquire how often the letter A must appear.

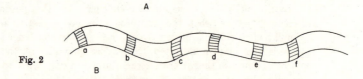

Fig. 2

11. If, however, the number of bridges leading to A is even, then one must distinguish whether or not the walk started in A. For if two bridges lead to A, and the walk starts in A, then the letter A must appear twice, once to indicate the leaving of A over one bridge, and a second time to indicate the return to A over the other bridge. But if the traveler starts his path in another region, then the letter A occurs only once, for in my notation the one appearance of A means entrance into A as well as exit out of A.

12. Let now four bridges lead into the region A, and the path begin in A. Then in the symbol of the completed path the letter A must appear three times, if he crosses each bridge only once. But if the path were to start in another region, then A would appear only twice. If six bridges lead into A, then A appears four times, if A is the initial region, otherwise only three times. And in general: if the number of bridges is even, then $\frac{1}{2}$ of it indicates how often A must make its appearance, if A is not the initial region; then one-half increased by one indicates how often A must appear if the walk starts in A.

13. As any walk has to start in some region, I define in the following way from the number of bridges leading to a region the number indicating how often the corresponding letter appears in the path symbol: If the number of bridges is odd, then I increase it by one and take half; if, however, it is even, then I take its half. If the sum of the numbers thus obtained is equal to the number of bridges increased by one, then we shall succeed in finding a path, but one must start in a region to which an odd number of bridges leads. If this sum happens to be smaller by one than the number of bridges increased by one, then the walk succeeds if one starts in a region to which an even number of bridges leads, for in this case our sum must still be increased by one.

Euler continues with a generalization to more regions A, B, C, ... and more bridges, and also deals with the question how, after it has been decided that a solution exists, the actual method of crossing the bridges can be found. Anyone who wishes to study these questions may read chap. IX of W. W. Rouse Ball, *Mathematical recreations and essays*, revised by H. S. M. Coxeter (Macmillan, New York, 1947).

CHAPTER IV ANALYSIS BEFORE NEWTON AND LEIBNIZ

During the Middle Ages mathematical meditation on the infinitely great and infinitesimally small usually took the form of speculation on ideas of Aristotle and Plato concerning the relation of point to line, the nature of the incommensurable, the paradoxes of Zeno, the existence of the indivisible, the potentially and the actually infinite. Occasionally an infinite series, or a simple "integration," appears. A good account of these speculations can be found in C. B. Boyer, *The history of the calculus and its conceptual development* (Dover, New York, 1959), chap. 3, which includes also direct quotations. On the study of Archimedes in the Middle Ages see M. Clagett, *Archimedes in the Middle Ages* (University of Wisconsin Press, Madison, 1964).

The mathematical technique of dealing with infinite processes improved greatly, when the study of Archimedes became possible on a more extensive scale. His writings first became generally accessible through the *editio princeps* of 1544 (Basel), which contained the original Greek text with a Latin translation, together with the important commentaries of Eutocius. Other, sometimes limited, editions appeared before and afterward. One of the first applications of Archimedean ideas can be found in Federigo Commandino's *Liber de centro gravitatis solidorum* (Bologna, 1565). He was followed by Stevin, Kepler, Valerio, Galileo, and later authors. Although inspired by Archimedes, these men developed their own methods of integration (differentiation came somewhat later; see Selection IV.9) and, as a rule, rejected the mathematical rigor of the Archimedean proof, which was based on a *reductio ad absurdum*. This indirect method of Archimedes (and of Euclid and other Greek authors) became known, during the seventeenth century, as the method of exhaustion, on which see, for instance, T. L. Heath, *Manual of Greek mathematics* (Clarendon Press, Oxford, 1931), 293–297. The Renaissance mathematicians were primarily out for new results, and often were not particularly worried about possible logical loopholes in their methods. They knew that their results were correct, and also knew that, if challenged, these results could be proved rigorously by Archimedean methods, that is, by showing that any supposition that the result was not true, would lead to an absurdity. But all this time-consuming indirect reasoning bored most of these authors. Wrote Johann Kepler in the preface to his *Nova stereometria doliorum* (*Opera omnia*, ed. C. Frisch (8 vols.; Meyder and Zinner, Frankfurt, Erlangen, 1858–1871), IV, 556; see Selection IV.2): "We could obtain absolute and in all respects perfect demonstrations from these books of Archimedes themselves, were we not repelled by the thorny reading thereof."

Later Christiaan Huygens (1629–1695), who had a keen sense of rigor, expressed this attitude in the following terms (*Oeuvres complètes*, XIV, 307): "In order to achieve the confidence of the experts it is not of great interest whether we give an absolute demonstration or such a foundation of it that after having seen it they do not doubt that a perfect demonstration can be given. I am willing to concede that it should appear in a clear, elegant, and ingenious form, as in all works of Archimedes. But the first and most important thing is the mode of discovery itself, which men of learning delight in knowing. Hence it seems that we must above all follow that method by which this can be understood and presented most concisely and clearly. We then save ourselves the labor of writing, and others that of reading—those others who have no time to take notice of the enormous quantity of geometrical inventions which increase from day to day and in this learned century seem to grow beyond bounds if they must use the prolix and perfect method of the Ancients." Similar statements occur in other works of this period and even later.

A penetrating study of the analysis of this period has been given by D. T. Whiteside, "Patterns of mathematical thought in the later seventeenth century," *Archive for History of Exact Sciences 1* (1961), 179–388.

We begin our selections with excerpts from Stevin and Kepler.

1   STEVIN. CENTERS OF GRAVITY

Simon Stevin (1548–1620), a Flemish-Dutch engineer and mathematician associated with Prince Maurice of Orange, was one of the first to open a new and productive period in the application of infinitesimals to mathematical problems. We show here how he determined the center of gravity of a triangle. The selection is Stevin's *Beghinselen der Weegconst* (Elements of the art of weighing; Leiden, 1585); the translation is taken (with some minor changes) from *The principal works of Simon Stevin* (Swets and Zeitlinger, Amsterdam), I (1955), 229–233, 251–255.

THEOREM II. PROPOSITION II

The center of gravity of any triangle is in the line drawn from the vertex to the middle point of the opposite side.

Supposition. Let *ABC* [Fig. 1] be a triangle of any form, in which from the angle *A* to *D*, the middle point of the side *BC*, there is drawn the line *AD*.

What is required to prove. We have to prove that the center of gravity of the triangle is in the line *AD*.

Preliminary. Let us draw *EF*, *GH*, *IK* parallel to *BC*, intersecting *AD* in *L*, *M*, *N*; after that *EO*, *GP*, *IQ*, *KR*, *HS*, *FT*, parallel to *AD*.

Proof. Since *EF* is parallel to *BC*, and *EO*, *FT* to *LD*, *EFTO* will be a parallelogram, in which *EL* is equal to *LF*, also *OD* and *DT*, in consequence of which the center of gravity of the quadrilateral *EFTO* is in *DL*, by the first proposition of this book.[1] And for the same reason the center of gravity of the

---

[1] Theorem I, Proposition I is: "The geometrical center of any plane figure is also its center of gravity." The proof is given for an equilateral triangle, a parallelogram, and a regular pentagon. The meaning of the proposition is that when a figure has a center of symmetry, it is its center of gravity. Then in Theorem II, Proposition II, Stevin discusses the case of an arbitrary triangle, and for this he needs infinitesimals.

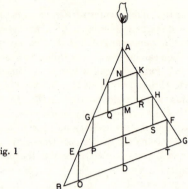

Fig. 1

parallelogram $GHSP$ will be in LM, and of $IKRQ$ in MN; and consequently the center of gravity of the figure $IKRHSFTOEPGQ$, composed of the aforesaid three quadrilaterals, will be in the line ND or AD. Now as here three quadrilaterals have been inscribed in the triangle, so an infinite number of such quadrilaterals can be inscribed therein, and the center of gravity of the inscribed figure will always be (for the reasons mentioned above) in the line AD. But the more such quadrilaterals there are, the less the triangle ABC will differ from the inscribed figure of the quadrilaterals. For if we draw lines parallel to BC through the middle points of AN, NM, ML, LD, the difference of the last figure will be exactly half of the difference of the preceding figure.[2] We can therefore, by infinite approximation, place within the triangle a figure such that the difference between the latter and the triangle shall be less than any given plane figure, however small. From which it follows that, taking AD to be the center line of gravity,[3] the apparent weight of the part ADC will differ less from the apparent weight of the part ADB than any plane figure that might be given, however small, from which I argue as follows:[4]

[2] It is obviously assumed that the side AB is divided into n equal segments (in the figure $n = 4$). The difference between the area Δ of the triangle ABC and that of the figure consisting of $(n - 1)$ parallelograms is Δ/n.

[3] The statement that AD is the center line of gravity seems to mean that AD is the vertical through the point of suspension of the triangle at rest and hence, by the rule of statics quoted by Stevin earlier in the book (Book I, Prop. 6: The center of gravity of a hanging solid is always in its center line of gravity), the center of gravity is in AD. (Notes 2 and 3 are based on footnotes in the *Principal works*, I).

[4] Stevin here uses the form of the syllogism known in ancient logic as $CAMESTRES$ (vowels AEE, A universal affirmation, as all P are Q, E universal negation, as no P are Q). He uses this formulation repeatedly (see *Principal works*, I, 143, note 2).

The reasoning amounts to this: When we know that the difference of two quantities A and B is smaller than a quantity that can be taken as small as we like, then $A = B$. The *reductio ad absurdum*, typical of the Greeks, is replaced by a syllogism.

It has been justly observed that Stevin's way of reasoning constitutes an important step in the evolution of the limit concept; see H. Bosmans, "Sur quelques exemples de la théorie des limites chez Simon Stevin," *Annales de la Société Scientifique de Bruxelles 37* (1913), 171–199; "L'Analyse infinitésimale chez Simon Stevin," *Mathesis 37* (1923), 12–18, 55–62, 105–109, summarized in sec. V of Bosmans, "Le mathématicien belge Simon Stevin de Bruges," *Periodico di matematiche* (ser. 4) *6* (1926), 231–261.

A. Beside any different apparent gravities there may be placed a gravity less than their difference;

O. Beside the present apparent gravities ADC and ADB there cannot be placed any gravity less than their difference;

O. Therefore the present apparent gravities ADC and ADB do not differ.

Therefore AD is the center line of gravity, and consequently the center of gravity of the triangle ABC is in it.

Conclusion. The center of gravity of any triangle therefore is in the line drawn from the vertex to the middle point of the opposite side, which we had to prove.

Problem I, Proposition III. Given a triangle: to find its center of gravity.

Supposition. Let ABC be a triangle [Fig. 2].

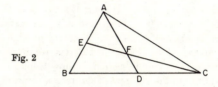

Fig. 2

What is required to find. We have to find its center of gravity.

Construction. There shall be drawn from A to the middle point of BC the line AD, likewise from C to the middle point of AB the line CE, intersecting AD in F. I say that F is the required center of gravity.

Proof. The center of gravity of the triangle ABC is in the line AD, and also in CE, by the second proposition. It is therefore F, which we had to prove.

Conclusion. Given therefore a triangle, we have found its center of gravity, as required.

The next propositions deal with centers of gravity of specific figures; for instance, Theorem V, Proposition VII, states that "the center of gravity of the quadrilateral with two parallel sides is in the line joining the middle points of those sides," and in Problem IV, Proposition XII, we find that the center of gravity of any parabolic segment is at three-fifths of its diameter. Stevin then finds the center of gravity of the parabolic segment at the point E on AD such that $AE : ED = 3 : 2$.

Before we pass to Kepler and his approach to the limit concept, we shall quote the formulation given by the Italian Luca Valerio (1552–1618) in his *De centro gravitatis solidorum* (Rome, 1604; 2nd ed., Bologna, 1659):

"If a quantity, either greater or smaller than a first quantity, has had a proportion to a quantity greater or smaller than a second quantity, with an excess or defect smaller than any arbitrary quantity [*excessu, vel defectu quantacumque proposita*], then the first quantity will have to the second the same proportion."

For instance, if we wish to prove that the areas of two circles C_1 and C_2 are as the squares of the diameters with the aid of inscribed (circumscribed) polygons, as Euclid does (*Elements*, XII, Proposition 2), then these areas are the first and second quantity, and the smaller (greater) quantities, differing from them by an arbitrary quantity, are the areas of the inscribed (circumscribed) polygons with the same number of sides.

Valerio's statement is closer to the modern limit concept than Stevin's. See Bosmans, cited in note 4, and E. J. Dijksterhuis, *De Elementen van Euclides* (2 vols.; Noordhoff, Groningen, 1929, 1930), II, 242.

2 KEPLER. INTEGRATION METHODS

Johann Kepler (1571–1630) is best known as an astronomer. He was a professor at Graz (1593–1599), assistant to Tycho Brahe and in 1601 became imperial mathematician at the court of the Austrian emperors at Prague. From 1612 to 1626 he taught at Linz. His work led him to many questions now solved by means of the calculus. In his *Astronomia nova* (Heidelberg, 1609), in which he announced the first and second "laws of Kepler," he has computations equivalent to our $\int_0^\varphi \sin \varphi \, d\varphi = 1 - \cos \varphi$, and attempted a solution of "Kepler's equation," $x = e \sin x + m$ (e, m constants). He dedicated his *Nova stereometria doliorum vinariorum* (New solid geometry of wine barrels; Linz, 1615) to the computation of the volume of solids obtained by the rotation of conics about a line in their plane, of which we select a simple section to show Kepler's nonrigorous but direct approach. In his computations of the volumes of surfaces of rotation, he took his axis of rotation in different directions (parallel to the principal axis, the other axis, a diameter, a tangent), either passing through the center, intersecting the conic section, not intersecting it, or tangent to it, and so arrived at 92 solids. Some of these solids already had names, such as sphere, conoid, and spheroid; to the others he gave names such as apple [*malus*], lemon [*citrium*], pear, plum, nut, and so forth.

Our selections are taken from the *Opera*, ed. Frisch (Heyder and Zimmer, Frankfurt, Erlangen, 1863), IV, 557–558, 582–584, or *Gesammelte Werke*, ed. M. Caspar (Beck, München, 1960), IX, 13–16, 47–49. There exists a German translation of the book in Ostwald's *Klassiker*, No. 165, ed. R. Klug (Engelmann, Leipzig, 1908).

Part I. *The Solid Geometry of Regular Bodies*

Theorem I. We first need the knowledge of the ratio between circumference and diameter. Archimedes taught:

The ratio of circumference to diameter is about 22 : 7. To prove it we use figures inscribed in and circumscribed about the circle. Since there is an infinite number of such figures, we shall, for the sake of convenience, use the hexagon [Fig. 1].

Fig. 1

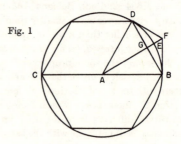

Let a regular hexagon CDB be inscribed in the circle; let its angles be C, D, B, its side DB, and F the point of intersection of the two tangents at D and B respectively. The line AF connects the center A with F, and intersects the line DB at G, the curve DB at E. But as DGB is a straight line, it is the shortest distance between D and B.

DEB, on the other hand, being a curve, is not the shortest distance between D and B. Hence DEB is longer than DGB. On the other hand, BF is tangent to the circle and therefore all parts of the curve EB are between FB and GB; therefore, if EB were straight, it would altogether be shorter than FB. For AEB, FEB are equivalent to a right angle, and, as EFB is an acute angle, EB, opposite the smaller angle EFB, must be smaller than FB, since this is opposite the larger angle. And we can consider EB a straight line, because in the course of the proof the circle is cut into very small arcs, which appear to be equal to straight lines.[1]

Now since, as can be observed, the curve DEB is contained in the triangle DBF, it must be smaller than the lines DF, FB, since it bends toward the angle DFB, and still has not the slightest part outside the lines DF, FB; but the containing, according to common sense, is greater than the contained.[2] This would be different, were the curve DEB winding and irregular.

But as DB is a side of the inscribed hexagon, and DF, FB are two halves of the circumscribed hexagon, arc DEB must be a sixth of the circle, since it was greater than DB and smaller than DF, FB; 6 DB is smaller than the circumference of the circle and 12 DF (or FB) is greater than the circumference.

But the side DB of the regular hexagon is equal to the radius AB. Therefore 6 radii AB, that is, three diameters CB or (if the diameter is divided by 7) $\frac{2}{7}$ CB are shorter than the circumference.

And again, since DG, GB are equal, GB is half of AB. The square of AB, however, is equal to the sum of the squares of AG and GB and is the quadruple of the square of GB. Therefore the square of AG is three times the square of GB. The ratio therefore of the squares of AB and AG is $\frac{4}{3}$ of the lines, therefore the ratio $AB:AG$ is $\sqrt{\frac{4}{3}}$, that is, the ratio of the numbers $100{,}000 : 86{,}603$.[3] But as $AG:AB = GB:BF$, then also $BF:GB$ is $\sqrt{\frac{4}{3}}$ and as GB is half of AB, for example, 50,000, BF must have about 57,737 of such parts. Twelvefold this total number therefore will be greater than the circumference of the circle. Computation gives the number 477,974 for those circles which have 200,000 for

[1] This statement of Kepler's was attacked by Paul Guldin, in his *Centrobaryca seu de centro gravitatis* (2 vols.; Vienna, 1635, 1641). There exists no geometric proof whatever, wrote Guldin, that a circular arc, be it as small as you like, may be equated to a straight line. Guldin (1577–1645), a Swiss-born Jesuit mathematician who taught in Rome, Vienna, and Graz, was critical not only of the methods of Kepler, but also of those of Cavalieri. His book also contains the "rules of Guldin"; see note 7.

[2] Here Guldin criticized again: if this were evident, then Archimedes would not have found it necessary (in *De sphaero et cylindro*) to prove that the circumference of a polygon circumscribed about a circle is larger than that of a circle: "In geometry we should not trust too much in what is evident."

[3] Kepler wrote "one-half of $\frac{4}{3}$" [*semi-sesquitertia*], expressing the square root by "one-half," a mode of expression that goes back to Boethius (sixth century A.D.) and even to Euclid. See Tropfke, *Geschichte*, II (1933), 81. This mode of expression, with its "logarithmic" flavor, has a relation to the ancient theory of music.

diameter. And those of diameter 7 have for twelve times BF the value 24 minus $\frac{1}{10}$. But this number is greater than the circumference itself; on the other hand the number 21 is smaller than the said circumference. And it is obvious that the curve BE is nearer to BG than the line BF. The circumference therefore is nearer the number 21 than $24 - \frac{1}{10}$.[4] We suppose it differs by 1 from 21, from the other by $2 - \frac{1}{10}$, and that it therefore doubtless is 22. This, however, Archimedes shows much more accurately by means of multisided figures of 12, 24, 48 sides; there it also becomes apparent how little the difference of the circumference from 22 is. Adrianus Romanus proved by the same method that when the diameter is divided into 20,000,000,000,000,000 parts, then about 62,831,853,071,795,862 of those parts make up the circumference.[5]

Remark [*Episagma*]. Of the three conical lines, which are called parabola, hyperbola, and ellipse, the ellipse is similar to the circle, and I showed in the *Commentary on the motions of Mars* that the ratio of the length of the elliptic line to the arithmetic mean of its two diameters (which are called the right and transversal axes) is about equal to $22:7$.[6]

Theorem II. *The area of a circle compared with the area of the square erected on the diameter has about the ratio* $11:14$.

Archimedes uses an indirect proof in which he concludes that if the area exceeds this ratio it is too large. The meaning of it seems to be this [Fig. 2].

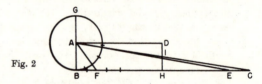

Fig. 2

The circumference of the circle BG has as many parts as points, namely, an infinite number; each of these can be regarded as the base of an isosceles triangle with equal sides AB, so that there are an infinite number of triangles in the area of the circle. They all converge with their vertices in the center A. We now straighten the circumference of circle BG out into the line BC, equal to it. The bases of these infinite triangles or sectors are therefore all supposed to be on the straight line BC, arranged one next to the other. Let BF be one of these bases, and CE any other, equal to it, and let the points F, E, C be connected with A. Since there are as many triangles ABF, AEC over the line BC as there are sectors in the area of the circle, and the bases BF, EC are equal, and all have the altitude BA in common (which is also one of the sectors), the triangles EAC, BAF will be equal, and equal to one of the circle sectors. As they all have their

[4] The actual value is $12 \cdot \frac{7}{8} \tan 30° = 24.25$. Kepler writes "24, minus decima."

[5] Adriaen Van Roomen (1561–1615) had published this in his *Ideae mathematicae* (Louvain, 1593).

[6] For this approximation of the circumference C of the ellipse of semiaxes a and b as $C = \frac{22}{7}(a + b)$, see Kepler's *Astronomia nova* (Heidelberg, 1609), *Gesammelte Werke*, ed. Caspar, III, 368. This statement of Kepler's was also criticized by Guldin. However, for planetary orbits, with small eccentricity, Kepler's approximation was not bad; if developed up to fourth powers of the eccentricity e, it is only $\frac{3}{64}\pi e^4$ greater than the circumference C (*ibid.*, III, 484; IV, 480).

bases on *BC*, the triangle *BAC*, consisting of all those triangles, will be equal to all the sectors of the circle and therefore equal to the area of the circle which consists of all of them. This is equivalent to Archimedes' conclusion by means of an absurdity.

If now we divide *BC* in half at *H*, then *ABHD* forms a parallelogram. Let *DH* intersect *AC* in *I*. This parallelogram is equal to the circle in area. Indeed, *CB* is to its half *CH* as *AB* (that is, *DH*) is to its half *IH*. Therefore $IH = ID$ and $HC = DA$ (equal to *BH*). The angles at *I* are equal, and those at *D* and *H* are right angles. The triangle *ICH*, which is outside the parallelogram, is equal to triangle *IAD* by which the parallelogram exceeds the trapezoid *AIHB*.

If now the diameter *GB* is 7 parts, then its square will be 49. And since the circumference consists of 22 such parts—hence also *BC*—its half *BH* will consist of 11, hardly more or less. Multiply it by the semidiameter $3\frac{1}{2}$, which is *AB*, and we get for the rectangle *AH* $38\frac{1}{2}$ [38 *semis*]. Therefore, if the square of the diameter is 49, the area of the circle is as twice 49 or 98 to 77. Dividing by 7 we obtain 14 to 11,   Q.E.D.

Corollary 1. The area of the sector of a circle (consisting of straight lines from the center intersecting the arc) is equal to the rectangle over the radius and half the arc.

Corollary 2 deals with the area of a segment of a circle.

The next theorems deal with the cone, cylinder, and sphere. In a supplement Kepler introduces conic sections and solids generated by these curves. Among the solids he discusses we find the torus, which he calls a ring [*annulus*].

Theorem XVIII. *Any ring with circular or elliptic cross section is equal to a cylinder whose altitude equals the length of the circumference which the center of the rotated figure describes, and whose base is the same as the cross section of the ring.*[7]

[7] In Theorem I and Theorem II Kepler had replaced Archimedes' *reductio ad absurdum* with a more direct proof, and in a vague way identified the points on the circumference with very small segments. His reasoning reminds us of Antiphon; see T. L. Heath, *Manual of Greek mathematics* (Clarendon Press, Oxford, 1931), 140. In Theorems XVIII and XIX we find the solid divided into very small disks. These theorems are special cases of the so-called Guldin or Pappus theorem, which in the version of Pappus runs as follows: "The ratio of two perfect [complete] surfaces of rotation is composed of the ratio of the rotated areas and of that of the straight lines drawn perpendicularly to the axes of rotation from the centers of gravity of the rotated areas of the axes"; Pappus, *Mathematical collection*, Book VII, trans. Ver Eecke (see Selection III.3, note 1), pp. 510–511. It should be pointed out that some scholars believe that this theorem is a later insertion. Kepler and Guldin probably found their theorems independently of Pappus.

The special case of the torus, which interested Kepler here, can be found in Heron's *Metrica* (c. A.D. 100), where it is attributed to a certain Dionysodoros (probably second century B.C.), author of a lost treatise on the torus (Heath, *Manual of Greek mathematics*, p. 385). Guldin, referring to Kepler's theorems on figures of rotation, stated his own rules and pointed out that Kepler had almost found them.

By cross section is meant the intersection of a plane through the center of the ring-shaped space and perpendicular to the ring-shaped surface. The proof of this theorem follows partly from theorem XVI[8] and can be established by the same means by which Archimedes taught as the principles of solid geometry.

Indeed, if we cut the ring GCD [Fig. 3] from its center A into an infinite number of very thin disks, any one of them will be the thinner toward the center A,

Fig. 3

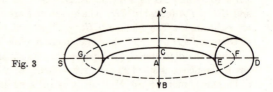

the nearer its part, such as E, lies to the center A than to F and the normal through F erected in the intersecting plane to the line ED. It also will be the thicker the nearer it is to the point D. At such two extreme points, such as D and E, the sum of the two thicknesses will be twice the one in the middle of the disk.

This consideration would not be valid if the parts at E and D of the disk on either side of the circumference FG and the perpendiculars through F and G were not equal and equally situated.

Corollary. This mode of measuring is valid for circular and for elliptical rings as well, high, narrow, or reclining, for open and closed rings alike, as indeed even for all rings whatever shape their cross section may have (instead of the circle ED)—so long as in the plane through AD perpendicular to the ring the parts on either side of F are equal and equally situated. We shall explore this in the case of a square section. Let the ring be of square shape and assume the square to be on ED. This ring can also be measured in another way. For it is the outer part of a cylinder whose base is a circle with AD as radius and whose height is DE. From this cylinder, according to Theorem XVI, the middle part has to be subtracted, that is, the cylinder whose base is the circle of radius AE and whose height is ED. The product, therefore, of ED and the circular area AD minus the circular area AE is equal to the volume of the ring with a square as cross section. And if ED is multiplied by the difference of the squares of AD and AE, then the ratio of this body to the fourth part of the ring would be as the square to the circle, therefore as 14 to 11. Let AE be equal to 2, AD equal to 4, then its square is 16; but the square of AE is 4, therefore the difference of the squares is 12; this number multiplied by the altitude 2 gives the volume as 24, of which the quadruple is 96. Since 14 is to 11 as $96:75\frac{3}{7}$, the volume of the square ring is $75\frac{3}{7}$. This is according to the computation of Theorem XVI. And according to the preceding method, if AF is 3, FG is 6. Since 7 is to 22 as 6 is to 19 minus $\frac{1}{7}$, this therefore will be the length of the circumference FG, the altitude of the cylinder. And since $ED = 2$, its square is 4. To obtain the base of the cylinder,

---

[8] Theorem XVI deals with the ratio of conical segments of the same height and different bases.

multiply therefore 4 by $(19 - \frac{1}{7})$. In this way also we see the truth of the theorem.

Theorem XIX *and Analogy. A closed ring is equal to a cylinder whose base is the circle of the cross section and whose height equals the circumference of the circle described by its center.*

As this method is valid for every ring, whatever the ratio of AE and AF may be, therefore it also holds for a closed ring, in which the center F of the circle ED describes the circle FG, where FG is equal to the rotated AD itself. This is because such a closed ring is intersected from A in disks that have no thickness at A and at D twice the thickness of that at F. Hence the circle through D is twice that through F.

Corollary. The cylindric body that is created by rotation of $MIKN$ [Fig. 4*a*], the four-sided figure of straight and curved lines, is according to the same consideration equal to a column with this figure as base and the length of the circle FG as height. But the outer fringe IKD that surrounds the cylindric body—as a wooden hoop surrounds a barrel—clearly does not yield to this theorem, and must be computed by other means.

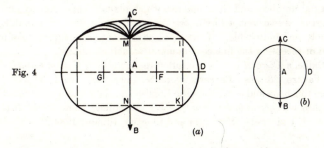

Fig. 4

(a)

(b)

Analogy. Moreoever, this method is valid for all cylindric bodies or parts of apples (or figs), no matter how slender, until I, K coincide with M, N, which happens in the formation of the sphere [Fig. 4*b*], where instead of the two lines MN and IK there exists only one, namely, BC. For this body the demonstration and use of this theorem fail for the first time.

Corollary. The ratio of the sphere to the closed ring created by the same circle is 7 to 33, since one-third of the radius multiplied by four times the area of the greatest circle, or two-thirds of the diameter multiplied by the area of the greatest circle, produce a cylinder equal to the sphere.[9] And a cylinder equal to the closed ring has the same base, and its altitude is the circumference [formed by the center]. Therefore as the circumference is to two-thirds of the diameter, that is, 33 : 7,[10] so is the ring to the sphere.[11]

---

[9] The text writes "cube."

[10] The ratio is $3\pi:2$.

[11] On Kepler there exists in English a symposium of the History of Science Society, *Johann Kepler 1571–1630, a tercentenary commemoration of his work* (Williams and Wilkins, Baltimore, 1931).

3 GALILEI. ON INFINITES AND INFINITESIMALS

Galileo Galilei (1564–1642) was from 1589 to 1610 professor, first at Pisa, then at Padua, and in 1610 became first mathematician at the grand-ducal court of Tuscany. With the newly invented telescope he discovered, in 1610, that Jupiter had four satellites; this and other startling facts led him to a defense of the Copernican system in his *Dialogo sopra due massimi sistemi del mondo* (Florence, 1632). Confined to Florence after his condemnation by the Holy Office, he published his ideas on kinematics and elasticity in the *Discorsi e dimostrazioni matematiche intorno a due nuove scienze attenenti alla mecanica e ai movimenti locali* (Leiden, 1638), which contain discussions on the infinite and the infinitesimal.

Galilei never wrote a book on the purely mathematical aspects of his work, but we find them discussed among other questions. His pupils Cavalieri and Torricelli later gave full elaboration to his ideas on problems dealing with infinitesimals (see Selections IV.5, 6, 9). In the *Discorsi* of 1638, in which Galilei laid the foundations of modern mechanics, we see him (as Salviati) discussing with Simplicio some of the fundamental difficulties one meets in the concepts of infinite and infinitesimal. Such difficulties were already widely discussed in antiquity, and equally among the scholastic writers. Galilei made some points clear, such as the difference between actual and potential infinity and the "equality" in number of the natural numbers and their squares, which play an important role in the modern theory of aggregates as developed by Georg Cantor. The text shows that Galilei was not afraid of the scholastic "indivisible," of letting a line segment be an aggregate of infinitely many points, and of accepting the line continuum as an actual infinite.

Galilei's *Discorsi* can be consulted in *Le opere di Galileo Galilei, edizione nazionale* (ed. A. Favaro, 20 vols.; Barbera, Florence, 1890–1909), VIII, 39–318. Our selection is taken from this book and is based on the English translation by H. Crew and A. De Salvio entitled: *Dialogues concerning two new sciences* (Macmillan, New York, 1914; reissued by North-western University, Evanston and Chicago, Illinois, 1939), 71–68 (*Opere*, VIII, 68–78).

SALVIATI. . . . Now since we have arrived at paradoxes let us see if we cannot prove that within a finite extent it is possible to discover an infinite number of vacua. At the same time we shall at least reach a solution of the most remarkable of all that list of problems which Aristotle himself calls wonderful; I refer to his *Questions in Mechanics*.[1] This solution may be no less clear and conclusive than

[1] The *Questions in mechanics* (also translated as *Mechanical problems*) is a collection of mechanical problems and their solutions; it is not a work of Aristotle (384–322 B.C.) but probably originated in his school, perhaps composed in the time of his successors, Theophrastus or Strato (322–269 B.C.). The book contains the parallelogram of velocities, and also the problem called that of the *rota Aristotelis*, the wheel that Galilei discusses in connection with Fig. 1. See *Aristotle, minor works, with an English translation by W. S. Hett* (Harvard University Press, Cambridge, Mass., 1936), 330–441. The *rota Aristotelis* is the subject of Problem 24, which begins as follows: "A difficulty arises as to how it is that the greater circle when it revolves traces out a path of the same length as a smaller circle, if the two are concentric. When they are revolved separately, then the paths along which they travel are in the same ratio as their respective sizes." We now say that when the larger circle *AB* (Fig. 1) rolls on *BF*, then the same smaller circle *AC*, fixed to circle *AB*, rolls and slides along *CE*. See further G. S. Klügel, *Mathematisches Wörterbuch* (Schwickert, Leipzig, 1823), IV, under "Rad, Aristotelisches." We meet the *rota Aristotelis* again when the cycloid is investigated; see Selection IV.10.

that which he himself gives and quite different also from that so cleverly expounded by the most learned Monsignor di Guevara.[2]

First it is necessary to consider a proposition, not treated by others, but one upon which depends the solution of the problem and from which, if I mistake not, we shall derive other new and remarkable facts. For the sake of clearness let us draw an accurate figure [Fig. 1]. About G as a center describe an equiangular and equilateral polygon of any number of sides, say the hexagon

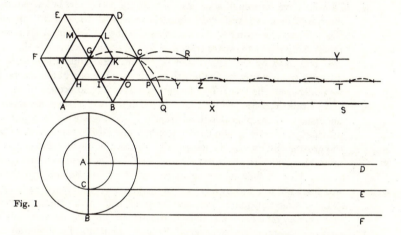

Fig. 1

$ABCDEF$. Similar to this and concentric with it, describe another smaller one which we shall call $HIKLMN$. Prolong the side AB, of the larger hexagon, indefinitely toward S; in like manner prolong the corresponding side HI of the smaller hexagon, in the same direction, so that the line HT is parallel to AS; and through the center draw the line GV parallel to the other two. This done, imagine the larger polygon to roll upon the line AS, carrying with it the smaller polygon. It is evident that, if the point B, the end of the side AB, remains fixed at the beginning of the rotation, the point A will rise and the point C will fall describing the arc CQ until the side BC coincides with the line BQ, equal to BC. But during this rotation the point I, on the smaller polygon, will rise above the line IT because IB is oblique to AS; and it will not again return to the line IT until the point C shall have reached the position Q. The point I, having described the arc IO above the line HT, will reach the position O at the same time the side IK assumes the position OP; but in the meantime the center G has traversed a path above GV and does not return to it until it has completed the arc GC. This step having been taken, the larger polygon has been brought to rest with its side BC coinciding with the line BQ while the side IK of the smaller polygon has been made to coincide with the line OP, having passed over the portion IO without touching it; also the center G will have reached the position C after having traversed all its course above the parallel line GV. And finally the

[2] Giovanni di Guevara (1561–1651), in later life bishop of Teano, was a correspondent of Galilei's. One of their points of discussion was Aristotle's *Questions in mechanics*.

entire figure will assume a position similar to the first, so that if we continue the rotation and come to the next step, the side DC of the larger polygon will coincide with the portion QX and the side KL of the smaller polygon, having first skipped the arc PY, will fall on YZ, while the center still keeping above the line GV will return to it at R after having jumped the interval CR. At the end of one complete rotation the larger polygon will have traced upon the line AS, without break, six lines together equal to its perimeter; the lesser polygon will likewise have imprinted six lines equal to its perimeter, but separated by the interposition of five arcs, whose chords represent the parts of HT not touched by the polygon: the center G never reaches the line GV except at six points. From this it is clear that the space traversed by the smaller polygon is almost equal to that traversed by the larger, that is, the line HT approximates the line AS, differing from it only by the length of one chord of one of these arcs, provided we understand the line HT to include the five skipped arcs.

Now this exposition which I have given in the case of these hexagons must be understood to be applicable to all other polygons, whatever the number of sides, provided only they are similar, concentric, and rigidly connected, so that when the greater one rotates the lesser will also turn, however small it may be. You must also understand that the lines described by these two are nearly equal provided we include in the space traversed by the smaller one the intervals which are not touched by any part of the perimeter of this smaller polygon.

Let a large polygon of, say, one thousand sides make one complete rotation and thus lay off a line equal to its perimeter; at the same time the small one will pass over an approximately equal distance, made up of a thousand small portions each equal to one of its sides, but interrupted by a thousand spaces which, in contrast with the portions that coincide with the sides of the polygon, we may call empty. So far the matter is free from difficulty or doubt.

But now suppose that about any center, say A, we describe two concentric and rigidly connected circles; and suppose that from the points C and B, on their radii, there are drawn the tangents CE and BF and that through the center A the line AD is drawn parallel to them; then if the large circle makes one complete rotation along the line BF, equal not only to its circumference but also to the other two lines CE and AD, tell me what the smaller circle will do and also what the center will do. As to the center it will certainly traverse and touch the entire line AD while the circumference of the smaller circle will have measured off by its points of contact the entire line CE, just as was done by the above-mentioned polygons. The only difference is that the line HT was not at every point in contact with the perimeter of the smaller polygon, but there were left untouched as many vacant spaces as there were spaces coinciding with the sides. But here in the case of the circles the circumference of the smaller one never leaves the line CE, so that no part of the latter is left untouched, nor is there ever a time when some point on the circle is not in contact with the straight line. How now can the smaller circle traverse a length greater than its circumference unless it go by jumps?

SAGREDO. It seems to me that one may say that just as the center of the circle, by itself, carried along the line AD is constantly in contact with it, although it is only a single point, so the points on the circumference of the smaller circle, carried along by the motion of the larger circle, would slide over some small parts of the line *CE*.

SALVIATI. There are two reasons why this cannot happen. First because there is no ground for thinking that one point of contact, such as that at *C*, rather than another, should slip over certain portions of the line *CE*. But if such slidings along *CE* did occur, they would be infinite in number since the points of contact (being mere points) are infinite in number: an infinite number of finite slips will however make an infinitely long line, while as a matter of fact the line *CE* is finite. The other reason is that as the greater circle, in its rotation, changes its point of contact continuously, the lesser circle must do the same because *B* is the only point from which a straight line can be drawn to *A* and pass through *C*. Accordingly the small circle must change its point of contact whenever the large one changes: no point of the small circle touches the straight line *CE* in more than one point. Not only so, but even in the rotation of the polygons there was no point on the perimeter of the smaller which coincided with more than one point on the line traversed by that perimeter; this is at once clear when you remember that the line *IK* is parallel to *BC* and that therefore *IK* will remain above *IP* until *BC* coincides with *BQ*, and that *IK* will not lie upon *IP* except at the very instant when *BC* occupies the position *BQ*; at this instant the entire line *IK* coincides with *OP* and immediately afterward rises above it.

SAGREDO. This is a very intricate matter. I see no solution. Pray explain it to us.

SALVIATI. Let us return to the consideration of the above-mentioned polygons whose behavior we already understand. Now in the case of polygons with 100,000 [3] sides, the line traversed by the perimeter of the greater, i.e., the line laid down by its 100,000 sides one after another, is equal to the line traced out by the 100,000 sides of the smaller, provided we include the 100,000 vacant spaces interspersed. So in the case of the circles, polygons having an infinitude of sides, the line traversed by the continuously distributed infinitude of sides is in the greater circle equal to the line laid down by the infinitude of sides in the smaller circle but with the exception that these latter alternate with empty spaces; and since the sides are not finite in number, but infinite, so also are the intervening empty spaces not finite but infinite. The line traversed by the larger circle consists then of an infinite number of points which completely fill it, while that which is traced by the smaller circle consists of an infinite number of points which leave empty spaces and only partly fill the line. And here I wish you to observe that after resolving and dividing a line into a finite number of parts, that is, into a number which can be counted, it is not possible to arrange them again into a greater length than that which they occupied when they formed a continuum [4] and were connected without the interposition of as many empty spaces. But if we consider the line resolved into an infinite number of infinitely small and indivisible parts, we shall be able to conceive the line extended indefinitely by the interposition, not of a finite, but of an infinite number of infinitely small indivisible empty spaces.

---

[3] Galilei uses words: *cento mila lati.*
[4] Italian: *stavano continuate.*

Now this which has been said concerning simple lines must be understood to hold also in the case of surfaces and solid bodies, it being assumed that they are made up of an infinite, not a finite, number of atoms. Such a body once divided into a finite number of parts, it is impossible to reassemble them so as to occupy more space than before unless we interpose a finite number of empty spaces, that is to say, spaces free from the substance of which the solid is made. But if we imagine the body, by some extreme and final analysis, resolved into its primary elements, infinite in number, then we shall be able to think of them as indefinitely extended in space, not by the interposition of a finite, but of an infinite number of empty spaces. Thus one can easily imagine a small ball of gold expanded into a very large space without the introduction of a finite number of empty spaces, always provided the gold is made up of an infinite number of indivisible parts.

After a short intermission, in which Simplicio praises the religious mind of Salviati, Simplicio continues:

But to return to our subject, your previous discourse leaves with me many difficulties which I am unable to solve. First among these is that, if the circumferences of the two circles are equal to the two straight lines, CE and BF, the latter considered as a continuum, the former as interrupted with an infinity of empty points, I do not see how it is possible to say that the line AD described by the center, and made up of an infinity of points, is equal to this center which is a single point. Besides, this building up of lines out of points, divisibles out of indivisibles, and finites out of infinites, offers me an obstacle difficult to avoid; and the necessity of introducing a vacuum, so conclusively refuted by Aristotle, presents the same difficulty.

SALVIATI. These difficulties are real; and they are not the only ones. But let us remember that we are dealing with infinities and indivisibles, both of which transcend our finite understanding, the former on account of their magnitude, the latter because of their smallness. In spite of this, men cannot refrain from discussing them, even though it must be done in a roundabout way.

To explain some of his ideas, Salviati goes on:

How can a single point be equal to a line? Since I cannot do more at present I shall attempt to remove, or at least diminish, one improbability by introducing a similar or a greater one, just as sometimes a wonder is diminished by a miracle.

And this I shall do by showing you two equal surfaces, together with two equal solids located upon these same surfaces as bases, all four of which diminish continuously and uniformly in such a way that their remainders always preserve

equality among themselves, and finally both the surfaces and the solids terminate their previous constant equality by degenerating, the one solid and the
one surface into a very long line, the other solid and the other surface into a
single point; that is, the latter to one point, the former to an infinite number of
points.

SAGREDO. This proposition appears to me wonderful, indeed; but let us hear
the explanation and demonstration.

SALVIATI. Since the proof is purely geometrical we shall need a figure [Fig. 2].
Let AFB be a semicircle with center at C; about it describe the rectangle $ADEB$

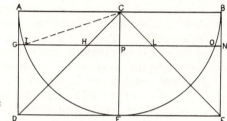

Fig. 2

and from the center draw the straight lines CD and CE to the points D and E.
Imagine the radius CF to be drawn perpendicular to either of the lines AB or
DE, and the entire figure to rotate about this radius as an axis. It is clear that
the rectangle $ADEB$ will thus describe a cylinder, the semicircle AFB a hemisphere, and the triangle CDE a cone. Next let us remove the hemisphere but
leave the cone and the rest of the cylinder, which, on account of its shape, we
will call a "bowl." First we shall prove that the bowl and the cone are equal;
then we shall show that a plane drawn parallel to the circle which forms the base
of the bowl and which has the line DE for diameter and F for a center—a plane
whose trace is GN—cuts the bowl in the points G, I, O, N, and the cone in the
points H, L, so that the part of the cone indicated by CHL is always equal to the
part of the bowl whose profile is represented by the triangles GAI and BON.
Besides this we shall prove that the base of the cone, i.e., the circle whose
diameter is HL, is equal to the circular surface which forms the base of this
portion of the bowl, or, as one might say, equal to a ribbon whose width is GI.
(Note by the way the nature of mathematical definitions which consist merely
in the imposition of names, or, if you prefer, abbreviations of speech established
and introduced in order to avoid the tedious drudgery which you and I now
experience simply because we have not agreed to call this surface a "circular
band" and that sharp solid portion of the bowl a "round razor.") Now call them
by what name you please, it suffices to understand that the plane, drawn at any
height whatever, so long as it is parallel to the base, i.e., to the circle whose
diameter is DE, always cuts the two solids so that the portion CHL of the cone
is equal to the upper portion of the bowl; likewise the two areas which are the
bases of these solids, namely, the band and the circle HL, are also equl. Here we
have the miracle mentioned above; as the cutting plane approaches the line AB
the portions of the solids cut off are always equal, so also the areas of their bases.

And as the cutting plane comes near the top, the two solids (always equal) as well as their bases (areas which are also equal) finally vanish, one pair of them degenerating into the circumference of a circle, the other into a single point, namely, the upper edge of the bowl and the apex of the cone. Now, since as these solids diminish equality is maintained between them up to the very last, we are justified in saying that, at the extreme and final end of this diminution, they are still equal and that one is not infinitely greater than the other. It appears, therefore, that we may equate the circumference of a large circle to a single point. And this which is true of the solids is true also of the surfaces which form their bases; for these also preserve equality between themselves throughout their diminution and in the end vanish, the one into the circumference of a circle, the other into a single point. Shall we not then call them equal, seeing that they are the last traces and remnants of equal magnitudes? Note also that, even if these vessels were large enough to contain immense celestial hemispheres, both their upper edges and the apexes of the cones therein contained would always remain equal and would vanish, the former into circles having the dimensions of the largest celestial orbits, the latter into single points. Hence in conformity with the preceding we may say that all circumferences of circles, however different, are equal to each other, and are each equal to a single point.

SAGREDO. This presentation strikes me as so clever and novel that, even if I were able, I would not be willing to oppose it; for to deface so beautiful a structure by a blunt pedantic attack would be nothing short of sinful. But for our complete satisfaction pray give us this geometrical proof that there is always equality between these solids and between their bases; for it cannot, I think, fail to be very ingenious, seeing how subtle is the philosophical argument based upon this result.

SALVIATI. The demonstration is both short and easy. Referring to the preceding figure, since IPC is a right angle the square of the radius IC is equal to the sum of the squares on the two sides IP, PC; but the radius IC is equal to AC and also to GP, while CP is equal to PH. Hence the square of the line GP is equal to the sum of the squares of IP and PH, or multiplying through by 4, we have the square of the diameter GN equal to the sum of the squares on IO and HL. And, since the areas of circles are to each other as the squares of their diameters, it follows that the area of the circle whose diameter is GN is equal to the sum of the areas of circles having diameters IO and HL, so that if we remove the common area of the circle having IO for diameter the remaining area of the circle GN will be equal to the area of the circle whose diameter is HL. So much for the first part. As for the other part, we leave its demonstration for the present, partly because those who wish to follow it will find it in the twelfth proposition of the second book of *De centro gravitatis solidorum* by the Archimedes of our age, Luca Valerio,[5] who made use of it for a different object, and partly because, for our purpose, it suffices to have seen that the above-mentioned surfaces are always equal and that, as they keep on diminishing uniformly, they degenerate, the one into a single point, the other into the circumference of a circle larger than any assignable; in this fact lies our miracle.

[5] See Selection IV.1.

SAGREDO. The demonstration is ingenious and the inferences drawn from it are remarkable.

After some further discussion, Salviati continues:

One of the main objections urged against this building up of continuous quantities out of indivisible quantities [*continuo d'indivisibili*] is that the addition of one indivisible to another cannot produce a divisible, for if this were so it would render the indivisible divisible. Thus if two indivisibles, say two points, can be united to form a quantity, say a divisible line, then an even more divisible line might be formed by the union of three, five, seven, or any other odd number of points. Since, however, these lines can be cut into two equal parts, it becomes possible to cut the indivisible which lies exactly in the middle of the line. In answer to this and other objections of the same type we reply that a divisible magnitude cannot be constructed out of two or ten or a hundred or a thousand indivisibles, but requires àn infinite number of them.

SIMPLICIO. Here a difficulty presents itself which appears to me insoluble. Since it is clear that we may have one line greater than another, each containing an infinite number of points, we are forced to admit that, within one and the same class, we may have something greater than infinity, because the infinity of points in the long line is greater than the infinity of points in the short line. This assigning to an infinite quantity a value greater than infinity is quite beyond my comprehension.

SALVIATI. This is one of the difficulties which arise when we attempt, with our finite minds, to discuss the infinite, assigning to it those properties which we give to the finite and limited; but this I think is wrong, for we cannot speak of infinite quantities as being the one greater or less than or equal to another. To prove this I have in mind an argument which, for the sake of clearness, I shall put in the form of questions to Simplicio, who raised this difficulty.

Here follows the argument that, when we compare the sequences 1, 2, 3, 4, ... and 1, 4, 9, 16, ..., a one-to-one correspondence can be established between the numbers of the first and those of the second sequence, $1 \leftrightarrow 1$, $2 \leftrightarrow 4$, $3 \leftrightarrow 9$, ... Then this argument is transferred to continua: "One line does not contain more or less or just as many points as another, but each line contains an infinite number." Salviati continues with the discussion of the *rota Aristotelis*, after remarks by Simplicio and Sagredo:

SIMPLICIO. Leaving this to one side for the moment, I should like to hear how the introduction of these indivisible quantities helps us to understand contraction and expansion, avoiding at the same time the vacuum and the penetrability of bodies.

SAGREDO. I also shall listen with keen interest to this same matter, which is far from clear in my mind; provided I am allowed to hear what, a moment ago, Simplicio suggested we omit, namely, the reasons which Aristotle offers against the existence of the vacuum and the arguments which you must advance in rebuttal.

SALVIATI. I will do both. And first, just as, for the production of expansion, we employ the line described by the small circle during one rotation of the large one—a line greater than the circumference of the small circle—so, in order to explain contraction, we point out that, during each rotation of the smaller circle, the larger one describes a straight line which is shorter than its circumference.

For the better understanding of this we proceed to the consideration of what happens in the case of polygons. Employing a figure similar to the earlier one [Fig. 3], construct the two hexagons, ABC and HIK, about the common center

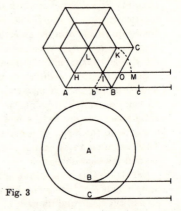

Fig. 3

L, and let them roll along the parallel lines HOM and ABc. Now holding the vertex I fixed, allow the smaller polygon to rotate until the side IK lies upon the parallel, during which motion the point K will describe the arc KM, and the side KI will coincide with IM. Let us see what, in the meantime, the side CB of the larger polygon has been doing. Since the rotation is about the point I, the terminal point B of the line IB, moving backward, will describe the arc Bb underneath the parallel cA so that, when the side KI coincides with the line MI, the side BC will coincide with bc, having advanced only through the distance Bc, but having retreated through a portion of the line BA which subtends the arc Bb. If we allow the rotation of the smaller polygon to go on it will traverse and describe along its parallel a line equal to its perimeter; while the larger one will traverse and describe a line less than its perimeter by as many times the length bB as there are sides less one; this line is approximately equal to that described by the smaller polygon, exceeding it only by the distance bB. Here now we see, without any difficulty, why the larger polygon, when carried by the smaller, does not measure off with its sides a line longer than that traversed by the smaller one; this is because a portion of each side is superposed upon its immediately preceding neighbor.

Let us next consider two circles, having a common center at A, and lying upon their respective parallels, the smaller being tangent to its parallel at the point B; the larger, at the point C. Here when the small circle commences to roll the point B does not remain at rest for a while so as to allow BC to move backward and carry with it the point C, as happened in the case of the polygons, where the point I remained fixed until the side KI coincided with MI and the line IB carried the terminal point B backward as far as b, so that the side BC fell upon bc, thus superposing upon the line BA the portion Bb, and advancing by an amount Bc, equal to MI, that is, to one side of the smaller polygon. On account of these superpositions, which are the excesses of the sides of the larger over the smaller polygon, each net advance is equal to one side of the smaller polygon and, during one complete rotation, these amount to a straight line equal in length to the perimeter of the smaller polygon.

But now reasoning in the same way concerning the circles, we must observe that, whereas the number of sides in any polygon is comprised within a certain limit, the number of sides in a circle is infinite; the former are finite and divisible; the latter infinite and indivisible. In the case of the polygon, the vertices remain at rest during an interval of time which bears to the period of one complete rotation the same ratio which one side bears to the perimeter; likewise, in the case of the circles, the delay of each of the infinite number of vertices is merely instantaneous, because an instant is such a fraction of a finite interval as a point is of a line which contains an infinite number of points. The retrogression of the sides of the larger polygon is not equal to the length of one of its sides but merely to the excess of such a side over one side of the smaller polygon, the net advance being equal to this smaller side; but in the circle the point or side C, during the instantaneous rest of B, recedes by an amount equal to its excess over the side B, making a net progress equal to B itself. In short, the infinite number of indivisible sides of the greater circle with their infinite number of indivisible retrogressions, made during the infinite number of instantaneous delays of the infinite number of vertices of the smaller circle, together with the infinite number of progressions, equal to the infinite number of sides in the smaller circle— all these, I say, add up to a line equal to that described by the smaller circle, a line which contains an infinite number of infinitely small superpositions, this bringing about a thickening or contraction without any overlapping or interpenetration of finite parts. This result could not be obtained in the case of a line divided into finite parts such as is the perimeter of any polygon, which when laid out in a straight line cannot be shortened except by the overlapping and interpenetration of its sides. This contraction of an infinite number of infinitely small parts without the interpenetration or overlapping of finite parts and the previously mentioned expansion of an infinite number of indivisible parts by the interposition of indivisible vacua is, in my opinion, the most that can be said concerning the contraction and rarefaction of bodies, unless we give up the impenetrability of matter and introduce empty spaces of finite size. If you find anything here that you consider worth while, pray use it; if not, regard it, together with my remarks, as idle talk; but this remember, we are dealing with the infinite and the indivisible.

4 GALILEI. ACCELERATED MOTION

An application of Galilei's theory of indivisibles is his derivation of the law for uniformly accelerated motion: if the acceleration is a, then $v = at$ and $s = \frac{1}{2}at^2 = (\frac{1}{2}at)t$, where $\frac{1}{2}at$ is the mean velocity between beginning and end. It will be seen that Galilei regards an area as generated by lines, or, we may say, as composed of lines—hence discarding the ancient difficulty that a sum of points can never be a line, and a sum of lines can never be an area. The text is again from the *Dialogues concerning two new sciences*, trans. H. Crew and A. de Salvio, 166–167; the original text is found in *Opere*, VIII, 208–209.

The theorem had already appeared in scholastic writings (Selection III.1). P. Duhem, *Etudes sur Léonard de Vinci* (Hermann, Paris, 1913), III, 388–398, called it "the rule of Oresme." See also C. B. Boyer, *History of the calculus* (Dover, New York, 1959), 83, 113; E. J. Dijksterhuis, *The mechanization of the world picture* (Clarendon Press, Oxford, 1961), 197–198; and A. Maier, *An der Grenze von Scholastik und Naturwissenschaft* (2nd ed.; Edizioni di Storia e Letteratura, Rome, 1953).

From Theorem I, Proposition I, Galilei could pass without infinitesimals to Theorem II, Proposition II, which states that $s_1 : s_2 = t_1^2 : t_2^2$.

THEOREM I, PROPOSITION I

The time in which any space is traversed by a body starting from rest and uniformly accelerated is equal to the time in which that same space would be traversed by the same body moving at a uniform speed whose value is the mean of the highest speed and the speed just before acceleration began.

Let us represent by the line AB [Fig. 1] the time in which the space CD is traversed by a body which starts from rest at C and is uniformly accelerated; let the final and highest value of the speed gained during the interval AB be represented by the line EB drawn at right angles to AB; draw the line AE, then all lines drawn from equidistant points on AB and parallel to BE will represent

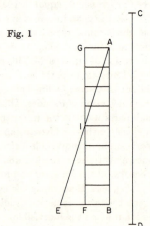

Fig. 1

the increasing values of the speed, beginning with the instant A. Let the point F bisect the line EB; draw FG parallel to BA, and GA parallel to FB, thus forming a parallelogram $AGFB$ which will be equal in area to the triangle AEB, since the side GF bisects the side AE at the point I; for if the parallel lines in the triangle AEB are extended to GI, then the sum of all the parallels contained in the quadrilateral is equal to the sum of those contained in the triangle AEB; for those in the triangle IEF are equal to those contained in the triangle GIA, while those included in the trapezium $AIFB$ are common. Since each and every instant of time in the time interval AB has its corresponding point on the line AB, from which points parallels drawn in and limited by the triangle AEB represent the increasing values of the growing velocity, and since parallels contained within the rectangle represent the values of a speed which is not increasing, but constant, it appears, in like manner, that the momenta assumed by the moving body may also be represented, in the case of the accelerated motion, by the increasing parallels of the triangle AEB, and, in the case of the uniform motion, by the parallels of the rectangle GB. For what the momenta may lack in the first part of the accelerated motion (the deficiency of the momenta being represented by the parallels of the triangle AGI) is made up by the momenta represented by the parallels of the triangle IEF.

Hence it is clear that equal spaces will be traversed in equal times by two bodies, one of which, starting from rest, moves with a uniform acceleration, while the momentum of the other, moving with uniform speed, is one-half its maximum momentum under accelerated motion. Q.E.D.

---

5 CAVALIERI. PRINCIPLE OF CAVALIERI

Bonaventura Cavalieri ($c.$ 1598–1647), a professor at Bologna, was a disciple of Galilei's. He wrote a treatise on the use of "indivisibles" that constituted a first textbook on what we now call integration methods. The *Geometria indivisibilibus continuorum* (Bologna, 1635; 2nd ed., 1653) considered areas as sums of indivisibles, the line segments of which it is composed, and volumes as sums of plane areas. It showed how to measure plane areas and solid volumes by comparing the indivisibles of one with the indivisibles of the other. By taking these indivisibles parallel to each other, Cavalieri arrived at the principle that is still called by his name. We present it in the translation by G. W. Evans, "Cavalieri's theorem in his own words," *American Mathematical Monthly* 24 (1917), 447–451, of Cavalieri's text on pp. 113–145 of his book. The lettering in the text and in Fig. 1 is modernized, as the facsimile, Fig. 2, shows.

In modern notation the principle of Cavalieri is as follows. Let the two lines $x = a, x = b$ together with the continuous curves $y = f_1(x)$, $y = f_2(x)$ enclose an area A_1, and let the same lines together with the continuous curves $y = \varphi_1(x)$, $y = \varphi_2(x)$ enclose another area A_2. Then if for all x, $a \leqslant x \leqslant b$,

$$f_2(x) - f_1(x) = \varphi_2(x) - \varphi_1(x),$$

$$A_1 = \int_a^b [f_2(x) - f_1(x)]\, dx = A_2 = \int_a^b [\varphi_2(x) - \varphi_1(x)]\, dx.$$

This theorem of Cavalieri's is still useful in the high-school teaching of the mensuration of solids, since it makes possible a considerable amount of integration without the formal apparatus of the calculus. The translator observed that his English version is "intended to give, as faithfully as possible, the verbal meaning of the Latin, while not, of course, following the prolix idiom of the time in all its ramifications." This is doing some injustice to the time, but no injustice to Cavalieri.

The Theorem. If between the same parallels any two plane figures are constructed, and if in them, any straight lines being drawn equidistant from the parallels, the included portions of any one of these lines are equal, the plane figures are also equal to one another; and if between the same parallel planes any solid figures are constructed, and if in them, any planes being drawn equidistant from the parallel planes, the included plane figures out of any one of the planes so drawn are equal, the solid figures are likewise equal to one another.

The figures so compared let us call analogues, the solid as well as the plane . . .

The Proof. Let any two plane figures ABC and XYZ [Fig. 1] be constructed between the same parallels PQ, RS; and let DN, OU, be drawn parallel to the

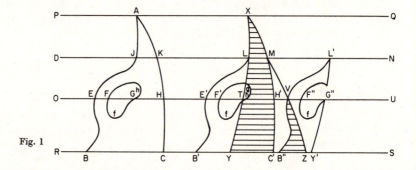

Fig. 1

aforesaid PQ, RS; and let the portions, for example of DN, included in the figures, namely JK, LM, be equal to each other; and again, in the line OU, let the portions EF, GH, taken together (for the figure ABC, for example, may be hollow within, according to the contour of FfG), be likewise equal to TV; and let this happen in all the other lines equidistant from PQ. I say that the figures, ABC, XYZ, are equal to each other.

Let either, then, of the two figures ABC, XYZ be taken, for example ABC itself, with the portions of the parallels PQ, RS coterminous with it, namely the portions PA, RB, and let it be superposed upon the other figure XYZ, but so that the lines PA, RB may fall upon AQ, CS; then either the whole figure ABC coincides with the whole figure XYZ (and thus, since they coincide with each other they are equal), or not; yet let there be some part which will coincide with some part, as $XMC'YThL$, part of the figure ABC, with $XMC'YThL$, part of the figure XYZ.

It is manifest, moreover, if the superposition of the figures is effected in such a way that portions of the parallels PQ, RS coterminous with our two figures are mutually superposed, that whatever straight lines (included in the figures) are in line remain in line; as, for example, since EF, GH are in line with TV, when the aforesaid superposition is made they will remain in line (namely $E'F'TH'$ in line with TV), for the distance of those lines EF, GH from PQ is equal to the distance of TV from PQ; whence, no matter how many times PA is placed over AQ, at any place, EF, GH will always remain in line[1] with TV, which is clearly apparent not only for this but for all other lines parallel to PQ in either figure.

In the case where part of one figure (as ABC) coincides of necessity with part of the figure XYZ, and not with the whole, granting that the superposition be made by such a rule as has been told, the demonstration will be as follows. For since when any parallels are drawn to PQ, the portions of them, included in the figures, which were in line, will still remain in line after superposition, and moreover since they were by hypothesis equal before superposition, therefore, after superposition the portions included in the figures will likewise be equal—as, for example, $E'F'$, TH' taken together will be equal to TV—therefore, if $E'F'$, TH' do not coincide with the whole of TV, then, one part [of one] coinciding with some part [of the other], as TH' with TH' itself, $E'F'$ will be equal to $H'V$, $E'H'$ being in the residuum of the figure ABC which is superposed, and $H'V$ in the figure XYZ upon which the other is superposed. In the same way we shall show that to any line whatever parallel to PQ, and included in the residuum of the superposed figure ABC (which may be $LB'YTF'$) corresponds an equal straight line, in line [with the former], which will be in the residuum of the figure XYZ on which ABC is superposed; therefore, the superposition being made by this rule, when anything of the superposed figure is left over and does not fall upon the figure, it must be that something of the other figure must also be left over, and have nothing superposed upon it.

Since, moreover, to each of the straight lines parallel to PQ and included in the residuum (or residua, for there may be several residual figures) of the superposed figure ABC (or $XB'C'$) there corresponds another straight line, in line [with the first] and included in the residuum (or residua) of the figure XYZ, it is manifest that these residual figures, or their aggregates, are between the same parallels; so since the residual figure $LB'YTF'$[2] is between the parallels DN, RS, likewise the residual figure (or aggregate of residual figures) of the figure XYZ (because it has the frusta Thg, $MC'Z$) will be between the same parallels DN, RS. For if it did not extend both ways to the parallels DN, RS, as for example if it extended up to DN, but not down to RS, only as far as OU, then to the straight lines included in the frustum $E'B'Yff'$, and parallel to PQ, there would not be found in the residuum of the figure XYZ (or in the aggregate of the residua) other corresponding lines as has been proved to be unavoidable. Therefore these residua, or their aggregates, are between the same parallels; and the portions of the lines parallel ot PQ, RS, included therein, are equal, as we

---

[1] Here begins the text in the facsimile, Fig. 2.
[2] Here ends the text in the facsimile, Fig. 2.

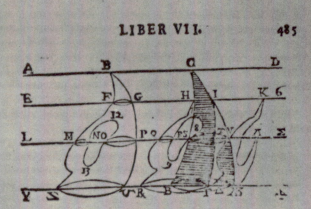

ipfi, SV , quod , & de cæteris quibufcunq; ipfi , AD , parallelis in
vtraque figura liquidò apparet . Quod vero pars vnius figuræ, vt ,
BZ&, congruat neceffario parti figuræ, (βA . & non toti , dum fit
fuperpofitio tali lege, quali dictum eft , fic demonftrabitur . Cum
enim ductis quibufcunq; ipfi , AD , parallelis conceptæ in figuris
ipfarum portiones , quæ erant fibi in directum , adhuc poft fuperpofitionem maneant fibi in directum, illæ vero ante fuperpofitionem effent ex hypotefi æquales , ergo poft fuperpofitionem portiones parallelarum ipfi, AD, in figuris fuperpofitis conceptæ erût
pariter æquales, vt ex.g. QR, ST , fimul fumptæ æquabuntur ipfi,
SV, ergo nifi vtræque, QR, ST, congruant toti , SV, congruente
parte alicui parti, vt, ST, ipfi, ST, erit, QR, æqualis ipfi, IV, &,
QR, quidem erit in refiduo figuræ, BZ& , fuperpofitæ , TV , verò
in refiduo figuræ, (βA, cui fit fuperpofitio . Eodem modo often-
demus cuicunq; parallelæ ipfi, AD, conceptæ in refiduo, figura, B
Z&, fuperpofitæ, quod fit, ti ℞ 5 97, refpondere in directum æqua-
lem rectam lineam, quæ erit in refiduo figuræ, (βA , cui fit fuperpofitio, ergo fuperpofitione hac lege facta , cum fupereft aliquid
de figura fuperpofita , quod non cadat fuper figuram , cui fit fuperpofitio, neceffe eft reliquæ figuræ aliquid etiam fupereffe, fuper
quod nihil fit fuperpofitum . Cum autem vnicuiq; rectæ lineæ
parallelæ, AD, conceptæ in refiduo, vel refiduis (quia poffunt effe
p'ures figuræ refiduæ) figuræ, BZ&, fiue, (℞ Г , fuperpofitæ , re-
fpondeat in directum in refiduo, vel refiduis figuræ ; (βA , alia recta linea , manifeftum eft has refiduas figuras , fiue rei duarum ag-
gregata, effe in eifdem parallelis, cum ergo refidua figura, H ℞ 597,

Fig. 2

have shown above; therefore the residua are subject to the same condition as has been assumed for ABC, XYZ; that is, they are analogues.

So let the residua be now superposed, but so that the parallels KL, CY may fall upon the parallels LN, YS, and the part $VB''Z$ of the frustum $LB'YTF'$ may coincide with the part $VB''Z$ of the frustum $MC'Z$; then we shall show, as above, that as long as there is found a residuum of one, there will be found also a residuum of the other, and these residua, or aggregates of residua, will be found within the same parallels. Let $L'VZY'G''F''$ be a residuum belonging to the figure ABC; and let $MC'B''V$, Thg, be residua belonging to the figure XYZ, whose aggregate is between the same parallels as the residuum $L'VZY'G''F''$, that is, between DN, RS. If now we superpose these residua again, but so that the parallels between which they lie be always superposed respectively, and this is supposed to be done continually, until the whole figure ABC shall have been superposed, I say the whole of it must coincide with XYZ; otherwise if there were any residuum of the figure XYZ, upon which nothing is superposed, there would be also some residuum of the figure ABC which would not have been superposed, as we have shown above to be unavoidable; but it is granted that the whole of ABC is superposed upon XYZ, therefore they are so superposed upon each other that there are no residua of either, therefore they are so superposed that they coincide, therefore the figures ABC, XYZ are equal to each other.

Now in the same diagram let ABC, XYZ be any two solid figures constructed between the same parallel planes PQ, RS; and let DN, OU be any planes drawn equidistant from the planes previously spoken of; and let the figures that lie in the same plane and that are included in the solids be equal to each other always; as JK equal to LM, and EF, GH, taken together (for a solid figure, for example ABC, may be hollow in any way within, according to the surface $FfGg$), equal to TV. I say that these solid figures are equal to each other.

For if we superpose the solid ABC, with the portions PA, RC of the planes PQ, RS, coterminous with it, upon the solid XYZ, in such a way that the plane PA be on the plane PQ, and the plane RC on the plane RS, we shall show (as we did above about the portions of the lines parallel to PQ included in the plane figures ABC, XYZ) that the figures included in the solids and lying in the same plane will also after superposition remain in the same plane; and therefore thus far the figures included in the superposed solids are equal—and parallel to PQ, RS.

Then unless the entire solid coincides with the other solid entire in the first superposition, residual solids will remain, or solids composed of residua, in either solid, which will not be superposed upon each other. Since for example the figures $E'F'$, TH' are equal to the figure TV, then when the common figure TH' is taken away, the remaining figure $E'F'$ will be equal to the remaining figure $H'V$; and this will happen in any plane whatever parallel to PQ and meeting the solids ABC, XYZ. Therefore whenever we have a residuum of one solid, we shall always have a residuum of the other also; and it will be evident, according to the method applied in the former part of this Proposition in the case of plane figures, that the residua of the solids, or the aggregates of residua, will always be between the same parallel planes (as the residua $LB'YTF'$, $MC'Z$, Thg are between the same parallels DN, RS) and will be analogues.

Now if these residua be superposed again, so that the plane DL will be placed on the plane LN, and RY on YS, and this is understood to be done continually, until ABC, which is being superposed, is entirely taken, the entire solid ABC will finally coincide with the entire solid XYZ. For when the entire solid ABC is superposed upon XYZ, unless they coincided there would be some residuum of one, as of the solid XYZ, therefore also some residuum of the solid $XB'C'$, or ABC, and this residuum would not be superposed; which is absurd, for it is already assumed that the entire solid ABC is superposed on XYZ. Therefore there will not be any residuum in these solids; therefore they will coincide; therefore the solid figures spoken of, ABC, XYZ, will be equal to each other, which was to be proved of them.

---

6 CAVALIERI. INTEGRATION

Cavalieri's method consisted in comparing the indivisibles of one figure with those of another. This was a dangerous procedure, because it was not always clear how and which indivisibles should be compared. Then he added these indivisibles, using such expressions as *omnes lineae* (o.l.) = *all lines* (a.l.). Summing correctly, he reached the important result that we write

$$\int_0^a x^n \, dx = \frac{a^{n+1}}{n+1} \qquad (n \text{ a positive integer}).$$

The abbreviation o.l. formed a kind of integration symbol, like Leibniz's use of $\int$, and then $\int \ldots dx$.

Cavalieri derived the formula in his *Geometry of indivisibles* (see the preceding selection) for the case $n = 2$, so that he did not get a new result: Archimedes had found it before him. In later work Cavalieri found the formula for $n = 3, 4, \ldots, 9$. We show here from his *Exercitationes geometricae sex* (Bologna, 1647), Part IV, Proposition 21, how he obtained the result for $n = 3$. To understand it we must know that Cavalieri first proved, in his prolix geometric way, the equivalent of the binomial expansion for $(a + b)^3$, later also for $(a + b)^n$, n a positive integer > 3. We give first a translation of the clumsy original version, then modify it somewhat by replacing Cavalieri's words by modern symbols. See on this subject H. Bosmans, "Un chapitre de l'oeuvre de Cavalieri. Les propositions XVI–XXVII de l'Exercitio quarta," *Mathesis 36* (1922), 365–373, 446–456.

Fig. 1

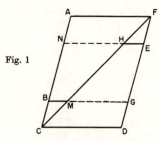

By the time Cavalieri's work was published, the integral of x^n was known to Torricelli, Fermat, Roberval, and possibly also to other mathematicians—each finding the result in his own way and some already generalizing it for fractional and negative n.

Proposition 21. All cubes of the parallelogram AD [Fig. 1] are the quadruple of all cubes of either triangles ACF or FDC.[1]

All cubes of parallelogram AD are equal to a.c. of [the line NH of] triangle ACF with a.s. of [the lines of] triangle FDC, together with three times a.l. of triangle FDC with a.s. of [the lines of] triangle ACF.[2]

[1] To understand this reasoning we reproduce first in our notation how Cavalieri shows that the area of a parallelogram $AFCD$ is twice that of one of the triangles into which it is divided by a diagonal FC. In Fig 1 draw $NE \parallel BG \parallel AF$ so that $HE = BM$, $NH = MG$. Take $HE = x$, $NH = y$, $AF = a$. Then $x + y = a$, hence $\sum x + \sum y = \sum a$ when we sum on all parallel lines NE from AF to CD, for to every NH there corresponds one and only one MG. But since $HE = BM$, $\sum x = \sum y$; hence $\sum x = \frac{1}{2} \sum a$. Here $\sum x$ is the area of triangle FCD, $\sum a$ that of the parallelogram. (If we introduce our present symbol Δx, we can write $(x + y) \Delta x = a \Delta x$, and thus we obtain $\sum x \Delta x = \frac{1}{2} \sum a \Delta x = \frac{1}{2} a \sum \Delta x = \frac{1}{2} a$, which shows that Cavalieri's formula is equivalent to $\int_0^a x \, dx = \frac{1}{2} a^2$.) Similarly, we obtain from $(x + y)^2 = a$:

$$\sum x^2 + 2 \sum xy + \sum y^2 = \sum a^2,$$
$$2 \sum x^2 + 2 \sum xy = \sum a^2.$$

To find $\sum xy$ we could write $x = \frac{1}{2}a + z$, $y = \frac{1}{2}a - z$. Then

$$\sum xy = \sum (\tfrac{1}{4}a^2 - x^2) = \tfrac{1}{4} \sum a^2 - \sum x^2.$$

From similar triangles we derive $\sum z^2 = \frac{1}{4} \sum x^2$. Hence

$$2 \sum x^2 + \tfrac{1}{2} \sum a^2 - \tfrac{1}{2} \sum x^2 = \sum a^2,$$

or
$$\sum x^2 = \tfrac{1}{3} \sum a^2.$$

[2] We follow the text step by step in modern notation. In Fig. 1 we take $AF = CD = a$, $HE = BM = x$, $NH = y$, $x + y = a$. Then Prop. 21 is that $\sum a^3 = 4 \sum x^3 = 4 \sum y^3$. The next paragraph states that

$$\sum (x + y)^3 = \sum x^3 + \sum y^3 + 3 \sum xy^2 + 3 \sum x^2 y$$

(a.l. = all lines, a.s. = all squares, a.c. = all cubes; "together with" means plus). Then

$$\sum (x + y)^3 : \sum (x + y)x^2 = \sum (x + y)^2 : \sum x^2 = 3 : 1,$$
$$\sum (x + y)^3 = 3 \sum (x + y)x^2,$$
$$\sum (x + y)x^2 = \sum x^2 y + \sum x^2 x = \sum x^2 y + \sum x^3,$$
$$\sum (x + y)^3 = 3 \sum x^3 + 3 \sum x^2 y,$$
$$\sum (x + y)^3 = \sum x^3 + \sum y^3 + 3 \sum xy^2 + 3 \sum x^2 y.$$

Subtracting $3 \sum x^2 y$ from both sides, we get

$$\sum x^3 + \sum y^3 + 3 \sum xy^2 = 3 \sum x^3;$$

but since
$$\sum x^3 + \sum y^3 = 2 \sum x^3, \quad \sum x^3 = \sum y^3,$$

we find
$$3 \sum xy^2 = 3 \sum x^2 y = \sum x^3,$$

hence
$$\sum x^3 + \sum y^3 + 3 \sum xy^2 + 3 \sum x^2 y = \sum (x + y)^3 = 4 \sum x^3 = 4 \sum y^3.$$

Now a.c. of AD are the product of a.l. AD by a.s. AD, and this is to the product of a.l. of AD by a.s. of triangle FDC as a.s. of the parallelogram AD is to a.s. of triangle FDC (because their altitude is the same, namely a.l. AD), and this ratio is 3. Hence a.c. of AD are equal to three times the product of a.l. of AD by a.s. of triangle FDC, and this is equal to the product of a.l. of triangle ACF by a.s. of triangle FDC plus the product of a.l. of triangle FDC by a.s. of the same triangle FDC, and this is equal to a.c. of triangle FDC. Hence a.c. of AD will be three times the sum of a.c. of triangle FDC and the product of a.l. of triangle ACF by a.s. of triangle FDC.

If now we resolve a.c. of AD into its parts, then we shall get a.c. of ACF plus a.c. of FDC plus three times the product of a.l. of triangle FDC by a.s. of triangle ACF plus three times the product of a.l. of triangle ACF by a.s. of triangle FDC. But three times the product of a.l. of triangle ACF by a.s. of triangle FDC is three times the same product. If we take it away three times of what we take away remains. So that a.c. of ACF plus a.c. of FDC and three times the product of a.l. of triangle FDC by a.s. of triangle FAC are three times a.c. of triangle FDC. Now a.c. of triangle ACF plus a.c. of triangle FCD are twice a.c. of triangle FCD, since a.c. of triangle ACF will be equal to a.c. of triangle FCD.

Hence three times the product of a.l. of triangle ACF by a.s. of triangle FCD together with three times the product of a.l. of triangle FCD by a.s. of triangle FAC and a.c. of the triangles ACF, FDC, that is a.c. of parallelogram AD, are equal to the quadruple of a.c. of triangle FDC (or triangle FAC).

This is so, since the product of a.l. of ACF by a.s. of FDC is equal to the product of a.l. of FDC by a.s. of ACF, and this is so because of the equality of the lines and their squares in those triangles FDC, ACF which alternately correspond. Hence three times the product of a.l. of ACF with a.s. of FDC are equal to three times the product of a.l. of FDC and a.s. of ACF. This makes the proof clear.[3]

---

[3] This reasoning can be much shortened in the following way. In order to find $\sum x^3$ we can proceed as before:

$$\sum (x + y)^3 = 2 \sum x^3 + 6 \sum x^2 y = \sum a^3.$$

But

$$\sum a^3 = a \sum a^2 = a\left(2 \sum x^2 + 2 \sum xy\right) = \tfrac{2}{3} \sum a^3 + 2a \sum xy$$

$$= \tfrac{2}{3} \sum a^3 + 2 \sum (x + y)xy = \tfrac{2}{3} \sum a^3 + 4 \sum x^2 y.$$

Hence

$$\sum x^2 y = \tfrac{1}{12} \sum a^3, \quad \text{and} \quad \sum x^3 = \tfrac{1}{4} \sum a^3.$$

We can in a similar way climb from x^3 to x^n, $n > 3$, to prove that

$$\sum x^n = \frac{1}{n + 1} a^n,$$

Proposition 22 proves that $\sum x^4 = \frac{1}{5} \sum a^4$. Proposition 23 is as follows:

Proposition 23. In any parallelogram such as BD [Fig. 2] with base CD we draw an arbitrary parallel EF to CD and the diagonal AC, intersecting EF in G. Then $DA : AF = (CD$ or $EF) : FG$. We call AC the prime diagonal. Then we construct point H on EF such that $DA^2 : AF^2 = EF : FH$, and so on all parallels to CD, so that all lines like this HF end on a curve CHA. In a similar way we construct a curve CIA, where $DA^3 : AF^3 = EF : FI$, a curve CLA such that $DA^4 : AF^4 = EF : FL$, etc. We call CHA the second diagonal, CIA the third, CLA the fourth, etc., and similarly $AGCD$ the first diagonal space of parallelogram BD, the trilinear figure $AHCD$ the second, $AICD$ the third, $ALCD$ the fourth, etc. Then I say that parallelogram BD is twice the first, three times the second, four times the third, five times the fourth space, etc.[4]

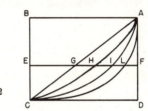

Fig. 2

Proposition 24. When a parallelogram and a parabola have the same base and the same axis, and are rotated about the base, then the cylinder generated by

---

which is equivalent to the integral

$$\int_0^a x^n \, dx = \frac{a^{n+1}}{n+1}.$$

We thus see how closely Cavalieri's method resembles our integration of polynomials. If we do not sum over all x, but only over the integers, we obtain (in our notation)

$$\lim_{a \to \infty} \frac{1^n + 2^n + \cdots + a^n}{a^{n+1}} = \frac{1}{n+1} \quad (a \text{ an integer}),$$

the generalization for integer $n > 2$ of the formula implied in Archimedes' quadrature of the parabola. For further information on this formula see G. Kowalewski, *Die klassischen Probleme der Analysis des Unendlichen* (Engelmann, Leipzig, 1910; 2nd ed., 1920), 44, 49, where this limit is named after Wallis. It is also implicit in the work of Roberval and Fermat around 1636–37.

[4] We give Cavalieri's demonstration in modern notation. We put $CE = a$, $DA = b$, $AF = z$, $FG = x$, $FH = x_1$, $FI = x_2$, $FL = x_3$, $GE = y$, $HE = y_1$ and the area of parallelogram $BD = \sum a = P$. Then the demonstration is as follows:

Parallelogram BD is twice the first space ACD, $P = \sum a = 2 \sum x$, as has been shown in previous propositions. Since $b^2 : z^2 = a^2 : x^2 = a : x_1$, we find that $a^2 : \sum x^2 = a : \sum x_1$, and, since according to a previous proposition $\sum x^2 = \frac{1}{3} \sum a^2 = \frac{1}{3} aP$, we conclude that $a^2 : \frac{1}{3}aP = a : \sum x_1$, or $P = 3 \sum x_1$; in other words, parallelogram BD is three times the space $AHCD$. In the same way we can show from $a^3 : x^3 = a : x_2$ and $\sum x^3 = \frac{1}{4} \sum a^3 = \frac{1}{4}a^2 P$ that $P = 4 \sum x_2$, or parallelogram BD is four times the space $AICD$. And so on.

the parallelogram is to the solid generated by the parabola (which Kepler called the parabolic spindle) as $15:8$.[5]

Cavalieri's summation of lines into areas and of areas into volumes can easily trip the unwary, as we have observed. Cavalieri was well aware of it, but expected that the difficulties would be removed in due time, that to cut the Gordian knot could be left to some later Alexander, as he put it. The correspondence of Cavalieri with his younger friend, Evangelista Torricelli (1608–1647), revealed some difficulties. For instance, in a letter from Cavalieri to Torricelli[6] the following paradox appears, paraphrased here in a modernized language:

Take a nonisosceles triangle ABC, of altitude AD. Draw an arbitrary line $PQ\|BC$, and draw PR and $QS\|AD$. Then $PR = QS$, hence $\sum PR = \sum QS$. But $\sum PR$, according to the theory of indivisibles, is equal to the area of $\triangle ABD$, and $\sum QS$ to that of $\triangle ACD$. These areas are, therefore, both equal and unequal.

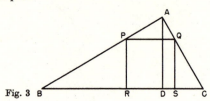

Fig. 3

---

[5] We use the figure of the previous propositions in which CHA is a parabolic curve with vertex A so that $AHCB$ is half a parabolic area. Let AB be the axis, and let parallelogram BD together with the half-parabolic area $AHCB$ be rotated about the half-base BC. Then BD generates a cylinder, and $AHCB$ half a parabolic spindle. We must prove that the cylinder is to the half-spindle (and twice the cylinder to the whole spindle) as $15:8$. For this purpose we start as in the previous proposition, with

$$a : x_1 = b^2 : x^2 = a^2 : x^2, \quad \text{so that} \quad a^2 : x_1^2 = a^4 : x^4.$$

Now

$$\sum a^2 : \sum ax_1 = \sum a : \sum x_1 = 3:1 = 15:5.$$

But

$$\sum ax_1 = \sum (x_1 + y_1) \sum x_1 = \sum x_1^2 + \sum x_1 y_1 = \tfrac{5}{15} \sum a^2,$$

and since

$$\sum a^2 : \sum x_1^2 = \sum a^4 : \sum x^4 = 5:1, \quad \text{or} \quad \sum x_1^2 = \tfrac{3}{15} \sum a^2,$$

we conclude that

$$\sum x_1 y_1 = \tfrac{2}{15} \sum a^2, \quad \text{or} \quad 2 \sum x_1 y_1 = \tfrac{4}{15} \sum a^2,$$

and

$$\sum x_1^2 + 2 \sum x_1 y_1 = \tfrac{7}{15} \sum a^2, \quad \text{or} \quad \sum y_1^2 = \tfrac{8}{15} \sum a^2.$$

The solid generated by BD is to the solid generated by $AHCB$ as $\sum a^2$ is to $\sum y_1^2$, and therefore as $15:8$.

[6] E. Torricelli, *Opere* (Montanari, Faenza, 1919), I, 170–171.

Cavalieri solves this paradox by considering the lines PR, QS as threads of a fabric. If $AB = 2AC$, and if AC contains 100 points, then AB contains 200 points, and hence there are 100 threads in ADC against 200 threads in ADB. He here exchanges his indivisibles of one dimension for infinitesimals (or finite differences) of two dimensions, an important step which he did not consistently follow up. It is exactly this type of pitfall that Leibniz avoided when he changed his original notation $\int_y = omnes\ linae\ y$ into $\int y\,dx$. See on this subject H. Bosmans, "Sur une contradiction reprochée à la théorie des 'indivisibles' de Cavalieri," *Annales de la Société Scientifique de Bruxelle 42* (1922), 82–89; E. Bortolotti, "I progressi del metodo infinitesimale nell'opera geometrica di Evangelista Torricelli," *Periodico de Matematiche [4] 8* (1928), 19–59.

7 FERMAT. INTEGRATION

Many mathematicians tried to generalize Cavalieri's integral $\int_0^a x^n\,dx = a^{n+1}/(n+1)$, n a positive integer, to n fractional or negative. Pierre Fermat and Evangelista Torricelli accomplished this about 1640 by an investigation of the area between an arc of the "hyperbola" $x^m y^n = k$ (m, n positive integers), an ordinate and an asymptote. Their approach was purely geometric. Fermat, of whose work on number theory we gave examples in our section on Arithmetic, solved the problem in the following way. We take our text from *Oeuvres*, I (1891), 255–259; III (1896), 216–219.

ON THE TRANSFORMATION AND SIMPLIFICATION OF THE EQUATIONS OF LOCI

APPLICATION TO THE COMPARISON OF ALL FORMS OF CURVILINEAR AREAS, EITHER AMONG THEMSELVES, OR WITH RECTILINEAR ONES

APPLICATIONS OF THE GEOMETRIC PROGRESSION TO THE QUADRATURE[1] OF PARABOLAS AND INFINITE HYPERBOLAS

Archimedes did not employ geometric progressions except for the quadrature of the parabola; in comparing various quantities he restricted himself to arithmetic progressions. Was this because he found that the geometric progression was less suitable for the quadrature? Was it because the particular device that he used to square the parabola by this progression can only with difficulty be applied to other cases? Whatever the reason may be, I have recognized and proved that this progression is very useful for quadratures, and I am willing to present to modern mathematicians my invention which permits us to square, by a method absolutely similar, parabolas as well as hyperbolas.

  The entire method is based on a well-known property of the geometric progression, namely the following theorem:

---

[1] Fermat uses the Greek term *tetragonizein* for "to perform a quadrature," a practice not uncommon in the seventeenth century.

*Given a geometric progression the terms of which decrease indefinitely, the dif-
ference between two consecutive terms of this progression is to the smaller of them as
the greater one is to the sum of all following terms.*[2]

This established, let us discuss first the quadrature of hyperbolas:

I define hyperbolas as curves going to infinity, which, like *DSEF* [Fig. 1],
have the following property. Let *RA* and *AC* be asymptotes which may be
extended indefinitely; let us draw parallel to the asymptotes any lines *EG*, *HI*,
NO, *MP*, *RS*, etc. We shall then always have the same ratio between a given
power of *AH* and the same power of *AG* on one side, and a power of *EG* (the
same as or different from the preceding) and the same power of *HI* on the other.

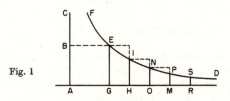

Fig. 1

I mean by powers not only squares, cubes, fourth powers, etc., the exponents of
which are 2, 3, 4, etc., but also simple roots the exponent of which is unity.[3]

I say that all these infinite hyperbolas except the one of Apollonius,[4] *or the first,
may be squared by the method of geometric progression according to a uniform and
general procedure.*

Let us consider, for example, the hyperbolas the property of which is defined
by the relations $AH^2/AG^2 = EG/HI$ and $AO^2/AH^2 = HI/NO$, etc. I say that
the indefinite area which has for base *EG* and which is bounded on the one side
by the curve *ES* and on the other side by the infinite asymptote *GOR* is equal
to a certain rectilinear area.

Let us consider the terms of an indefinitely decreasing geometric progression;
let *AG* be the first term, *AH* the second, *AO* the third, etc. Let us suppose that
those terms are close enough to each other that following the method of Archimedes we could adequate [*adégaler*] according to Diophantus,[5] that is, equate
approximately the rectilinear parallelogram *GE × GH* and the general quadrilateral *GHIE*; in addition we shall suppose that the first intervals *GH*, *HO*, *OM*,
etc. of the consecutive terms are sufficiently equal that we can easily employ
Archimedes' method of exhaustion by circumscribed and inscribed polygons.
It is enough to make this remark once and we do not need to repeat it and insist
constantly upon a device well known to mathematicians.

---

[2] This is Fermat's way of expressing that the sum of a convergent series $a + ar + ar^2
+ \cdots + ar^n + \cdots = a/(1 - r)$.

[3] This may mean "exponents that are unit fractions."

[4] The hyperbola of Apollonius is the ordinary hyperbola, of which, if its equation is
$xy = a^2$, the integral $\int_0^\infty y \, dx$ diverges.

[5] The term *adequatio* is a Latin translation of the Greek term *parisótēs*, by which Diophantus denoted an approximation to a certain number as closely as possible. See T. L.
Heath, *Manual of Greek mathematics* (Clarendon Press, Oxford, 1931), 493. Fermat uses the
term to denote what we call a limiting process.

Now, since $AG/AH = AH/AO = AO/AM$, we have also $AG/AH = GH/HO = HO/OM$, for the intervals. But for the parallelograms,

$$\frac{EG \times GH}{HI \times HO} = \frac{HI \times HO}{ON \times OM}.$$

Indeed, the ratio $EG \times GH/HI \times HO$ of the parallelograms consists of the ratios EG/HI and GH/HO; but, as indicated, $GH/HO = AG/AH$; therefore, the ratio $EG \times GH/HI \times HO$ can be decomposed into the ratios EG/HI and AG/AH. On the other hand, by construction, $EG/HI = AH^2/AG^2$ or AO/AG, because of the proportionality of the terms; therefore, the ratio $GE \times GH/HI \times HO$ is decomposed into the ratios AO/AG and AG/GH; now AO/AH is decomposed into the same ratios; we find consequently for the ratio of the parallelograms: $EG \times GH/HI \times HO = AO/AH = AH/AG$.

Similarly we prove that $HI \times HO/NO \times MO = AO/AH$.

But the lines AO, AH, AG, which form the ratios of the parallelograms, define by their construction a geometric progression; hence the infinitely many parallelograms $EG \times GH$, $HI \times HO$, $NO \times OM$, etc., will form a geometric progression, the ratio of which will be AH/AG. Consequently, according to the basic theorem of our method, GH, the difference of two consecutive terms, will be to the smaller term AG as the first term of the progression, namely, the parallelogram $GE \times GH$, to the sum of all the other parallelograms in infinite number. According to the adequation of Archimedes, this sum is the infinite figure bounded by HI, the asymptote HR, and the infinitely extended curve IND.

Now if we multiply the two terms by EG we obtain $GH/AG = EG \times GH/EG \times AG$; here $EG \times GH$ is to the infinite area the base of which is HI as $EG \times GH$ is to $EG \times AG$. Therefore, the parallelogram $EG \times AG$, which is a given rectilinear area, is adequated to the said figure; if we add on both sides the parallelogram $EG \times GH$, which, because of infinite subdivisions, will vanish and will be reduced to nothing, we reach a conclusion that would be easy to confirm by a more lengthy proof carried out in the manner of Archimedes, namely, that for this kind of hyperbola the parallelogram AE is equivalent to the area bounded by the base EG, the asymptote GR, and the curve ED infinitely extended.

It is not difficult to extend this idea to all the hyperbolas defined above except the one that has been indicated.

Fermat then extends his method to parabolas. His reasoning can be translated as follows.

Divide the interval $0 \leqslant x < a$ into parts by the points $x_1 = a$, $x_2 = ar$, $x_3 = ar^2, \ldots, r < 1$, which are separated by the intervals $l_1 = a(1 - r)$, $l_2 = ar(1 - r)$, $l_3 = ar^2(1 - r), \ldots$. If $y = x^n$ ($n = p/q$, $p, q \gtrless 0$) is the equation of the "hyperbola" or "parabola," then the values of y corresponding to $x_1, x_2, x_3, \ldots$ are $y_1 = a^n$, $y_2 = a^n r^n$, $y_3 = a^n r^{2n}, \ldots$. Then the sum S of the rectangles $l_1 x_1 + l_2 x_2 + l_3 x_3 + \cdots$ is

$$S = a(1 - r)a^n + ar(1 - r)a^n r^n + ar^2(1 - r)a^n r^{2n} + \cdots$$

$$= (1 - r)a^{n+1}(1 + r^{n+1} + r^{2n+2} + \cdots) = \frac{1 - r}{1 - r^{n+1}} a^{n+1}.$$

When $r = s^q$ ($s < 1$) and $n \neq -1$, then

$$\int_0^a x^n \, dx = a^{n+1} \lim \frac{1 - r}{1 - r^{n+1}} = a^{n+1} \lim \frac{1 - s^q}{1 - s^{p+q}}$$

$$= \frac{q a^{n+1}}{p + q} = \frac{a^{n+1}}{n + 1}.$$

As we see, this procedure holds for n positive and negative, but it fails for $n = -1$.

This method approaches our modern method of limits; it uses the concept of the limit of an infinite geometric series.

8   FERMAT. MAXIMA AND MINIMA

Modern textbooks on calculus take up first the differential and then the integral calculus. It may therefore come as a surprise to find that up to the middle of the seventeenth century the whole theory of infinitesimals concentrated on the computation of areas, volumes, and centers of gravity, that is, on what we now call the integral calculus. Tangent constructions were, until that period, based on the property that the tangent has only one point in common with the curve, as we can see in Euclid or Apollonius. Archimedes, in his book on spirals, found tangents by a method that seems to have been inspired by kinematic considerations. Even Torricelli, when determining the tangent at a point of the "hyperbola" $x^m y^n = k$, still used the ancient method (A. Agostini, "Il metodo delle tangenti fondato sopra la dottrina dei moti nelle opere di Torricelli," *Periodico di matematica* [4] 28 (1950), 141–158), and Descartes sought the normal prior to the tangent, and found it in some cases of algebraic curves by asking for double roots of a certain equation that expresses the abscissa of the intersections of the curve with a circle.

The beginning of the differential calculus, in which the tangent appears as the limit of a secant, can be studied in considerations concerning maxima and minima, as in Kepler's *Nova stereometria doliorum vinariorum* (Linz, 1615; see Selection IV.2). Here we read that "near a maximum the decrements on both sides are in the beginning only imperceptible" (decrementa habet insitio insensibilia; *Opere*, IV (1863), 612).

With Fermat we obtain an algorithm based on this fact. To understand his approach and its subsequent development into the method of the "characteristic triangle" (dx, dy, ds) we must take notice of the fact that Fermat and Descartes were among the first to apply the new algebra developed by Cardan, Bombelli, and Viète to the geometry of the ancients. This was, as we have seen, the beginning of the coordinate method. Descartes published his method in 1637, but Fermat's discovery was known only through his correspondence until 1679, the year of the publication of his works. Here is Fermat's approach, from his *Oeuvres*, III (1896), 121–123. It is followed by a paper in which he applied his method to the finding of a center of gravity (*Ibid.*, 124–126).

(1) ON A METHOD FOR THE EVALUATION OF MAXIMA AND MINIMA[1]

The whole theory of evaluation of maxima and minima presupposes two unknown quantities and the following rule:

Let a be any unknown of the problem (which is in one, two, or three dimensions, depending on the formulation of the problem). Let us indicate the maximum or minimum by a in terms which could be of any degree. We shall now replace the original unknown a by $a + e$ and we shall express thus the maximum or minimum quantity in terms of a and e involving any degree. We shall adequate [*adégaler*], to use Diophantus' term,[2] the two expressions of the maximum or minimum quantity and we shall take out their common terms. Now it turns out that both sides will contain terms in e or its powers. We shall divide all terms by e, or by a higher power of e, so that e will be completely removed from at least one of the terms. We suppress then all the terms in which e or one of its powers will still appear, and we shall equate the others; or, if one of the expressions vanishes, we shall equate, which is the same thing, the positive and negative terms. The solution of this last equation will yield the value of a, which will lead to the maximum or minimum, by using again the original expression.

Here is an example:

To divide the segment AC [Fig. 1] at E so that AE × EC may be a maximum.

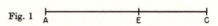

Fig. 1

We write $AC = b$; let a be one of the segments, so that the other will be $b - a$, and the product, the maximum of which is to be found, will be $ba - a^2$. Let now $a + e$ be the first segment of b; the second will be $b - a - e$, and the product of the segments, $ba - a^2 + be - 2ae - e^2$; this must be adequated with the preceding: $ba - a^2$. Suppressing common terms: $be \sim 2ae + e^2$. Suppressing e: $b = 2a$.[3] To solve the problem we must consequently take the half of b.

We can hardly expect a more general method.

ON THE TANGENTS OF CURVES

We use the preceding method in order to find the tangent at a given point of a curve.

Let us consider, for example, the parabola BDN [Fig. 2] with vertex D and of diameter DC; let B be a point on it at which the line BE is to be drawn tangent to the parabola and intersecting the diameter at E.

---

[1] This paper was sent by Fermat to Father Marin Mersenne, who forwarded it to Descartes. Descartes received it in January 1638. It became the subject of a polemic discussion between him and Fermat (*Oeuvres*, I, 133). On Mersenne, see Selection I.6, note 1.

[2] See Selection IV.7, note 5.

[3] Our notation is modern. For instance, where we have written (following the French translation in *Oeuvres*, III,122) $be \sim 2ae + e^2$, Fermat wrote: B in E adaequabitur A in E bis $+ Eq$ (Eq standing for E quadratum). The symbol $\sim$ is used for "adequates."

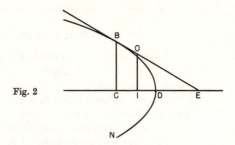

Fig. 2

We choose on the segment BE a point O at which we draw the ordinate OI; also we construct the ordinate BC of the point B. We have then: $CD/DI > BC^2/OI^2$, since the point O is exterior to the parabola. But $BC^2/OI^2 = CE^2/IE^2$, in view of the similarity of triangles. Hence $CD/DI > CE^2/IE^2$.

Now the point B is given, consequently the ordinate BC, consequently the point C, hence also CD. Let $CD = d$ be this given quantity. Put $CE = a$ and $CI = e$; we obtain

$$\frac{d}{d-e} > \frac{a^2}{a^2 + e^2 - 2ae}. \quad^4$$

Removing the fractions:

$$da^2 + de^2 - 2dae > da^2 - a^2e.$$

Let us then adequate, following the preceding method; by taking out the common terms we find:

$$de^2 - 2dae \sim -a^2e,$$

or, which is the same,

$$de^2 + a^2e \sim 2dae.$$

Let us divide all terms by e:

$$de + a^2 \sim 2da.$$

On taking out de, there remains $a^2 = 2da$, consequently $a = 2d$.

Thus we have proved that CE is the double of CD—which is the result.

This method never fails and could be extended to a number of beautiful problems; with its aid, we have found the centers of gravity of figures bounded by straight lines or curves, as well as those of solids, and a number of other results which we may treat elsewhere if we have time to do so.

I have previously discussed at length with M. de Roberval[5] the quadrature of areas bounded by curves and straight lines as well as the ratio that the solids which they generate have to the cones of the same base and the same height.

---

[4] Fermat wrote: D ad $D - E$ habebit majorem proportionem quam $Aq.$ ad $Aq. + Eq. - A$ in E bis (D will have to $D - E$ a larger ratio than A^2 to $A^2 + E^2 - 2AE$).

[5] See the letters from Fermat to Roberval, written in 1636 (*Oeuvres*, III, 292–294, 296–297).

Now follows the second illustration of Fermat's "*e*-method," where Fermat's e = Newton's o = Leibniz' dx.[6]

---

(2) CENTER OF GRAVITY OF PARABOLOID OF REVOLUTION, USING THE SAME METHOD[7]

Let $CBAV$ (Fig. 3) be a paraboloid of revolution, having for its axis IA and for its base a circle of diameter CIV. Let us find its center of gravity by using the same method which we applied for maxima and minima and for the tangents of curves; let us illustrate, with new examples and with new and brilliant applications of this method, how wrong those are who believe that it may fail.

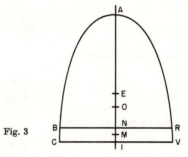

Fig. 3

In order to carry out this analysis, we write $IA = b$. Let O be the center of gravity, and a the unknown length of the segment AO; we intersect the axis IA by any plane BN and put $IN = e$, so that $NA = b - e$.

It is clear that in this figure and in similar ones (parabolas and paraboloids) the centers of gravity of segments cut off by parallels to the base divide the axis in a constant proportion (indeed, the argument of Archimedes can be extended by similar reasoning from the case of a parabola to all parabolas and paraboloids of revolution[8]). Then the center of gravity of the segment of which NA is the axis and BN the radius of the base will divide AN at a point E such that $NA/AE = IA/AO$, or, in formula, $b/a = (b - e)/AE$.

---

[6] The gist of this method is that we change the variable x in $f(x)$ to $x + e$, e small. Since $f(x)$ is stationary near a maximum or minimum (Kepler's remark), $f(x + e) - f(x)$ goes to zero faster than e does. Hence, if we divide by e, we obtain an expression that yields the required values for x if we let e be zero. The legitimacy of this procedure remained, as we shall see, a subject of sharp controversy for many years. Now we see in it a first approach to the modern formula: $f'(x) = \lim_{e \to 0} \dfrac{f(x + e) - f(x)}{e}$, introduced by Cauchy (1820–21).

[7] This paper seems to have been sent in a letter to Mersenne written in April 1638, for transmission to Roberval. Mersenne reported its contents to Descartes. Fermat used the term "parabolic conoid" for what we call "paraboloid of revolution."

[8] "All parabolas" means "parabolas of higher order," $y = kx^n$, $n > 2$. The reference is to Archimedes' *On floating bodies*, II, Prop. 2 and following; see T. L. Heath, *The works of Archimedes* (Cambridge University Press, Cambridge, England, 1897; reprint, Dover, New York), 264ff.

The portion of the axis will then be $AE = (ba - ae)/b$ and the interval between the two centers of gravity, $OE = ae/b$.

Let M be the center of gravity of the remaining part $CBRV$; it must necessarily fall between the points N, I, inside the figure, in view of Archimedes' postulate 9 in *On the equilibrium of planes*, since $CBRV$ is a figure completely concave in the same direction.[9]

But

$$\frac{\text{Part } CBRV}{\text{Part } BAR} = \frac{OE}{OM},$$

since O is the center of gravity of the whole figure CAV and E and M are those of the parts.

Now in the paraboloid of Archimedes,

$$\frac{\text{Part } CAV}{\text{Part } BAR} = \frac{IA^2}{NA^2} = \frac{b^2}{b^2 + e^2 - 2be};$$

hence by dividing,

$$\frac{\text{Part } CBRV}{\text{Part } BAR} = \frac{2be - e^2}{b^2 + e^2 - 2be}.$$

But we have proved that

$$\frac{\text{Part } CBRV}{\text{Part } BAR} = \frac{OE}{OM}.$$

Then in formulas,

$$\frac{2be - e^2}{b^2 + e^2 - 2be} = \frac{OE \ (= ae/b)}{OM};$$

hence

$$OM = \frac{b^2ae + ae^3 - 2bae^2}{2b^2a - be^2}.$$

From what has been established we see that the point M falls between points N and I; thus $OM < OI$; now, in formula, $OI = b - a$. The question is then prepared from our method, and we may write

$$b - a \sim \frac{b^2ae + ae^3 - 2bae^2}{2b^2e - be^2}.$$

Multiplying both sides by the denominator and dividing by e:

$$2b^3 - 2b^2a - b^2e + bae \sim b^2a + ae^2 - 2bae.$$

[9] This is postulate 7 in the modern Heiberg edition, and is translated in Heath, p. 190, as follows: "In any figure whose perimeter is concave in (one and) the same direction the center of gravity must be within the figure." (On the term "concave in the same direction," see Heath, p. 2.)

Since there are no common terms, let us take out those in which e occurs and let us equate the others:

$$2b^3 - 2b^2a = b^2a, \quad \text{hence} \quad 3a = 2b.$$

Consequently

$$\frac{IA}{AO} = \frac{3}{2}, \quad \text{and} \quad \frac{AO}{OI} = \frac{2}{1},$$

and this was to be proved.[10]

The same method applies to the centers of gravity of all the parabolas ad infinitum as well as those of paraboloids of revolution. I do not have time to indicate, for example, how to look for the center of gravity in our paraboloid obtained by revolution about the ordinate;[11] it will be sufficient to say that, in this conoid, the center of gravity divides the axis into two segments in the ratio 11/5.

---

9 TORRICELLI. VOLUME OF AN INFINITE SOLID

Evangelista Torricelli (1608–1647) succeeded Galilei at Florence as mathematician to the grand duke of Tuscany. He was well acquainted with the works of Archimedes, Galilei, and Cavalieri, and corresponded with Mersenne, Roberval, and other mathematicians. He computed many areas, volumes, and tangents, discussed the cycloid, performed what we now see as partial integration, and had an idea of the inverse character of tangent and area problems. He was aware of the logical difficulties in the method of indivisibles (see Selection IV.6). Torricelli is best known as a physicist (we speak of the "vacuum of Torricelli" in the mercury barometer), but his *Opere* (ed. G. Loria and G. Vassura, 3 vols.; Montanari, Faenza, 1919) show his ingenuity also in mathematics. From the *Opere* his manuscript "De infinitis spiralibus" (c. 1646) has been republished (with improved text) with an Italian translation by E. Carruccio (Domus Galilaeana, Pisa, 1955). Our selection is from *De solido hyperbolico acuto* (c. 1643), not published until 1919 in the *Opere*, vol. I, part 1, pp. 191–221. Here we see how he integrated, by a purely geometric method, an integral with an infinite range of integration, but yet finite, something quite remarkable in those days. The method used is that of indivisibles, in this case formed by circles in parallel planes.

ON THE ACUTE HYPERBOLIC SOLID

Consider a hyperbola of which the asymptotes AB, AC enclose a right angle [Fig. 1]. If we rotate this figure about the axis AB, we create what we shall call

---

[10] These relations were known to Archimedes (see note 8). But Fermat solved this problem on centers of gravity, hence a problem in the integral calculus, with what we might call an application of the principle of virtual variations.

[11] Here ACI of Fig. 3 is rotated about CI.

an acute hyperbolic solid, which is infinitely long in the direction of B. Yet this solid is finite. It is clear that there are contained within this acute solid rectangles through the axis AB, such as $DEFG$. I claim that such a rectangle is equal to the square of the semiaxis of the hyperbola.[1]

We draw from A, the center of the hyperbola, the semiaxis AH, which bisects the angle BAC. This gives us the rectangle $AIHC$, which is certainly a square (it is a rectangle and the angle at A is bisected by the axis AH). Therefore the square of AH is twice the square $AIHC$, or twice the rectangle AF, and therefore equal to the rectangle $DEFG$, as claimed.[2]

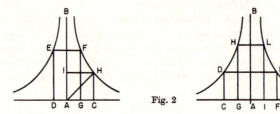

Fig. 1 Fig. 2

Lemma 2. All cylinders described within the acute hyperbolic solid and constructed about the common axis are isoperimetric (I always mean without their bases). Consider the acute solid with axis AB [Fig. 2] and visualize within it the arbitrary cylinders $CDEF$, $GHLI$, drawn about the common axis AB. The rectangles through the axes CE, GL are equal and so the curved surfaces of the cylinders will be equal. Q.E.D.[3]

Lemma 3. All isoperimetric cylinders (for instance, those that are drawn within the acute hyperbolic solid) are to each other as the diameters of their bases. Indeed, in Fig. 2, the rectangles AE, AL are equal, hence $FE : IL = AI : AF$. The cylinder CE has to cylinder GL a ratio composed of $AF^2 : AI^2$ and of $FE : IL$, or of $FA : IA$, or of $FA^2 : AI$ times AF. The cylinders CE, GL are therefore to each other as FA^2 is to AI times AF, and thus as line FA is to line AI. Q.E.D.[4]

Lemma 4. Let ABC [Fig. 3] be an acute body with axis DB, D the center of the hyperbola (where the asymptotes meet), and DF the axis of the hyperbola.

---

[1] Torricelli speaks of the *latus versum* where we speak of the real axis. The term *latus versum*, or *latus transversum*, is a translation of a Greek term used by Apollonius; see T. L. Heath, *Manual of Greek mathematics* (Clarendon Press, Oxford, 1931), 359. In the present case, taking the rectangular asymptotes as X- and Y-axes (AB the axis of positive Y), the equation of the hyperbola is $xy = \frac{1}{4}a^2$, if the length of the *latus versum* is $2a$.

[2] The theorem used is $xy = $ const., which Torricelli takes (as he remarks in the margin) from Apollonius' *Conics*, II, Prop. 12; see T. L. Heath, *Apollonius of Perga* (Cambridge University Press, Cambridge, England, 1896).

[3] Here Torricelli quotes Archimedes, *On the square and cylinder*, I, Prop. 6; see T. L. Heath, *The works of Archimedes* (Cambridge University Press, Cambridge, England, 1897; reprint Dover, New York).

[4] This reasoning seems rather clumsy to us, since we see immediately that $x_1^2 y_1 : x_2^2 y_2 = x_1 : x_2$, when $x_1 y_1 = x_2 y_2$ (= const.). However, to restate this reasoning in the geometric form usual in the seventeenth century (comparing and transforming parallelepipeds) would take as much space as Torricelli needs. The phrase "composed of $AF^2 : AI^2$ and of $FE : IL$" means $(AF^2 : AI^2) \times (FE : IL)$. The text has $IA : AF$, which should be $AF : IA$.

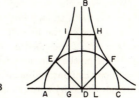

Fig. 3

We construct the sphere $AEFC$ with center D and radius DF. This is the largest sphere with center D that can be described in the acute body. We take an arbitrary cylinder contained in the acute body, say $GIHL$. I claim that the surface of cylinder GH is one-fourth that of the sphere $AEFC$.

Indeed, since the rectangle GH through the axis of the cylinder is equal to DF^2, hence to the circle $AEFC$, therefore this cylindrical surface $GIHL = \frac{1}{4}$ the surface of the sphere $AEFC$, of which the great circle $AEFG$ is also one-fourth.

Lemma 5. The surface of any cylinder $GHIL$ described in the acute solid (the surface without bases) is equal to the circle of radius DF, which is the semiaxis, or half the latus versum of the hyperbola, for this is proved in the demonstration of the preceding lemma.

Theorem. An acute hyperbolic solid, infinitely long, cut by a plane [perpendicular] to the axis, together with the cylinder of the same base, is equal to that right cylinder of which the base is the latus versum (that is, the axis) of the hyperbola, and of which the altitude is equal to the radius of the basis of this acute body.

Consider a hyperbola of which the asymptotes AB, AC [Fig. 4] enclose a right angle. We draw from an arbitrary point D of the hyperbola a line DC parallel to AB, and DP parallel to AC. Then the whole figure is rotated about AB as axis, so that the acute hyperbolic solid EBD is formed together with a cylinder $FEDC$ with the same base. We extend BA to H, so that AH is equal to the entire axis, that is, the latus versum of the hyperbola. And on the diameter AH we imagine a circle [in the plane] constructed perpendicular to the asymptote AC, and over the base AH we conceive a right cylinder $ACGH$ of altitude

Fig. 4

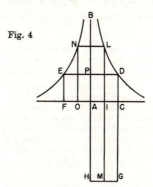

AC, which is the radius of the base of the acute solid. I claim that the whole body $FEBDC$, though long without end, yet is equal to the cylinder $ACGH$.

We select on the line AC an arbitrary point I and we form the cylindrical surface $ONLI$ inscribed in the acute solid about the axis AB, and likewise the circle IM on the cylinder $ACGH$ parallel to the base AH. Then we have, according to our lemma: (cylindrical surface $ONLI$) is to (circle IM) as (rectangle OL through the axis) is to (square of the radius of circle OM), hence as (rectangle OL) is to (square of the semiaxis of the hyperbola).

And this will always be true no matter where we take point I. Hence all cylindrical surfaces together, that is, the acute solid EBD itself, plus the cylinder of the base $FEDC$, will be equal to all the circles together, that is, to the cylinder $ACGH$. Q.E.D.[5]

Scholium. It might seem incredible that, though this body has an infinite length, yet none of those cylindrical surfaces which we have considered has infinite length. Each of them is limited, as is obvious to anybody who is even moderately familiar with the theory of conics.

The truth of the preceding theorem is sufficiently clear in itself, and it is, I think, sufficiently confirmed by the examples at the beginning of this paper. However, I shall, in order to satisfy in this also the reader who has his doubts about the indivisibles, repeat the same demonstration at the end of this work, in the accustomed way of demonstration as used by the ancient geometers, a way longer, but to me therefore not necessarily safer.[6]

But before we do this, first something else. Since we have given demonstrations about that acute solid of which the asymptotes of the generating hyperbola form a right angle, we shall here in passing state without demonstration to which figures the acute solids are equal, when the asymptotes are at an obtuse, or an acute, angle.

We omit the proofs to avoid ballast; the industrious reader will be able to supply them with little effort.

Let a hyperbola be given of which the asymptotes AB, AC form an obtuse angle. Revolve the figure around the axis AB. Then we will obtain an acute solid, infinitely long toward B, which we cut by a plane DE perpendicular to the axis. Then [Fig. 5] the acute body DBE is equal to the cylinder $DILE$ plus the cone IAL.[7] In Fig. 6, where the intersecting plane is DE, the whole acute solid that stands on the circle DE minus the cone OAV is equal to the cylinder

---

[5] The reasoning amounts to the evaluation of $\int_0^c 2\pi xy\, dx = \pi a^2 \int_0^c dx = \pi a^2 c$, where $OC = c$. The fact that astonished Torricelli, that the infinite extent of the solid does not imply infinite volume, can be expressed in our language by saying that $\int^\infty dy/y^2$ converges.

[6] The second part of the paper is entitled "On the dimension of the acute hyperbolic solid according to the methods of the ancients." Torricelli says that because of the infinite extent of the solid it is impossible to comprehend it between inscribed and circumscribed solids. Yet he had been able to find a proof with Archimedean methods, and, so he said, had Roberval. The proof takes much more space than that with indivisibles.

[7] If we take AB as the positive Y-axis and the X-axis perpendicular to AB at A, then the equation of the hyperbola is $mx^2 - xy + a^2 = 0$, where $y = mx$ is the equation of the asymptote AL. Then, if $m = -p$, and $DE = 2c$, solid $DBE = 2\pi \int_0^c x(y - y_0)\, dx$, where $-pc^2 - cy_0 + a^2 = 0$, or solid $DBE = \pi(a^2 c + \frac{1}{3}pc^3)$. The other theorems of Torricelli can be verified in a similar way.

IE taken together with the cone *IAC*. Let us now assume that the asymptotes meet at an acute angle, and let plane *CD* be as in Fig. 7. Then the acute solid *CHD* together with the cone *EAI* will be equal to the cylinder *CEDI*. And in Fig. 8 the whole acute solid consisting of the rotation of the mixed infinite quadrangle *ABCDA* will be twice the cylinder *IEDC*.

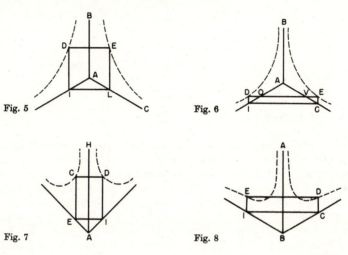

Fig. 5

Fig. 6

Fig. 7

Fig. 8

Then follow 29 corollaries, dealing with special properties of these figures.

Torricelli's paper contains discoveries that he announced to several mathematicians in letters written during 1643. The result was that before the end of the year Roberval, Fermat, and Mersenne were acquainted with them; see E. Bortolotti, *Archivio di storia della scienze 5* (1924), 212–213. Torricelli also discovered ("De infinitis spiralibus") that the arc length of a logarithmic spiral remains finite when it winds an infinite number of times around its asymptotic point; this paper is also reprinted in E. Torricelli, *Opere* (1919). See also on Torricelli's work A. Agostini, "Il problema inverso delle tangenti nelle opere di Torricelli," *Archeion 12* (1930), 33–37, E. Bortolotti, "Le 'Coniche' di Apollonio e il problema inverso delle tangenti de Torricelli," *ibid.*, 267–271, and Selection IV.6, note 6.

The existence of finite areas of infinite extent was already known to scholastic writers and can be found in the so-called sophismata literature of the fourteenth century (Suisseth, Oresme). The subject of this literature was the discussion of logical, mathematical, or physical antinomies (contradictions), which easily involved questions concerning the infinite and the infinitesimal; see A. Maier, *An der Grenze von Scholastik und Naturwissenschaften* (2nd ed.; Edizioni di Storia e Letteratura, Rome, 1953), 264–269, 336–338. An example of such a surface is a step figure (Fig. 9), of which the first step consists of a unit square ABB_1A_1, the second step of a rectangle $B_1C_1C_2B_2$, where C_1 is the center of A_1B_1 and $B_1B_2 = BB_1$, the third step of a rectangle $B_2D_2D_3B_3$, where D_2 is the center of B_2C_2 and $B_2B_3 = B_1B_2 = BB_1$, and so on. The total area is $(1 + \frac{1}{2} + \frac{1}{4} + \frac{1}{8} + \cdots) = 2$, since all rectangles above A_1B_1 can be placed inside the unit square and "exhaust" it.

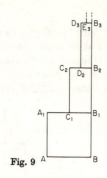

Fig. 9

10 ROBERVAL. THE CYCLOID

The cycloid has been traced back to the French theologian and mathematician Charles Bouvelles (*c.* 1470–*c.* 1553). Galilei was attracted by it, and wrote in a letter to Cavalieri of February 24, 1640 (*Opere*, Edizione nazionale (Barbera, Florence, 1890–1909), XVIII, 153–154): "More than 50 years ago this curved line came to my mind and I wanted to describe it, admiring it because of its most gracious curvature, adaptable to the arches of a bridge. I made several tentative calculations on it and on the space comprised between it and its chord, in order to demonstrate some property. And it seemed at first that such space may be three times the circle which it describes, but it was not that." Galilei gave the curve its name.

About 1630 Father Marin Mersenne (1588–1648), a correspondent of Descartes, Fermat, and many other mathematicians, suggested the cycloid as a test curve for the different methods of dealing with infinitesimals. It soon became one of the most discussed curves of the period, the discussion occasionally leading to acrimonious remarks, so that the curve has been compared to an apple of discord or called the Helen of the geometers. Among those who took up the challenge of Mersenne was Gilles Personne de Roberval (1602–1675), a professor of mathematics in Paris at the Collège du Roi (now Collège de France). From his *Traité des indivisibles* (1634; first published Paris, 1693; reprinted Paris, 1730; Amsterdam, 1736) we present here a section on the cycloid, translated rather freely (the original is somewhat prolix) by E. Walker in *A study of the Traité des Indivisibles* (Teachers College, New York, 1932). It shows how Roberval handled indivisibles,[1] and how he introduced the so-called companion of the cycloid, that is, the sine curve, which was long known under this name, even in the days of Euler. Roberval usually called the cycloid a *roulette*, a custom followed by Pascal; another name was *trochoid* (after Greek *trochos*, wheel). We have, with Walker, used the now customary term cycloid. The interest in this curve was also connected with the age-old speculation concerning the *rota Aristotelis* (see Selection IV.3).

On Roberval see further L. Auger, *Un savant méconnu, G. P. de Roberval* (Blanchard, Paris, 1962). On his mathematics see also C. B. Boyer, *The history of the calculus* (Dover,

---

[1] It is clear that Roberval, like Cavalieri, uses the method of indivisibles, of which he may have been an independent discoverer (Walker, *A study of the Traité des indivisibles*, 15, 142), but his view was somewhat different. He made clear in his *Traité* that the phrase "the infinite number of points" stands for the infinity of little lines which make up the whole line; see Boyer, *History of the calculus*, 141–142.

New York, 1949). On the cycloid see E. A. Whitman, "Some historical notes on the cycloid," *American Mathematical Monthly 50* (1943), 309–315.

We follow here, with some modifications, the text in Walker, 174–177, 219–222, corresponding to pp. 209ff of the 1736 edition of Roberval's *Traité*. The Walker version also introduces some modern symbolism, and the division into Propositions 1, 2, . . . is Walker's. Her book also contains translations and paraphrases of other sections of the *Traité*.

To Generate the Cycloid. Let the diameter AB [Fig. 1] of the circle $AEGB$ move along the tangent AC, always remaining parallel to its original position, until it takes the position CD, and let AC be equal to the semicircle AGB. At the same time, let the point A move on the semicircle AGB, in such a way that the speed of AB along AC may be equal to the speed of A along the semicircle AGB. Then, when AB has reached the position CD, the point A will have reached the position D. The point A is carried along by two motions—its own on the semicircle $AEGB$, and that of the diameter along AC. The path of the point A, due to these two motions, is the half cycloid $A \cdots D$, the second half being symmetrical with the first.

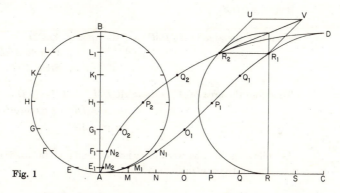

Fig. 1

The Nature of the Cycloid. Let the line AC and the semicircle AGB be divided into an infinite number of parts such that arc AE = arc EF = $\cdots$

$$= \text{line } AM = \text{line } MN = \text{line } NO = \cdots.$$

Draw the sine EE_1 perpendicular to the diameter AB, and the versed sine AE_1 is the altitude of A when it has come to E. Similarly draw FF_1, GG_1, etc.

Let MM_1 be parallel and equal to AE_1, NN_1 parallel and equal to AF_1, etc. Let M_1M_2 be parallel to AC and equal to EE_1, N_1N_2 parallel to AB and equal to FF_1, etc. [Roberval's notation for $M_1, N_1, \ldots$ is 1, 2, $\ldots$; for $M_2, N_2, \ldots$ is 8, 9, $\ldots$.]

When the diameter has reached the point M, the point A will have reached the position E, the distance of A above AC will be $MM_1 = AE_1$, and the distance of A from the diameter AB will be $EE_1 = M_1M_2$, hence when the diameter is at M the point A is at M_2. In the same way, when the diameter is at N the point A is at N_2, etc. We thus get two curves, one $AM_2N_2\cdots R_2D$, and the other $AM_1N_1\cdots R_1D$. The first of these is the path of the point A, which is the first half of the cycloid.

The Companion of the Cycloid. The curve drawn through the points $AM_1N_1\cdots R_1D$, is known as the companion of the cycloid.[2]

Proposition 1. The area of the figure included between the cycloid and the companion of the cycloid is equal to the area of half of the generating circle.

Proof. Within the figure $AM_2N_2\cdots D\cdots N_1M_1\cdots A$ we have $M_1M_2 = EE_1$, $N_1N_2 = FF_1$, $O_1O_2 = GG_1$, etc.

Now M_1M_2, N_1N_2, O_1O_2 divide this figure into strips whose altitudes are AE_1, E_1F_1, $F_1G_1,\ldots$, while EE_1, FF_1, $GG_1,\ldots$ divide the semicircle AHB into strips having the same altitudes. Hence the corresponding infinitesimal strips are equal. Therefore the area of the figure $AM_2N_2\cdots D\cdots N_1M_1\cdots A$ is equal to the area of the semicircle AHB.[3]

Proposition 2. The area of the figure included between the cycloid and its base is equal to three times the area of the generating circle.

Proof. The companion of the cycloid, the curve $AM_1N_1\cdots D$, bisects the parallelogram $ABCD$, since to each line in $ACDM_1$ there corresponds an equal line in $ABDM_1$.

Therefore the area of $ACDM_1 = \frac{1}{2}$ the area of $ABCD$
$$= \frac{1}{2} \quad ,, \quad ,, \quad ,, \; 2\cdot\text{circle } AGB$$
$$= \quad ,, \quad ,, \quad ,, \; \text{circle } AGB.$$

Therefore the area of $ACDM_2 = ACDM_1 + AM_2\cdots D\cdots M_1$
$$= \text{circle } AGB + \tfrac{1}{2} \text{ circle } AGB$$
$$= \tfrac{3}{2} \text{ circle } AGB.$$

Doubling, the area between the whole cycloid and its base is equal to three times the area of the generating circle.

Proposition 3. To construct a tangent to the cycloid.

Construction. Let R_2 be the given point at which the tangent is to be drawn. Draw R_2R_1 parallel to AC. Draw R_2U tangent to the generating circle RR_2 and make $R_2U = R_2R_1$. Complete the parallelogram R_2UVR_1, and draw the diagonal R_2V. Then R_2V is the required tangent.

Proof. The direction of the motion of the point R_2 which is due to the motion of AB along AC is R_2R_1; the direction of the motion of the point R_2 which is due to the motion of the point A on the circumference is R_2U, and since these motions are always equal, it follows that R_2R_1 must equal R_2U. Therefore R_2V is the tangent to the cycloid at R_2, since it is the resultant of the two motions.

---

[2] The "companion of the cycloid" is a sine curve. If AC is taken as the X-axis, AB as the Y-axis, its equation is, in our notation, $y = 1 - \cos x$.

[3] When arc $AE = \varphi$ [radius $R = 1$], then the equation of the cycloid is $x = \varphi - \sin\varphi$, $y = 1 - \cos\varphi$, and the area $AM_2DM_1A = \int_0^\pi (\varphi - \varphi + \sin\varphi)^2\, dy = \int_0^\pi \sin^2\varphi\, d\varphi = \pi/2$.

Addendum. If, instead of being equal, the magnitudes of the two motions had been in some other ratio, the parallelogram would have been constructed with its sides in that ratio.[4]

The next 22 Propositions deal with a number of area computations and other integrations. Curves discussed include the parabola, the limaçon of Pascal (which Roberval calls the conchoid of the circle), ring surfaces, the hyperboloid of revolution (which Roberval calls the hyperbolic conoid), cones, spheroids, the conchoid of Nicomedes, and the curve introduced as follows, which is known as the hippopede of Eudoxus.[5]

Proposition 26. On the surface of a right cylinder draw a line enclosing an area equal to the area of a given square, and that with a single stroke of the compasses.

Construction. Let AB [Fig. 2] be the side of the given square. Bisect AB at C. Describe a circle FME whose diameter FE is equal to AC. Construct a right cylinder whose midsection is the circle FME, and whose altitude is equal at

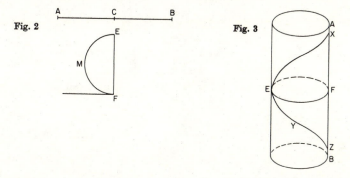

Fig. 2

Fig. 3

least to $2FE$. With F as a fixed point, and with an opening of the compasses equal to FE, draw a closed curve XYZ on the cylinder [Fig. 3]. Then XYZ is the required curve enclosing an area equal to the square on AB.

---

[4] The tangent construction uses kinematic concepts and is related to the method of Archimedes in his book *On spirals*; see T. L. Heath, *The works of Archimedes* (Cambridge University Press, Cambridge, England, 1897; reprint, Dover, New York), 151ff, esp. Props. 16–20. While Roberval used Greek methods to find tangents, his contemporary Fermat was laying the foundations of the present method, based on the derivative (see Selection IV.8).

[5] This curve, which plays a role in the planetary model constructed by Eudoxus (fourth century B.C.), was one of the curves discussed by Mersenne, Fermat, Roberval, and other mathematicians of their day, including the Toulouse mathematician Antoine de Lalouvère, who called the curve cyclocylindrique; *Veterum geometria promota in septem de cycloide libras* (Toulouse, 1660). The curve is the intersection of a sphere with a cylinder.

The proof is based on dividing the circumference FME into an indefinite number of equal parts. Then the text continues with the integration of the squares of sines and cosines, as follows.

Proposition 32. The sum of the squares of the sines on a semicircle is equal to one-eighth of the square of the diameter taken as many times as there are sines.

Proof. Let the circumference FR_1E (center $D = S_1$) be divided into an infinite number of equal parts, and let all the lines RS represent the sines of the successive arcs [Fig. 4, where the arc R_1F is divided into $n = 10$ equal sections by $R_1R_2 \cdots R_9$, and similarly for the arc R_1E]. Now

$$DR_1^2 = R_1S_1^2 + DS_1^2,$$
$$DR_2^2 = R_2S_2^2 + DS_2^2,$$

. . . . .

Fig. 4

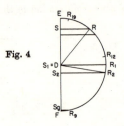

But

$$R_1S_1 = \sin(ER_1)$$
$$R_2S_2 = \sin(ER_2)$$

and

$$DS_2 = \sin(90° - ER_2)$$
$$= R_9S_9.$$

Likewise,

$$R_3S_3 = \sin(ER_3)$$

and

$$DS_3 = R_8S_8.$$

. . . . .

Thus each line DS is equal to a corresponding line RS which is the sine of one arc. Therefore, adding,

$$n \cdot DR^2 = 2 \sum RS^2 \text{ (from zero to } DR_1) = n \cdot \tfrac{1}{4} \cdot EF^2,$$

where n represents the measure of the quadrant arc. But in the semicircle there will be twice as many sines as we have here. Hence

$$\frac{2 \cdot \sum RS^2}{2 \cdot n \cdot EF^2} = \frac{1}{8},$$

or, since $2n$ is the measure of the semicircle, the sum of the squares of the sines in a semicircle is one-eighth of the square of the diameter multiplied by the number of units in the semicircle.[6]

Proposition 33. The sum of the squares of the versed sines in a semicircle is three-eighths of the square of the diameter taken as many times as there are versed sines.[7]

Proof. $FE^2 = FS_9^2 + S_9E^2 + 2FS_9 \cdot S_9E$

$$= FS_9^2 + S_9E^2 + 2R_9S_9^2,$$

$$FE^2 = FS_8^2 + S_8E^2 + 2R_8S_8^2,$$

Adding:

$$2n \cdot FE^2 = \sum FS^2 + \sum SE^2 + 2\left(2 \sum{}' RS\right),$$

where $2n$ = semicircle FRE; the sum $\sum$ is taken from 0 to FE and $\sum'$ from 0 to RD. Now by Proposition 32, $n \cdot FE^2 = 8 \sum' RS^2$; therefore

$$8\left(2 \sum{}' RS^2\right) = 2 \sum FS^2 + 2 \cdot \left(2 \sum{}' RS^2\right),$$

whence

$$6\left(2 \sum{}' RS^2\right) = 2 \sum FS^2,$$

or

$$2 \sum{}' RS^2 = \tfrac{1}{3} \sum FS^2.$$

But by Proposition 32,

$$2 \sum{}' RS^2 = \tfrac{1}{8}(2n \cdot EF^2).$$

---

[6] Hence $\int_0^\pi \sin^2 \varphi \, d\varphi = \frac{1}{8} \cdot 4 \cdot \pi = \pi/2$. Compare this result with that of Pascal (Selection III.7). In those days sines were taken as line segments, whose length depended on the radius R of the circle. The custom of taking $R = 1$, and hence of regarding sines (and cosines, tangents, etc.) as ratios, begins with Euler, *Introductio in analysin infinitorum* (Lausanne, 1748).

[7] The versed sines were introduced by versed sin $\alpha = R - \cos \alpha$. The companion of the cycloid is a versed-sine curve with respect to BA and BC as axes.

Therefore

$$\tfrac{1}{8}(2n \cdot EF^2) = \tfrac{1}{3} \sum FS^2$$

or

$$\sum FS^2 = \tfrac{3}{8}(2n \cdot EF^2).$$

or, in words, the sum of the squares of the versed sines in a semicircle is three-eighths of the square of the diameter multiplied by the number of units in the length of the semicircle.[8]

Proposition 34. The volume of the solid generated by the cycloid as it revolves about its base line as an axis is equal to five-eighths of the volume of the circumscribed cylinder.

In Fig. 2 the lines MM_1, $NN_1, \ldots$, are versed sines, hence it follows from Proposition 33 that

the solid generated by $AN_1DC = \tfrac{3}{8}$ the cylinder $ABDC$,

but

the solid generated by $AN_1DN_2 = \tfrac{1}{4}$,,   ,, $ABDC$,

and therefore, by addition,

the solid generated by $AN_2DC = \tfrac{5}{8}$,,   ,, $ABDC$.

But AN_2DC is only one-half of the cycloid, therefore the solid generated by the whole cycloid is five-eighths of the whole circumscribed cylinder.[9]

Notice that the solid generated by AN_2DN_1 is equal to the solid generated by the semicircle DC, because these two plane figures have their corresponding lines equal each to each and at the same distance from the axis AC; and the semicircle $DC = \tfrac{1}{4}$ of the parallelogram $ABDC$, hence the solid $AN_2DN_1 = \tfrac{1}{4}$ of the cylinder $ABDC$.

---

11 PASCAL. THE INTEGRATION OF SINES

Roberval was one of the men who influenced Blaise Pascal (on Desargues's influence see Selection III.7), who in his turn wrote a treatise on the cycloid, which he called a roulette (1658). The following paper, which still uses to a certain extent the notion of indivisibles, shows how Pascal integrated $\sin^n \varphi$, $n = 1, 2, 3, 4, \ldots$, making use thereby of a "characteristic triangle," though not yet that of (dx, dy, ds) which we often use now. It is entitled

[8] $\int_0^\pi (1 - \cos \varphi)^2 \, d\varphi = \tfrac{3}{8} \cdot 4 \cdot \pi = \dfrac{3\pi}{2}.$

[9] $\pi \int_0^\pi (1 - \cos \varphi)^3 \, d\varphi = \tfrac{5}{8} \cdot 4\pi \cdot \pi = \dfrac{5\pi^2}{2}.$

Traité des sinus du quart de cercle (1659); *Oeuvres*, ed. L. Brunschwicg and P. Boutroux (Hachette, Paris, 1914–1921), IX, 60–76.

Pascal's paper differs from that of Roberval: (*a*) in his partial rejection of indivisibles; (*b*) in his more general choice of the limits of the integration interval, so that here we may see a transition to the indefinite integral:

$$\int_{\varphi}^{\pi/2} \sin \varphi \, d\varphi = \cos \varphi, \qquad \int_{\varphi}^{\pi/2} \sin^2 \varphi \, d\varphi = \frac{\pi}{4} - \frac{\varphi}{2} + \frac{1}{2} \sin \varphi \cos \varphi,$$

and so forth.

ON THE SINES OF A QUADRANT OF A CIRCLE

Let ABC [Fig. 1] be a quadrant of a circle of which the radius AB will be considered the axis and the perpendicular radius AC the base; let D be any point on the arc from which the sine DI will be drawn to the radius AC; and let DE be the tangent on which we choose the points E arbitrarily, and from these points we draw the perpendiculars ER to the radius AC.[1]

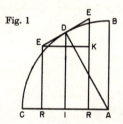

Fig. 1

I say that the rectangle formed by the sine[2] DI and the tangent EE is equal to the rectangle formed by a portion of the base (enclosed between the parallels) and the radius AB.

For the radius AD is to the sine DI as EE is to RR, or to EK, which is clear because of the similarity of the right-angled triangles DIA, EKE, the angle EEK or EDI being equal to the angle DAI.

Proposition I. The sum of the sines of any arc of a quadrant is equal to the portion of the base between the extreme sines, multiplied by the radius.[3]

---

[1] The triangle EEK of this figure led Leibniz to his early researches into the calculus; it gave him the idea of the "characteristic" triangle, when EE is small. The segment EE is a tangent in Pascal's essay. With Leibniz it became a chord. For the different forms of this triangle see D. Mahnke, "Neue Einblicke in die Entdeckungsgeschichte der höheren Analysis," *Abhandlungen der preussischen Akademie der Wissenschaften, Kl. Math. Phys. 1* (1925), 1–64.

[2] As with all authors up to the eighteenth century, Pascal's sine of an angle φ is a line, and not a ratio. It is what we now write $R \sin \varphi$, R being the radius of the circle.

[3] This is equivalent to our formula $\int_{\varphi_0}^{\varphi_1} \sin \varphi \, d\varphi = \cos \varphi_0 - \cos \varphi_1$.

Proposition II. The sum of the squares of those sines is equal to the sum of the ordinates[4] of the quadrant that lie between the extreme sines, multiplied by the radius.[5]

Proposition III. The sum of the cubes of the same sines is equal to the sum of the squares of the same ordinates between the extreme sines, multiplied by the radius.[6]

Proposition IV. The sum of the fourth powers of the same sines is equal to the sum of the cubes of the same ordinates between the extreme sines, multiplied by the radius.

And so on to infinity.

Preparation for the proof. Let any arc BP be divided into an infinite number of parts by the points D [Fig. 3] from which we draw the sines PO, DI, etc. ... ; let us take in the other quadrant of the circle the segment AQ, equal to AO (which measures the distance between the extreme sines of the arc, BA, PO); let AQ be divided into an infinite number of equal parts by the points H, at which the ordinates HL will be drawn.

Fig. 3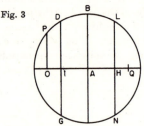

[4] To understand the difference between "ordinates" and "sines" we must take Pascal's *triligne* ("trilinear figure"), formed by two perpendicular lines AB and AC and a (convex) curve BLC (Fig. 2); see *Oeuvres*, VIII, 369). The points D divide AB into equal parts; the

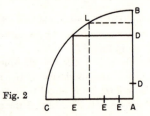

Fig. 2

points E do the same for AC, and the points L for the arc BLC (every arc is of the same length). Then: (a) the perpendiculars to AB, from D to the curve, are the "ordinates to the axis" (*ordonnées à l'axe*); (b) the perpendiculars to AC, from E to the curve, are the "ordinates of the base" (*ordonnées à la base*); (c) the perpendiculars from L to AC are the "sines on the base"; (d) the perpendiculars from L to AB are the "sines on the axis." The difference between "ordinates" and "sines" is expressed at present by the change in the variable of integration.

[5] Proposition II is equivalent to our formula $\int_{\varphi_0}^{\varphi_1} \sin^2 \varphi \, d\varphi = -\int_{\varphi_0}^{\varphi_1} \sin \varphi \, d(\cos \varphi)$, and therefore expresses a change of variable. The area is equal to $\frac{1}{2}(\varphi_1 - \varphi_0) - \frac{1}{2}(\sin \varphi_1 \cos \varphi_1) + \frac{1}{2}(\sin \varphi_0 \cos \varphi_0)$.

[6] Hence $\int_{\varphi_0}^{\varphi_1} \sin^3 \varphi \, d\varphi = -\int_{\varphi_0}^{\varphi_1} \sin^2 \varphi \, d(\cos \varphi)$.

Proof of Proposition I. I say that the sum of the sines *DI* (each of them multiplied of course by one of the equal small arcs *DD*) is equal to the segment *AO* multiplied by the radius *AB*.

Indeed, let us draw at all the points *D* the tangents *DE* [Fig. 1], each of which intersects its neighbor at the points *E*; if we drop the perpendiculars *ER* it is clear that each sine *DI* multiplied by the tangent *EE* is equal to each distance *RR* multiplied by the radius *AB*. Therefore, all the quadrilaterals formed by the sines *DI* and their tangents *EE* (which are all equal to each other) are equal to all the quadrilaterals formed by all the portions *RR* with the radius *AB*; that is (since one of the tangents *EE* multiplies each of the sines, and since the radius *AB* multiplies each of the distances), the sum of the sines *DI*, each of them multiplied by one of the tangents *EE*, is equal to the sum of the distances *RR*, each multiplied by *AB*. But each tangent *EE* is equal to each one of the equal arcs *DD*. Therefore the sum of the sines multiplied by one of the equal small arcs is equal to the distance *AO* [Fig. 3] multiplied by the radius.

Note. It should not cause surprise when I say that all the distances *RR* are equal to *AO* and likewise that each tangent *EE* is equal to each of the small arcs *DD*, since it is well known that, even though this equality is not true when the number of the sines is finite, nevertheless the equality is true when the number is infinite; because then the sum of all the equal tangents *EE* differs from the entire arc *BD*, or from the sum of all the equal arcs *DD*, by less than any given quantity: similarly the sum of the *RR* from the entire *AO*.

Proof of Proposition II. I say that the sum of the squares of the sines *DI* (each of them multiplied by one of the equal small arcs *DD*) is equal to the sum of the *HL*, or to the area *BHQL*, multiplied by the radius *AB*.

For if the sines *DI* as well as the ordinates *HL* are extended to the circumference on the other side of the base, intersecting them at the points *G* and *N*, it is clear that each *DI* will be equal to each *IG*, and *HN* to *HL*.

In order to prove the proposition that all the squares of the *DI* times *DD* are equal to all the *HL* times *AB*, it is enough to prove that the sum of all the *HL* times *AB*, or all the *HN* times *AB*, or the area *QNN* multiplied by *AB*, is equal to all the *GI* times *ID* times *EE*, or to all the *GI* times *RR* times *AB* (since *ID* times *EE* is equal to each *RR* times *AB*). Then, by taking out the common quantity *AB* we have to prove that the area *AQNN* is equal to the sum of the rectangles *GI* times *RR*: this is clear, since the sum of the rectangles formed by each *GI* and each *RR* differs only by less than any given quantity from the area *AOGN* or, what is the same, *AQNN*, since the segment *AQ* was constructed to be equal to *AO*: and this was to be proved.

---

12 PASCAL. PARTIAL INTEGRATION

From the previous selection we can see how well Pascal understood the importance of the change of variable. This led him to operations equivalent to our partial integration, as we can see in the following fragment of the *Traité des trilignes rectangles et de leurs anglets*; *Oeuvres*, ed. L. Brunschwicg and P. Boutroux (Hachette, Paris, 1914–1921), IX, 3–44. The *triligne* is defined in the previous selection (note 4). The present selection is taken from pp. 5–8.

TREATISE ON TRILINEAR RECTANGULAR FIGURES

Let ABC [Fig. 1] be a rectangular trilinear figure, of which the ordinates to the axis are DF, and the ordinates to the base are EG. These intersect the curve in G, from which perpendiculars GR are dropped to the axis; these perpendiculars I call *counterordinates* [*contr'ordonnées*]. Let GR and FD be continued indefinitely. And let there be on the axis AB at the other side of the trilinear figure an arbitrary figure $BKOA$ in the same plane bounded by the extreme parallels CA, BK (this figure will be called the *adjunct* of the trilinear figure). Let this adjunct figure be intersected by the ordinates FD in the points O and by the counterordinates GR in the points I.

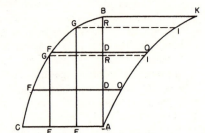

Fig. 1

I say that the sum of the rectangles on FD and DO, bounded by every ordinate of the trilinear figure and every ordinate of the adjunct figure, is equal to the sum of the areas ARI which are the portions of the adjunct between every one of the counterordinates and the end point of the adjunct at A.

Indeed, let us suppose that the trilinear figure BCA is multiplied into the figure $BAOK$, so that thus a certain solid figure is formed.[1] This means that at all points of ABC perpendiculars to its plane are erected, forming an infinite prismatic solid with ABC as base. Let also the figure $BAOK$ be rotated about the axis BA into a position perpendicular to the plane of ABC, and finally let the base AC be lifted parallel to itself in such a way that the point A always stays on the contour of the lifted figure $AOIKB$ till it falls back to B; that portion of the infinite prismatic solid cut out by the surface described by the line CA in its motion will be the solid that is considered here. It is bounded by four surfaces, among which the trilinear figure serves as base.

---

[1] This multiplication of one figure into another is the integration principle introduced as "ductus plani in planum" by Grégoire de Saint Vincent, a Jesuit mathematician of Antwerp, in his *Opus geometricum quadraturae circuli* (Antwerp, 1647); see J. E. Hofmann, "Das Opus geometricum des Gregorius a S. Vincentio und seine Einwirkung auf Leibniz," *Abhandlungen der preussischen Akademie der Wissenschaften, Math.-Naturw. Klasse* (1941, publ. 1942), no. 3, 1–80. In our present notation it can be expressed as the evaluation of $\int_a^b y(x)z(x)\,dx$ when $y(x)$ and $z(x)$ are given.

Let us now introduce two types of plane perpendicular to the plane of ABC: the first passing through the ordinates DF (which produce as sections of the solid the rectangles on FD and DO, bounded by the ordinate DF and the ordinate DO of the adjunct figure); the second type passing through the ordinates GE, parallel to the adjunct $BAOK$ (in its rotated position) and intersecting the same solid in sections all equal and similar to the portions RIA lying between any counterordinate RI and the extreme point A of the figure (this can be understood by the parallelisms, each of these planes being parallel to the rotated adjunct, and AC being parallel to itself in its motion). Now it can be seen that the sum of the sections made by each type of plane is equal to the solid, and they are consequently equal to each other (since the indefinite parts AE, EE, etc. of the base are equal both among themselves and to the equal and indefinite parts AD, DD, etc. of the axis).[2] Hence the sum of all rectangles on FD and DO is equal to the sum of all the portions RIA: which was to be proved.

After a lemma on the area of a parabolic segment (parabola of any integral order) the text continues:

Proposition I. The sum of the ordinates to the base is the same as the sum of the ordinates to the axis.

Indeed, each of them is equal to the area of the trilinear figure.

Proposition II. The sum of the squares of the ordinates to the base is twice the sum of the rectangles formed by each ordinate to the base and its distance to the axis. That is, the sum of all the EG squared is twice the sum of all the rectangles on FD and DA.

If we take AC as the positive X-axis and AB as the positive Y-axis and if the curve AOK is given by $x = f(y)$, then Pascal's theorem can be expressed by

$$\int_{y=A}^{y=B} x\, dF(y) = xF(y)\Big]_{y=A}^{y=B} - \int_{y=A}^{y=B} F(y)\, dx = \int_{x=C}^{x=D} F(y)\, dx,$$

where $F(y) = \int_0^R f(y)\, dy$, and $x = 0$, $y = 0$ at A. It is therefore a partial integration. In Proposition I, $F(y) = y$, so that

$$\int_A^B x\, dy = \int_C^D y\, dx.$$

[2] This equipartition is necessary to avoid contradictions in the theory of indivisibles; see the letter of Torricelli to Cavalieri in Selection IV.6.

In Proposition II, $f(y) = y$ (the line AOK is a straight line), $F(y) = \frac{1}{2}y^2$, and

$$\int_0^B xy \, dy = \frac{1}{2} \int_0^D y^2 \, dx.$$

Pascal has several more examples based on this change of variables.

It is here that we meet one of Pascal's references to a fourth dimension, when he generalizes his *trilignes* from plane to space and beyond: "La quatrième dimension n'est point contre la pure géométrie" (The fourth dimension is not against pure geometry). See H. Bosmans, "Sur l'interprétation géométrique donnée par Pascal à l'espace à quatre dimensions," *Annales de la Société Scientifique de Bruxelles 42* (1923), 337–345.

13 WALLIS. COMPUTATION OF π BY SUCCESSIVE INTERPOLATIONS

After 1650, analytic methods began to receive more attention and to replace geometric methods based on the writings of the ancients. This was due partly to the acceptance into geometry of those algebraic methods that Descartes and Fermat had introduced, and partly to the still very active interest in numerical work—interpolation, approximation, logarithms —a heritage of the sixteenth and early seventeenth centuries. This tradition was strong in England, where Napier and Briggs had labored.

This analytic method advanced rapidly through the efforts of John Wallis (1616–1703), of Emmanuel College, Cambridge, who in 1649 became the Savilian professor of geometry at Oxford. He was one of the founders of the Royal Society and, through his work, influenced Newton, Gregory, and other mathematicians. In his *Arithmetica infinitorum* (Oxford, 1655), he led explorations into the realms of the infinite with daring analytic methods, using interpolation and extrapolation to obtain new results. The title of the book shows the difference between Wallis' method—he called it "arithmetica"; we would say (with Newton) "analysis"—and the geometric method of Cavalieri. First Wallis derived Cavalieri's integral in an original way. Thereupon, he plunged into a maelstrom of numerical work and, with fine mathematical intuition to guide him in his interpolations, arrived at the infinite product for π that bears his name. See J. F. Scott, *The mathematical work of John Wallis* (Taylor and Francis, Oxford, 1938); also A. Prag, "John Wallis," *Quellen und Studien zur Geschichte der Mathematik (B)* I (1931), 381–412.

Proposition 39.[1] Given a series of quantities that are the cubes of a series of numbers continuously increasing in arithmetic proportion (like the series of cubic numbers), which begin from a point or zero (say 0, 1, 8, 27, 64, . . .); we ask for the ratio of this series to the series of just as many numbers equal to the highest number of the first series.

[1] In previous propositions Wallis has derived the limit

$$\lim_{n \to \infty} \frac{\sum\limits_{i=1}^n i^k}{n^{k+1}} = \frac{1}{k+1}$$

for $k = 1, 2$. This Proposition 39 prepares for the case $k = 3$; it shows Wallis's typical inductive and analytic method.

The investigation is carried out by the inductive method, as before. We have

$$\frac{0+1=1}{1+1=2} = \frac{2}{4} = \frac{1}{4} + \frac{1}{4};$$

$$\frac{0+1+8=9}{8+8+8=24} = \frac{3}{8} = \frac{1}{4} + \frac{1}{8};$$

$$\frac{0+1+8+27=36}{27+27+27+27=108} = \frac{4}{12} = \frac{1}{4} + \frac{1}{12};$$

$$\frac{0+1+8+27+64=100}{64+64+64+64+64=320} = \frac{5}{16} = \frac{1}{4} + \frac{1}{16};$$

$$\frac{0+1+\cdots+125=225}{125+\cdots+125=750} = \frac{6}{20} = \frac{1}{4} + \frac{1}{20};$$

$$\frac{0+\cdots+125+216=441}{216+\cdots+216=1512} = \frac{7}{24} = \frac{1}{4} + \frac{1}{24};$$

and so forth.

The ratio obtained is always greater than one-fourth, or $\frac{1}{4}$. But the excess decreases constantly as the number of terms increases; it is $\frac{1}{4}, \frac{1}{8}, \frac{1}{12}, \frac{1}{16}, \frac{1}{20}, \frac{1}{24}, \ldots$ There is no doubt that the denominator of the fraction increases with every consecutive ratio by a multiple of 4, so that the excess of the resulting ratio over $\frac{1}{4}$ is the same as $1:4$ times the number of terms after 0, etc.

Proposition 40. *Theorem.* Given a series of quantities that are the cubes of a series of numbers continuously increasing in arithmetic proportion beginning, for instance, with 0, then the ratio of this series to the series of just as many numbers equal to the highest number of the first series will be greater than $\frac{1}{4}$. The excess will be 1 divided by four times the number of terms after 0, or the cube root of the first term after 0 divided by four times the cube root of the highest term.

The sum of the series $0^3 + 1^3 + \cdots + l^3$ is $\dfrac{l+1}{4} l^3 + \dfrac{l+1}{4l} l^3$, or, if m is the number of terms, $\dfrac{m}{4} l^3 + \dfrac{m}{4l} l^3 = \dfrac{1}{4} ml^3 + \dfrac{1}{4} ml^2$. This is apparent from the previous reasoning.

If, with increasing number of terms, this excess over $\frac{1}{4}$ diminishes continuously, so that it becomes smaller than any given number (as it clearly does), when it goes to infinity, then it must finally vanish. Therefore:

Proposition 41. *Theorem.* If an infinite series of quantities which are the cubes of a series of continuously increasing numbers in arithmetic progression, beginning, say, with 0, is divided by the sum of numbers all equal to the highest and equal in number, then we obtain $\frac{1}{4}$. This follows from the preceding reasoning.

Proposition 42. *Corollary.* The complement AOT [Fig. 1] of half the area of the cubic parabola therefore is to the parallelogram TD over the same arbitrary base and altitude as 1 to 4.

Indeed, let AOD be the area of half the parabola AD (its diameter AD, and the corresponding ordinates DO, DO, etc.) and let AOT be its complement.

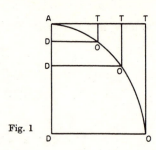

Fig. 1

Since the lines DO, DO, etc., or their equals AT, AT, etc. are the cube roots[2] of AD, AD, ..., or their equals TO, TO, ..., these TO, TO, etc. will be the cubes of the lines AT, AT, ... The whole figure AOT therefore (consisting of the infinite number of lines TO, TO, etc., which are the cubes of the arithmetically progressing lines AT, AT, ...) will be to the parallelogram ATD (consisting of just as many lines, all equal to the greatest TO), as 1 to 4, according to our previous theorem. And the half-segment AOD of the parabola (the residuum of the parallelogram) is to the parallelogram itself as 3 is to 4.

In Proposition 44 the result of these considerations on the quotient of the two series $\sum_{i=1}^{n} i^k$ and $\sum n^k$ ($n + 1$ terms) is laid down in a table for $k = 0, 1, 2, ..., 10$. Wallis discriminates for i^k between the series of equals ($k = 0$), of the first order ($k = 1$), of the second order ($k = 2$), and so forth (*series aequalium, primanorum, secundanorum*, and so forth).

Proposition 54. *Theorem*.[3] If we consider an infinite series of quantities beginning with a point or 0 and increasing continuously as the square, cube, biquadratic, etc. roots of numbers in an arithmetic progression (which I call the series of order $k = \frac{1}{2}, \frac{1}{3}, \frac{1}{4}, ...$), then the ratio of the whole series to the series of all numbers equal to the highest number is expressed in the following table:

k	Result		
$\frac{1}{2}$	$\frac{1}{3}$		$1\frac{1}{2}$
$\frac{1}{3}$	$\frac{1}{4}$	, or as 1 to	$1\frac{1}{3}$
$\vdots$	$\vdots$		$\vdots$
$\frac{1}{10}$	$\frac{1}{11}$		$1\frac{1}{10}$

---

[2] Wallis uses the terms *ratio subduplicata, subtriplicata* etc., to denote square, cubic, etc., roots; the *ratio subduplicata* of A^2/B^2 is A/B. These terms are not classical, and may be medieval. Wallis uses them here and in his *Mathesis universalis* (Oxford, 1657), chap. 30. The term *duplicate ratio* is classical; see Euclid, *Elements*, Book V, Definition 9: if $a/b = b/c$, then a/c has the duplicate ratio of a/b, hence $a/c = a^2/b^2$. Similarly, *triplicate ratio* in Definition 10 means the ratio of cubes. See G. Eneström, "Ueber den Ursprung des Termes 'ratio subduplicata'," *Bibliotheca mathematica* [3] *4* (1903), 210–211; *6* (1905), 410; *12* (1911–12), 180–181.

[3] Propositions 54 and 59 are supplementary, with the tabulation of $\int_0^1 x^k \, dx = 1/(k + 1)$ for all positive rational k.

Wallis calls these series of order $\frac{1}{2}, \frac{1}{3}, \ldots$ *series subsecundanorum, subtertianorum*, etc. Proposition 59 gives the full table for $k = p/q$ (Table 1).

	q \ p	0	1	2	3	⋮	10
	1	$\frac{1}{1}$	$\frac{1}{2}$	$\frac{1}{3}$	$\frac{1}{4}$	⋮	$\frac{1}{11}$
Quadraticae	2	$\frac{2}{2}$	$\frac{2}{3}$	$\frac{2}{4}$	$\frac{2}{5}$	⋮	$\frac{2}{12}$
Cubicae	3	$\frac{3}{3}$	$\frac{3}{4}$	$\frac{3}{5}$	$\frac{3}{6}$	⋮	$\frac{3}{13}$
	…	…	…	…	…	⋱	…
Decimanae	10	$\frac{10}{10}$	$\frac{10}{11}$	$\frac{10}{12}$	$\frac{10}{13}$	⋮	$\frac{10}{20}$
		Aequalium	Primorum	Secundorum	Tertiorum		

Table 1

Wallis's table uses for p the terms *aequalium, primanorum,* etc., and for q the terms *quadraticae, cubicae,* etc. if $q = 2$, $q = 3$, etc.

Proposition 64. If we take an infinite series of quantities, beginning with a point or 0, continuously increasing in the ratio of any power, an integer or a rational fraction, then the ratio of the whole to the series of as many numbers equal to the highest number is 1 divided by the index of this power $+1$.[4]

At the end of the explanation Wallis adds: "If we suppose the index irrational, say $\sqrt{3}$, then the ratio is as 1 to $1 + \sqrt{3}$, etc."

In Prop. 87 we have the analogous result for negative powers (the term "negative" is used).

Proposition 108. If two series be given, one that of equals, the other of the first order, and if the first term of the latter series is subtracted from the first term of the series of equals, the second term from the second term, etc., then the differences give one-half of the total first series. However, when we add the term, the aggregates are found to be $\frac{3}{2}$ of the series of equals.[5]

For instance, let R be the arbitrary term of the series of equals and the highest term of the series of the first order. Let its infinitely small part be denoted by $a = R/\infty$, and let A be the number of all terms (or the altitude of the figure); this number will go to infinity. Then the sum of the aggregates is:

---

[4] Here the theorem of note 3 is explicitly formulated as "Theorema universalis."

[5] This means, in our notation, $\int_0^1 (1 - x)\, dx = \frac{1}{2}$, $\int_0^1 (1 + x)\, dx = \frac{3}{2}$.

$$R - 0a \qquad R + 0a$$
$$R - 1a \qquad R + 1a$$
$$R - 2a \qquad R + 2a$$
$$R - 3a \qquad R + 3a$$
$$\text{etc.} \qquad \text{etc.}$$
$$\underline{R - R \qquad\qquad R + R}$$
$$AR - \tfrac{1}{2}AR \qquad AR + \tfrac{1}{2}AR$$

The sum of all equal terms is clearly AR. The sum of the terms of the series of the first order is half of it: $\tfrac{1}{2}AR$. Now $AR - \tfrac{1}{2}AR = \tfrac{1}{2}AR, AR + \tfrac{1}{2}AR = \tfrac{3}{2}AR$. This means that the former series is to the series of equals as $\tfrac{1}{2}$ to 1, and the latter as $\tfrac{3}{2}$ to 1.

Proposition 111. *Theorem.* If from a series of equals are subtracted, term by term, the terms of a series of the second, third, fourth, etc. order, these differences give $\tfrac{2}{3}, \tfrac{3}{4}, \tfrac{4}{5}$ of the total series of equals. If we add, the aggregates are $\tfrac{4}{3}, \tfrac{5}{4}, \tfrac{6}{5}$, etc. of this total sum.[6] Indeed, take the terms

$$R^2 \mp 0a^2 \qquad R^3 \mp 0a^3 \qquad R^4 \mp 0a^4$$
$$R^2 \mp 1a^2 \qquad R^3 \mp 1a^3 \qquad R^4 \mp 1a^4$$
$$R^2 \mp 4a^2 \qquad R^3 \mp 8a^3 \qquad R^4 \mp 16a^4$$
$$R^2 \mp 9a^2 \qquad R^3 \mp 27a^3 \qquad R^4 \mp 81a^4$$
$$\cdots\cdots\cdots \qquad \cdots\cdots\cdots \qquad \cdots\cdots\cdots$$

until
$$R^2 \mp R^2 \qquad R^3 \mp R^3 \qquad R^4 \mp R^4$$

Then the sums are (Prop. 44)

$$AR^2 \mp \tfrac{1}{3}AR^2, \quad AR^3 \mp \tfrac{1}{4}AR^3, \quad AR^4 \mp \tfrac{1}{5}AR^4.$$

Hence the sum of the differences gives

$$1 - \tfrac{1}{3} = \tfrac{2}{3}, \quad 1 - \tfrac{1}{4} = \tfrac{3}{4}, \quad 1 - \tfrac{1}{5} = \tfrac{4}{5}, \quad \text{etc.}$$

and the sum of the aggregates gives

$$1 + \tfrac{1}{3} = \tfrac{4}{3}, \quad 1 + \tfrac{1}{4} = \tfrac{5}{4}, \quad 1 + \tfrac{1}{5} = \tfrac{6}{5}, \quad \text{etc.}$$

Proposition 117. ...We replace the $1a, 2a, 3a$, etc. of previous propositions by a, b, c, etc., to show better the procedure of the operation:[7]

Series	Squares	Cubes
$R - 0$	$R^2 - 0R + 00$	$R^3 - 0R^2 + 00R - 000$
$R - a$	$R^2 - 2aR + a^2$	$R^3 - 3aR^2 + 3aR - a^3$
$R - b$	$R^2 - 2bR + b^2$	$\cdots \cdots \cdots \cdots$
$R - c$	$R^2 - 2cR + c^2$	
etc.	etc.	
$R - R$	$R^2 - 2RR + R^2$	$R^3 - 3RR^2 + 3R^2R - R^3$
$AR - \tfrac{1}{2}$	$AR^2 - \tfrac{2}{2}AR^2 + \tfrac{1}{3}AR^2$	$AR^3 - \tfrac{3}{2}AR^3 + \tfrac{3}{3}AR^3 - \tfrac{1}{4}AR^3$

[6] In our notation, $\int_0^1 (1 \pm x^n)\, dx = 1 \pm 1/(n+1)$, $n \geqslant 0$.

[7] In our notation, $\int_0^1 (1 - x)^k\, dx = 1/(k+1) = k!/(k+1)!$, $k \geqslant 0$.

Hence

$1 - \frac{1}{2} = \frac{1}{2};$ $\qquad$ $1 - \frac{2}{2} + \frac{1}{3} = \frac{1}{3};$ $\qquad$ $1 - \frac{3}{2} + \frac{3}{3} - \frac{1}{4} = \frac{1}{4};$

or $\frac{1}{2};$ $\qquad\qquad$ $\dfrac{1 \times 2}{2 \times 3};$ $\qquad\qquad\qquad$ $\dfrac{1 \times 2 \times 3}{2 \times 3 \times 4},$

and so on. We multiply continuously the numbers in arithmetic progression by each other (as many as agree with the value of the power), beginning with 1 and 2 and then regularly increasing with 1.

Proposition 121. *Corollary.* The ratio of the [area of a] circle to the square of the diameter (or of an ellipse to any of its circumscribed parallelograms) is as the series of square roots of the term-by-term differences of the infinite series of equals and the series of the second order to this series of equals.[8]

Indeed, if we call R [Fig. 2] the radius of the circle (of which $a = R/\infty$ is the infinitesimally small part), and if we construct an infinite number of per-

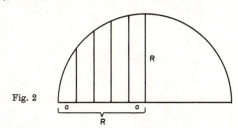

Fig. 2

pendiculars or *sinus recti* in order to complete the quadrant, then these perpendiculars are the mean proportionals between the segments of the diameters (as is well known), or

between	$R + 0,$	$R + 1a,$	$R + 2a,$	$R + 3a,$ etc.
and	$R - 0,$	$R - 1a,$	$R - 2a,$	$R - 3a,$ etc.
whose rectangles are	$R^2 - 00,$	$R^2 - 1a^2,$	$R^2 - 4a^2,$	$R^2 - 9a^2,$ etc.
the mean prop. are	$\sqrt{R^2 - 00},$	$\sqrt{R^2 - 1a^2},$	$\sqrt{R^2 - 4a^2},$	$\sqrt{R^2 - 9a^2},$ etc.

Hence, whatever the ratio of the sum of these roots is to that of their maximum (the radius), such is also the ratio of a quadrant of the circle (which consists of these roots) to the square of the radius (which consists of these maxima). Therefore it is also the ratio of the whole circle to the square of the diameter.

For this ratio Wallis writes $1 : \square$ (in our notation $\pi/4$).[9]

Proposition 132 [Fig. 3]. If we subtract term by term from the infinite series of equals the series of the first order (or, if we like, of $\frac{1}{1}$th order), of order

---

[8] In our notation, $\int_0^1 \sqrt{1 - x^2}\, dx = \pi/4$.

[9] The standard notation π for $4 : \square$ is due to William Jones (1675–1749), a friend of Newton, who assisted him in having some of his manuscripts published (see Selection V.4). In a textbook of 1706 he wrote π for 3.14159 etc. Euler adopted it and provided for its universal acceptance through his *Introductio in analysin infinitorum* (Lausanne, 1748).

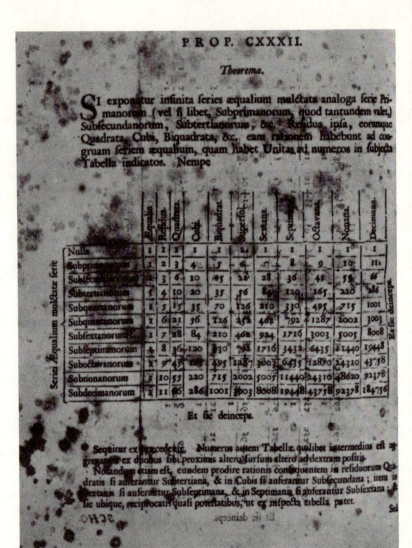

Fig. 3

$[1/p =] \frac{1}{2}, \frac{1}{3}$, etc., then the series of differences and the series of order $[k =] 2$, 3, 4, etc. formed from them have the same ratio to the series of equals as has 1 to the numbers in [Table 2].[10]

k \ p	0	1	2	3	4	:	10
0	1	1	1	1	1	:	1
1	1	2	3	4	5	:	11
2	1	3	6	10	15	:	66
3	1	4	10	20	35	:	
...	...	...	...	...	...	:	...
10	1	11	66	206	1001	:	184754

Table 2

k \ p	$\frac{1}{2}$	0	$\frac{1}{2}$	1	$\frac{3}{2}$	2	$\frac{5}{2}$	A ($l = k+1$)
$-\frac{1}{2}$	∞	1	$\frac{1}{2}$ □	$\frac{1}{2}$	$\frac{1}{3}$ □	$\frac{3}{8}$	$\frac{4}{15}$ □	A ($l=k+1$)
0	1	1	1	1	1	1	1	1
$\frac{1}{2}$	$\frac{1}{2}$ □	1	□	$\frac{3}{2}$	$\frac{4}{3}$ □	$\frac{15}{8}$	$\frac{8}{5}$ □	$A \times \frac{2l-1}{1}$
1	$\frac{1}{2}$	1	$1\frac{1}{2}$	2	$2\frac{1}{2}$	3	$\frac{7}{2}$	$\frac{2l+0}{2}$
$\frac{3}{2}$	$\frac{1}{3}$ □	1	$\frac{4}{3}$ □	$\frac{5}{2}$	$\frac{8}{3}$ □	$\frac{35}{8}$	$\frac{64}{15}$ □	$A \times \frac{2l-1}{1} \cdot \frac{2l+1}{3}$
2	$\frac{3}{8}$	1	$1\frac{7}{8}$	3	$4\frac{3}{8}$	6	$\frac{63}{8}$	$\frac{2l+0}{2} \cdot \frac{2l+1}{2}$

Table 3

This follows from the preceding. Any intermediate number in this table is the sum of the two numbers next to it, one above and the other to the left.

Proposition 184. In the preceding table we can interpolate in the following way [Fig. 4].

Proposition 189. We can now interpolate other series in the preceding table [as in Table 3].[11]

Proposition 191. Problem. It is proposed to determine this term □ as closely as possible in absolute numbers.

Wallis finds, by further interpolation (see the row for $p = \frac{1}{2}$ in Table 3, from which can be derived

$$\frac{\frac{15}{8}}{\frac{4}{3}\square} < \frac{\frac{4}{3}\square}{\frac{3}{2}} < \frac{\frac{3}{2}}{\square}, \quad \text{etc.}),$$

[10] In our notation, $\int_0^1 (1 - x^{1/p})^k \, dx = \frac{l(l+1)\ldots(l+p-1)}{1\cdot 2\cdots p}, l = k+1, k, p \geqslant 0$. The listing $(2l + 0)/2$, etc. is from Proposition 184.

[11] The interpolation is by means of the expressions A, 1, $A(2l - 1)/1$, etc., with the insertion of fractional values for 1. Since

$$\int_0^1 (1 - x^{1/p})^k \, dx = p \int_0^1 (1 - y)^k y^{p-1} \, dy = pB(p, k + 1),$$

$$\int_0^1 (1 - x^{1/k})^p \, dx = k \int_0^1 (1 - y)^p y^{k-1} \, dy = kB(k, p + 1),$$

the symmetry of the table expresses the symmetry of the B-function. Both integrals are equal to

$$\frac{kp}{k + p}\frac{\Gamma(k)\Gamma(p)}{\Gamma(k + p)} = \frac{\Gamma(k + 1)\Gamma(p + 1)}{\Gamma(k + p + 1)}.$$

The values for k and p are positive integers and multiples of $\frac{1}{2}$. If $y = z^2$, we find

$$\int_0^1 (1 - x^{1/p})^k \, dx = 2p \int_0^1 (1 - z^2)^k z^{2p-1} \, dz,$$

which is a multiple of the integral $\int_0^1 x^m y^k \, dx$, $x^2 + y^2 = 1$. We can say that Wallis computed this integral for integral values of m and k. The symbol ∞ for "infinite" is due to Wallis. See also T. P. Nunn, "The arithmetic of infinities," *Mathematical Gazette 5* (1909–1911), 345–356, 377–386, with a paraphrase of the book.

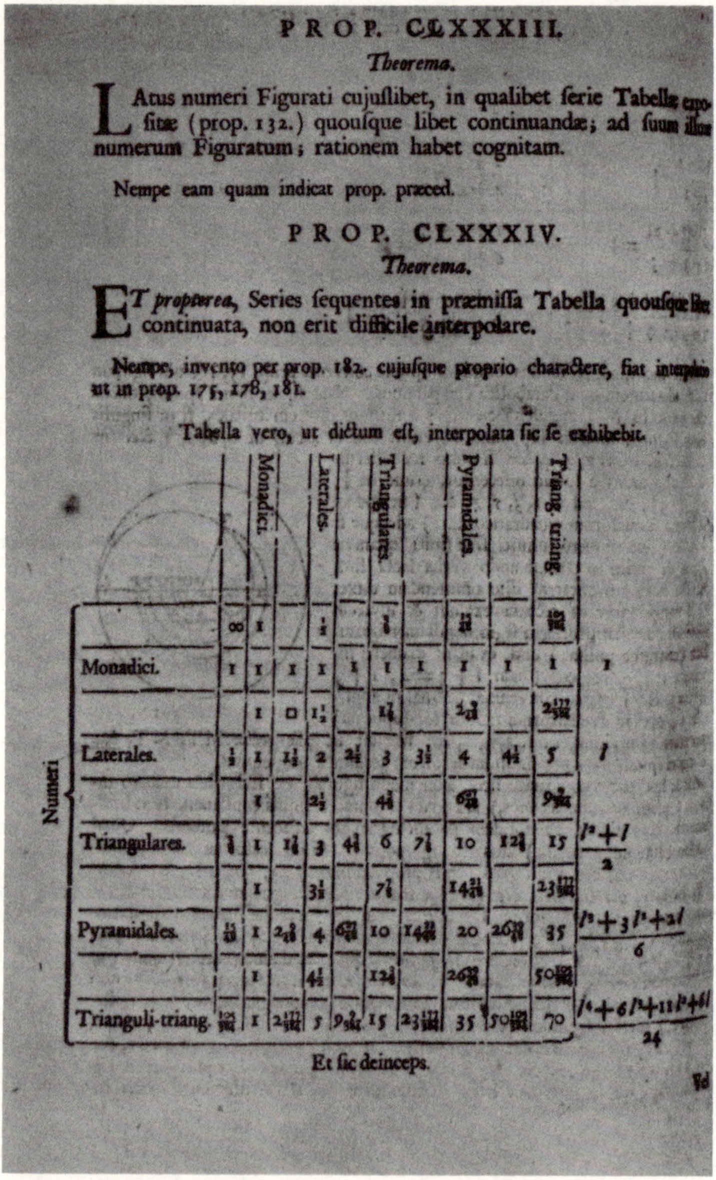

PROP. CLXXXIII.
Theorema.

LAtus numeri Figurati cujuslibet, in qualibet serie Tabellæ oppo-
sitæ (prop. 132.) quousque libet continuandæ; ad suum illum
numerum Figuratum; rationem habet cognitam.

Nempe eam quam indicat prop. præced.

PROP. CLXXXIV.
Theorema.

ET propterea, Series sequentes in præmissa Tabella quousque libet
continuata, non erit difficile interpolare.

Nempe, invento per prop. 182. cujusque proprio charactere, fiat interpolatio
ut in prop. 175, 178, 181.

Tabella vero, ut dictum est, interpolata sic se exhibebit.

Numeri		Monadici		Laterales		Triangulares		Pyramidales		Triang. triang.	
	∞	1		$\frac{1}{2}$		$\frac{1}{3}$		$1\frac{1}{4}$		$2\frac{1}{11}$	
Monadici.	1	1	1	1	1	1	1	1	1	1	1
		1	0	$1\frac{1}{2}$		$1\frac{1}{4}$		$2\frac{1}{11}$		$2\frac{177}{...}$	
Laterales.	$\frac{1}{2}$	1	$1\frac{1}{2}$	2	$2\frac{1}{2}$	3	$3\frac{1}{2}$	4	$4\frac{1}{2}$	5	l
		1		$2\frac{1}{2}$		$4\frac{1}{2}$		$6\frac{1}{11}$		$9\frac{1}{...}$	
Triangulares.	$\frac{1}{8}$	1	$1\frac{3}{4}$	3	$4\frac{1}{2}$	6	$7\frac{1}{2}$	10	$12\frac{3}{4}$	15	$\dfrac{l^2+l}{2}$
		1		$3\frac{1}{2}$		$7\frac{1}{2}$		$14\frac{11}{...}$		$23\frac{11}{...}$	
Pyramidales.	$\frac{1}{16}$	1	$2\frac{1}{16}$	4	$6\frac{11}{...}$	10	$14\frac{11}{...}$	20	$26\frac{11}{...}$	35	$\dfrac{l^3+3l^2+2l}{6}$
		1		$4\frac{1}{2}$		$12\frac{1}{4}$		$26\frac{11}{...}$		$50\frac{11}{...}$	
Trianguli-triang.	$\frac{1}{...}$	1	$2\frac{11}{...}$	5	$9\frac{1}{...}$	15	$23\frac{11}{...}$	35	$50\frac{11}{...}$	70	$\dfrac{l^4+6l^3+11l^2+6l}{24}$

Et sic deinceps.

Fig. 4

that $\square$ is

$$\text{less than} \quad \frac{3 \times 3 \times 5 \times 5 \cdots 13 \times 13}{2 \times 4 \times 4 \times 6 \cdots 12 \times 14} \sqrt{1\tfrac{1}{13}}$$

and

$$\text{greater than} \quad \frac{3 \times 3 \times 5 \times 5 \cdots 13 \times 13}{2 \times 4 \times 4 \times 6 \cdots 12 \times 14} \sqrt{1\tfrac{1}{14}},$$

and so forth to as close an approximation as we like.[12]

14 BARROW. THE FUNDAMENTAL THEOREM OF THE CALCULUS

The so-called inverse-tangent problem consisted in finding the curve, given a law concerning the behavior of the tangent. An early example was the search for loxodromes on the sphere, which are curves intersecting the meridians at a given angle; this problem was originated by Pedro Nuñez and Simon Stevin in the sixteenth century. A later example of importance was contained in a letter to Descartes written by Florimond De Beaune in 1639, which led to the search for the curve of constant subtangent; see Descartes, *Oeuvres*, ed. C. Adam and P. Tannery, *Correspondance*, II (Paris, 1898), 510–519, and Selection V.1. The next step was the recognition that finding quadratures and solving inverse-tangent problems were identical propositions—in other words, the discovery that the integral calculus is the inverse of the differential calculus. Torricelli came to this understanding in his case of generalized parabolas and hyperbolas, satisfying the equation $x \, dy = ky \, dx$; see E. Bortolotti, *Archeion 12* (1930), 60–64. James Gregory (1638–1675), the great Scottish mathematician who died so young, seems to have been the first to see the proposition in its generality, though still in a geometric manner. This was in his *Geometriae pars universalis* (Padua, 1668); see *James Gregory tercentenary memorial volume*, ed. H. W. Turnbull (London, 1939), where Gregory's work can be enjoyed in an English paraphrase. See also M. Dehn and E. D. Hellinger, "Certain mathematical achievements of James Gregory," *American Mathematical Monthly 50* (1943), 149–163. We then find the fundamental theorem in the *Lectiones geometricae* (London, 1670) by Isaac Barrow (1630–1677), in his day a famous theologian and from 1662 to 1670 professor of mathematics at Cambridge, where he was the first to occupy the Lucasian chair. His most famous disciple was Isaac Newton, who succeeded him in his chair. See P. C. Osmond, *Isaac Barrow: his life and times* (Society for Promoting Christian Knowledge, London, 1944).

The *Lectiones geometricae* present, in 13 lectures, a curious collection of theorems, mostly concerned with the finding of tangents, areas, and lengths of arcs. Barrow himself says in the preface that he did not find the presentation very satisfactory, but instead of editing his lectures he chose rather to send them forth "in Nature's garb," just as they were born. His starting point is motion, and his early method of finding tangents is thus kinematic. He then begins to use indivisibles, but with some caution, and at the end he arrives at the method of differentiation, as used by Fermat, and at that of the characteristic (or differential) triangle (dx, dy, ds). The method is thoroughly geometrical, and this makes it not easy to recognize the importance of Barrow's results. On the (partial) translation by

---

[12] We now write $\dfrac{\pi}{2} = \lim\limits_{n \to \infty} \dfrac{2 \times 2 \times 4 \times 4 \times \ldots (2n)(2n)}{3 \times 3 \times 5 \times 5 \times \ldots (2n-1)(2n+1)}$.

J. M. Child, *The geometrical lectures of Isaac Barrow* (Open Court, Chicago, London, 1916) we base certain sections of Lectures X and XI which contain theorems equivalent to $ds^2 = dx^2 + dy^2$ (rectification) and $(d/dx)\int_0^x y\,dx = y$ (the fundamental theorem). The notation is slightly modernized; see footnote 11.

LECTURE X

1. Let AEG [Fig. 1] be any curve whatever, and AFI another curve so related to it that, if any straight line EF is drawn parallel to a straight line given in

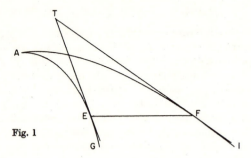

Fig. 1

position (which cuts AEG in E and AFI in F), EF is always equal to the arc AE of the curve AEG, measured from A; also let the straight line ET touch the curve AEG at E, and let ET be equal to the arc AE; join TF; then TF touches the curve AFI.

The proof follows.

---

2. Moreover, if the straight line EF always bears the same ratio to the arc AE, in just the same way FT can be shown to touch the curve AFI.[1] ...

3. Let AGE [Fig. 2] be any curve, D a fixed point, and AIF another curve such that, if any straight line DEF is drawn through D, the intercept EF is always equal to the arc AE; and let the straight line ET touch the curve AGE; make TE equal to the arc AE; let TKF be a curve such that, if any straight

---

[1] This is one of the many theorems by which Barrow passes from the knowledge of the tangent of one curve to that of another by means of methods which originally are based on motion (EF is moving parallel to itself), but eventually can be interpreted purely geometrically. If $E(x, y)$ and the y-axis are in the EF direction, then $EF = y + s$ ($s = $ arc AE), and $F(x, y + s)$. Hence the slope of FT is

$$\frac{1 + \sin \varphi}{\cos \varphi} = \tan \varphi + \sec \varphi = \frac{dy}{dx} + \frac{ds}{dx}, \quad \text{if } \frac{ds}{dx} = \sec \varphi,$$

hence $\frac{ds}{dx} = \sqrt{1 + \left(\frac{dy}{dx}\right)^2}$. Here we have taken $\frac{dy}{dx} = \tan \varphi$, hence the X-axis is $\perp AF$.

line DHK is drawn through D, cutting the curve TKF in K and the straight
line TE in H, $HK = HT$; then let FS be drawn to touch TKF at F; FS will
touch the curve AIF also.[2]

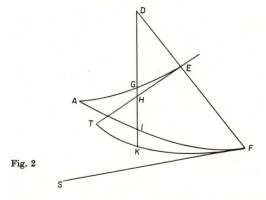

Fig. 2

4. Moreover, if the straight line EF always bears the same ratio to the arc
AE, the tangent to it can easily be found from the above and Lect. VIII, §8.

A number of similar theorems follow, and applications to some special curves (from a
straight line to a hyperbola, from a circle to a quadratix). Then Barrow says: "I add one
or two theorems, which it will be seen are of great generality, and not lightly to be passed
over." Here they are:

---

11. Let ZGE [Fig. 3] be any curve of which the axis is VD and let there be
perpendicular ordinates to this axis (VZ, PG, DE) continually increasing from
the initial ordinate VZ; also let VIF be a line such that, if any straight line EDF
is drawn perpendicular to VD, cutting the curves in the points E, F, and VD in
D, the rectangle contained by DF and a given length R is equal to the intercepted space $VDEZ$; also let $DE : DF = R : DT$, and join [T and F]. Then TF
will touch the curve VIF.[3] For, if any point I is taken in the line VIF (first on
the side of F towards V), and if through it IG is drawn parallel to VZ, and $IŁ$ is
parallel to VD, cutting the given lines as shown in the figure; then $LF : LK = DF : DT = DE : R$, or $R \times LF = LK \times DE$.

[2] This is similar to Art 1, but now in polar form.
[3] If the curve ZGE is given by $y = f(x)$ and curve AIF by $z = g(x)$, then $Rz = \int_0^x y\,dx$,
and $y : z = R : DT$. The theorem that DT is tangent to the curve AIF gives $y : z = R : z\dfrac{dx}{dz}$,
hence $y = R\dfrac{dz}{dx}$ or $y = \dfrac{d}{dx}\int_0^x y\,dx$. This therefore is, in geometrical form, the fundamental
theorem of the calculus. Figure 4 gives the text in facsimile.

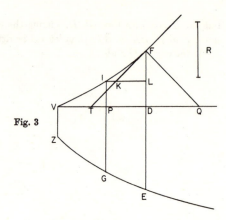

Fig. 3

But, from the stated nature of the lines DF, LK, we have $R \times LF =$ area $PDEG$: therefore $LK \times DE =$ area $PDEG < DP \times DE$; hence $LK < DP < LI$.

Again, if the point I is taken on the other side of F, and the same construction is made as before, plainly it can be easily shown that $LK > DP > LI$.

From which it is quite clear that the whole of the line TKF lies within or below the curve $VIFI$.

Other things remaining the same, if the ordinates, VZ, PG, DE, continually decrease, the same conclusion is attained by similar argument; only one distinction occurs, namely, in this case, contrary to the other, the curve VIF is concave to the axis VD.

Corollary. It should be noted that $DE \times DT = R \times DF =$ area $VDEZ$.

12. From the preceding we can deduce the following theorem.

Let ZGE, VIE be any two lines so related that, if any straight line EDF is applied to a common axis VD, the square on DF is always equal to twice the space $VDEZ$; also take DQ, along VD produced, equal to DE, and join FQ; then FQ is perpendicular to the curve VIF.[4]

I will also add the following kindred theorems.

13. Let $AGEZ$ be any curve [Fig. 5], and D a certain fixed point such that the radii, DA, DG, DE, drawn from D, decrease continually from the initial radius DA; then let DKE be another curve intersecting the first in E and such that, if any straight line DKG is drawn through D, cutting the curve AEZ in G and the curve DKE in K, the rectangle contained by DK and a given length R is equal to the area ADG; also let DT be drawn perpendicular to DE, so that $DT = 2R$; join TE. Then TE touches the curve DKE.

. . . .

Moreover, if any point, K say, is taken in the curve DKE, and through it DKG is drawn, and $DG : DK = R : P$; then, if DT is taken equal to $2P$ and TG is joined, and also KS is drawn parallel to GT; KS will touch the curve DKE.

---

[4] This shows how to construct the normal to the figure of Art. 11. Arts. 13 and 14 show how the argument runs in terms of what we would call polar coordinates.

7ᴱ

L ᴇ ᴄ ᴛ. X.

Hujufmodi plura quædam cogitaram hic inferere ; verùm hæc ex-
iftimo fufficere fubindicando modo, juxta quem, citra *Calculi moleſti-*
am, curvarum tangentes exquirere licet, unaque conftructiones de-
monftrare. Subjiciam tamen unum aut alterum non afpernanda, ut vi-
detur *Theoremata* perquam generalia.

XI. Sit linea quæpiam Z G E, cujus axis V D ; ad quam impri-
mis applicatæ perpendiculares (V Z, P G, D E) ab initio V Z con-
tinuè utcunque crefcant ; fit item linea V I F talis, ut ductâ quâcunq;
rectâ E D F ad V D perpendiculari (quæ *curvam* fecet punctis E, F,
ipfam V D in D) fit femper *rectangulum* ex D F, & defignatâ quâ-
dam R æquale *fpatio* refpectivè *intercepto* V D E Z ; fiat autem D E.
D F :: R . D T ; & connectatur recta T F ; hæc curvam V I F
continget.

Sumatur enim in linea V I F punctum quodpiam I (illud primò fu-
pra punctum F, verfus initium V) & per hoc ducantur rectæ I G ad
V Z, ac K L ad V D parallelæ (quæ lineas expofitas fecent, ut vides)
eſtque tum L F . L K :: (D F . D T ::) D E . R ; adeóque L F ×
R = L K × D E. Eſt autem (ex præſtituta linearum iſtarum natura)
L F × R æquale fpatio P D E G ; ergò L K × D E = P D E G =
D P × D E. Unde eſt L K = D P, vel L K = L I.

Rurfus accipiatur quodvis punctum I, infra punctum F, reliquaq;
fiant, uti prius ; fimilique jam plane difcurfu conſtabit fore L K × D E
= P D E G = D P × D E, unde jam erit L K = D P, vel L I. E
quibus liquidò patet totam rectam T K F K intra (feu extra) curvam
V I F I exiftere.

Iifdem quoad cætera pofitis, fi *ordinatæ* V Z, P G, D E, &c. con-
tinuè decrefcant, eadem conclufio fimili ratiocinio colligetur ; unicum obvenit *Difcrimen*, quòd in hoc cafu (contra quam in priore)
linea V I F concavas fuas axi V D obvertat.

Corol. Notetur D E × D T æquari fpatio V D E Z.

XII. Exindè deducitur hoc *Theorema*: Sint duæ lineæ quævis
Z G E, V K F ita relatæ, ut ad communem ipfarum axem V D ap-
plicatâ quâvis rectâ E D F, fit femper quadratum ex D E æquale *du-*
plo fpatio V D E Z, fumatur autem D Q = D E, & connectatur F Q ;
hæc curvæ V K F perpendicularis erit.

Concipiatur enim linea V I F, per F tranfiens, talis qualem mox
attigimus (cujus fcilicet ad V D applicatæ fe habeant ut fpatia VDEZ,
hoc eſt ut quadrata ex applicatis à curva V K F in præfente hypothefi)
lineamque

Fig. 109.

Fig. 110.

Fig. 111.

Fig. 4

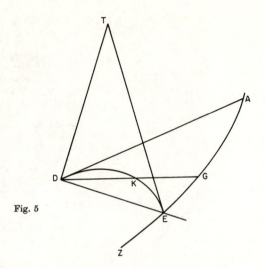

Fig. 5

Observe that Square on DG : Square on $DK = 2R : DS$.

Now, the above theorem is true, and can be proved in a similar way, even if the radii drawn from D, DA, DG, DE, are equal (in which case the curve $AGEZ$ is a circle and the curve DKE is the Spiral of Archimedes), or if they continually increase from A.

14. From this we may easily deduce the following theorem.

Let AGE, DKE [Fig. 6] be two curves so related that, if straight lines DA, DG are drawn from some fixed point D in the curve DKE (of which the latter cuts

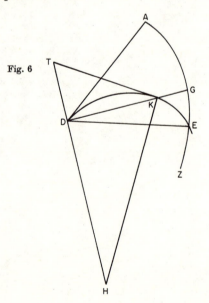

Fig. 6

the curve DKE in K), the square on DK is equal to four times the area ADG; then, if DH is drawn perpendicular to DG, and $DK : DG = DG : DH$; and HK is joined; then HK is perpendicular to the curve DKE.

We have now finished in some fashion the first part, as we declared, of our subject. Supplementary to this we add, in the form of appendices, a method for finding tangents by calculation frequently used by us. Although I hardly know, after so many well-known and well-worn methods of the kind above, whether there is any advantage in doing so. Yet I do so on the *advice of a friend*;[5] and all the more willingly, because it seems to be more profitable and general than those which I have discussed.

Let AP, PM be two straight lines given in position [Fig. 7] of which PM cuts a given curve in M, and let MT be supposed to touch the curve at M, and to cut the straight line at T.

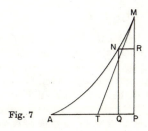

Fig. 7

In order to find the length of the straight line PT, I set off an indefinitely small arc, MN, of the curve; then I draw NQ, NR parallel to MP, AP; I call $MP = m$, $PT = t$, $MR = a$, $NR = e$, and other straight lines, determined by the special nature of the curve, useful for the matter in hand, I also designate by name; also I compare MR, NR (and through them, MP, PT) with one another by means of an equation obtained by calculation; meantime observing the following rules.[6]

Rule 1. In the calculation, I omit all terms containing a power of a or e, or products of these (for these terms have no value).[7]

Rule 2. After the equation has been formed, I reject all terms consisting of letters denoting known or determined quantities or terms which do not contain a or e (for these terms, brought over to one side of the equation, will always be equal to zero).

Rule 3. I substitute m (or MP) for a, and t (or PT) for e. Hence at length the quantity of PT is found.

Moreover, if any indefinitely small arc of the curve enters the calculation, an indefinitely small part of the tangent, or of any straight line equivalent to it (on

---

[5] This friend probably is Newton, to whom Barrow refers by name in the preface, saying that Newton has helped him in preparing the book, adding some things from his own work.

[6] This introduces the "characteristic triangle" (NR, RM, NM) or (dx, dy, ds), on the advice, it seems, of Newton.

[7] This neglecting of terms of higher order reminds us of Fermat (Selections IV.7, 8) and also of Newton's fluxion theory (Selection V.7).

account of the indefinitely small size of the arc) is substituted for the arc. But these points will be made clearer by the following examples.

Barrow gives five examples of this method of the characteristic triangle. Two of them are the folium of Descartes $x^3 + y^3 = axy$ (written by Barrow AP cub $+ PM$ cub $= AX \times AP \times PM$; he calls the curve La Galande) and the quadratrix; the others are the curves $x^3 + y^3 = a^3$, $r = a \tan \theta$, and $y = a \tan x$. The result of the differentiation of $y = a \tan x$ is shown to be (in our notation, of course), $dy/dx = a \sec^2 x$.

The next lecture deals with integration.

LECTURE XI

1. If VH [Fig. 8] is a curve whose axis is VD, and HD is an ordinate perpendicular to VD, and $\varphi Z\psi$ is a line such that, if from any point chosen at random on the curve, say E, a straight line EP is drawn normal to the curve, and a straight line EAZ perpendicular to the axis, AZ is equal to the intercept AP; then the area $VD\psi\varphi$ will be equal to half the square on the line DH.

For if the angle HDO is half a right angle, and the straight line VD is divided into an infinite number of equal parts at A, B, C, and if through these points

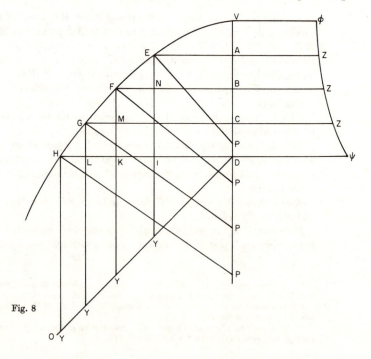

Fig. 8

straight lines EAZ, FBZ, GCZ, are drawn parallel to HD, meeting the curve in E, F, G; and if from these points are drawn straight lines EIY, FKY, GLY, parallel to VD (or HO); and if also EP, FP, GP, HP are normals to the curve, the lines intersecting as in the figure; then the triangle HLG is similar to the triangle PDH (for, on account of the infinite section, the small arc HG can be considered as a straight line).

Hence, $HL:LG = PD:DH$, or $HL \times DH = LG \times PD$; that is,

$$HL \times HO = DC \times D\psi.$$

By similar reasoning it may be shown that, since the triangle GMF is similar to the triangle PCG, $LK \times LY = CB \times CZ$; and in the same way,

$$KI \times KY = BA \times BZ, \quad ID \times IY = AV \times AZ.$$

Hence it follows that the triangle DHO (which differs in the slightest degree only from the sum of the rectangles $HL \times HO + LK \times LY + KI \times KY + ID \times IY$) is equal to the space $VD\psi\varphi$ (which similarly differs in the least degree only from the sum of the rectangles $DC \times D\psi + CB \times CZ + BA \times BZ + AV \times AZ$); that is,

$$DH^2/2 = \text{area } VD\psi\varphi.$$

A lengthier indirect argument may be used; but what advantage is there?[8]

2. With the same data and construction as before, the sum of the rectangles $AZ \times AE$, $BZ \times BF$, $CZ \times CG$, etc., is equal to one-third of the cube on the base DH.[9]

For, since $HL:LG = PD:DH = PD \times DH:DH^2$; therefore $HL \times DH^2 = LG \times PD \times DH$ or $LH \times HO^2 = DC \times D\psi \times DH$; and, similarly, $LK \times LY^2 = CB \times CZ \times CG$, $KI \times KY^2 = BA \times BZ \times BF$, etc.

But the sum $HL \times HO^2 + LK \times LY^2 + KI \times KY^2 +$ etc. $= DH^3/3$; and the proposition follows at once.

3. By similar reasoning, it follows that

the sum of $AZ \times AE^2$, $BZ \times BF^2$, $CZ \times CG^2$, etc. $= DH^4/4$;
the sum of $AZ \times AE^3$, $BZ \times BF^3$, $CZ \times CG^3$, etc. $= DH^5/5$;

and so on.

4. Hence we may deduce the following important theorems.

Let $VD\psi\varphi$ be any space of which the axis VD is equally divided [as in Fig. 7];

[8] If we measure x along VP and y in the direction of EA, then $AP = AZ = y\,dy/dx$ and the theorem states that

$$\int_0^{x_0} y \frac{dy}{dx}\,dx = \int_0^{y_0} y\,dy = \frac{y^2}{2}\Big]_0^{y_0} = \frac{y_0^2}{2},$$

when D has the coordinates (x_0, y_0). This is a form of change of independent variable from x to y.

[9] That is, $\int_0^{x_0} y^2 \frac{dy}{dx}\,dx = \int_0^{y_0} y^2\,dy = \frac{y_0^3}{3}$.

then if we imagine that each of the spaces $VAZ\varphi$, $VBZ\varphi$, $VCZ\varphi$, etc., is multiplied by its own ordinate AZ, BZ, CZ, etc., respectively, the sum which is produced will be equal to half the square of the space $VD\psi\varphi$.[10]

Several more examples are given, concerning the area of a quadrant of a circle and of a parabolic segment and the volume of a surface of rotation, after which comes the following theorem:

10. Again, if VH [Fig. 9] is a curve whose axis is VD and base DH, and DZZ is a curve such that, if any point such as E is taken on the curve VH and ET is drawn to touch the curve, and a straight line EIZ is drawn parallel to the axis, then IZ is always equal to AT; in that case, I say, the space DHO is equal to the space VHD.

This extremely useful theorem is due to that most learned man, Gregory of Aberdeen: we will add some deductions from it . . .[11]

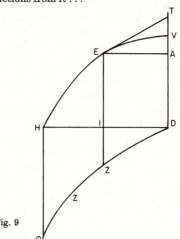

Fig. 9

[10] That is,

$$\int_0^{y_0} y \, dy \int_0^x y \frac{dy}{dx} dx = \int_0^{y_0} \tfrac{1}{2} y^3 \, dy = \frac{y_0^4}{8} = \tfrac{1}{2} \left(\frac{y_0^2}{2}\right)^2.$$

Art. 5 shows that

$$\int_0^{y_0} y^{1/2} \, dy = \tfrac{2}{3} y_0^{3/2}.$$

[11] When D is taken as origin, $DI = x$, $DA = y$, then $AT = IZ = x \, dy/dx$, and if we write $H(x_0, 0)$, $V(0, y_0)$, then

$$\int_0^{y_0} x \, dy = \int_0^{x_0} y \, dx.$$

Barrow refers to James Gregory, who, in 1668, had arrived independently at the fundamental theorem.

Barrow goes on to give more examples and Art. 19 arrives again at the fundamental theorem, now in the form converse to that given in Lecture X, Art. 11: if the curve AMB is given by $z = f(x)$, and the curve KZL by $y = f_1(x)$, $z\,dx/dz:z = R:y$, then $\int y\,dx = R\int dz$, or $\int y\,dx = Rz$.

We list here some of Barrow's notations which we have modified: $A\ \boxed{}\ B$, A is greater than B; $A\ \boxed{}\ B$, A is less than B; $A.B::C.D$, $A:B = C:D$; Aq, the square of A, for instance, in Lecture X, Art. 13, square on DG is written DGq; Ac or A cub, the cube of A; $DHqq$ the fourth power of DH. We have kept his symbol of multiplication, $A \times B$.

We end with a word of caution. Despite the fact that, in order to understand these seventeenth-century mathematicians, we are inclined to translate their reasoning into the notation and language with which we are familiar, we must constantly be aware that our point of view is not equivalent to theirs. They saw geometric theorems in the sense of Euclid, where we see operations and calculating processes. At the same time, just because these mathematicians applied their geometric notions in an attempt to transcend the static character of classical mathematics, their geometric thought has a richness that may easily escape observation in the modern transcription. If we were to rewrite Euclid in the notation of analytic geometry we would obtain a body of knowledge with a character different from that of Euclid and, despite all the advantages that the algebraic computations would bring, we would lose some of the more subtle and esthetic qualities of Euclid.

15 HUYGENS. EVOLUTES AND INVOLUTES

The search for reliable clocks, a necessity for scientific navigation and geography as well as for theoretical astronomy, led Christiaan Huygens (1629–1695), a Dutch patrician and a founding member of the French Academy of Sciences (1666), to the invention of the pendulum clock (the idea of which seems to have already occurred to Galilei). Huygens described this invention in the *Horologium oscillatorium* (Paris, 1673; reprinted, with French translation, in *Oeuvres complètes de Christiaan Huygens*, XVIII, 68–368). This book, in its five parts, contains a number of important discoveries in mechanics and mathematics, so that, with the books of Cavalieri and Wallis (see Selections IV.5, 6, 13), it is a landmark on the path that led to the invention of the calculus.

After describing his pendulum clock in Part I, Huygens deals in Part II with "The fall of heavy bodies and their cycloidal movement." Here we find a theory of the cycloid and, based on it, the following theorem on a heavy point moving on a cycloid in a field of gravity:

Proposition XXV. On a cycloid with a vertical axis whose vertex is below, the times of descent in which a mobile point, starting from rest at an arbitrary point of the curve, reaches the lowest point, are all equal, and have to the times of the vertical fall along the total axis of the cycloid a ratio equal to that of the semicircumference of a circle to that of the diameter [in our terms, as $\pi:2$].

In other words, the cycloid is a *tautochrone*. From this theorem Huygens obtains the tautochronic pendulum, which has a period independent of its amplitude. This property of the cycloid leads him to the discovery that the evolute of the cycloid is also a cycloid, and then, in Part III of the book, to the general theory of evolutes and involutes of plane curves.

The *Horologium oscillatorium* can also be studied in a German translation in Ostwald's *Klassiker*, No. 192 (Engelmann, Leipzig, 1913).

Here follows a translation of the beginning of Part III, entitled "On the evolution and dimension of curved lines."

Definition I. A curve is said to be curved [*inflexa*] to one side, if all its tangents touch it just on that side. If it has some parts straight, then these, continued at both ends, are themselves regarded as tangents.

Definition II. When two curves of this kind pass through the same point, and when the convexity of one is directed toward the concavity of the other, as the curves *ABC* and *ADE* [Fig. 1], then we shall call both "concave [*cavae*] to the same side."

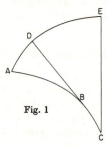

Fig. 1

Definition III. When we consider that a thread or flexible line is laid along a curve concave to one side, and when we remove one end from it while the other end of the thread stays on the curve in such a way that the developed part remains taut, then this end of the thread will clearly describe another curve; this curve is called an involute [*descripta ex evolutione*].[1]

Definition IV. The curve, however, along which the thread has been laid may be called the evolute [*evoluta*]. In the figure *ABC* is the evolute, *ADE* the involute of *ABC*, for if the end of the thread has come from *A* to *D*, then the straight part *DB* of the thread will be taut, while the other part *BC* still lies along the curve. It is clear that *DB* is tangent to the evolute at *B*.

Proposition I. Every tangent of the evolute intersects the involute at right angles:

Let *AB* [Fig. 2] be the evolute, *AH* its involute. Let the straight line *FDC*, tangent to curve *ADB* at *D*, intersect the curve *ACH* at *C*. I claim that it

---

[1] The term *involute*, a curious English construction for the more natural "evolvent" (as in German: *Evolute, Evolvente*; in French: *développée, développante*) appears first in Charles Hutton's *Mathematical dictionary* (London, 1796), according to the *Oxford English Dictionary*. This gives for the first English appearance of the term *evolute* 1730–36.

intersects the curve at right angles, that is, if we construct on CD the perpendicular CE, then this line should touch the curve ACH at C. Indeed, since the straight line DC is tangent to the evolute at D, it clearly represents the position of the thread at the moment when its end has come to C. When therefore we prove that the thread while describing the whole curve ACH can reach the line CE only at the point C, we shall have proved that CE is tangent to the curve ACH at the point C.

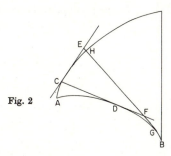

Fig. 2

Let us take on AC another point H different from C, and let us consider first the case in which H is farther removed than C from the starting point A of the evolution. Let the free part of the thread have the position HG, when its end is at H. The line HG is therefore tangent to the curve AB at G. While the end of the thread describes the arc CH, the thread evolves itself away from arc DG. Hence CD will intersect the line HG if extended beyond D; say at F. Let GH intersect the line CE at E. We then have[2]

$$DF + FG > DG,$$

whether DG be a straight or a curved line. If we add to both sides the straight segment DC, then we obtain

$$CF + FG > CD + DG.$$

In connection with the evolution we have

$$CD + DG = HG.$$

Hence the sum $CG + FG$ will also be $> HG$, and if we subtract from both sides the segment FG, then we find that

$$CF > HF.$$

But we have

$$FE > FC,$$

[2] Huygens does not use the Harriot symbol $>$, but uses words: "Quia igitur duae simul DF, FG, majores sunt quam DG."

since in the triangle FCE the angle C is right. Hence we have a fortiori

$$FE > FH.$$

From this it follows that the thread on this side of the point C no longer intersects the line CE.

Now let the point H [Fig. 3] be closer to the starting point A than the point C. Let HG be the position of the thread at the moment when its end is at H. Let

Fig. 3

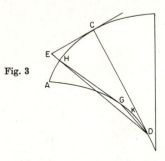

us draw the lines DG and DH, of which the last one meets the straight line CE at E. It is clear that the straight line DG cannot be on the continuation of HG, and that HGD is therefore a triangle. Now, since

$$DG \leqslant DKG,$$

the sign $=$ holding for the case where the part DG of the evolute is straight, we will find, adding GH on both sides, that

$$DG + GH \leqslant DKG + GH,$$

or

$$DG + GH \leqslant DC.$$

But

$$DH < DG + GH,$$

hence DH is a fortiori $< DC$. But $DE > DC$, since in triangle DCE the angle C is right. Hence DH is much more $< DE$. The point H, the end of the thread GH, is therefore situated inside the angle DCE. From this it follows that between A and C the end of H never gets as far as CE. Hence CE touches the curve AC at C, so that DC, to which CE has been constructed as a perpendicular, cuts the curve at right angles.   Q.E.D.[3]

---

[3] We see that Huygens proves that a line is tangent to the curve by using the method of the ancients; see our commentary to Selections IV.8, 10.

We give the next three propositions without the proof, which is similar in character to that given above for Proposition I.

Proposition II. Every curved line segment, concave in one direction, as *ABD* [Fig. 4], can be divided in so many parts that if we draw the chords that subtend every one of the arcs, as *AB*, *BC*, and *CD*, and then draw the tangents *AN*,

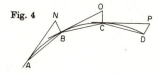

Fig. 4

BO, *CP* from every one of the points of division and also from the end of the curve, till each of them meets the normal to the curve at the next point of division (*BN*, *CO*, *DP* are the normals), then every chord will have to the corresponding normal (*AB* to *BN*, *BC* to *CO*, *CD* to *PD*) a ratio superior to any given ratio.[4]

Proposition III. Two curved lines curved both in the same direction and concave in that same direction cannot issue from the same point in such a mutual position that every line normal to the one is also normal to the other.

Huygens proves that if *ACE* and *AGK* [Fig. 5] are such curves, and *KE* is a common normal, then the proposition that all normals at the points *G* of *AGK* are also normals to *ACE* leads to an absurdity.

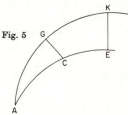

Fig. 5

Proposition IV. If from a point pass two curved lines curved both in the same direction and concave in the same direction, and in such a mutual position that the tangents to the one of them meet those of the other at right angles, then this other curve will be the evolute of the first from the common point on.

---

[4] In our words: the segments on the normals are of higher order of infinity than the chords. See Selection V.5.

Proposition V. If a straight line touches a cycloid at its vertex and we construct on this straight line as base another cycloid, similar and equal to the first, then, beginning at the vertex mentioned, an arbitrary tangent to the inferior cycloid will be normal to the superior cycloid.

Let us suppose that the straight line AG [Fig. 6] touches the cycloid at its vertex A and that on this line as base is constructed another cycloid AEF with

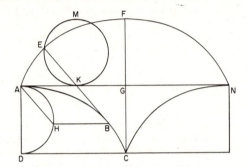

Fig. 6

vertex F. Let BK be a tangent to the cycloid ABC. I claim that this tangent, continued to the cycloid AEF, will meet it at right angles. Indeed, let us describe about AD, the axis of the cycloid ABC, the generating circle AHD, which intersects BH, parallel to the base, at H. Let us draw the line HA. Since BK is tangent to the cycloid at B, it is parallel to this line HA, hence $AHBK$ is a parallelogram and $AK = HB$, that is, equal to the arc AH.[5] Let us now describe the circle KM, equal to the generating circle AHD, tangent to the base AG at K and intersecting the continued line BK at E. Since BKE is parallel to AH, and hence $EKA = KAH$, it is clear that the continued line BK cuts from the circle KM an arc equal to the arc which AH cuts from the circle AHD. The arc KE is therefore equal to the arc AH, that is, to the line HB, hence to the line KA. But it follows from this equality, from a property of the cycloid (since the generating circle MK is tangent to line AG at K), that the point which describes the cycloid [AEF] has passed through E. The line KE therefore meets the cycloid at E at right angles, that is, KE is no other line than the continuation of BK. Q.E.D.

Proposition VI. By the evolution of a half-cycloid, beginning at the summit, another half-cycloid is described equal and similar to the first, whose base coincides with the straight line that is tangent to the cycloid evolved at its vertex.

In the following propositions of Part III Huygens investigates many other evolutes, notably those of conic sections, and uses this information for some computations of length

[5] In this proof Huygens uses several properties of the cycloid that he has established in Part II of his book. See also Selection IV.10.

and area. This leads to the general theorem on the construction of evolutes for "geo-metrical" curves.[6]

Part IV of Huygens' book, "On centers of oscillation," contains the theory of the oscillating bodies. The short Part V deals with centrifugal force and another clock construction.

[6] The establishment of this theorem is the beginning of a series of investigations on curves that are congruent or similar to their evolutes. The search leads to logarithmic spirals, epicycloids, and hypocycloids. See *Oeuvres complètes de Christiaan Huygens*, XVIII, 40–41, and C. A. Crommelin and W. van der Woude, *Simon Stevin 30* (1954), 17–24. As to "geo-metric curves," Descartes, in the second book of his *Géométrie*, called curves "geometric" if they "admit of precise and exact measurement," so that all their points must bear a definite relation to all points of a straight line, a relation to be expressed by means of a single equation, which then can be of different degrees (see Selection III.4). For such curves Huygens expresses, in geometric form, the formula for the radius of curvature which later Jakob Bernoulli would express by $z = ds^3 : dy\, d\, dx$, a formula equivalent to the one familiar to us in our calculus texts; see "Curvatura laminae elasticae . . . Radii circulorum oscula-tium in terminis simplicissimis exhibiti . . .," *Acta Eruditorum* (June 1694), 262–266 (*Opera* II, 576–600). See further Selection V.23.

CHAPTER V NEWTON, LEIBNIZ, AND THEIR SCHOOL

The final discovery of the calculus required the assimilation of the geometric methods of Cavalieri and Barrow with the analytic methods of Descartes, Fermat, and Wallis; it also required the understanding of the relation between the search for tangent constructions and quadratures. This fundamental step was taken by Isaac Newton (1643–1727) and Gottfried Wilhelm Leibniz (1646–1716). Newton's "golden period" of discovery fell between 1664 and 1668 (see Selection III.8); Leibniz also had such a "golden period," from 1672 to 1676, when he resided in Paris, met with the twenty-one-years older Huygens, and twice made trips to London to confer with British scientists. In those years he discovered his form of the calculus. Newton's discovery of the calculus thus came before that of Leibniz, but Leibniz published his first (1684 and after). Newton, after hints in his *Principia* (1687) and in a new edition of Wallis's *Algebra* in the *Opera mathematica* of Wallis (Oxford, 1693), did not publish his calculus until 1704 (see Selection V.7), after Leibniz's discoveries had already been presented in L'Hôpital's textbook (1696).

The development of the calculus in Leibniz's sense proceeded fast through the work of Leibniz himself, the brothers Bernoulli, Euler, and other continental mathematicians. Newton's calculus, the so-called theory of fluxions, had a much less spectacular progress in the works of Taylor, Maclaurin, and other British mathematicians. The struggle over priority, started during the lifetime of Newton and Leibniz (the opening shot in this war was fired by the mathematician-adventurer-mystic Nicolas Fatio de Duillier in 1699), did not help to reconcile the two points of view and their adherents. Not until the early years of the nineteenth century was the conflict resolved in the sense that leading mathematicians in the English-speaking countries began to adopt the Leibniz notation, primarily through the impact of the work done by Laplace (though the notation $\dot{x}$ for the time derivative of x has persisted to the present day). From that time on the priority struggle has become a priority question, and thus a comparatively minor chapter in the whole history of the origin of the calculus. New light is constantly being shed on this history through the publication of documents hitherto available only in archives, such as papers and letters of Newton, Oldenburg, and Mersenne. Literature on this question is found in J. E. Hofmann, *Geschichte der Mathematik*, III (Sammlung Göschen 882; De Gruyter, Berlin, 1957), 38; a short survey of the struggle is given in J. Q. Fleckenstein, *Der Prioritätsstreit zwischen Leibniz und Newton* (*Elemente der Mathematik*, Beiheft 12; Birkhaüser, Basel, Stuttgart, 1956). The bibliography in R. C. Archibald, "Outline of the History of Mathematics," *American Mathematical Monthly 56*, No. 1, Part II (January 1949), 114 pp., is very helpful, not only on Newton and Leibniz, but also on the whole of the history of mathematics.

1 LEIBNIZ. THE FIRST PUBLICATION OF HIS DIFFERENTIAL CALCULUS

Gottfried Wilhelm Leibniz (1646–1716), born in Leipzig, studied philosophy and law at Leipzig and Jena, and became a diplomatic counselor in the service of the elector of Mayence. This allowed him to spend the years 1672–1676 in Paris, where he developed many of his mathematical ideas, including the calculus, under the personal influence of Huygens and by studying Descartes, Pascal, and British mathematicians. From 1676 to his death he lived most of the time in Hanover as a librarian in the service of the kings of Hanover.

The manuscripts in which Leibniz wrote down his discoveries during his Paris stay were published by Gerhardt and have been translated into English by J. M. Child, *The early mathematical manuscripts of Leibniz* (Open Court, Chicago, London, 1920), with critical and historical notes. Very interesting is the note, dated October 26, 1675, with the first utilization of the signs $\int$, signifying Cavalieri's "omnes lineae," and d, originally placed in the denominator to balance dimensions. Contact with British mathematicians, who considered Leibniz somewhat of an upstart, remained difficult, but during 1676 Leibniz succeeded, through the intermediation of Henry Oldenburg (1615–1677), German-born secretary of the Royal Society, in obtaining two letters from Newton, containing many of his results on series, but little on the calculus (see Selection V.4).

After settling in Hanover, Leibniz became a collaborator on the new Leipzig periodical *Acta Eruditorum* (founded 1682, after 1732 entitled *Nova Acta Eruditorum*, discontinued 1782; 117 volumes), in which in 1682 he announced the existence of his calculus and in 1684 published the first account.

On Leibniz's formative years see J. E. Hofmann, *Die Entwicklungsgeschichte der Leibnizschen Mathematik während des Aufenthaltes in Paris* (Leibnizens Verlag, Munich, 1949).

Here follows a translation from the Latin of Leibniz's paper of 1684, which opens the modern period in the history of the calculus. The original can be found in *Acta Eruditorum 3* (1684), 467–473; it has been reprinted in Leibniz, *Mathematische Schriften*, Abth. 2, Band III (1863); this is also Dritte Folge, Band VII, of *Leibnizens gesammelte Werke*, ed. G. H. Pertz, so that this book is sometimes quoted as Band III and sometimes as Band VII. The significant title is *Nova methodus pro maximis et minimis, itemque tangentibus, quae nec fractas nec irrationales quantitates moratur, et singulare pro illi calculi genus* (A new method for maxima and minima as well as tangents, which is impeded neither by fractional nor by irrational quantities, and a remarkable type of calculus for this).

The paper contains the general rules for differentiation, and it uses differentials (called *differentiae*, that is, differences) rather than derivatives. It has the d-notation. The differentials dx, dv are defined as finite increments, and it is not explained why in $d(xv) = x\,dv + v\,dx$ the term $dx\,dv$ is neglected. The paper contains the condition $dv = 0$ for a maximum or minimum, and $d\,dv = 0$ for a point of inflection. It introduces the term "differential calculus," but, although v, or y, is taken as a function of x, the term "function" does not appear in this paper. To find this term we must consult a paper by Leibniz of 1692; see note 1 and Selections V.6, 9, 16.

Before introducing the term *calculus differentialis*, Leibniz used the expression *methodus tangentium directa*. For the *methodus tangentium inversa* or *calculus summatorius* (the $\int$ is derived from S for *summatio*) Leibniz and Johann Bernoulli, during 1698, introduced the term *calculus integralis*; the term *integral* appears already in a paper by Jakob Bernoulli of 1690 (see Selection V.2).

There exists a German translation of Leibniz's paper of 1684 by G. Kowalewski in Ostwald's *Klassiker*, No. 162 (Engelmann, Leipzig, 1908). A partial English translation by Evelyn Walker is given in Smith, *Source Book*, 619–626, and an Italian translation in G. Castelnuovo, *Le origini del calcolo infinitesimale nell' era moderna* (Zanichelli, Bologna, 1938), 147–160.

A NEW METHOD FOR MAXIMA AND MINIMA AS WELL AS TANGENTS, WHICH IS NEITHER IMPEDED BY FRACTIONAL NOR IRRATIONAL QUANTITIES, AND A REMARKABLE TYPE OF CALCULUS FOR THEM, BY G.W.L. [FIG. 1]

Let an axis AX [Fig. 2; simplified from Leibniz's figure] and several curves such as VV, WW, YY, ZZ be given, of which the ordinates VX, WX, YX, ZX, perpendicular to the axis, are called v, w, y, z respectively. The segment AX, cut off from the axis [*abscissa ab axe*[1]] is called x. Let the tangents be VB, WC, YD, ZE, intersecting the axis respectively at B, C, D, E. Now some straight line selected arbitrarily is called dx, and the line which is to dx as v (or w, or y, or z) is to XB (or XC, or XD, or XE) is called dv (or dw, or dy, or dz),[2] or the difference[3] of these v (or w, or y, or z). Under these assumptions we have the following rules of the calculus.

If a is a given constant, then $da = 0$, and $d(ax) = a\,dx$. If $y = v$ (that is, if the ordinate of any curve YY is equal to any corresponding ordinate of the curve VV), then $dy = dv$. Now *addition* and *subtraction*: if $z - y + w + x = v$, then $d(z - y + w + x) = dv = dz - dy + dw + dx$. Multiplication: $d(xv) = x\,dv + v\,dx$, or, setting $y = xv$, $dy = x\,dv + v\,dx$. It is indifferent whether we take a formula such as xv or its replacing letter such as y. It is to be noted that x and dx are treated in this calculus in the same way as y and dy, or any other indeterminate letter with its difference. It is also to be noted that we cannot always move backward from a differential equation without some caution, something which we shall discuss elsewhere.

[1] Note the Latin term *abscissa*. This term, which was not new in Leibniz's day, was made by him into a standard term, as were so many other technical terms. In the article "De linea ex lineis numero infinitis ordinatim ductis inter se concurrentibus formata...," *Acta Eruditorum 11* (1692), 168–171 (Leibniz, *Mathematische Schriften*, Abth. 2, Band I (1858), 266–269), in which Leibniz discusses evolutes, he presents a collection of technical terms. Here we find *ordinata, evolutio, differentiare, parameter, differentiabilis, functio,* and *ordinata* and *abscissa* together designated as *coordinatae.* Here he also points out that ordinates may be given not only along straight but also along curved lines. The term *ordinate* is derived from *rectae ordinatim applicatae*, "straight lines designated in order," such as parallel lines. The term *functio* appears in the sentence: "the tangent and some other functions depending on it, such as perpendiculars from the axis conducted to the tangent."

[2] When the subtangent—a term Leibniz used in a paper in the *Acta Eruditorum* (1694; *Mathematische Schriften*, Abth. 2, Band I, 306), though it may be older—is denoted by s, Leibniz defines $dy:dx = y:s$, or $s = y:dy/dx$. We may express this by saying that Leibniz takes the derivative (geometrically, in the form of the tangent) without further definition, and defines the differentials in terms of the derivative.

[3] Leibniz uses the term *differentia* and conceives it as a finite line segment. What we now call *differential* would long after Leibniz often be called *difference.* Leibniz also uses other terms. As to the meaning of the differentials, see the end of this selection.

MENSIS OCTOBRIS A. M DC LXXXIV. 467

NOVA METHODVS PRO MAXIMIS ET MI.

nimis, itemque tangentibus, quæ nec fractas, nec irrati-
onales quantitates moratur, & singulare pro
illis calculi genus, per G.G.L.

TAB. XII.

Sit axis AX, & curvæ plures, ut VV, WW, YY, ZZ, quarum ordinatæ, ad axem normales, VX, WX, YX, ZX, quæ vocentur respective, v, vv, y, z; & ipsa AX abscissa ab axe, vocetur x. Tangentes sint
VB, WC, YD, ZE axi occurrentes respective in punctis B, C, D, E.
Jam recta aliqua pro arbitrio assumta vocetur dx, & recta quæ sit ad
dx, ut v (vel vv, vel y, vel z) est ad VB (vel WC, vel YD, vel ZE) vocetur dv (vel d vv, vel dy vel dz) sive differentia ipsarum v (vel ipsarum vv, aut y, aut z) His positis calculi regulæ erunt tales:

Sit a quantitas data constans, erit da æqualis o, & d ax erit æqu-
a dx: si sit y æqu. v (seu ordinata quævis curvæ YY, æqualis cuivis ordinatæ respondenti curvæ VV) erit dy æqu. dv. Jam *Additio & Sub-*
tractio: si sit z -y † vv † x æqu. v, erit dz -y † vv † x seu dv, æqu.
dz -dy † dvv † dx. *Multiplicatio,* $\overline{d x v}$ æqu. x dv † v dx, seu posito
y æqu. xv, fiet dy æqu. x dv † v dx. In arbitrio enim est vel formulam,
ut xv, vel compendio pro ea literam, ut y, adhibere.   Notandum & x
& dx eodem modo in hoc calculo tractari, ut y & dy, vel aliam literam
indeterminatam cum sua differentiali.   Notandum etiam non dari
semper regressum a differentiali Æquatione, nisi cum quadam cautio-
ne, de quo alibi.   Porro *Divisio,* d $\stackrel{v}{-}$ vel (posito z æqu. $\stackrel{v}{-}$) dz æqu.

† v dy † y dv
 $\overline{}$
                                          y   y
 yy

Quoad *Signa* hoc probe notandum, cum in calculo pro litera
substituitur simpliciter ejus differentialis, servari quidem eadem signa,
& pro † z scribi † dz, pro -z scribi -- dz, ut ex additione & subtra-
ctione paulo ante posita apparet; sed quando ad exegesin valorum
venitur, seu cum consideratur ipsius z relatio ad x, tunc apparere, an
valor ipsius dz sit quantitas affirmativa, an nihilo minor seu negativa:
quod posterius cum sit, tunc tangens ZE ducitur a puncto Z non versus A, sed in partes contrarias seu infra X, id est tunc cum ipsæ ordinatæ

N n n 3   z decre-

Fig. 1. The first page of Leibniz's paper, *Acta Eruditorum 3* (1684), 467.

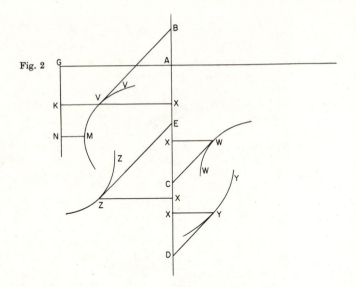

Fig. 2

Now *division*: $d \dfrac{v}{y}$ or $\left(\text{if } z = \dfrac{v}{y}\right)$ $dz = \dfrac{\pm v\,dy \mp y\,dv}{yy}$.

The following should be kept well in mind about the *signs*.[4] When in the calculus for a letter simply its differential is substituted, then the signs are preserved; for z we write dz, for $-z$ we write $-dz$, as appears from the previously given rule for addition and subtraction. However, when it comes to an explanation of the values, that is, when the relation of z to x is considered, then we can decide whether dz is a positive quantity or less than zero (or negative). When the latter occurs, then the tangent ZE is not directed toward A, but in the opposite direction, down from X. This happens when the ordinates z decrease with increasing x. And since the ordinates v sometimes increase and sometimes decrease, dv will sometimes be positive and sometimes be negative; in the first case the tangent VB is directed toward A, in the latter it is directed in the opposite sense. None of these cases happens in the intermediate position at M, at the moment when v neither increases nor decreases, but is stationary. Then $dv = 0$, and it does not matter whether the quantity is positive or negative, since $+0 = -0$. At this place v, that is, the ordinate LM, is *maximum* (or, when the convexity is turned to the axis, *minimum*), and the tangent to the curve at M is directed neither in the direction from X up to A, to approach the axis, nor down to the other side, but is parallel to the axis. When dv is infinite with respect to dx, then the tangent is perpendicular to the axis, that is, it is the ordinate itself. When $dv = dx$, then the tangent makes half a right angle with the axis. When with increasing ordinates v its increments or differences dv also

---

[4] The ambiguity in signs is due to the fact that s is taken positive. Systematic discrimination between positive and negative senses in analytic geometry came only with Monge and Möbius in the early nineteenth century.

increase (that is, when dv is positive, $d\,dv$, the difference of the differences, is also positive, and when dv is negative, $d\,dv$ is also negative), then the curve turns toward the axis its *concavity*, in the other case its *convexity*.[5] Where the increment is maximum or minimum, or where the increments from decreasing turn into increasing, or the opposite, there is a *point of inflection*.[6] Here concavity and convexity are interchanged, provided the ordinates too do not turn from increasing into decreasing or the opposite, because then the concavity or convexity would remain. However, it is impossible that the increments continue to increase or decrease, but the ordinates turn from increasing into decreasing, or the opposite.[7] Hence a point of inflection occurs when $d\,dv = 0$ while neither v nor $dv = 0$. The problem of finding inflection therefore has not, like that of finding a maximum, two equal roots, but three. This all depends on the correct use of the signs.

Sometimes it is better to use *ambiguous signs*, as we have done with the division, before it is determined what the precise sign is. When with increasing x v/y increases (or decreases), then the ambiguous signs in $d\dfrac{v}{y} = \dfrac{\pm v\,dy \mp y\,dv}{yy}$ must be determined in such a way that this fraction is a positive (or negative) quantity. But $\mp$ means the opposite of $\pm$, so that when one is $+$ the other is $-$ or vice versa. There also may be several ambiguities in the same computation, which I distinguish by parentheses. For example, let $\dfrac{v}{y} + \dfrac{y}{z} + \dfrac{x}{v} = w$; then we must write

$$\frac{\pm v\,dy \mp y\,dv}{yy} + \frac{(\pm)y\,dz\,(\mp)\,z\,dy}{zz} + \frac{((\pm))x\,dv\,((\mp))\,v\,dx}{vv} = dv,$$

so that the ambiguities in the different terms may not be confused. We must take notice that an ambiguous sign with itself gives $+$, with its opposite gives $-$, while with another ambiguous sign it forms a new ambiguity depending on both.

Powers. $dx^a = ax^{a-1}\,dx$; for example, $dx^3 = 3x^2\,dx$. $d\dfrac{1}{x^a} = -\dfrac{a\,dx}{x^{a-1}}$; for example, if $w = \dfrac{1}{x^3}$, then $dw = -\dfrac{3\,dx}{x^4}$.

Roots. $d\sqrt[b]{x^a} = \dfrac{a}{b}\,dx\sqrt[b]{x^{a-b}}$ (hence $d\sqrt[2]{y} = \dfrac{dy}{2\sqrt[2]{y}}$, for in this case $a = 1, b = 2$), therefore $\dfrac{a}{b}\sqrt[b]{x^{a-b}} = \tfrac{1}{2}\sqrt[2]{y^{-1}}$, but y^{-1} is the same as $\dfrac{1}{y}$; from the nature of the exponents in a geometric progression, and $\sqrt[2]{\dfrac{1}{y}} = \dfrac{1}{\sqrt[2]{y}}$, $d\dfrac{1}{\sqrt{x^a}} = \dfrac{-a\,dx}{b\sqrt[b]{x^{a+b}}}$. The law for integral powers would have been sufficient to cover the case of fractions

---

[5] Leibniz has "concavity" and "convexity" interchanged.

[6] Leibniz' term is *punctum flexii contrarii* (point of opposite flection). On this term see T. F. Mulcrone, *The Mathematics Teacher 61* (1968), 475–478.

[7] There seems to be something wrong here: when $y = x^2$, $dy = 2x\,dx$; then, when x passes from negative to positive ($dx > 0$), dy increases while y first decreases and then increases. However, see note 4.

as well as roots, for a power becomes a fraction when the exponent is negative, and changes into a root when the exponent is fractional. However, I prefer to draw these conclusions myself rather than relegate their deduction to others, since they are quite general and occur often. In a matter that is already complicated in itself it is preferable to facilitate the operations.

Knowing thus the *Algorithm* (as I may say) of this calculus, which I call *differential calculus*, all other differential equations can be solved by a common method. We can find maxima and minima as well as tangents without the necessity of removing fractions, irrationals, and other restrictions, as had to be done according to the methods that have been published hitherto. The demonstration of all this will be easy to one who is experienced in these matters and who considers the fact, until now not sufficiently explored, that dx, dy, dv, dw, dz can be taken proportional to the momentary differences, that is, increments or decrements, of the corresponding x, y, v, w, z. To any given equation we can thus write its differential equation. This can be done by simply substituting for each *term* (that is, any part which through addition or subtraction contributes to the equation) its differential quantity. For any other quantity (not itself a term, but contributing to the formation of the term) we use its differential quantity, to form the differential quantity of the term itself, not by simple substitution, but according to the prescribed Algorithm. The methods published before have no such transition. They mostly use a line such as DX or of similar kind, but not the line dy which is the fourth proportional to DX, DY, dx—something quite confusing. From there they go on removing fractions and irrationals (in which undetermined quantities occur). It is clear that our method also covers transcendental[8] curves—those that cannot be reduced by algebraic computation, or have no particular degree—and thus holds in a most general way without any particular and not always satisfied assumptions.

We have only to keep in mind that to find a *tangent* means to draw a line that connects two points of the curve at an infinitely small distance, or the continued side of a polygon with an infinite number of angles, which for us takes the place of the *curve*. This infinitely small distance can always be expressed by a known differential like dv, or by a relation to it, that is, by some known tangent. In particular, if y were a transcendental quantity, for instance the ordinate of a cycloid, and it entered into a computation in which z, the ordinate of another curve, were determined, and if we desired to know dz or by means of dz the tangent of this latter curve, then we should by all means determine dz by means of dy, since we have the tangent of the cycloid. The tangent to the cycloid itself, if we assume that we do not yet have it, could be found in a similar way from the given property of the tangent to the circle.

Now I shall propose an example of the calculus, in which I shall indicate division by $x:y$, which means the same as x divided by y, or $\frac{x}{y}$.[9] Let the *first* or

---

[8] This may be the first time that the term "transcendental" in the sense of "nonalgebraic" occurs in print.

[9] From this suggestion by Leibniz dates the general adoption of this notation; see J. Tropfke, *Geschichte*, 3rd ed., II (1933), 30. See also the reference to Mengoli in G. Castelnuovo, *Le origini del calcolo infinitesimale nell'era moderna*, 153.

given equation be[10] $x : y + (a + bx)(c - xx) : (ex + fxx)^2 + ax\sqrt{gg + yy} + yy :$ $\sqrt{hh + lx + mxx} = 0$. It expresses the relation between x and y or between AX and XY, where a, b, c, e, f, g, h are given. We wish to draw from a point Y the line YD tangent to the curve, or to find the ratio of the line DX to the given line XY. We shall write for short $n = a + bx$, $p = c - xx$, $q = ex + fxx$, $r = gg + yy$, and $s = hh + lx + mxx$. We obtain $x : y + np :$ $qq + ax\sqrt{r} + yy : \sqrt{s} = 0$, which we call the *second* equation. From our calculus it follows that

$$d(x : y) = (\pm x \, dy \mp y \, dx) : yy,^{11}$$

and equally that

$$d(np : qq) = [(\pm) \, 2np \, dq \, (\mp) \, q(n \, dp + p \, dn)] : q^3,$$

$$d(ax\sqrt{r}) = +ax \, dr : 2\sqrt{r} + a \, dx\sqrt{r},$$

$$d(yy : \sqrt{s}) = ((\pm))yy \, ds \, ((\mp)) \, 4ys \, dy : 2s\sqrt{s}.$$

All these differential quantities from $d(x : y)$ to $d(yy : \sqrt{s})$ added together give 0, and thus produce a *third* equation, obtained from the terms of the second equation by substituting their differential quantities. Now $dn = b \, dx$ and $dp = -2x \, dx$, $d = e \, dx + 2fx \, dx$, $dr = 2y \, dy$, and $ds = l \, dx + 2mx \, dx$. When we substitute these values into the third equation we obtain a *fourth* equation, in which the only remaining differential quantities, namely dx, dy, are all outside of the denominators and without restrictions. Each term is multiplied either by dx or by dy, so that the law of homogeneity always holds with respect to these two quantities, however complicated the computation may be. From this we can always obtain the value of $dx : dy$, the ratio of dx to dy, or the ratio of the required DX to the given XY. In our case this ratio will be (if the fourth equation is changed into a proportionality):

$$\mp x : yy - axy : \sqrt{r} \, (\mp) \, 2y : \sqrt{s}$$

divided by

$$\mp 1 : y \, (\pm) \, (2npe + 2fx) : q^3 \, (\mp) \, (-2nx + pb) : qq$$

$$+ a\sqrt{r} \, ((\pm)) \, yy(l + 2mx) : 2s\sqrt{s}.$$

Now x and y are given since point Y is given. Also given are the values of n, p, q, r, s expressed in x and y, which we wrote down above. Hence we have obtained what we required. Although this example is rather complicated we

---

[10] We have retained Leibniz' notation : but substituted parentheses for superscript bars: Leibniz writes

$x : y + \overline{a + bx} \, \overline{c \, xx} : \text{quadrat.} \ ex + fxx + ax\sqrt{gg + yy} + yy : \sqrt{hh + lx + mxx} \ \text{aequ. 0.}$
[11] Leibniz writes $d, x : y$.

have presented it to show how the above-mentioned rules can be used even in a more difficult computation. Now it remains to show their use in cases easier to grasp.

Let two points C and E [Fig. 3] be given and a line SS in the same plane. It is required to find a point F on SS such that when E and C are connected with F

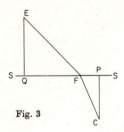

Fig. 3

the sum of the rectangle of CF and a given line h and the rectangle of FE and a given line r are as small as possible.[12] In other words, if SS is a line separating two media, and h represents the density of the medium on the side of C (say water), r that of the medium on the side of E (say air), then we ask for the point F such that the path from C to E via F is the shortest possible. Let us assume that all such possible sums of rectangles, or all possible paths, are represented by the ordinates KV of curve VV perpendicular to the line GK [Fig. 2]. We shall call these ordinates w. Then it is required to find their minimum NM. Since C and E [Fig. 3] are given, their perpendiculars to SS are also given, namely CP (which we call c) and EQ (which we call e); moreover PQ (which we call p) is given. We denote $QF = GN$ (or AX) by x, CF by f, and EF by g. Then $FP = p - x, f = \sqrt{cc + pp - 2px + xx}$ or $= \sqrt{l}$ for short; $g = \sqrt{ee + xx}$ or $= \sqrt{m}$ for short. Hence

$$w = h\sqrt{l} + r\sqrt{m}.$$

The differential equation (since $dw = 0$ in the case of a minimum) is, according to our calculus,

$$0 = +h \, dl : 2\sqrt{l} + r \, dm : 2\sqrt{m}.$$

But $dl = -2(p - x) \, dx$, $dm = 2x \, dx$; hence

$$h(p - x) : f = rx : g.$$

When we now apply this to dioptrics, and take f and g, that is, CF and EF, equal to each other (since the refraction at the point F is the same no matter how long the line CF may be), then $h(p - x) = rx$ or $h : r = x : (p - x)$, or $h : r = QF : FP$; hence the sines of the angles of incidence and of refraction,

---

[12] For this problem, due to Fermat (*Oeuvres*, II (1844), 457), see note 13.

FP and QF, are in inverse ratio to r and h, the densities of the media in which the incidence and the refraction take place. However, this density is not to be understood with respect to us, but to the resistance which the light rays meet. Thus we have a demonstration of the computation exhibited elsewhere in these *Acta* [1682], where we presented a general foundation of optics, catoptrics, and dioptrics.[13] Other very learned men have sought in many devious ways what someone versed in this calculus can accomplish in these lines as by magic.

This I shall explain by still another example. Let *13* [Fig. 4] be a curve of such a nature that, if we draw from one of its points, such as *3*, six lines *34, 35, 36*,

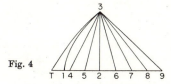

Fig. 4

37, 38, 39 to six fixed points *4, 5, 6, 7, 8, 9* on the axis, then their sum is equal to a given line. Let *T14526789* be the axis, *12* the abscissa, *23* the ordinate, and let the tangent *3T* be required. Then I claim that *T2* is to *23* as $\dfrac{23}{34} + \dfrac{23}{35} + \dfrac{23}{36}$ $+ \dfrac{23}{37} + \dfrac{23}{38} + \dfrac{23}{39}$ to $-\dfrac{24}{34} - \dfrac{25}{35} + \dfrac{26}{36} + \dfrac{27}{37} + \dfrac{28}{38} + \dfrac{29}{39}$. The same rule will hold if we increase the number of terms, taking not six but ten or more fixed points. If we wanted to solve this problem by the existing tangent methods, removing irrationals, then it would be a most tedious and sometimes insuperable task; in this case we would have to set up the condition that the rectangular planes and solids which can be constructed by means of all possible combinations of two or three of these lines are equal to a given quantity.[14] In all these cases and even in more complicated ones our methods are of astonishing and unequaled facility.

And this is only the beginning of much more sublime Geometry, pertaining to even the most difficult and most beautiful problems of applied mathematics, which without our differential calculus or something similar no one could attack with any such ease. We shall add as appendix the solution of the problem which De Beaune proposed to Descartes and which he tried to solve in Vol. 3 of the *Letters*, but without success.[15] It is required to find a curve *WW* such that, its

[13] In this paper, "Unicum opticae, catoptricae et dioptricae principium," *Acta Eruditorum 1* (1683), 186–190, dealing with the laws of refraction and reflection, Leibniz makes known' for the first time in print that he has his own *methodus de maximis et minimis*.

[14] If the coordinates of point i are a_i, $i = 4, 5, \ldots$, and those of point 3 are x, y, then this result can immediately be obtained by differentiating $\sum_i \sqrt{(x - a_i)^2 + y^2}$. Leibniz writes $-24, -25$ because his segments are all positive.

[15] This is an inverse-tangent problem; Leibniz quotes *Les lettres de René Descartes* (3 vols.; Paris, ed. C. de Clerselier, 1657–1667). The problem is part of a long series of investigations that begins with the invention of logarithms by Napier by comparing an arithmetic and a geometric series and leads up to the full recognition of the inverse relation of the two functions $y = \log x$ and $x = e^y$ by Euler. Florimond De Beaune (1601–1652), a jurist at Blois, had written to Descartes about some curves; Descartes's answer of 1639 exists (Descartes,

(footnote continued)

tangent WC being drawn to the axis, XC is always equal to a given constant line a. Then XW or w is to XC or a as dw is to dx. If dx (which can be chosen arbitrarily) is taken constant, hence always equal to, say, b, that is, x or AX increases uniformly, then $w = \dfrac{a}{b}\, dw$. Those ordinates w are therefore proportional to their dw, their increments or differences, and this means that if the x form an arithmetic progression, then the w form a geometric progression. In other words, if the w are numbers, the x will be logarithms, so that the curve WW is logarithmic.

Leibniz (like Newton) was never very consistent in his explanation of differentials. For instance, in his reply to his critic, the Dutch physician Bernard Nieuwentijt (1654–1718), who rebuked him for rejecting infinitely small quantities as if they were nothing at all (Cantor, *Geschichte*, III, 244–247), he answered that it was correct to consider quantities of which the difference is incomparably small to be equal; a line is not lengthened by adding a point. It was only a question of words, he added, whether one rejected such an equality: "Responsio ad nunnullos difficultates a Dn. Bernardo Niewentiit...motas," *Acta Eruditorum* 7 (1695), 310–316; *Mathematische Schriften*, Abth. 2, Band I, 320–328. And later he pointed out: "There are different degrees of infinity or of infinitely small, just as the globe of the Earth is estimated as a point in proportion to the distance of the fixed stars, and a play ball is still a point as compared to the radius of the terrestrial sphere, so that the distance of the fixed stars is an infinitely infinite or infinite of the infinite with respect to the diameter of the ball" ("Mémoire de Mr G. G. Leibniz touchant son sentiment sur le calcul différentiel," *Mémoires pour l'histoire des sciences et des beaux arts* = *Journal de Trévoux* (Nov.–Dec. 1701), 270–272; also Leibniz, *Mathematische Schriften*, Abth. 2, Band I, 350. No wonder that in his reply to Nieuwentijt he referred to the indirect method of Archimedes as the final test of the truth. See also C. B. Boyer, *The history of the calculus* (Dover, New York, 1949), 213–222.

Oeuvres, II, 510–519); it was printed in the above-mentioned seventeenth-century edition of Descartes's letter and was studied by Leibniz. One of the curves was defined by a geometric description equivalent to the equation $dy/dx = (x - y)/b$. By means of the substitution $x' = b - x + y$, $y' = y$ this equation is transformed into $dy'/dx' = -y'/b$, the differential equation of what we now call the logarithmic curve. Descartes comes to the equivalent of this result; without mentioning logarithms he derives an inequality that can be written in our notation

$$\frac{1}{n} + \frac{1}{n+1} + \cdots + \frac{1}{m-1} > \log\frac{m}{n} > \frac{1}{n+1} + \frac{1}{n+2} + \cdots + \frac{1}{m}$$

($n < m - 1$; m, n positive integers); see C. J. Scriba, "Zur Lösung des 2. Debeauneschen Problems durch Descartes," *Archive for History of Exact Sciences 1* (1961), 406–419. Descartes, like Napier, lets the logarithms grow when the argument decreases, while Briggs, who introduces 10 as base, lets argument and function grow at the same time. The next important steps, known to Leibniz, were Grégoire De Saint Vincent's determination of the area enclosed by a hyperbola, two ordinates, and an asymptote (1647), which Alfons Anton De Sarasa (1649) interpreted with the aid of logarithms, and Nikolaus Mercator's series (1667) for this area of the hyperbola: $a - \dfrac{a^2}{2} + \dfrac{a^3}{3} - \dfrac{a^4}{4} + \cdots$.

2 LEIBNIZ. THE FIRST PUBLICATION OF HIS INTEGRAL CALCULUS

Two years after Leibniz had published his first account of the differential calculus, he published a paper on the inverse tangent problem in which the symbol $\int$ appears. This was done in a rather casual way, since the paper was a review of a book by the Scottish pupil of Newton, John Craig. Leibniz used the occasion to illustrate two fundamental points at the same time: (a) the power of the integration symbol in combination with that of differentiation (if $p \, dy = x \, dx$, then $\int p \, dy = \int x \, dx$, and conversely, which is the expression of the inverse character of the differential and the integral calculus), and (b) the power of the method to represent that still poorly explored type of relation, the "transcendental" quantities.

Leibniz's paper "De geometria recondita et analysi indivisibilium atque infinitorum" (On a deeply hidden geometry and the analysis of indivisibles and infinities) appeared in the *Acta Eruditorum 5* (1686) and was reprinted in Leibniz, *Mathematische Schriften*, Abth. 2, Band III, 226–235. A German translation by G. Kowalewski can be found in Ostwald's *Klassiker*, No. 162 (Engelmann, Leipzig, 1908). Here follows a translation of that part of the paper dealing with the introduction of the integral calculus.

For transcendental problems,[1] wherever dimensions and tangents occur that have to be found by computation, there can hardly be found a calculus more useful, shorter, and more universal than my *differential calculus, or analysis of indivisibles and infinites*, of which only a small sample or corollary is contained in my Method of Tangents published in the *Acta* of October 1684.[2] It has been much praised by Dr. Craig, who has also suspected that there is more to it, and on p. 29 of his little book[3] has made an attempt to prove Barrow's theorem (that the sum of the intervals between the ordinates and perpendiculars to a curve taken on the axis and measured in it is equal to half of the square of the final ordinate).[4] In trying this he deviates a bit from his goal, which does not surprise me in the new method: so that I believe that I may oblige him and others by publishing here an addition to a subject that seems to have so wide a use. From it flow all the admirable theorems and problems of this kind with such ease that there is no more need to teach and retain them than for him who knows our present algebra[5] to memorize many theorems of ordinary geometry.

[1] Leibniz repeatedly stressed (even in the title of his paper of 1684) that his method was also valid for nonalgebraic quantities. In those days, with an undeveloped function concept, there were algebraic expressions and curves, "mechanical" curves such as the cycloid, arbitrary curves, and tabulated rows of numbers such as sines and logarithms. Leibniz's point is that his calculus can deal in a unified way not only with algebraic expressions but also with others.

[2] This is the preceding Selection V.1.

[3] J. Craig, *Methodus figurarum lineis rectis et curvis comprehensarum quadraturas determinandi* (London, 1685). This book contained a reference to Leibniz's calculus. The Leibniz form of the calculus was therefore announced in an English book before Newton published his theory. The Scottish theologian John Craig (1660 ?–1731) was at Cambridge around 1680; he loved to apply mathematics to matters divine.

[4] I. Barrow, *Lectiones geometricae* (Cambridge, 1670), Lecture XI, no. 1; see Selection IV.14. When p is the subnormal of the curve $y = f(x)$, then $\int_0^x p \, dx = \int_0^y y \, dy = \frac{1}{2} y^2$.

[5] Latin "illi qui Speciosam tenet," which we think may mean: "to him who knows his speciosa." The "logistica speciosa" of François Viète (1591; see our Selection II.5) was the art of writing equations with letters as coefficients instead of numbers.

I proceed to this subject in the following way. Let the ordinate be x, the abscissa y, and the interval between perpendicular and ordinate, described before, p. Then according to my method it follows immediately that $p\,dy = x\,dx$, as Dr. Craig has also found. When we now subject this differential equation to summation we obtain $\int p\,dy = \int x\,dx$ (like powers and roots in ordinary calculations, so here sum and difference, or $\int$ and d, are each other's converse). Hence we have $\int p\,dy = \frac{1}{2}xx$, which was to be demonstrated. Now I prefer to use dx and similar symbols rather than special letters, since this dx is a certain modification of the x and by virtue of this it happens that—when necessary— only the letter x with its powers and differentials enters into the calculus, and transcendental relations are expressed between x and some other quantity.[6] Transcendental curves can therefore also be expressed by an equation, for example, if a is an arc, and the versed sine x, then $a = \int dx : \sqrt{2x - x^2}$ and if the ordinate of a cycloid is y, then $y = \sqrt{2x - xx} + \int dx : \sqrt{2x - xx}$, which equation perfectly expresses the relation between the ordinate y and the abscissa x. From it all properties of the cycloid can be demonstrated. The analytic calculus is thus extended to those curves that hitherto have been excluded for no better reason than that they were thought to be unsuited to it. Wallis's interpolations and innumerable other questions can be derived from this.

---

3 LEIBNIZ. THE FUNDAMENTAL THEOREM OF THE CALCULUS

From the many papers that Leibniz wrote on the calculus we reproduce a part of the "Supplementum geometriae dimensoriae[1] . . . similiterque multiplex constructio lineae ex data tangentium conditione," *Acta Eruditorum* (1693), 385–392, translated from Leibniz, *Mathematische Schriften*, Abth. 2, Band I, 294–301. Here he expresses by means of a figure the inverse relation of integration and differentiation. There exists a German translation in Ostwald's *Klassiker*, No. 162 (Engelmann, Leipzig, 1908), 24–34.

I shall now show *that the general problem of quadratures can be reduced to the finding of a line that has a given law of tangency (declivitas)*, that is, for which the sides of the characteristic triangle have a given mutual relation. Then I shall show how this line can be described by a motion that I have invented. For this

[6] The meaning of this obscure Latin sentence seems to be (a) that the introduction of the dx under the integral sign makes it easier to see (what we would call) the functional character of the expression, as is clear when we pass from one variable to another and (b) that therefore we have a way of writing operating symbols expressing transcendental quantities such as $\int dx : x$, $\int dx : \sqrt{1 - x^2}$, symbols that express the nature of the quantity (contrary to such expressions as $\log x$ or $\sin x$, which in themselves do not express a property). Leibniz is often quite obscure when he wants to tell us about a really exciting discovery he has made.

[1] Leibniz distinguishes here between *geometria dimensoria*, which deals with quadratures, and *geometria determinatrix*, which can be reduced to algebraic equations.

purpose [Fig. 1] I assume for every curve $C(C')$ a *double characteristic triangle*,[2] one, TBC, that is assignable, and one, GLC, that is inassignable,[3] and these two are similar. The inassignable triangle consists of the parts GL, LC, with the elements of the coordinates CF, CB as sides, and GC, the element of arc, as the base or hypotenuse. But the assignable triangle TBC consists of the axis, the ordinate, and the tangent, and therefore contains the angle between the direction of the curve (or its tangent) and the axis or base, that is, the inclination of

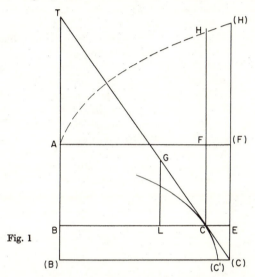

Fig. 1

the curve at the given point C. Now let $F(H)$, the region of which the area has to be squared,[4] be enclosed between the curve $H(H)$, the parallel lines FH and $(F)(H)$, and the axis $F(F)$; on that axis let A be a fixed point, and let a line AB, the conjugate axis, be drawn through A perpendicular to AF. We assume that point C lies on HF (continued if necessary); this gives a new curve $C(C')$ with the property that, if from point C to the conjugate axis AB (continued if necessary) both its ordinate CB (equal to AF) and tangent CT are drawn, the part TB of the axis between them is to BC as HF to a constant [segment] a, or a times BT is equal to the rectangle AFH (circumscribed about the trilinear figure

[2] In Fig. 1 Leibniz assigns the symbol (C) to two points which we denote by (C) and (C'). If, with Leibniz, we write $CF = x$, $BC = y$, $HF = z$, then $E(C) = dx$, $CE = F(F) = dy$, and $H(H)(F)F = z\,dy$. First Leibniz introduces curve $C(C')$ with its characteristic triangle, and then later reintroduces it as the squareing curve [*curva quadratrix*] of curve $AH(H)$.

[3] For want of anything better we use Leibniz's terms *assignabilis* and *inassignabilis*. G. Kowalewski, *Leibniz über die Analysis des Unendlichen*, Ostwald's *Klassiker*, No. 162 (Engelmann, Leipzig, 1908), 30, uses the German *angebbar* and *unangebbar*, "indicable" and "unindicable." For "differential" Leibniz in our text uses the term "element." Observe also the use of the term "coordinates" (Latin *coordinatae*).

[4] The Latin is here a little more expressive than the English. From the Latin *quadrare* we can derive *quadrans, quadrandus, quadratrix, quadratura*, which can be translated by "to square," "squaring," "to be squared," "squaring curve" or "quadratrix," and "quadrature."

$AFHA$).[5] This being established, I claim that the rectangle on a and $E(C)$ (we must discriminate between the ordinates FC and $(F)(C)$ of the curve) is equal to the region $F(H)$. When therefore I continue line $H(H)$ to A, the trilinear figure $AFHA$ of the figure to be squared is equal to the rectangle with the constant a and the ordinate FC of the squaring curve as sides. This follows immediately from our calculus. Let $AF = y$, $FH = z$, $BT = t$, and $FC = x$; then $t = zy : a$, according to our assumption; on the other hand, $t = y\, dx : dy$ because of the property of the tangents expressed in our calculus. Hence $a\, dx = z\, dy$ and therefore $ax = \int z\, dy = AFHA$. Hence the curve $C(C')$ is the quadratrix with respect to the curve $H(H)$, while the ordinate FC of $C(C')$, multiplied by the constant a, makes the rectangle equal to the area, or the sum of the ordinates $H(H)$ corresponding to the corresponding abscissas AF. Therefore, since $BT : AF = FH : a$ (by assumption), and the relation of this FH to AF (which expresses the nature of the figure to be squared) is given, the relation of BT to FH or to BC, as well as that of BT to TC, will be given, that is, the relation between the sides of triangle TBC.[6] Hence, all that is needed to be able to perform the quadratures and measurements is to be able to describe the curve $C(C')$ (which, as we have shown, is the quadratrix), when the relation between the sides of the assignable characteristic triangle TBC (that is, the law of inclination of the curve) is given.

Leibniz continues by describing an instrument that can perform this construction.

4 NEWTON AND GREGORY. BINOMIAL SERIES

Isaac Newton started to work on what is now called the calculus in 1664 under Barrow at Cambridge (Selection IV.14). One of his early sources was the Latin edition by F. van Schooten of the *Géométrie* of Descartes, which also had contributions to the infinitesimal calculus. Newton's first manuscript notes date from 1665. Here we see emerge his "pricked" letters, such as $\dot{x}$ for our dx/dt. Studying Wallis's *Arithmetica infinitorum* he also discovered the binomial series. Then, in 1669, having studied Nicolas Mercator's *Logarithmotechnia* (London, 1668) and James Gregory's *De vera circuli et hyperbolae quadratura* (Padua, 1667), he composed the manuscript later published as *De analysi per aequationes numero terminorum infinitas* (ed. W. Jones, London, 1711). Expanding on his fluxional methods, he wrote another text in 1671, entitled *Methodus fluxionum et serierum infinitorum*, first published, in English translation, as *The method of fluxions and infinite series*, ed. John Colson (London, 1736); the original was first published by Samuel Horsley in the *Opera omnia* (London, 1779–1785), under the title *Geometria analytica*.

Then, in 1676, in two letters to Henry Oldenburg, the secretary of the Royal Society and, like Mersenne at an earlier date, a man whose scientific contacts connected him with prac-

---

[5] This is Pascal's expression; see Selections IV.11, 12.

[6] This reasoning is still very much like that of Barrow, Gregory, and Torricelli, but because Leibniz possesses the converse relation $a\, dx = x\, dy \leftrightarrow \int a\, dx = \int x\, dy$ he needs only one demonstration, where Barrow needed two (Lecture X, 11; XI, 19; Selection IV.14).

tically all who worked in the exact sciences, Newton presented some of his results, especially on the binomial series and on fluxions. The letters were destined for Leibniz, then in his early struggles for the discovery of his own calculus (see Selection V.1). After some time, Newton's attention was directed toward mechanics and astronomy; the result was the immortal *Philosophiae naturalis principia mathematica* (London, 1687; *Principia* for short), with its exposition of the planetary theory on the basis of the law of universal gravity. Newton did not explain his theory of fluxions in this book, preferring to give his proofs in classical geometric form as Huygens had done. However, some of his lemmas and propositions present, in carefully chosen language, a few products of his meditations on the calculus, and we reprint them here as Selections V.5, 6. Then, finally, in an attempt to collect his thoughts on fluxions, Newton produced in 1693 a manuscript that was eventually published as *Tractatus de quadratura curvarum* (London, 1704), which we have chosen for Selection V.7, being, as it seems, part of the last formulation that Newton gave to his theory of fluxions.

The *Analysis per aequationes*, the *Quadratura curvarum*, and the *Methodus fluxionum* have been republished, in their eighteenth-century English translations, by D. T. Whiteside, *The mathematical works of Isaac Newton* (Johnson Reprint Co., New York, London, 1964).

Our first selection of Newton's work gives essential parts of his two letters of 1676 to Oldenburg, dealing in the main with the binomial series. By applying Wallis's methods of interpolation and extrapolation to new problems, Newton had taken the concept of negative and fractional exponents from Wallis, and so had been able to generalize the binomial theorem, already known for a long time for positive integral exponents (see Selection I.5 on the Pascal triangle), to these more generalized exponents, by which a polynomial expression was changed into an infinite series. He then was able to show how a great many series that already existed in the literature could be regarded as special cases, either directly or by differentiation or integration.

Here follow the two letters from Newton to Oldenburg; they are taken from *The correspondence of Isaac Newton*, ed. H. W. Turnbull (Cambridge University Press, New York, 1959), vol. 1.

LETTER OF JUNE 13, 1676

Most worthy Sir,

Though the modesty of Mr. Leibniz, in the extracts from his letter which you have lately sent me, pays great tribute to our countrymen for a certain theory of infinite series, about which there now begins to be some talk, yet I have no doubt that he has discovered not only a method for reducing any quantities whatever to such series, as he asserts, but also various shortened forms, perhaps like our own, if not even better. Since, however, he very much wants to know what has been discovered in this subject by the English, and since I myself fell upon this theory some years ago, I have sent you some of those things which occurred to me in order to satisfy his wishes, at any rate in part.

Fractions are reduced to infinite series by division; and radical quantities by extraction of the roots, by carrying out those operations in the symbols just as they are commonly carried out in decimal numbers. These are the foundations of these reductions: but extractions of roots are much shortened by this theorem,

$$(P + PQ)^{m/n} = P^{m/n} + \frac{m}{n} AQ + \frac{m - n}{2n} BQ + \frac{m - 2n}{3n} CQ$$

$$+ \frac{m - 3n}{4n} DQ + \text{etc.},$$

where $P + PQ$ signifies the quantity whose root or even any power, or the root of a power, is to be found; P signifies the first term of that quantity, Q the remaining terms divided by the first, and m/n the numerical index of the power of $P + PQ$, whether that power is integral or (so to speak) fractional, whether positive or negative. For as analysts, instead of aa, aaa, etc., are accustomed to write a^2, a^3, etc., so instead of $\sqrt{a}$, $\sqrt{a^3}$, $\sqrt{c:a^5}$, etc. I write $a^{\frac{1}{2}}$, $a^{\frac{3}{2}}$, $a^{\frac{5}{3}}$, and instead of $1/a$, $1/aa$, $1/a^3$, I write a^{-1}, a^{-2}, a^{-3}.[1] And so for

$$\frac{aa}{\sqrt{c:(a^3 + bbx)}}$$

I write $aa(a^3 + bbx)^{-\frac{1}{3}}$, and for

$$\frac{aab}{\sqrt{c:\{(a^3 + bbx)(a^3 + bbx)\}}}$$

I write $aab(a^3 + bbx)^{-\frac{2}{3}}$: in which last case, if $(a^3 + bbx)^{-\frac{2}{3}}$ is supposed to be $(P + PQ)^{m/n}$ in the Rule, then P will be equal to a^3, Q to bbx/a^3, m to -2, and n to 3. Finally, for the terms found in the quotient in the course of the working I employ A, B, C, D, etc., namely, A for the first term, $P^{m/n}$; B for the second term, $(m/n)AQ$; and so on. For the rest, the use of the rule will appear from the examples.

Example 1.

$$\sqrt{(c^2 + x^2)} \quad \text{or} \quad (c^2 + x^2)^{\frac{1}{2}} = c + \frac{x^2}{2c} - \frac{x^4}{8c^3} + \frac{x^6}{16c^5} - \frac{5x^8}{128c^7} + \frac{7x^{10}}{256c^9} + \text{etc.}$$

For in this case, $P = c^2$, $Q = x^2/c^2$, $m = 1$, $n = 2$, $A (= P^{m/n} = (cc)^{\frac{1}{2}}) = c$, $B (= (m/n)AQ) = x^2/2c$, $C \left(= \frac{m - n}{2n} BQ \right) = -\frac{x^4}{8c^3}$; and so on.

---

[1] Newton had learned this method of broken and negative exponents from Wallis, but the idea goes back as far as Oresme and Chuquet; see Selection II.2. Through the influence of Wallis and Newton the method was gradually adopted by other mathematicians. The notation $\sqrt{c:(\ \ )}$ indicates the cube root.

Other examples give the solution of the equations $y^3 - 2y - 5 = 0$ and $y^3 + axy + a^2y - x^3 - 2a^3 = 0$, series for $\sin x$ and for $\sin^2 x$, the solution of Kepler's problem (to divide a semicircle by a line through a given point on the diameter into two sections of which the areas are in given proportion[2]) for an ellipse, the rectification of the arc of an ellipse and a hyperbola, the area of a hyperbola with the aid of the series for the logarithm, the quadrature of the quadratrix $x = y \tan x/a$, and the volume of a segment of an ellipsoid of rotation. Newton, careful not to give too much away, selected these examples from results that were already known.

Leibniz answered in his letter of August 17 with an account of several of his own results in finding quadratures, hinting at his possession of a general method. He also offered several series, among them $1 - \frac{1}{3} + \frac{1}{5} - \frac{1}{7} + \cdots$, as the ratio of the area of a circle to the circumscribed square—a series which Leibniz had already mentioned to friends in 1673, but which James Gregory had found before.[3]

Newton was interested, and answered as follows:

LETTER OF OCTOBER 24, 1676

Cambridge October 24 1676

Most worthy Sir,

I can hardly tell with what pleasure I have read the letters of those very distinguished men Leibniz and Tschirnhaus.[4] Leibniz's method for obtaining convergent series is certainly very elegant, and it would have sufficiently revealed the genius of its author, even if he had written nothing else. But what he has scattered elsewhere throughout his letter is most worthy of his reputation—it leads us also to hope for very great things from him. The variety of ways by which the same goal is approached has given me the greater pleasure, because three methods of arriving at series of that kind had already become known to me, so that I could scarcely expect a new one to be communicated to us. One of mine I have described before; I now add another, namely, that by which I first chanced on these series—for I chanced on them before I knew the divisions and extractions of roots which I now use. And an explanation of this will serve to

---

[2] Kepler, in his *Astronomia nova stellae Martis* (Heidelberg, 1609), showed that the problem of finding the position of a planet in its elliptical orbit leads to this other problem.

[3] Newton's letters were first published in J. Wallis, *Opera mathematica*, III (Oxford, 1699), 622–629; the letter of Gregory was first published in *Correspondence of scientific men of the XVIIth century*, ed. S. J. Rigaud (2 vols.; Oxford University Press, Oxford, 1841), II, 209. Gregory, like Newton, used his result to expand many functions into infinite series. He also discovered Taylor's theorem; see Selection V.11. In a letter to Collins of February 15, 1671, we find as one of his results the series for $\tan^{-1} x$, or, more precisely (since the tangent for Gregory is a line), the expansion

$$a = t - \frac{t^3}{3r^2} + \frac{t^5}{5r^4} - \frac{t^7}{7r^6} + \frac{t^9}{9r^8} + \cdots,$$

if $r \tan a/r = t$, $a = $ arc, $r = $ radius (*Gregory Memorial Volume*, p. 170), first published in *Commercium epistolicum J. Collins et aliorum de analysi promota* (London, 1712), 25–26. For $a = \pi/4$, $r = 1$, $t = 1$, we obtain the series of Leibniz. On Collins see p. 290.

[4] Tschirnhaus (see Selection II.11) wrote a letter to Oldenburg, dated September 1, 1676, of which Oldenburg informed Newton.

lay bare, what Leibniz desires from me, the basis of the theorem set forth near the beginning of the former letter.

At the beginning of my mathematical studies, when I had met with the works of our celebrated Wallis, on considering the series by the intercalation of which he himself exhibits the area of the circle and the hyperbola, the fact that, in the series of curves whose common base or axis is x and the ordinates

$$(1 - x^2)^{\frac{0}{2}}, \quad (1 - x^2)^{\frac{1}{2}}, \quad (1 - x^2)^{\frac{2}{2}}, \quad (1 - x^2)^{\frac{3}{2}}, \quad (1 - x^2)^{\frac{4}{2}}, \quad (1 - x^2)^{\frac{5}{2}}, \quad \text{etc.},$$

if the areas of every other of them, namely

$$x, \quad x - \tfrac{1}{3}x^3, \quad x - \tfrac{2}{3}x^3 + \tfrac{1}{5}x^5, \quad x - \tfrac{3}{3}x^3 + \tfrac{3}{5}x^5 - \tfrac{1}{7}x^7, \quad \text{etc.}$$

could be interpolated, we should have the areas of the intermediate ones, of which the first $(1 - x^2)^{\frac{1}{2}}$ is the circle: in order to interpolate these series I noted that in all of them the first term was x and that the second terms $\tfrac{0}{3}x^3$, $\tfrac{1}{3}x^3$, $\tfrac{2}{3}x^3$, $\tfrac{3}{3}x^3$, etc., were in arithmetical progression, and hence that the first two terms of the series to be intercalated ought to be $x - \tfrac{1}{3}(\tfrac{1}{2}x^3)$, $x - \tfrac{1}{3}(\tfrac{3}{2}x^3)$, $x - \tfrac{1}{3}(\tfrac{5}{2}x^3)$, etc. To intercalate the rest I began to reflect that the denominators 1, 3, 5, 7, etc. were in arithmetical progression, so that the numerical coefficients of the numerators only were still in need of investigation. But in the alternately given areas these were the figures of powers of the number 11, namely of these, 11^0, 11^1, 11^2, 11^3, 11^4, that is, first 1; then 1, 1; thirdly, 1, 2, 1; fourthly 1, 3, 3, 1; fifthly 1, 4, 6, 4, 1, etc. And so I began to inquire how the remaining figures in these series could be derived from the first two given figures, and I found that on putting m for the second figure, the rest would be produced by continual multiplication of the terms of this series,

$$\frac{m - 0}{1} \times \frac{m - 1}{2} \times \frac{m - 2}{3} \times \frac{m - 3}{4} \times \frac{m - 4}{5}, \quad \text{etc.}$$

For example, let $m = 4$, and $4 \times \tfrac{1}{2}(m - 1)$, that is 6 will be the third term, and $6 \times \tfrac{1}{3}(m - 2)$, that is 4 the fourth, and $4 \times \tfrac{1}{4}(m - 3)$, that is 1 the fifth, and $1 \times \tfrac{1}{5}(m - 4)$, that is 0 the sixth, at which term in this case the series stops. Accordingly, I applied this rule for interposing series among series, and since, for the circle, the second term was $\tfrac{1}{3}(\tfrac{1}{2}x^3)$, I put $m = \tfrac{1}{2}$, and the terms arising were

$$\frac{1}{2} \times \frac{\tfrac{1}{2} - 1}{2} \quad \text{or} \quad -\tfrac{1}{8}, \qquad -\frac{1}{8} \times \frac{\tfrac{1}{2} - 2}{3} \quad \text{or} \quad +\tfrac{1}{16}, \qquad \frac{1}{16} \times \frac{\tfrac{1}{2} - 3}{4} \quad \text{or} \quad -\tfrac{5}{128},$$

and so to infinity. Whence I came to understand that the area of the circular segment which I wanted was

$$x - \frac{\tfrac{1}{2}x^3}{3} - \frac{\tfrac{1}{8}x^5}{5} - \frac{\tfrac{1}{16}x^7}{7} - \frac{\tfrac{5}{128}x^9}{9} \quad \text{etc.}$$

And by the same reasoning the areas of the remaining curves, which were to be inserted, were likewise obtained: as also the area of the hyperbola and of the other alternate curves in this series $(1 + x^2)^{\frac{1}{2}}$, $(1 + x^2)^{\frac{3}{2}}$, $(1 + x^2)^{\frac{5}{2}}$, $(1 + x^2)^{\frac{7}{2}}$, etc. And the same theory serves to intercalate other series, and that through intervals of two or more terms when they are absent at the same time. This was my first entry upon these studies, and it had certainly escaped my memory, had I not a few weeks ago cast my eye back on some notes.

But when I had learnt this, I immediately began to consider that the terms

$$(1 - x^2)^{\frac{1}{2}}, \quad (1 - x^2)^{\frac{3}{2}}, \quad (1 - x^2)^{\frac{5}{2}}, \quad (1 - x^2)^{\frac{7}{2}}, \quad \text{etc.,}$$

that is to say,

$$1, \quad 1 - x^2, \quad 1 - 2x^2 + x^4, \quad 1 - 3x^2 + 3x^4 - x^6, \quad \text{etc.}$$

could be interpolated in the same way as the areas generated by them: and that nothing else was required for this purpose but to omit the denominators $1, 3, 5, 7$, etc., which are in the terms expressing the areas; this means that the coefficients of the terms of the quantity to be intercalated $(1 - x^2)^{\frac{1}{2}}$, or $(1 - x^2)^{\frac{3}{2}}$, or in general $(1 - x^2)^m$, arise by the continued multiplication of the terms of this series

$$m \times \frac{m - 1}{2} \times \frac{m - 2}{3} \times \frac{m - 3}{4}, \quad \text{etc.,}$$

so that (for example)

$$(1 - x^2)^{\frac{1}{2}} \quad \text{was the value of} \quad 1 - \tfrac{1}{2}x^2 - \tfrac{1}{8}x^4 - \tfrac{1}{16}x^6, \quad \text{etc.,}$$

$$(1 - x^2)^{\frac{3}{2}} \quad \text{of} \quad 1 - \tfrac{3}{2}x^2 + \tfrac{3}{8}x^4 + \tfrac{1}{16}x^6, \quad \text{etc.,}$$

and

$$(1 - x^2)^{\frac{1}{3}} \quad \text{of} \quad 1 - \tfrac{1}{3}x^2 - \tfrac{1}{9}x^4 - \tfrac{5}{81}x^6, \quad \text{etc.}$$

So then the general reduction of radicals into infinite series by that rule, which I laid down at the beginning of my earlier letter, became known to me, and that before I was acquainted with the extraction of roots. But once this was known, that other could not long remain hidden from me. For in order to test these processes, I multiplied

$$1 - \tfrac{1}{2}x^2 - \tfrac{1}{8}x^4 - \tfrac{1}{16}x^6, \quad \text{etc.}$$

into itself; and it became $1 - x^2$, the remaining terms vanishing by the continuation of the series to infinity. And even so $1 - \tfrac{1}{3}x^2 - \tfrac{1}{9}x^4 - \tfrac{5}{81}x^6$, etc. multiplied twice into itself also produced $1 - x^2$. And as this was not only sure proof of these conclusions so too it guided me to try whether, conversely, these series, which it thus affirmed to be roots of the quantity $1 - x^2$, might not be extracted out of it in an arithmetical manner. And the matter turned out well. This was the form of the working in square roots.

$$1 - x^2(1 - \tfrac{1}{2}x^2 - \tfrac{1}{8}x^4 - \tfrac{1}{16}x^6, \text{ etc.}$$

$$\frac{1}{0 - x^2}$$

$$\frac{- x^2 + \tfrac{1}{4}x^4}{- \tfrac{1}{4}x^4}$$

$$\frac{- \tfrac{1}{4}x^4 + \tfrac{1}{8}x^6 + \tfrac{1}{64}x^8.}{0 \quad - \tfrac{1}{8}x^6 - \tfrac{1}{64}x^8}$$

Newton continues by mentioning in a guarded way, by means of an anagram, that he has a general method for finding tangents and quadratures, which is not limited by irrationalities. He gives a series representation of the binomial integral

$$J_\theta = \int z^\theta (e + fz^\eta)^\lambda \, dz,$$

but without explanation (which Leibniz had no trouble in finding, as a marginal note to the letter shows). Among examples Newton gives the rectification of the cissoid $xy^2 = (a - x^3)$, a formula in series which we can write

$$\frac{\pi}{4} = \tan^{-1} \tfrac{1}{2} + \tfrac{1}{2} \tan^{-1} \tfrac{4}{7} + \tfrac{1}{2} \tan^{-1} \tfrac{1}{8},$$

and the solution of x from

$$y = x + \frac{x^2}{2} + \frac{x^3}{3} + \frac{x^4}{4} + \cdots$$

in the form

$$x = y - \frac{y^2}{2} + \frac{y^3}{6} - \frac{y^4}{24} + \cdots.$$

A few years after Newton's discovery of the binomial series James Gregory (1638–1675), the Scottish mathematician, rediscovered it independently. We know it from letters of Gregory to the London mathematician John Collins (1625–1683), a correspondent of Newton's and of many other mathematicians. We give here a translation from Gregory's letter of November 20, 1670, taken from *James Gregory tercentenary memorial volume*, ed. H. W. Turnbull (Bell, London, 1939), 131–133.

TO FIND THE NUMBER OF A LOGARITHM

Given b, $\log b = e$, $b + d$, $\log (b + d) = e + c$, it is required to find the number whose logarithm is $e + a$.

Take a series of continual proportions b, d, d^2/b, d^3/b^2, etc., and another series a/c, $(a - c)/2c$, $(a - 2c)/3c$, $(a - 3c)/4c$, etc., and let f/c be the product of the first two terms of the second series, g/c that of the first three, h/c that of the first four, i/c that of the first five, etc. The number whose logarithm is $e + a$ will be

$$e + a = b + \frac{ad}{c} + \frac{fd^2}{cb} + \frac{gd^3}{cb^2} + \frac{hd^4}{cb^3} + \frac{id^5}{cb^4} + \frac{kd^6}{cb^5} + \text{etc.}$$

Hence with a little work but without difficulty any pure equation whatever may be solved.

This statement gives for $x = e + a$

$$x = b + \frac{a}{c}d + \frac{a(a - c)}{c \cdot 2c}\frac{d^2}{b} + \frac{a(a - c)(a - 2c)}{c \cdot 2c \cdot 3c}\frac{d^3}{b^2} + \cdots,$$

which we recognize as the binomial expansion of $b(1 + d/b)^{a/c}$, whose logarithm is

$$\log b + \frac{a}{c}[\log (b + d) - \log b] = e + a.$$

By pure equation Gregory means an equation of the form $ax^n = b$, n rational. He adds an example which gives the daily rate percent at compound interest equivalent to 6 percent per annum; he takes

$$b = 100, \quad d = 6, \quad a = 1, \quad c = 365.$$

Then

$$x = 100\left(1 + \frac{6}{100}\right)^{1/365} = 100.0160101;$$

the correct value is 100.0159919.

On the atmosphere of intellectual tension typical of the period, see for example the accounts in J. E. Hofmann, *Die Entwicklungsgeschichte der Leibnizschen Mathematik* (Leibnizens Verlag, Munich, 1949), 60–87, 194–205, or J. F. Scott, *The mathematical work of John Wallis* (Taylor and Francis, London, 1938), chaps. 9, 10. See further H. W. Turnbull, *The mathematical discoveries of Newton* (Blackie and Son, Glasgow, 1945).

5 NEWTON. PRIME AND ULTIMATE RATIOS

In some sections of the *Principia*, Newton gave some information on his discoveries in the field that Leibniz, in 1684, had called the differential calculus. After opening his work with his three *Axioms, or Laws of Motion*, he presented in Book I his dynamics of bodies moving under central forces, ending with his theories of the moon and of the attraction of a sphere.

The demonstrations are all geometric in the ancient Greek way, but instead of the indirect proof Newton uses his theory of prime and ultimate ratios, in which we now recognize the theory of limits, although he explained it in a manner difficult to understand, especially by his contemporaries. Then, in Book II, which contains applications to bodies moving in a resisting medium, Newton uses certain principles of his theory of fluxions, but without introducing its notation, nor even the term fluxion (see Selection V.6).

Book III, the last one of the *Principia*, describes the system of the world.

Here we present the lemmas dealing with Newton's theory of prime and ultimate (or first and last) ratios. They can be found in Book I, Section I. We take them from the translation of the second edition of the *Principia* (1713) by Andrew Motte (London, 1729), as revised by F. Cajori in *Sir Isaac Newton's Mathematical principles of natural philosophy* (University of California Press, Berkeley, 1945). The sections marked by * are absent or different in the first edition.

Section I is called, *The method of prime and ultimate ratios of quantities, by the help of which we demonstrate the propositions that follow.* Then follows:

LEMMA I

Quantities, and the ratios of quantities, which in any finite time converge continually to equality, and before the end of that time approach nearer to each other than bt any given difference, become ultimately equal.

If you deny it, suppose them to be ultimately unequal, and let D be their ultimate difference. Therefore they cannot approach nearer to equality than by that given difference D; which is contrary to the supposition.

LEMMA II

If in any figure AacE [Fig. 1], terminated by the right lines Aa, AE, and the curve acE, there be inscribed any number of parallelograms Ab, Bc, Cd, &c., comprehended under equal bases AB, BC, CD, &c., and the sides, Bb, Cc, Dd, &c., parallel to one side Aa of the figure; and the parallelograms aKbl, bLcm, cMdn,

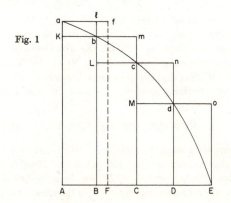

Fig. 1

&c., are completed: then if the breadth of those parallelograms be supposed to be diminished, and their number to be augmented in infinitum, I say, that the ultimate ratios which the inscribed figure $AKbLcMdD$, the circumscribed figure $AalbmcndoE$, and curvilinear figure $AabcdE$, will have to one another, are ratios of equality.

For the difference of the inscribed and circumscribed figures is the sum of the parallelograms Kl, Lm, Mn, Do, that is (from the equality of all their bases), the rectangle under one of their bases Kb and the sum of their altitudes Aa, that is, the rectangle $ABla$. But this rectangle, because its breadth AB is supposed diminished in infinitum, becomes less than any given space. And therefore (by Lem. 1) the figures inscribed and circumscribed become ultimately equal one to the other; and much more will the intermediate curvilinear figure be ultimately equal to either.   Q.E.D.

LEMMA III

The same ultimate ratios are also ratios of equality, when the breadths AB, BC, DC, &c., of the parallelograms are unequal, and are all diminished in infinitum.

For suppose AF equal to the greatest breadth, and complete the parallelogram $FAaf$. This parallelogram will be greater than the difference of the inscribed and circumscribed figures; but, because its breadth AF is diminished in infinitum it will become less than any given rectangle.   Q.E.D.

Corollary I. Hence the ultimate sum of those evanescent parallelograms will in all parts coincide with the curvilinear figure.

Corollary II. Much more will the rectilinear figure comprehended under the chords of the evanescent arcs ab, bc, cd, &c., ultimately coincide with the curvilinear figure.

Corollary III. And also the circumscribed rectilinear figure comprehended under the tangents of the same arcs.

Corollary IV. And therefore these ultimate figures (as to their perimeters acE) are not rectilinear, but curvilinear limits of rectilinear figures.

LEMMA IV

If in two figures $AacE$, $PprT$ [Fig. 2], there are inscribed (as before) two series of parallelograms, an equal number in each series, and, their breadths being diminished in infinitum, if the ultimate ratios of the parallelograms in one figure to those in the other, each to each respectively, are the same: I say, that those two figures, $AacE$, $PprT$, are to each other in that same ratio.

Fig. 2

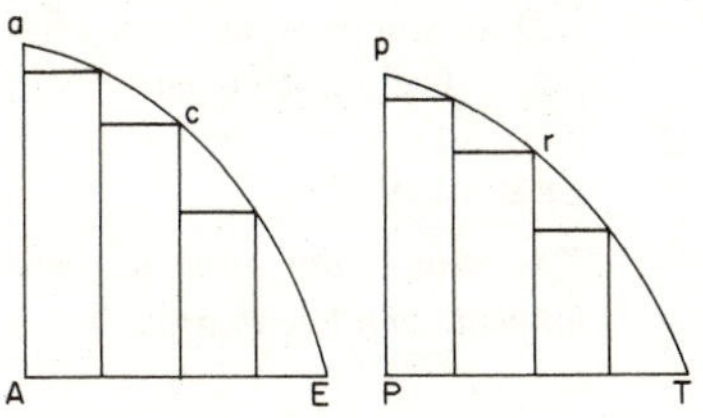

For as the parallelograms in the one are severally to the parallelograms in the other, so (by composition) is the sum of all in the one to the sum of all in the other; and so is the one figure to the other; because (by Lem. III) the former figure to the former sum, and the latter figure to the latter sum, are both in the ratio of equality. Q.E.D.

Corollary. Hence if two quantities of any kind are divided in any manner into an equal number of parts, and those parts, when their number is augmented, and their magnitude diminished *in infinitum,* have a given ratio to each other, the first to the first, the second to the second, and so on in order, all of them taken together will be to each other in that same given ratio. For if, in the figures of this Lemma, the parallelograms are taken to each other in the ratio of the parts, the sum of the parts will always be as the sum of the parallelograms; and therefore supposing the number of the parallelograms and parts to be augmented, and their magnitudes diminished *in infinitum,* those sums will be in the ultimate ratio of the parallelogram in the one figure to the correspondent parallelogram in the other; that is (by the supposition), in the ultimate ratio of any part of the one quantity to the correspondent part of the other.

LEMMA V

All homologous sides of similar figures, whether curvilinear or rectilinear, are proportional; and the areas are as the squares of the homologous sides.

LEMMA VI

If any arc ACB [Fig. 3], *given in position, is subtended by its chord AB, and in any point A, in the middle of the continued curvature, is touched by a right line AD, produced both ways; then if the points A and B approach one another and meet, I say, the angle BAD, contained between the chord and the tangent, will be diminished* in infinitum, *and ultimately will vanish.*

Fig. 3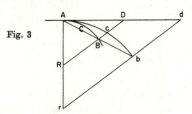

For if that angle does not vanish, the arc *ACB* will contain with the tangent *AD* an angle equal to a rectilinear angle; and therefore the curvature at the point *A* will not be continued, which is against the supposition.

LEMMA VII

The same things being supposed, I say that the ultimate ratio of the arc, chord, and tangent, any one to any other, is the ratio of equality.[1]

[1] In our notation, lim arc *ACB*/chord *AB* = 1, *B* → *A*.

For while the point B approaches towards the point A, consider always AB and AD as produced to the remote points b and d; and parallel to the secant BD draw bd; and let the arc Acb be always similar to the arc ACB. Then, supposing the points A and B to coincide, the angle dAb will vanish, by the preceding Lemma; and therefore the right lines Ab, Ad (which are always finite), and the intermediate arc Acb, will coincide, and become equal among themselves. Wherefore, the right lines AB, AD, and the intermediate arc ACB (which are always proportional to the former), will vanish, and ultimately acquire the ratio of equality.   Q.E.D.

Corollary I. Whence if through B we draw BF parallel to the tangent, [Fig. 4], always cutting any right line AF passing through A in F, this line BF will be ultimately in the ratio of equality with the evanescent arc ACB; because, completing the parallelogram $AFBD$, it is always in a ratio of equality with AD.

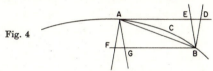

Fig. 4

Corollary II. And if through B and A more right lines are drawn, as BE, BD, AF, AG, cutting the tangent AD and its parallel BF; the ultimate ratio of all the abscissas AD, AE, BF, BG, and of the chord and arc AB, any one to any other, will be the ratio of equality.

Corollary III. And therefore in all our reasoning about ultimate ratios, we may freely use any one of those lines for any other.

LEMMA VIII

If the right lines AR, BR [Fig. 3], with the arc ACB, the chord AB, and the tangent AD, constitute three triangles RAB, $RACB$, RAD, and the points A and B approach and meet: I say, that the ultimate form of these evanescent triangles is that of similitude, and their ultimate ratio that of equality.

We omit this proof.

LEMMA IX

If a right line AE [Fig. 5], and a curved line ABC, both given in position, cut each other in a given angle, A; and to that right line, in another given angle, BD, CE

are ordinately applied,[2] *meeting the curve in B, C; and the points B and C together approach towards and meet in the point A: I say, that the areas of the triangles ABD, ACE, will ultimately be to each other as the squares of homologous sides.*

Fig. 5

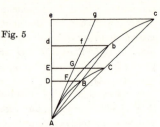

For while the points B, C, approach towards the point A, suppose always AD to be produced to the remote points d and e, so as Ad, Ae may be proportional to AD, AE; and the ordinates db, ec, to be drawn parallel to the ordinates DB and EC, and meeting AB and AC produced in b and c. Let the curve Abc be similar to the curve ABC, and draw the right line Ag so as to touch both curves in A, and cut the ordinates DB, EC, db, ec, in F, G, f, g. Then, supposing the length Ae to remain the same, let the points B and C meet in the point A; and the angle cAg vanishing, the curvilinear areas Abd, Ace will coincide with the rectilinear areas Afd, Age; and therefore (by Lemma V) will be one to the other in the duplicate ratio of the sides Ad, Ae. But the areas ABD, ACE are always proportional to these areas; and so the sides AD, AE are to these sides. And therefore the areas ABD, ACE are ultimately to each other as the squares of the sides AD, AE.   Q.E.D.

LEMMA X

The spaces which a body describes by any finite force urging it, whether that force is determined and immutable, or is cŏntinually augmented or continually diminished, are in the very beginning of the motion to each other as the squares of the times.

We omit the proof and the corollaries.

---

*SCHOLIUM

If in comparing with each other indeterminate quantities of different sorts, any one is said to be directly or inversely as any other, the meaning is, that the former is augmented or diminished in the same ratio as the latter, or as its

---

[2] Meaning "applied in order." From this expression, common at the time, our term *ordinate* is derived. See Selection V.1, note 1; also Selection IV.11, note 4.

reciprocal. And if any one is said to be as any other two or more, directly or inversely, the meaning is, that the first is augmented or diminished in the ratio compounded of the ratios in which the others, or the reciprocals of the others, are augmented or diminished. Thus, if A is said to be as B directly, and C directly, and D inversely, the meaning is, that A is augmented or diminished in the same ratio as $B \cdot C \cdot \frac{1}{D}$, that is to say, that A and $\frac{BC}{D}$ are to each other in a given ratio.

LEMMA XI

The evanescent subtense of the angle of contact, in all curves which at the point of contact have a finite curvature, is ultimately as the square of the subtense of the conterminous arc.[3]

Case 1. Let AB [Fig. 6] be that arc, AD its tangent, BD the subtense of the angle of contact perpendicular on the tangent, AB the subtense of the arc.

Fig. 6

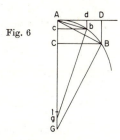

Draw BG perpendicular to the subtense AB, and AG perpendicular to the tangent AD, meeting in G; then let the points D, B, and G approach to the points d, b, and g, and suppose I to be the ultimate intersection of the lines BG, AG, when the points D, B have come to A. It is evident that the distance GI may be less than any assignable distance. But (from the nature of the circles passing through the points A, B, G, and through A, b, g),

$$AB^2 = AG \cdot BD, \quad \text{and}$$

$$Ab^2 = Ag \cdot bd.$$

But because GI may be assumed of less length than any assignable, the ratio of AG to Ag may be such as to differ from unity by less than any assignable difference; and therefore the ratio of AB^2 to Ab^2 may be such as to differ from the ratio of BD to bd by less than any assignable difference. Therefore, by Lem. I, ultimately,

$$AB^2 : Ab^2 = BD : bd. \qquad \text{Q.E.D.}$$

[3] A way of expressing this is that if $AD = x$, $BD = y$, the equation of the curve near A can be written in the form $y = \alpha x^2 + \beta x^3 + \cdots$.

Case 2. Now let BD be inclined to AD in any given angle, and the ultimate ratio of BD to bd will always be the same as before, and therefore the same with the ratio of AB^2 to Ab^2. Q.E.D.

Case 3. And if we suppose the angle D not to be given, but that the right line BD converges to a given point, or is determined by any other condition whatever; nevertheless the angles D, d, being determined by the same law, will always draw nearer to equality, and approach nearer to each other than by any assigned difference, and therefore, by Lemma I, will at last be equal; and therefore the lines BD, bd are in the same ratio to each other as before. Q.E.D.

Corollary I. Therefore since the tangents AD, Ad, the arcs AB, Ab, and their sines, BC, bc, become ultimately equal to the chords AB, Ab, their squares will ultimately become as the subtenses BD, bd.

*$Corollary$ II. Their squares are also ultimately as the versed sines of the arcs, bisecting the chords, and converging to a given point. For those versed sines are as the subtenses BD, bd.

*$Corollary$ III. And therefore the versed sine is as the square of the time in which a body will describe the arc with a given velocity.

*$Corollary$ IV. The ultimate proportion,

$$\triangle ADB : \triangle Adb = AD^3 : Ad^3 = DB^{\frac{3}{2}} : db^{\frac{3}{2}},$$

is derived from

$$\triangle ADB : \triangle Adb = AD \cdot DB : Ad \cdot db$$

and from the ultimate proportion

$$AD^2 : Ad^2 = DB : db.$$

So also is obtained ultimately

$$\triangle ABC : \triangle Abc = BC^3 : bc^3.$$

Corollary V. And because DB, db are ultimately parallel and as the squares of the lines AD, Ad, the ultimate curvilinear areas ADB, Adb will be (by the nature of the parabola) two-thirds of the rectilinear triangles ADB, Adb, and the segments AB, Ab will be one-third of the same triangles. And thence those areas and those segments will be as the squares of the tangents AD, Ad, and also of the chords and arcs AB, AB.

SCHOLIUM

But we have all along supposed the angle of contact to be neither infinitely greater nor infinitely less than the angles of contact made by circles and their tangents; that is, that the curvature at the point A is neither infinitely small nor infinitely great, and that the interval AJ is of a finite magnitude.

Now follows a discussion of these angles of contact, between which other intermediate angles of contact can be established, and so forth.[4] The scholium continues:

These Lemmas are premised to avoid the tediousness of deducing involved demonstrations *ad absurdum*, according to the method of the ancient geometers. For demonstrations are shorter by the method of indivisibles; but because the hypothesis of indivisibles seems somewhat harsh, and therefore that method is reckoned less geometrical, I chose rather to reduce the demonstrations of the following Propositions to the first and last sums and ratios of nascent and evanescent quantities, that is, to the limits of those sums and ratios, and so to premise, as short as I could, the demonstrations of those limits. For hereby the same thing is performed as by the method of indivisibles; and now those principles being demonstrated, we may use them with greater safety. Therefore if hereafter I should happen to consider quantities as made up of particles, or should use little curved lines for right ones, I would not be understood to mean indivisibles, but evanescent divisible quantities; not the sums and ratios of determinate parts, but always the limits of sums and ratios; and that the force of such demonstrations always depends on the method laid down in the foregoing Lemmas.

Perhaps it may be objected, that there is no ultimate proportion of evanescent quantities; because the proportion, before the quantities have vanished, is not the ultimate, and when they are vanished, is none.[5] But by the same argument it may be alleged that a body arriving at a certain place, and there stopping, has no ultimate velocity; because the velocity, before the body comes to the place, is not its ultimate velocity; when it has arrived, there is none. But the answer is easy; for by the ultimate velocity is meant that with which the body is moved, neither before it arrives at its last place and the motion ceases, nor after, but at the very instant it arrives; that is, that velocity with which the body arrives at its last place, and with which the motion ceases. And in like manner, by the

---

[4] Newton touches here upon the different types of angles of contact made by straight lines, tangent circles, and so on. Here he touches the age-old problem of the nature of hornlike angles (*anguli cornuti*). It arose in connection with Proposition 16 of the third book of Euclid's *Elements*: "The straight line drawn at right angles to the diameter of a circle from its extremity will fall outside the circle, and into the space between the straight line and the circumference another straight line cannot be interposed; further the angle of the semicircle is greater, and the remaining angle less, than every acute rectilineal angle." See T. L. Heath, *The thirteen books of Euclid's Elements* (2nd ed.; Dover, New York, 1956), II, 39–43, where we find an account of the contributions to this question by Proclus, Cardan, Peletier, Clavius, Viète, and Wallis. See also F. Klein, *Elementary mathematics from an advanced point of view*, II (Macmillan, New York, 1939), 204, where it is shown that we are dealing here with non-Archimedean quantities. See also E. Kasner, "The recent theory of the horn angle," *Scripta Mathematica 11* (1945), 263–267.

[5] Here we see Newton wrestling with the concept of "ultimate ratio of evanescent quantities" and Zeno's reasoning. It was about this type of argument that Berkeley wrote his *Analyst* (Selection V.12). On Zeno's argumentation and the debate around them during the centuries see F. Cajori, "The history of Zeno's arguments on motion," *American Mathematical Monthly 22* (1915), nine articles, and "The purpose of Zeno's arguments on motion," *Isis 3* (1920), 7–20. We now know—and it is even implicit in Newton—that the concept of limit can be established without that of "ultimate ratio of evanescent quantities."

ultimate ratio of evanescent quantities is to be understood the ratio of the quantities not before they vanish, nor afterwards, but with which they vanish. In like manner the first ratio of nascent quantities is that with which they begin to be. And the first or last sum is that with which they begin and cease to be (or to be augmented or diminished). There is a limit which the velocity at the end of the motion may attain, but not exceed. This is the ultimate velocity. And there is the like limit in all quantities and proportions that begin and cease to be. And since such limits are certain and definite, to determine the same is a problem strictly geometrical. But whatever is geometrical we may use in determining and demonstrating any other thing that is also geometrical.

It may also be objected, that if the ultimate ratios of evanescent quantities are given, their ultimate magnitudes will be also given: and so all quantities will consist of indivisibles, which is contrary to what *Euclid* has demonstrated concerning incommensurables, in the tenth Book of his *Elements*. But this objection is founded on a false supposition. For those ultimate ratios with which quantities vanish are not truly the ratios of ultimate quantities, but limits towards which the ratios of quantities decreasing without limit do always converge; and to which they approach nearer than by any given difference, but never go beyond, nor in effect attain to, till the quantities are diminished *in infinitum*. This thing will appear more evident in quantities infinitely great. If two quantities, whose difference is given, be augmented *in infinitum*, the ultimate ratio of these quantities will be given, namely, the ratio of equality: but it does not from thence follow, that the ultimate or greatest quantities themselves, whose ratio that is, will be given. Therefore if in what follows, for the sake of being more easily understood, I should happen to mention quantities as least, or evanescent, or ultimate, you are not to suppose that quantities of any determinate magnitude are meant, but such as are conceived to be always diminished without end.

---

6 NEWTON. GENITA AND MOMENTS

There is a place in the *Principia* (Book II, Sec. II, between Propositions 7 and 8) where differentials are introduced by the name of "moments," which were produced by variable quantities called "genita." This was an approach to the concept of function, and we take it from the same Motte-Cajori translation of the *Principia* as we used before (pp. 249–251).

It is not easy to understand the meaning attached to moments, whether they are just "nascent principles" or equivalent to "finite quantities proportional to velocities." Since the text from which we have taken our translation is the second edition of the *Principia*, it is interesting that the first edition reads: "Moments, as soon as they are of finite magnitude, cease to be moments. To be given finite bounds is in some measure contradictory to their continuous increase or decrease," where our second edition reads: "Finite particles are not moments, but the very quantities generated by the moments." Newton continued to struggle with the problem of the meaning of his moments, as Leibniz did with that of his differences and "differentials."

Here, for the first time, Newton explains some of the principles of his calculus of fluxions, in a section dealing with the motion of bodies that move against a resistance. These propositions deal with problems resulting from the integration of (in our notation) differential equations such as $\ddot{x} = g - k\dot{x}$ and $\ddot{x} = g - k(\dot{x})^2$. It will be seen, however, that Newton does not use the term "fluxion."

LEMMA II

The moment of any genitum *is equal to the moments of each of the generating sides multiplied by the indices of the powers of those sides, and by their coefficients continually.*

I call any quantity a *genitum*[1] which is not made by addition or subtraction of divers parts, but is generated or produced in arithmetic by the multiplication, division, or extraction of the root of any terms whatsoever; in geometry by the finding of contents and sides, or of the extremes and means of proportionals. Quantities of this kind are products, quotients, roots, rectangles, squares, cubes, square and cubic sides, and the like. These quantities I here consider as variable and indetermined, and increasing or decreasing, as it were, by a continual motion or flux; and I understand their momentary increments or decrements by the name of moments; so that the increments may be esteemed as added or affirmative moments; and the decrements as subtracted or negative ones. But take care not to look upon finite particles as such. Finite particles are not moments, but the very quantities generated by the moments. We are to conceive them as the just nascent principles of finite magnitudes. Nor do we in this Lemma regard the magnitude of the moments, but their first proportion, as nascent. It will be the same thing, if, instead of moments, we use either the velocities of the increments and decrements (which may also be called the motions, mutations, and fluxions of quantities), or any finite quantities proportional to those velocities. The coefficient of any generating side is the quantity which arises by applying the genitum to that side.

Wherefore the sense of the Lemma is, that if the moments of any quantities A, B, C, &c., increasing or decreasing by a continual flux, or the velocities of the mutations which are proportional to them, be called a, b, c, &c., the moment or mutation of the generated rectangle AB will be $aB + bA$; the moment of the generated content ABC will be $aBC + bAC + cAB$; and the moments of the generated powers A^2, A^3, A^4, $A^{\frac{1}{2}}$, $A^{\frac{3}{2}}$, $A^{\frac{1}{3}}$, $A^{\frac{2}{3}}$, A^{-1}, A^{-2}, $A^{-\frac{1}{2}}$ will be $2aA$, $3aA^2$, $4aA^3$, $\frac{1}{2}aA^{-\frac{1}{2}}$, $\frac{3}{2}aA^{\frac{1}{2}}$, $\frac{1}{3}aA^{-\frac{2}{3}}$, $\frac{2}{3}aA^{-\frac{1}{3}}$, $-aA^{-2}$, $-2aA^{-3}$, $-\frac{1}{2}aA^{-\frac{3}{2}}$ respectively;

and, in general, that the moment of any power $A^{n/m}$ will be $\frac{n}{m}aA^{(n-m)/m}$. Also, that the moment of the genitum A^2B will be $2aAB + bA^2$; the moment of the generated quantity $A^3B^4C^2$ will be $3aA^2B^4C^2 + 4bA^3B^3C^2 + 2cA^3B^4C$; and the moment of the generated quantity $\frac{A^3}{B^2}$ or A^3B^{-2} will be $3aA^2B^{-2} - 2bA^3B^{-3}$; and so on. The Lemma is thus demonstrated.

---

[1] A *genitum* is therefore an expression of one term which is dependent on one variable. The Motte-Cajori translation is *genitum*, making the term neuter, but in Latin it is *quantitas genita*, a generated quantity.

Case 1. Any rectangle, as AB, augmented by a continual flux, when, as yet, there wanted of the sides A and B half their moments $\frac{1}{2}a$ and $\frac{1}{2}b$, was $A - \frac{1}{2}a$ into $B - \frac{1}{2}b$, or $AB - \frac{1}{2}aB - \frac{1}{2}bA + \frac{1}{4}ab$; but as soon as the sides A and B are augmented by the other half-moments, the rectangle becomes $A + \frac{1}{2}a$ into $B + \frac{1}{2}b$, or $AB + \frac{1}{2}aB + \frac{1}{2}bA + \frac{1}{4}ab$. From this rectangle subtract the former rectangle, and there will remain the excess $aB + bA$. Therefore with the whole increments a and b of the sides, the increment $aB + bA$ of the rectangle is generated. Q.E.D.

Case 2. Suppose AB always equal to G, and then the moment of the content ABC or GC (by Case 1) will be $gC + cG$, that is (putting AB and $aB + bA$ for G and g), $aBC + bAC + cAB$. And the reasoning is the same for contents under ever so many sides. Q.E.D.

Case 3. Suppose the sides A, B, and C, to be always equal among themselves; and the moment $aB + bA$, of A^2, that is, of the rectangle AB, will be $2aA$; and the moment $aBC + bAC + cAB$ of A^3, that is, of the content ABC, will be $3aA^2$. And by the same reasoning the moment of any power A^n is naA^{n-1}. Q.E.D.

Case 4. Therefore since $\frac{1}{A}$ into A is 1, the moment of $\frac{1}{A}$ multiplied by A, together with $\frac{1}{A}$ multiplied by a, will be the moment of 1, that is, nothing. Therefore the moment of $\frac{1}{A}$, or of A^{-1}, is $\frac{-a}{A^2}$. And generally since $\frac{1}{A^n}$ into A^n is 1, the moment of $\frac{1}{A^n}$ multiplied by A^n together with $\frac{1}{A^n}$ into naA^{n-1} will be nothing. And, therefore, the moment of $\frac{1}{A^n}$ or A^{-n} will be $-\frac{na}{A^{n+1}}$. Q.E.D.

Case 5. And since $A^{\frac{1}{2}}$ into $A^{\frac{1}{2}}$ is A, the moment of $A^{\frac{1}{2}}$ multiplied by $2A^{\frac{1}{2}}$ will be a (by Case 3); and, therefore, the moment of $A^{\frac{1}{2}}$ will be $\frac{a}{2A^{\frac{1}{2}}}$ or $\frac{1}{2}aA^{-\frac{1}{2}}$. And generally, putting $A^{m/n}$ equal to B, then A^m will be equal to B^n, and therefore maA^{m-1} equal to nbB^{n-1}, and maA^{-1} equal to nbB^{-1}, or $nbA^{-m/n}$; and therefore $\frac{m}{n}aA^{(n-m)/n}$ is equal to b, that is, equal to the moment of $A^{m/n}$. Q.E.D.

Case 6. Therefore the moment of any genitum $A^m B^n$ is the moment of A^m multiplied by B^n, together with the moment of B^n multiplied by A^m, that is, $maA^{m-1}B^n + nbB^{n-1}A^m$; and that whether the indices m and n of the powers be whole numbers or fractions, affirmative or negative. And the reasoning is the same for higher powers. Q.E.D.

Corollary I. Hence in quantities continually proportional, if one term is given, the moments of the rest of the terms will be as the same terms multiplied by the number of intervals between them and the given term. Let A, B, C, D, E, F be continually proportional; then if the term C is given, the moments of the rest of the terms will be among themselves as $-2A$, $-B$, D, $2E$, $3F$.[2]

[2] When $A:B = B:C = C:D = D:E = E:F$ and $c = 0$, then C is constant and
$$a:b:d:e:f = -2A:-B:D:2E:3F.$$

Corollary II. And if in four proportionals the two means are given, the moments of the extremes will be as those extremes. The same is to be understood of the sides of any given rectangle.

Corollary III. And if the sum or difference of two squares is given,[3] the moments of the sides will be inversely as the sides.

Now follows in the text that famous Scholium which is different in the first and second editions of the *Principia*. In the first edition (1687) Newton mentions the letters which he and "the very excellent G. W. Leibnizius" had exchanged ten years before (see Selection V.4), and praises him as coinventor of the calculus. In the second edition (1713) the Scholium mentions several men who had contributed to the invention, but omits Leibniz. By this time the priority struggle between Newton and Leibniz was in full swing.

7 NEWTON. QUADRATURE OF CURVES

Newton published the first full exposition of this theory of fluxions in his *Tractatus de quadratura curvarum* as an appendix to his *Opticks* (London, 1704); it was written in 1693. He has hardly any moments, that is, infinitesimals, here, and uses fluxions and fluents, as well as his principle of prime and ultimate ratios. He demands great care in their use: "errores quam minimi in rebus mathematicis non sunt contemnendi" (see no. 6 of the Introduction). We present here a section of the *Tractatus* in the translation by J. Stewart, *Two treatises on the quadrature of curves and analysis by equations of an infinite number of terms, explained* (London, 1745).

In no. 5 of the Introduction the characteristic triangle is introduced, and the definition of a tangent as the limit of a secant.

INTRODUCTION TO THE QUADRATURE OF CURVES

1. I consider mathematical quantities in this place not as consisting of very small parts; but as described by a continued motion. Lines are described, and thereby generated not by the apposition of parts, but by the continued motion of points; superficies by the motion of lines; solids by the motion of superficies; angles by the rotation of the sides; portions of time by a continual flux: and so in other quantities. These geneses really take place in the nature of things, and are daily seen in the motion of bodies. And after this manner the ancients, by drawing moveable right lines along immoveable right lines, taught the genesis of rectangles.

---

[3] When $A^2 \pm B^2 = $ const., then $a : b = \pm B : A$. Newton, who in Book I has used for his proofs only his theorem of prime and ultimate ratios, in the sections of Book II following his lemma on genita uses occasionally that part of his fluxional calculus which is expressed by the lemma, that is, the formula $d(A^m B^n C^p \cdots) = mA^{m-1}B^nC^p \cdots dA + nA^mB^{n-1}C^p \cdots dB + pA^mB^nC^{p-1} \cdots dC + \cdots$. Newton's notation is usually like our own, but occasionally he writes for A^2 either Aq, or A quad.

2. Therefore considering that quantities, which increase in equal times, and by increasing are generated, become greater or less according to the greater or less velocity with which they increase and are generated; I sought a method of determining quantities from the velocities of the motions or increments, with which they are generated; and calling these velocities of the motions or increments *fluxions*, and the generated quantities *fluents*. I fell by degrees upon the method of fluxions, which I have made use of here in the quadrature of curves, in the years 1665 and 1666.

3. Fluxions are very nearly as the augments of the fluents generated in equal but very small particles of time, and, to speak accurately, they are in the *first ratio* of the nascent augments; but they may be expounded by any lines which are proportional to them.

4. Thus if the areas ABC, $ABDG$ [Fig. 1] be described by the ordinates BC, BD moving along the base AB with an uniform motion, the fluxions of these

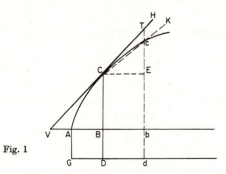

Fig. 1

areas shall be to one another as the describing ordinates BC and BD, and may be expounded by these ordinates, because that these ordinates are as the nascent augments of the areas.

5. Let the ordinate BC advance from its place into any new place bc. Complete the parallelogram $BCEb$, and draw the right line VTH touching the curve in C, and meeting the two lines bc and BA produced in T and V: and Bb, Ec, and Cc will be the augments now generated of the absciss AB, the ordinate BC and the curve line ACc; and the sides of the triangle CET are in the *first ratio* of these augments considered as nascent, therefore the fluxions of AB, BC, and AC are as the sides CE, ET, and CT of that triangle CET, and may be expounded by these same sides, or, which is the same thing, by the sides of the triangle VBC, which is similar to the triangle CET.

6. It comes to the same purpose to take the fluxions in the *ultimate ratio* of the evanescent parts. Draw the right line Cc, and produce it to K. Let the ordinate bc return into its former place BC, and when the points C and c coalesce, the right line CK will coincide with the tangent CH, and the evanescent triangle CEc in its ultimate form will become similar to the triangle CET, and its evanescent sides CE, Ec, and Cc will be *ultimately* among themselves as the sides CE, ET, and CT of the other triangle CET are, and therefore the fluxions of the lines AB, BC, and AC are in this same ratio. If the points C and c are distant

from one another by any small distance, the right line CK will likewise be distant from the tangent CH by a small distance. That the right line CK may coincide with the tangent CH, and the ultimate ratios of the lines CE, Ec, and Cc may be found, the points C and c ought to coalesce and exactly coincide. The very smallest errors in mathematical matters are not to be neglected.

7. By the like way of reasoning, if a circle described with the center B and radius BC be drawn at right angles along the absciss AB, with an uniform motion, the fluxion of the generated solid ABC will be as that generating circle, and the fluxion of its superficies will be as the perimeter of that circle and the fluxion of the curve line AC jointly. For in whatever time the solid ABC is generated by drawing that circle along the length of the absciss, in the same time its superficies is generated by drawing the perimeter of that circle along the length of the curve AC. You may likewise take the following examples of this method.

8. *Let the right line PB* [Fig. 2], *revolving about the given pole P, cut another right line AB given in position: it is required to find the proportion of the fluxions of these right lines AB and PB.*

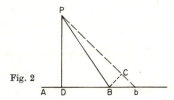

Fig. 2

Let the line PB move forward from its place PB into the new place Pb. In Pb take PC equal to PB, and draw PD to AB in such manner that the angle bPD may be equal to the angle bBC; and because the triangles bBC, bPD are similar, the augment Bb will be to the augment Cb as Pb to Db. Now let Pb return into its former place PB, that these augments may evanish, then the ultimate ratio of these evanescent augments, that is the ultimate ratio of Pb to Db, shall be the same with that of PB to DB, PDB being then a right angle, and therefore the fluxion of AB is to the fluxion of PB in that same ratio.

9. *Let the right line PB, revolving about the given pole P, cut other two right lines given in position, viz. AB and AE in B and E: the proportion of the fluxions of these right lines AB and AE is sought.*

Let the revolving right line PB [Fig. 3] move forward from its place PB into

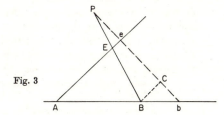

Fig. 3

the new place Pb, so as to cut the lines AB, AE in the points b and e: and draw BC parallel to AE meeting Pb in C, and it will be $Bb:BC::Ab:Ae$, and $BC:Ee::PB:PE$, and by joining the ratios, $Bb:Ee::Ab \times PB:Ae \times PE$.[1] Now let Pb return into its former place PB, and the evanescent augment Bb will be to the evanescent augment Ee as $AB \times PB$ to $AE \times PE$; and therefore the fluxion of the right line AB is to the fluxion of the right line AE in the same ratio.

10. Hence if the revolving right line PB cut any curve lines given in position in the points B and E, and the right lines AB, AE now becoming moveable, touch these curves in the points of section B and E: the fluxion of the curve, which the right line AB touches, shall be to the fluxion of the curve, which the right line AE touches, as $AB \times PB$ to $AE \times PE$. The same thing would happen if the right line PB perpetually touched any curve given in position in the moveable point P.

11. *Let the quantity x flow uniformly, and let it be proposed to find the fluxion of x^n.*

In the same time that the quantity x, by flowing, becomes $x + o$, the quantity x^n will become $(x + o)^n$, that is, by the method of infinite series, $x^n + nox^{n-1} + \dfrac{n^2 - n}{2} oox^{n-2} +$ &c. And the augments o and $nox^{n-1} + \dfrac{n^2 - n}{2} oox^{n-2} +$ &c. are to one another as 1 and $nx^{n-1} + \dfrac{n^2 - n}{2} ox^{n-2} +$ &c.

Now let these augments vanish, and their ultimate ratio will be 1 to nx^{n-1}.

12. By like ways of reasoning, the fluxions of lines, whether right or curve in all cases, as likewise the fluxions of superficies, angles, and other quantities, may be collected by the method of *prime* and *ultimate* ratios. Now to institute an analysis after this manner in finite quantities and investigate the *prime* or *ultimate* ratios of these finite quantities when in their nascent or evanescent state, is consonant to the geometry of the ancients: and I was willing to show that, in the method of fluxions, there is no necessity of introducing figures infinitely small into geometry. Yet the analysis may be performed in any kind of figures, whether finite or infinitely small, which are imagined similar to the evanescent figures; as likewise in these figures, which, by the method of indivisibles, use to be reckoned as infinitely small, provided you proceed with due caution.

From the fluxions to find the fluents, is a much more difficult problem, and the first step of the solution is equivalent to the quadrature of curves; concerning which I wrote what follows some considerable time ago.

13. In what follows I consider indeterminate quantities as increasing or decreasing by a continued motion, that is, as flowing forwards, or backwards, and I design them by the letters z, y, x, v, and their fluxions or celerities of increasing I denote by the same letters pointed $\dot{z}$, $\dot{y}$, $\dot{x}$, $\dot{v}$. There are likewise fluxions or mutations more or less swift of these fluxions, which may be called the second fluxions of the same quantities z, y, x, v, and may be thus designed $\ddot{z}$, $\ddot{y}$, $\ddot{x}$, $\ddot{v}$: and the first fluxions of these last, or the third fluxions of z, y, x, v,

---

[1] This is Barrow's notation; see Selection IV.14.

are thus denoted $\ddot{z}$, $\ddot{y}$, $\ddot{x}$, $\ddot{v}$: and the fourth fluxions thus $\dddot{z}$, $\dddot{y}$, $\dddot{x}$, $\dddot{v}$. And after the same manner that $\ddot{z}$, $\ddot{y}$, $\ddot{x}$, $\ddot{v}$ are the fluxions of the quantities $\ddot{z}$, $\ddot{y}$, $\ddot{x}$, $\ddot{v}$, and these the fluxions of the quantities $\dot{z}$, $\dot{y}$, $\dot{x}$, $\dot{v}$; and these last the fluxions of the quantities z, y, x, v: so the quantities z, y, x, v may be considered as the fluxions of others, which I shall design thus $\dot{z}$, $\dot{y}$, $\dot{x}$, $\dot{v}$; and these as the fluxions of others $\ddot{z}$, $\ddot{y}$, $\ddot{x}$, $\ddot{v}$; and these last still as the fluxions of others $\dddot{z}$, $\dddot{y}$, $\dddot{x}$, $\dddot{v}$. Therefore $\ddot{z}$, $\dot{z}$, z, $\dot{z}$, $\ddot{z}$, $\dot{\ddot{z}}$, $\ddot{z}$, $\dddot{z}$, &c. design a series of quantities whereof every one that follows is the fluxion of the one immediately preceding, and every one that goes before, is a flowing quantity having that which immediately succeeds, for its fluxion. The like is the series $\sqrt{az \overset{''}{-} zz}$, $\sqrt{az \overset{'}{-} zz}$, $\sqrt{az - zz}$, $\sqrt{az \overset{\cdot}{-} zz}$, $\sqrt{az \overset{\cdot\cdot}{-} zz}$, $\sqrt{az \overset{\cdot\cdot\cdot}{-} zz}$; as likewise the series $\dfrac{az + zz}{a - z}$, $\dfrac{az + zz}{a - z}$, $\dfrac{az + zz}{a - z}$, $\dfrac{az + zz}{a - z}$, $\dfrac{az + zz}{a - z}$, $\dfrac{az + zz}{a - z}$, &c.

14. And it is to be remarked that any preceding quantity in these series is as the area of a curvilinear figure of which the succeeding is the rectangular ordinate, and the absciss is z: as $\sqrt{ax \overset{'}{-} zz}$ the area of a curve, whose ordinate is $\sqrt{az - zz}$, and absciss z. The design of all these things will appear in the following propositions.

PROPOSITION I. PROBLEM I

15. *An equation being given involving any number of flowing quantities, to find the fluxions.*[1]

Solution. Let every term of the equation be multiplied by the index of the power[2] of every flowing quantity that it involves, and in every multiplication change the side or root of the power into its fluxion, and the aggregate of all the products with their proper signs, will be the new equation.

16. *Explication.* Let a, b, c, d, &c. be determinate and invariable quantities, and let any equation be proposed involving the flowing quantities z, y, x, &c. as $x^3 - xy^2 + a^2z - b^3 = 0$. Let the terms be first multiplied by the indexes of the powers of x, and in every multiplication for the root, or x of one dimension write $\dot{x}$, and the sum of the factors will be $3\dot{x}x^2 - \dot{x}y^2$. Do the same in y, and there arises $-2xy\dot{y}$. Do the same in z, and there arises $a^2\dot{z}$. Let the sum of these products be put equal to nothing, and you'll have the equation $3\dot{x}x^2 - \dot{x}y^2 - 2xy\dot{y} + a^2\dot{z} = 0$. I say the relation of the fluxions is defined by this equation.

17. *Demonstration.* For let o be a very small quantity, and let $o\dot{z}$, $o\dot{y}$, $o\dot{x}$ be the moments, that is the momentaneous synchronal increments of the quantities z, y, x. And if the flowing quantities are just now z, y, x, then after a moment of time, being increased by their increments $o\dot{z}$, $o\dot{y}$, $o\dot{x}$, these quantities shall become $z + o\dot{z}$, $y + o\dot{y}$, $x + o\dot{x}$: which being wrote in the first equation for z, y, and x, give this equation $x^3 + 3x^2o\dot{x} + 3xoo\dot{x}\dot{x} + o^3\dot{x}^3 - xy^2 - o\dot{x}y^2 - 2xoy\dot{y} - 2\dot{x}o^2\dot{y}y - xo^2\dot{y}\dot{y} - \dot{x}o^3\dot{y}\dot{y} + a^2z + a^2o\dot{z} - b^3 = 0$.

---

[1] Newton prefers to differentiate equations, but later also differentiates functions, often given as areas. See the remark of D'Alembert, Selection V.14, p. 343.

[2] [Footnote by the translator, Stewart] The word translated here *power* is *dignitas*, dignity, by which must be understood not only perfect, but also imperfect powers or surd roots, which are expressed in the manner of perfect powers, as is well known, by fractional indexes. In which sense $x^{1/2}$, $x^{2/3}$, etc. are powers; $\frac{1}{2}$ and $\frac{2}{3}$ their indexes, and x the side or root. I use the word power, because dignity is seldom used in English in this sense.

Subtract the former equation from the latter, divide the remaining equation by o, and it will be $3\dot{x}x^2 + 3\dot{x}\dot{x}ox + \dot{x}^3o^2 - \dot{x}y^2 - 2x\dot{y}y - 2\dot{x}o\dot{y}y - xo\dot{y}\dot{y} - \dot{x}o^2\dot{y}\dot{y} + a^2\dot{z} = 0$. Let the quantity o be diminished infinitely, and neglecting the terms which vanish, there will remain $3\dot{x}x^2 - \dot{x}y^2 - 2x\dot{y}y + a^2\dot{z} = 0$. Q.E.D.

18. *A fuller explication.* After the same manner if the equation were $x^3 - xy^2 + aa\sqrt{ax - y^2} - b^3 = 0$, thence would be produced $3x^2\dot{x} - \dot{x}y^2 - 2xy\dot{y} + aa\overset{\displaystyle\cdot}{\sqrt{ax - y^2}} = 0$. Where if you would take away the fluxion $\sqrt{ax - y^2}$, put $\sqrt{ax - y^2} = z$, and it will be $ax - y^2 = z^2$, and by this proposition $a\dot{x} - 2\dot{y}y = 2\dot{z}z$, or $\dfrac{a\dot{x} - 2\dot{y}y}{2z} = \dot{z}$, that is $\dfrac{a\dot{x} - 2\dot{y}y}{2\sqrt{ax - yy}} = \overset{\displaystyle\cdot}{\sqrt{ax - yy}}$. And thence $3x^2\dot{x} - \dot{x}y^2 - 2xy\dot{y} + \dfrac{a^3\dot{x} - 2a^2\dot{y}y}{2\sqrt{ax - yy}} = 0$.

19. And by repeating the operation, you proceed to second, third, and subsequent fluxions. Let $zy^3 - z^4 + a^4 = 0$ be an equation proposed, and by the first operation it becomes $\dot{z}y^3 + 3z\dot{y}y^2 - 4\dot{z}z^3 = 0$; by the second $\ddot{z}y^3 + 6z\dot{y}y^2 + 3z\ddot{y}y^2 + 6z\dot{y}^2y - 4\ddot{z}z^3 - 12\dot{z}^2z^2 = 0$, by the third, $\dot{z}y^3 + 9\ddot{z}\dot{y}y^2 + 18\dot{z}\dot{y}^2y + 3z\ddot{y}y^2 + 18z\ddot{y}\dot{y}y + 6z\dot{y}^3 - 4\dot{z}z^3 - 36\ddot{z}\dot{z}z^2 - 24\dot{z}^3z = 0$.

20. But when one proceeds thus to second, third, and following fluxions, it is proper to consider some quantity as flowing uniformly, and for its first fluxion to write unity, for the second and subsequent ones, nothing. Let there be given the equation $zy^3 - z^4 + a^4 = 0$, as above; and let z flow uniformly, and let its fluxion be unity: then by the first operation it shall be $y^3 + 3z\dot{y}y^2 - 4z^3 = 0$; by the second $6\dot{y}y^2 + 3z\ddot{y}y^2 + 6z\dot{y}^2y - 12z^2 = 0$; by the third $9\ddot{y}y^2 + 18\dot{y}^2y + 3z\ddot{y}y^3 + 18z\ddot{y}\dot{y}y + 6z\dot{y}^3 - 24z = 0$.

But in equations of this kind it must be conceived that the fluxions in all the terms are of the same order, i.e., either all of the first order $\dot{y}$, $\dot{z}$; or all of the second $\ddot{y}$, $\dot{y}^2$, $\dot{y}\dot{z}$, $\dot{z}^2$; or all of the third $\dddot{y}$, $\ddot{y}\dot{y}$, $\ddot{y}\dot{z}$, $\dot{y}^3$, $\dot{y}^2\dot{z}$, $\dot{y}\dot{z}^2$, $\dot{z}^3$, &c. And where the case is otherwise the order is to be completed by means of the fluxions of a quantity that flows uniformly, which fluxions are understood. Thus the last equation, by completing the third order, becomes $9\dot{z}\ddot{y}y^2 + 18\dot{z}\dot{y}^2y + 3z\dddot{y}y^2 + 18z\ddot{y}\dot{y}y + 6z\dot{y}^3 - 24z\dot{z}^3 = 0$.[3]

PROPOSITION II. PROBLEM II

22. *To find such curves as can be squared.*

Let ABC [Fig. 4] be the figure to be found, BC the rectangular ordinate, and AB the absciss. Produce CB to E, so that $BE = 1$, and complete the parallelogram $ABED$: and the fluxions of the areas ABC, $ABED$ shall be as BC and BE. Assume therefore any equation, by which the relation of the areas may be defined, and thence the relation of the ordinates BC and BE will be given by the first proposition. Q.E.I.

The two following propositions afford examples of this.

---

[3] Newton insists on homogeneity, which requires that each term of the equation has the same number of "pricks."

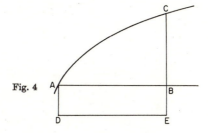

Fig. 4

PROPOSITION III. THEOREM I

23. If for the absciss AB and area AE or $AB \times 1$ you write z promiscuously, and if for $e + fz^\eta + gz^{2\eta} + hz^{3\eta} + \&c.$ you write R: let the area of the curve be $z^\theta R^\lambda$, the ordinate BC shall be equal to

$$[\theta e + (\theta + \lambda\eta)fz^\eta + (\theta + 2\lambda\eta)gz^{2\eta} + (\theta + 3\lambda\eta)hz^{3\eta} + \&c.]z^{\theta-1}R^{\lambda-1}.^4$$

24. *Demonstration.* For let $z^\theta R^\lambda = v$, it will be (by Prop. I) $\theta \dot{z}z^{\theta-1}R^\lambda + \lambda z^\theta \dot{R}R^{\lambda-1} = \dot{v}$. For R^λ in the first term of the equation and z^θ in the second write $RR^{\lambda-1}$ and $zz^{\theta-1}$, and it will become $(\theta\dot{z}R + \lambda z\dot{R})z^{\theta-1}R^{\lambda-1} = \dot{v}$. But it was $R = e + fz^\eta + gz^{2\eta} + hz^{3\eta} + \&c.$: and thence (by Prop. I) it becomes $\dot{R} = \eta f\dot{z}z^{\eta-1} + 2\eta g\dot{z}z^{2\eta-1} + 3\eta h\dot{z}z^{3\eta-1} + \&c.$ which being substituted, and BE or 1 wrote for $\dot{z}$, it becomes

$$[\theta e + (\theta + \lambda\eta)fz^\eta + (\theta + 2\lambda\eta)gz^{2\eta} + (\theta + 3\lambda\eta)hz^{3\eta} + \&c.]z^{\theta-1}R^{\lambda-1} = \dot{v} = BC.$$

Q.E.D.

In Proposition V Newton indicates interest in the convergence and divergence of series, and shows how to apply the method of partial fractions in integration, as Leibniz did in 1702 (see Selection II.10, note 1), but Newton's solution is not as exhaustive as that of Leibniz, in particular where he writes that an integral of the form $\int dx : (x + a)$ cannot be computed. Newton, like Johann Bernoulli later in his *Integral calculus* (Selection V.10), published without change results obtained many years before the date of publication.

Among other propositions is Proposition IV, Theorem II, to find the ordinate BC when $AB = z$ is the abscissa of a curve with area $z^\theta R^\lambda s^\mu$, when $R = e + fz^\eta + gz^{2\eta} + \cdots$ and $S = k + lz^\eta + mz^{2\eta} + \cdots$. Proposition IX, Theorem VII, is: "The areas of those curves are equal among themselves, whose ordinates are reciprocally as the fluxions of their abscisses,"[5] with many corollaries. On pp. 24–28 is a tabulation of integrals, divided into two sets. We reproduce (Fig. 5) the first set, of ten examples; d is a coefficient (not a Leibniz symbol d) and t is the area $\int y\,dz$, without a constant of integration.

The second set, with 33 examples, is entitled "A table of the more simple kind of curves which may be compared with the ellipsis and hyperbola." This is similar to what Leibniz and Bernoulli did, as we can see, for instance, in Selection V.9; we now say that the integral involves radicals and inverse circular functions.

[4] We have modernized Newton's notation somewhat. He writes $^{+\theta}_{+\lambda\eta} \times fz^\eta$ for what we write as $(\theta + \lambda\eta)z^\eta$ and instead of the square brackets he uses an overbar.
[5] From $y\,dx = y_1\,dx_1$ it follows only that the areas differ by a constant. Newton's examples are obtained by substitutions such as $x = z^s$ in algebraic expressions.

A TABLE

Of the more simple kind of Curves which may be squared.

Forms of Curves.	Areas of the Curves.
I $dz^{n-1}=y$	$\frac{d}{n}z^n = t.$
II $\dfrac{dz^{n-1}}{e^2+2efz^n+f^2z^{2n}}=y$	$\dfrac{dz^n}{ne^2+nefz^n}=t.$ Or $\dfrac{-d}{nef+nf^2z^n}=t$
III 1 $dz^{n-1}\sqrt{e+fz^n}=y$	$\frac{2d}{3nf}R^3 = t.$ Where $R=\sqrt{e+fz^n}$
2 $dz^{2n-1}\sqrt{e+fz^n}=y$	$\dfrac{-4e+6fz^n}{15nf^2}dR^3=t$
3 $dz^{3n-1}\sqrt{e+fz^n}=y$	$\dfrac{16e^2-24efz^n+30f^2z^{2n}}{105nf^3}dR^3=t$
4 $dz^{4n-1}\sqrt{e+fz^n}=y$	$\dfrac{-96e^3+144e^2fz^n-180ef^2z^{2n}+210f^3z^{3n}}{945nf^4}=t$
IV 1 $\dfrac{dz^{n-1}}{\sqrt{e+fz^n}}=y$	$\frac{2d}{nf}R = t.$
2 $\dfrac{dz^{2n-1}}{\sqrt{e+fz^n}}=y$	$\dfrac{-4e+2fz^n}{3nf^2}dR=t$
3 $\dfrac{dz^{3n-1}}{\sqrt{e+fz^n}}=y$	$\dfrac{16e^2-8efz^n+6f^2z^{2n}}{15nf^3}dR=t$
4 $\dfrac{dz^{4n-1}}{\sqrt{e+fz^n}}=y$	$\dfrac{-96e^3+48e^2fz^n-36ef^2z^{2n}+30f^3z^{3n}}{105nf^4}dR=t$

Fig. 5

Proposition XI contains an evaluation of moments (in our present sense). First Newton forms a sequence:

$$f_1(x) = \int_0^x f(x)\,dx, \quad f_2(x) = \int_0^x f_1(x)\,dx, \quad f_3(x) = \int_0^x f_2(x)\,dx, \ldots,$$

then a second sequence:

$$A = \int_0^t f(x)\,dx, \quad B = \int_0^t x f(x)\,dx, \quad C = \int_0^t x^2 f(x)\,dx, \ldots,$$

and a third sequence:

$$P = \int_0^t f(x)\,dx, \quad Q = \int_0^t (t-x)f(x)\,dx, \quad R = \int_0^t (t-x)^2 f(x)\,dx, \ldots,$$

and establishes relations between the terms of these last two sequences; for example,

$$A = P, \quad tA - B = Q, \quad \frac{t^2 A - 2tB + C}{2} = \tfrac{1}{2}R, \quad \text{etc.}$$

A Scholium was added when the *Quadrature of curves* was published. It opens with a remark which seems to show that Newton was on the way to the discovery of the Taylor series. He also shows that differential equations of the form $y^{(n)} = f(x), f(y^{(n)}, y^{(n+1)}) = 0$ can be integrated by quadratures.[6] See further H. W. Turnbull, *The mathematical discoveries of Newton* (Blackie, London and Glasgow, 1945), chap. IV. An Italian translation of the *Quadratura curvarum* is found in G. Castelnuovo, *Le origini del calcolo infinitesimale* (Zanichelli, Bologna, 1930), 113–145.

There exists, as we have seen (Selection V.4), another text by Newton on the same subject, the *Methodus fluxionum*, written in 1671, but published only in 1736 in English translation as *Method of fluxions*. It represents an earlier phase in Newton's development, differing from that reflected in the *Quadrature of curves*. In the *Method of fluxions* we find the moments again (pp. 32–33):

The moments of flowing quantities (that is, their indefinitely small parts, by the accession of which, in indefinitely small portions of time, they are continually increased) are as the velocities of their flowing or increasing. Wherefore, if the moment of any one, as x, be represented by the product of its celerity $\dot{x}$ into an infinitely small quantity o (that is, by $\dot{x}o$), the moments of the others, v, y, z will be represented by $\dot{v}o, \dot{y}o, \dot{z}o$, because $\dot{v}o, \dot{x}o, \dot{y}o$, and $\dot{z}o$ are to each other as $\dot{v}, \dot{x}, \dot{y}$, and $\dot{z}$.

---

[6] Example: $a^2\dot{v} = av + v^2$, homogenized $a^2\dot{v} = av\dot{z} + v^2\dot{z}$; with $\dot{v} = 1$ it becomes $\dot{z} = a^2/(av + v^2)$.

Newton considered the equation $x^3 - ax^2 + axy - y^3 = 0$ and changed x into $x + \dot{x}o$, y into $y + \dot{y}o$. He obtained (after division by o) $3\dot{x}x^2 + 3\dot{x}^2ox + x^2oo - 2a\dot{x}x - a\dot{x}^2o + a\dot{x}y + a\dot{y}x + a\dot{x}\dot{y}o - 3\dot{y}y^2 - 3\dot{y}^2oy - y^3oo = 0$ and continued:

But whereas o is supposed to be indefinitely little, that it may represent the moments of quantities, the terms that are multiplied by it will be nothing in respect to the rest [*termini in eam ducti pro nihilo possunt haberi cum aliis collato*]; therefore I reject them, and there remains

$$3x^2\dot{x} - 2ax\dot{x} + a\dot{x}y + a\dot{y}x - 3\dot{y}y^2 = 0.$$

The *Method of Fluxions* has been republished in the *Mathematical works of Isaac Newton*, ed. D. T. Whiteside (Johnson Reprint Co., New York, London, 1964), I, 29–137.

8 L'HÔPITAL. THE ANALYSIS OF THE INFINITESIMALLY SMALL

In the development of the calculus Leibniz received invaluable aid from the brothers Jakob (James, Jacques) Bernoulli (1654–1705) and Johann (John, Jean) Bernoulli (1667–1748). They belonged to a prominent merchant family in Basel, Switzerland. Jakob, in 1687, became professor of mathematics in the university of his home town; Johann, in 1695, professor of mathematics in the university of Groningen in the Netherlands. On his brother's death Johann was elected to fill his place at Basel, and here he stayed until his death, in his later years enjoying the reputation of an elder statesman in the field of mathematics, and proud of the achievements of his pupil Leonhard Euler.

The result of the collaboration of the brothers with Leibniz, which began in 1685 after Leibniz's first paper on his calculus in the *Acta Eruditorum* of 1684 (see Selection V.1), was the creation of almost the whole of the present elementary calculus before the end of the seventeenth century. Many of the problems they solved led to ordinary differential equations and even into the calculus of variations (see Selections V.19, 20, 21). A first exposition of the calculus can be found in lectures given by Johann during 1691–92, which were not published until 1922: *Lectiones de calculo differentialium*, ed. P. Schafheitlin (Basel, 1922); German translation in Ostwald's *Klassiker*, No. 211 (Engelmann, Leipzig, 1924). The integral calculus was published during Johann's lifetime in his *Opera omnia* (Lausanne, Geneva), III (1742); German translation by G. Kowalewski in Ostwald's *Klassiker*, No. 194 (1914). However, many of Bernoulli's and Leibniz's ideas on the differential calculus were made public in a book entitled *Analyse des infiniment petits* (Paris, 1696) by Guillaume François Antoine de l'Hôpital (1661–1701). The author, a French nobleman and amateur mathematician, freely acknowledged his indebtedness to his teachers: "I have made free use of their discoveries [Je me suis servi sans façon de leurs découvertes], so that I frankly return to them whatever they please to claim as their own." The book is best known because of its rule for expressions of the form $\frac{0}{0}$. Johann Bernoulli, after L'Hôpital's death, claimed this rule for his own (*Acta Eruditorum*, August 1704).

The exact relation between L'Hôpital and Bernoulli long remained a subject of conjecture, but it has recently been clarified through the publication of the correspondence between the two men: Johann Bernoulli, *Briefwechsel*, ed. O. Spiess (Birkhäuser, Basel), I (1955), 235–236. L'Hôpital, a good mathematician in his own right, found young Johann Bernoulli willing to tutor him in the new calculus, and, for a financial allowance, to communicate to him exclusively some of his (Bernoulli's) discoveries. The rule for $\frac{0}{0}$ is contained in a letter of Bernoulli's to L'Hôpital of July 22, 1694; see D. J. Struik, *Mathematics Teacher 56* (1963), 257–260. On L'Hôpital see also J. L. Coolidge, *The mathematics of great amateurs* (Clarendon Press, Oxford, 1949; Dover, New York, 1963), chap. 12, and C. B. Boyer, *Mathematics Teacher 39* (1946), 159–167.

Here follow certain sections of L'Hôpital's book in the translation by E. Stone, *The method of fluxions both direct and inverse* (London, 1730). However, where Stone changed to the Newton notation we have changed back to the original. This means that where Stone has "fluxion" we have "differential" (the original has "différence"), and Stone's $\dot{x}$ has been changed back to dx. L'Hôpital's word for "ordinate" is, as was customary for a long time, "appliquée." For "abscissa" he uses "coupée," the French translation of the term.

PART I. SECTION I. OF FINDING THE DIFFERENTIALS OF QUANTITIES

1. *Definition* I. *Variable quantities are those that continually increase or decrease; and constant or standing quantities, are those that continue the same while others vary.* As the ordinates and abscisses of a parabola are variable quantities, but the parameter is a constant or standing quantity.

Definition II. *The infinitely small part whereby a variable quantity is continually increased or decreased, is called the differential of that quantity.*

For example: let there be any curve line AMB [Fig. 1] whose axis or diameter is the line AC, and let the right line PM be an ordinate, and the right line pm another infinitely near to the former.

Now if you draw the right line MR parallel to AC, and the chords AM, Am; and about the centre A with the distance AM, you describe the small circular arch MS: then shall Pp be the differential of PA; Rm the differential of Pm; Sm the differential of AM; and Mm the differential of the arch AM. In like manner, the little triangle MAm, whose base is the arch Mm, shall be the differential of the segment AM; and the small space $MPpm$ will be the differential of the space contained under the right lines AP, PM, and the arch AM.

Corollary. It is manifest, that the differential of a constant quantity (which is always one of the initial letters a, b, c, etc. of the alphabet) is 0: or (which is all one) that constant quantities have no differentials.

Scholium. The differential of a variable quantity is expressed by the note or characteristic d, and to avoid confusion this note d will have no other use in the sequence of this calculus. And [Fig. 1] if you call the variable quantities AP, x; PM, y; AM, z; the arch AM, u; the mixtlined space APM, s; and the segment AM, t: then will dx express the value of Pp, dy the value of RM, dz the value of Sm, du the value of the small arch Mm, ds the value of the little space $MPpm$, and du the value of the small mixtlined triangle MAm.

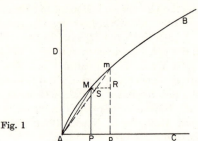

Fig. 1

2. *Postulate* I. Grant that two quantities, whose difference is an infinitely small quantity, may be taken (or used) indifferently for each other: or (which is the same thing) that a quantity, which is increased or decreased only by an infinitely small quantity, may be considered as remaining the same.

For example: grant that Ap may be taken for AP; pm for PM; the space Apm for APM; the small space $MPpm$ for the small rectangle $MPpR$; the small sector AMS for the small triangle AMm; the angle pAm for the angle PAM, etc.

3. *Postulate* II. Grant that a curve line may be considered as the assemblage of an infinite number of infinitely small right lines: or (which is the same thing) as a polygon of an infinite number of sides, each of an infinitely small length, which determine the curvature of the line by the angles they make with each other [Fig. 2].

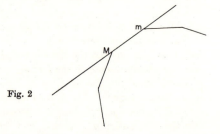

Fig. 2

For example: grant that the part Mm of the curve, and the circular arch MS, may be considered as straight lines, on account of their being infinitely small, so that the little triangle mSM may be looked upon as a right-lined triangle.

4. *Proposition* I. *To find the differentials of simple quantities connected together with the signs + and −.*

It is required to find the differentials of $a + x + y − z$. If you suppose x to increase by an infinitely small part, viz. till it becomes $x + dx$; then will y become $y + dy$; and z, $z + dz$: and the constant quantity a will still be the same a. So that the given quantity $a + x + y − z$ will become $a + x + dx + y + dy − z − dz$; and the differential of it (which will be had in taking it from this last expression) will be $dx + dy − dz$; and so of others. From whence we have the following

Rule I. For finding the differentials of simple quantities connected together with the signs + and −.

Find the differential of each term of the quantity proposed; which connected together by the same respective signs will give another quantity, which will be the differential of that given.

5. *Proposition* II. *To find the differentials of the product of several quantities multiplied, or drawn into each other.*

The differential of xy is $y\,dx + x\,dy$: for y becomes $y + dy$, when x becomes $x + dx$; and therefore xy then becomes $xy + y\,dx + x\,dy + dx\,dy$. Which is the product of $x + dx$ into $y + dy$, and the differential thereof will be $y\,dx + x\,dy + dx\,dy$, that is, $y\,dx + x\,dy$: because $dx\,dy$ is a quantity infinitely small, in respect of the other terms $y\,dx$ and $x\,dy$: For if, for example, you divide $y\,dx$ and $dx\,dy$ by dx, we shall have the quotients y and dx, the latter of which is infinitely less than the former.

Whence it follows, that the differential of the product of two quantities, is equal to the product of the differential of the first of those quantities into the second plus the product of the differential of the second into the first.

Now follows the application of these propositions: rules of differentiation and integration for elementary functions, tangency, maxima and minima, curvature, and envelopes, applied to many curves of the day. The radius of curvature is given by the formula $(dx^2 + dy^2)^{\frac{3}{2}}/ - dx\,d\,dy$. Then comes the rule for $\frac{0}{0}$ as follows:

SECTION IX. THE SOLUTION OF SOME PROBLEMS DEPENDING ON THE METHODS AFOREGOING

163. *Proposition* I. *Let AMD* [Fig. 3] *be a curve* $(AP = x,\ PM = y,\ AB = a)$ *of such a nature, that the value of the ordinate y is expressed by a fraction, the numerator and denominator of which, do each of them become 0 when $x = a$, viz.*

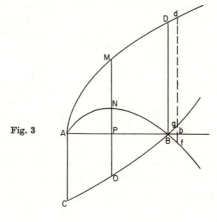

Fig. 3

*when the point P coincides with the given point B. It is required to find what will
then be the value of the ordinate BD.*

Let ANB, COB, be two curves (having the line AB as a common axis) of such
a nature, that the ordinate PN expresses the numerator, and the ordinate PO
the denominator of the general fraction representing any ordinate PM: so that
$PM = \dfrac{AB \times PN}{PO}$. Then it is manifest, that these two curves will meet one
another in the point B; since by the supposition PN, PO do each become 0 when
the point P falls in B. This being supposed, if an ordinate bd be imagined
infinitely near to BD, cutting the curves ANB, COB, in the points f, g; then will
$bd = \dfrac{AB \times bf}{bg}$, which will be equal to BD. Now our business is only to find the
relation of bg to bf. In order thereto it is manifest, when the absciss AP becomes
AB, the ordinates PN, PO will be 0, and when AP becomes Ab, they do
become bf, bg. Whence it follows, that the said ordinates bf, bg, themselves, are
the differentials of the ordinates in B and b, with regard to the curves ANB,
COB; and consequently, if the differential of the numerator be found, and that
be divided by the differential of the denominator, after having made $x = a =
Ab$ or AB, we shall have the value of the ordinates bd or BD sought. Which
was to be found.

164. *Example* I. Let $y = \dfrac{\sqrt{2a^3x - x^4} - a\sqrt[3]{aax}}{a - \sqrt[4]{ax^3}}.$ [1] Now it is manifest, when
$x = a$, that the numerator and denominator of the fraction will each be equal
to 0. Therefore, we must assume the differential of the numerator $\dfrac{a^3\, dx - 2x^3\, dx}{\sqrt{2a^3x - x^4}}$
$- \dfrac{aa\, dx}{\sqrt[3]{aax}}$, and divide it by the differential of the denominator $- \dfrac{3a\, dx}{4\sqrt[4]{a^2x}}$, after
having made $x = a$, viz. divide $-\tfrac{4}{3}a\, dx$ by $-\tfrac{1}{4}\, dx$; and there comes out
$\tfrac{16}{9}a = BD$.

165. Let $y = \dfrac{aa - ax}{a - \sqrt{ax}}.$ We find $y = 2a$, when $x = a$. We could solve this
example without need of the calculus of differentials in this way.

Having taken away the incommensurables, we shall have $aaxx + 2aaxy
- axyy - 2a^3x + a^4 + aayy - 2a^3y = 0$, which being divided by $x - a$, re-
duces to $aax - a^3 + 2aay - ayy = 0$, and substituting a for x, we obtain as
before $y = 2a$.

---

9 JAKOB BERNOULLI. SEQUENCES AND SERIES

The early success of Leibniz's calculus was due to his collaboration with the brothers
Bernoulli, who were among the earliest students of Leibniz's mathematical ideas and

[1] This example was communicated to L'Hôpital by Johann Bernoulli in his letter of July
22, 1694. Another example of Bernoulli's is $y = \dfrac{a\sqrt{ax} - xx}{a - \sqrt{ax}}$ for $x = a$. Then $y = 3a$.

developed, with Leibniz, such a productivity that by 1700 they had obtained most of what we now call the elementary differential and integral calculus, with the beginnings of ordinary differential equations and the calculus of variations. Most of their work appeared with that of Leibniz, in the *Acta Eruditorum*; Jakob's papers were collected in the *Opera*, ed. G. Cramer (2 vols.; Geneva, 1744), and Johann's in the *Opera omnia*, ed. G. Cramer (4 vols.; Lausanne, 1742). Jakob also wrote a book on probability, called *Ars conjectandi* (Basel, 1713), in which we find not only the "Bernoulli theorem" in probability, but also (pp. 97–99) the "Bernoulli numbers." He derived them as an extension of the work of Wallis, who had given the value of $1^c + 2^c + \cdots + n^c$ (c a positive integer) for large n (Selection IV.13) as close to $[1/(c + 1)]n^{c+1}$. Bernoulli then computes the precise value of $1^c + 2^c + \cdots + n^c$ as follows (in the translation of J. Ginsburg in Smith, *Source book*, 85–90):

Let the series of natural numbers 1, 2, 3, 4, 5, etc. up to n be given, and let it be required to find their sum, the sum of the squares, cubes, etc. Since in the table of combinations the general term of the second column is $n - 1$ and the sum of all terms, that is, all $n - 1$, or $\int \overline{n - 1}$ in consequence of above is[1]

$$\frac{n \cdot n - 1}{1 \cdot 2} = \frac{nn - n}{2}.$$

The sum $\int \overline{n - 1}$ or

$$\int n - \int 1 = \frac{nn - n}{n}.$$

Therefore

$$\int n = \frac{nn - n}{2} + \int 1.$$

But $\int 1$ (the sum of all units) $= n$. Therefore the sum of all n or

$$\int n = \frac{nn - n}{2} + n = \tfrac{1}{2}nn + \tfrac{1}{2}n.$$

A term of the third column is generally taken to be

$$\frac{n - 1 \cdot n - 2}{1 \cdot 2} = \frac{nn - 3n + 2}{2}$$

and the sum of all terms (that is, of all $(nn - 3n + 2)/2$) is

$$\frac{n \cdot n - 1 \cdot n - 2}{1 \cdot 2 \cdot 3} = \frac{n^3 - 3nn + 2n}{6}.$$

---

[1] Bernoulli has shown before that $1 + 2 + \cdots + n = \tfrac{1}{2}n(n - 1)$. He uses the symbol $\int$ where we use the symbol Σ.

We will have then that

$$\int \frac{n^2 - 3n + 2}{2}$$

or

$$\int \tfrac{1}{2}nn - \int \tfrac{3}{2}n + \int 1 = \frac{n^3 - 3nn + 2n}{6}$$

and

$$\int \tfrac{1}{2}nn = \frac{n^3 - 3nn + 2n}{6} + \int \tfrac{3}{2}n - \int 1 \, ;$$

but

$$\int \tfrac{3}{2}n = \tfrac{3}{2}\int n = \tfrac{3}{4}nn + \tfrac{3}{4}n$$

and

$$\int 1 = n.$$

Substituting, we have

$$\int \tfrac{1}{2}nn = \frac{n^3 - 3nn + 2n}{6} + \frac{3nn + 3n}{4} - n = \tfrac{1}{6}n^3 + \tfrac{1}{4}nn + \tfrac{1}{12}n,$$

of which the double $\int nn$ (the sum of the squares of all n) $= \tfrac{1}{3}n^3 + \tfrac{1}{2}nn + \tfrac{1}{6}n$. A term of the fourth column is generally

$$\frac{n - 1 . n - 2 . n - 3}{1 . 2 . 3} = \frac{n^3 - 6nn + 11n - 6}{6},$$

and the sum of all terms is

$$\frac{n . n - 1 . n - 2 . n - 3}{1 . 2 . 3 . 4} = \frac{n^4 - 6n^3 + 11nn - 6n}{24}.$$

It must certainly be that

$$\int \frac{n^3 - 6nn + 11n - 6}{6} \, ;$$

that is

$$\int \tfrac{1}{6}n^3 - \int nn + \int \tfrac{11}{6}n - \int 1 = \frac{n^4 - 6n^3 + 11nn - 6n}{24}.$$

Hence

$$\int \tfrac{1}{6}n^3 = \frac{n^4 - 6n^3 + 11nn - 6n}{24} + \int nn - \int \tfrac{11}{6}n + \int 1.$$

And before it was found that $\int nn = \tfrac{1}{3}n^3 + \tfrac{1}{2}nn + \tfrac{1}{6}n$, $\int \tfrac{11}{6}n$ or $\tfrac{11}{6}\int n = \tfrac{11}{12}nn + \tfrac{11}{12}n$, and $\int 1 = n$. When all substitutions are made, the following results:

$$\int \tfrac{1}{6}n^3 = \frac{n^4 - 6n^3 + 11nn - 6n}{24} + \tfrac{1}{3}n^3 + \tfrac{1}{2}nn + \tfrac{1}{6}n - \tfrac{11}{12}nn - \tfrac{11}{12}n + n$$

$$= \tfrac{1}{24}n^4 + \tfrac{1}{12}n^3 + \tfrac{1}{24}nn;$$

or, multiplying by 6,

$$\int n^3 = \tfrac{1}{4}n^4 + \tfrac{1}{2}n^3 + \tfrac{1}{4}nn.$$

Thus we can step by step reach higher and higher powers and with slight effort form the following table.[2]

Sums of Powers

$$\int n \ = \tfrac{1}{2}nn + \tfrac{1}{2}n,$$
$$\int nn \ = \tfrac{1}{3}n^3 + \tfrac{1}{2}nn + \tfrac{1}{6}n,$$
$$\int n^3 \ = \tfrac{1}{4}n^4 + \tfrac{1}{2}n^3 + \tfrac{1}{4}nn,$$
$$\int n^4 \ = \tfrac{1}{5}n^5 + \tfrac{1}{2}n^4 + \tfrac{1}{3}n^3 \ast - \tfrac{1}{30}n,$$
$$\int n^5 \ = \tfrac{1}{6}n^6 + \tfrac{1}{2}n^5 + \tfrac{5}{12}n^4 \ast - \tfrac{1}{12}nn,$$
$$\int n^6 \ = \tfrac{1}{7}n^7 + \tfrac{1}{2}n^6 + \tfrac{1}{2}n^5 \ast - \tfrac{1}{6}n^3 \ast + \tfrac{1}{42}n,$$
$$\int n^7 \ = \tfrac{1}{8}n^8 + \tfrac{1}{2}n^7 + \tfrac{7}{12}n^6 \ast - \tfrac{7}{24}n^4 \ast + \tfrac{1}{12}nn,$$
$$\int n^8 \ = \tfrac{1}{9}n^9 + \tfrac{1}{2}n^8 + \tfrac{2}{3}n^7 \ast - \tfrac{7}{15}n^5 \ast + \tfrac{2}{9}n^3 \ast - \tfrac{1}{30}n,$$
$$\int n^9 \ = \tfrac{1}{10}n^{10} + \tfrac{1}{2}n^9 + \tfrac{3}{4}n^8 \ast - \tfrac{7}{10}n^6 \ast + \tfrac{1}{2}n^4 \ast - \tfrac{1}{12}nn,$$
$$\int n^{10} = \tfrac{1}{11}n^{11} + \tfrac{1}{2}n^{10} + \tfrac{5}{6}n^9 \ast - 1n^7 \ast + 1n^5 \ast - \tfrac{1}{2}n^3 \ast + \tfrac{5}{66}n.$$

Whoever will examine the series as to their regularity may be able to continue the table. Taking c to be the power of any exponent, the sum of all n^c or

$$\int n^c = \frac{1}{c+1} n^{c+1} + \tfrac{1}{2}n^c + \frac{c}{2} An^{c-1} + \frac{c.c - 1.c - 2}{2.3.4} Bn^{c-3}$$

$$+ \frac{c.c - 1.c - 2.c - 3.c - 4}{2.3.4.5.6} Cn^{c-5}$$

$$+ \frac{c.c - 1.c - 2.c - 3.c - 4.c - 5.c - 6}{2.3.4.5.6.7.8} Dn^{c-7},$$

[2] The symbol ∗ indicates that there is a term with coefficient zero.

and so on, the exponents of n continually decreasing by 2 until n or nn is reached. The capital letters A, B, C, D denote in order the coefficients of the last terms in the expressions for $\int nn, \int n^4, \int n^6, \int n^8$, namely, A is equal to $\frac{1}{6}$, B is equal to $-\frac{1}{30}$, C is equal to $\frac{1}{42}$, D is equal to $-\frac{1}{30}$.

These coefficients are such that each one completes the others in the same expression to unity. Thus D must have the value $-\frac{1}{30}$ because $\frac{1}{9} + \frac{1}{2} + \frac{2}{3} - \frac{7}{15} + \frac{2}{9} + (+D) - \frac{1}{30} = 1$.

With the help of this table it took me less than half of a quarter of an hour to find that the tenth powers of the first 1000 numbers being added together will yield the sum

$$91,409,924,241,424,243,424,241,924,242,500$$

From this it will become clear how useless was the work of Ismael Bullialdus[3] spent on the compilation of his voluminous *Arithmetica infinitorum* in which he did nothing more than compute with immense labor the sums of the first six powers, which is only a part of what we have accomplished in the space of a single page.

Between 1689 and 1704, Bernoulli published five papers on infinite series, a subject on which much had already been written. His papers contain new results, or results he perhaps considered new. They were republished in the *Opera* (1744), and German translation by G. Kowalewski exists in Ostwald's *Klassiker*, No. 171 (Engelmann, Leipzig, 1909). We translate here a section on the harmonic series, taken from the first paper, *Proportiones arithmeticae de seriebus infinitis earumque summa finita* (Basel, 1689; *Opera*, I, 375–402), together with the verses with which it opens. The divergence of the harmonic series was demonstrated as early as the fourteenth century by Nicole Oresme in his *Questiones super geometriam Euclidis* (ed. H. L. L. Busard; Brill, Leiden, 1961), p. 6; *Paraphrase*, p. 76 (see Selection III.1), and a proof was published by P. Mengoli, *Novae quadraturae arithmeticae* (Bologna, 1650). Oresme's discussion is as follows (it follows the discussion of other infinite series):

Upon addition of nonproportional parts in a *proportio minoris inaequalitatis* [decrease of terms in ratio $< \frac{1}{2}$] the whole may become infinite. Add to a magnitude of 1 foot: $\frac{1}{2}$, $\frac{1}{3}$, $\frac{1}{4}$ foot, etc.; the sum of which is infinite. In fact, it is possible to form an infinite number of groups of terms with a sum greater than $\frac{1}{2}$. Thus $\frac{1}{3} + \frac{1}{4}$ is greater than $\frac{1}{2}$ [*quia 4ᵃ et 3ᵃ sunt plus quam una medietas*], similarly $\frac{1}{5} + \frac{1}{6} + \frac{1}{7} + \frac{1}{8}$ is greater than $\frac{1}{2}$, $\frac{1}{9} + \frac{1}{10} + \frac{1}{11} + \cdots + \frac{1}{16}$ is greater than $\frac{1}{2}$, and so *in infinitum*.

---

[3] I. Bullialdus or Boulliou (1605–1694), a French astronomer, wrote *Opus novum ad arithmeticum infinitorum* (Paris, 1682).

This way of reasoning is like the modern one; that of the Bernoullis is somewhat different. Here follows part of Jakob's text.

Ut non finitam seriem finita coercet
 Summula et in nullo limite limes adest:
Sic modico immensi vestigia Numinis haerent
 Corpore et augusto limite limes abest.
Cernere in immense parvum, dic, quanta voluptas!
 In parvo immensum cernere, quanta, Deum![4]

The sum of an infinite series of harmonic proportionals

$$\tfrac{1}{1} + \tfrac{1}{2} + \tfrac{1}{3} + \tfrac{1}{4} + \tfrac{1}{5} \quad \text{etc.}$$

is infinite.

My brother was the first to observe it:[5] because after having found, in the way explained before, the sum

$$\tfrac{1}{2} + \tfrac{1}{6} + \tfrac{1}{12} + \tfrac{1}{20} + \tfrac{1}{30} \quad \text{etc.}$$

he also wanted to see what becomes of the series

$$\tfrac{1}{2} + \tfrac{2}{6} + \tfrac{3}{12} + \tfrac{4}{20} + \tfrac{5}{30} \quad \text{etc.}$$

if it is solved with the method of Art. XIV.[6] Then he found the truth of the theorem from the clear contradiction which follows if it is assumed that the harmonic series has a finite sum.

---

[4] Just as a finite little sum embraces the infinite series, and a limit exists where there is no limit: so the vestiges of the immense Mind cling to the modest body, and there exists no limit within the narrow limit. O say, what glory it is to recognize the small in the immense! What glory to recognize in the small the immensity of God!

[5] The reasoning of Johann Bernoulli can be found in his *Opera omnia*, IV, 8, in a section entitled "De seriebus varia, Corollarium III," translated in Ostwald's *Klassiker*, No. 171, p. 116.

[6] The reasoning of Jakob's Art. XIV, applied to the case of a harmonic series, is as follows. Let

$$N = \tfrac{1}{1} + \tfrac{1}{2} + \tfrac{1}{3} + \tfrac{1}{4} + \tfrac{1}{5} + \cdots;$$

then

$$P = N - 1 = \tfrac{1}{2} + \tfrac{1}{3} + \tfrac{1}{4} + \tfrac{1}{5} + \tfrac{1}{6} + \cdots;$$

hence

$$N - P = 1 = (\tfrac{1}{1} - \tfrac{1}{2}) + (\tfrac{1}{2} - \tfrac{1}{3}) + (\tfrac{1}{3} - \tfrac{1}{4}) + (\tfrac{1}{4} - \tfrac{1}{5}) + (\tfrac{1}{5} - \tfrac{1}{6})$$

$$= \frac{1}{1.2} + \frac{1}{2.3} + \frac{1}{3.4} + \frac{1}{5.6} + \cdots;$$

(footnote continued)

Indeed, he observed that the

$$\text{Series } A = \tfrac{1}{2} + \tfrac{1}{3} + \tfrac{1}{4} + \tfrac{1}{5} + \tfrac{1}{6} + \tfrac{1}{7} \quad \text{etc.}$$

changes into

$$\text{Series } B = \tfrac{1}{2} + \tfrac{2}{6} + \tfrac{3}{12} + \tfrac{4}{20} + \tfrac{5}{20} + \tfrac{6}{42} \quad \text{etc.} = C + D + E + F \quad \text{etc.}$$

when the separate fractions are changed into others with the numerators 1, 2, 3, 4, etc.

Here

$$C = \tfrac{1}{2} + \tfrac{1}{6} + \tfrac{1}{12} + \tfrac{1}{20} + \tfrac{1}{30} + \tfrac{1}{42} \quad \text{etc.} = 1 \text{ (as proved before)}$$

when the separate fractions are changed into others with the numerators $1, 2, 3, 4$, etc.

$$
\begin{aligned}
D &= \phantom{+\tfrac{1}{6}} + \tfrac{1}{6} + \tfrac{1}{12} + \tfrac{1}{20} + \tfrac{1}{30} + \tfrac{1}{42} \quad \text{etc.} = C - \tfrac{1}{2} = \tfrac{1}{2}, \\
E &= \phantom{+\tfrac{1}{6} + \tfrac{1}{6}} + \tfrac{1}{12} + \tfrac{1}{20} + \tfrac{1}{30} + \tfrac{1}{42} \quad \text{etc.} = B - \tfrac{1}{6} = \tfrac{1}{3}, \\
F &= \phantom{+\tfrac{1}{6} + \tfrac{1}{6} + \tfrac{1}{12}} \tfrac{1}{20} + \tfrac{1}{30} + \tfrac{1}{42} \quad \text{etc.} = E - \tfrac{1}{12} = \tfrac{1}{4}, \\
& \text{etc.} = \text{etc.}
\end{aligned}
$$

from which it follows that

$$B = 1 + \tfrac{1}{2} + \tfrac{1}{3} + \tfrac{1}{4} \quad \text{etc.} = A,$$

so that a part would be equal to the whole if the sum were finite.

When he showed it to me, I gave a proof by showing in the following way that the sum of the infinite harmonic series

$$\tfrac{1}{1} + \tfrac{1}{2} + \tfrac{1}{3} + \tfrac{1}{4} \quad \text{etc.}$$

surpasses any given number, and hence is infinite, after Proposition II.[7] [To prove it] let N be a given number as large as you like. Take a group of terms away from the beginning of the series, of which the sum is equal to or larger than one, which is a unity contained in the number N. From the remaining terms take

---

We cannot accept this reasoning, since N is not a finite number, but the proof can be made acceptable by the following modification. Let

$$s_n = \frac{1}{1} + \frac{1}{2} + \frac{1}{3} + \cdots + \frac{1}{n},$$

then

$$s_n - (s_n - 1) = 1 = \frac{1}{1 \cdot 2} + \frac{1}{2 \cdot 3} + \frac{1}{3 \cdot 4} + \cdots + \frac{1}{n(n-1)} + \frac{1}{n};$$

hence

$$1 = \lim_{n \to \infty} \left[\frac{1}{2} + \frac{1}{6} + \frac{1}{12} + \cdots + \frac{1}{n(n-1)} \right].$$

[7] Proposition II: That which is greater than any given quantity is infinite.

away again a group of terms of which the sum is larger than another unity, and repeat this procedure, if possible, as often as there are unities in the number N. Then all the separated terms together will surpass the given number, and the whole series will a fortiori surpass it.

When you deny that after some terms have been taken away the other terms can surpass unity, then let $1 : a$ be the first term remaining after the last separation. The terms following $1 : a$ are $1 : (a + 1)$, $1 : (a + 2)$, $1 : (a + 3)$, etc. Now form a geometric progression starting with the first two members $1 : a$ and $1 : (a + 1)$. The terms in this progression following the second term are then smaller than the corresponding terms in the harmonic progression since they have the larger denominator (cf. Proposition IV).[8] Continue the geometric progression for so long as the terms are larger than $1 : a^2$ (this can be done by a finite number of steps, since a is a finite number). Then this finite geometric series becomes equal to 1, according to VIII.[9] The harmonic series with as many terms will therefore surpass unity.   Q.E.D.

Corollary 1. If it be permitted to take a jump into geometry, then it also follows that the area between a hyperbola and its asymptotes is infinite. Let us divide [Fig. 1] an asymptote into an infinite number of equal parts, beginning at the center A. Let the points dividing the parts be $B, C, D, E, \ldots$ Through them let us draw to the curve the lines BM, CN, DO, EP, etc. parallel to the other asymptote and let us complete the parallelograms AM, BN, CO, DP, etc. Since their bases are equal they will be in the same proportion as their altitudes,

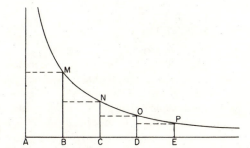

Fig. 1

---

[8] Proposition IV states that if the numbers $A, B, C, D, E, \ldots$ form a geometric progression, and $A, B, F, G, H, \ldots$ an arithmetic one with the same initial terms A, B, then $C > F$, $D > G$, $E > H$, and so on. This is shown by applying the theorem in Euclid, *Elements*, Book V, Prop. 25, that from $a : b = c : d$ it follows that $a + d > b + c$ if a is the largest, d the smallest number of the proportion (all positive). Hence from $A : B = B : C = C : D = D : E$ it follows that $A + C > 2B = A + F$, hence $C > F$, and so on.

[9] Proposition VIII gives the sum of n terms of a geometric series. The reasoning is as follows, in our present notation: We replace $\dfrac{1}{a} + \dfrac{1}{a + 1} + \dfrac{1}{a + 2} + \cdots$ by $\dfrac{1}{a} + \dfrac{r}{a} + \dfrac{r^2}{a} + \cdots$, where $r = \dfrac{a}{a + 1}$, and continue the sum until the last term of the geometric series $\dfrac{r^{n-1}}{a} \leqslant \dfrac{1}{a^2}$, or $ar^{n-1} \leqslant 1$; the sum s_n of the n terms of this series is

$$s_n = \frac{1}{a}\frac{1 - r^n}{1 - r} = \frac{a + 1}{a}(1 - r^n) \geqslant \frac{a + 1}{a}\left(1 - \frac{r}{a}\right) = 1.$$

that is, as the segments BM, CN, OD, EP, etc. or as $\frac{1}{1}$, $\frac{1}{2}$, $\frac{1}{3}$, $\frac{1}{4}$ etc., as follows from the nature of the hyperbola. But still, since we have shown that the sum $\frac{1}{1} + \frac{1}{2} + \frac{1}{3} + \frac{1}{4}$ etc. is infinite, the sum of the parallelograms AM, BN, CO, DP, etc., will also be infinite, and hence for even stronger reason the hyperbolic area in which lie these parallelograms.[10]

---

10   JOHANN BERNOULLI. INTEGRATION

L'Hôpital's book (see Selection V.8) can serve to show how Johann Bernoulli taught the differential calculus in 1691/92. From Bernoulli's *Opera omnia*, III, 305–558, we give here a translation of some of his lessons on integral calculus. It is clear from the whole text that Bernoulli, when he prepared his own *Opera*, simply took his manuscript of *c.* 1691 and left it practically unchanged, adding some modifications. This explains how at first he did not know what to do with the integral of $1/x$, but later in the text corrected himself. The title of the work is *Lectiones mathematicae de methodo integralium, aliisque conscriptae in usum Ill. Marchionis Hospitalii cum auctor Parisiis ageret annis 1691 et 1692* (Mathematical lectures on the method of integrals, and on other subjects written for the use of the Marquis de L'Hôpital, as the author gave them in Paris during 1691 and 1692). A partial German translation exists in Ostwald's *Klassiker*, No. 194 (Engelmann, Leipzig, 1914).

FIRST LECTURE. ON THE METHOD OF INTEGRALS AND OTHER MATTERS

We have seen how to find the *differentials* of quantities.[1] Now we shall show inversely how the integrals of the differentials are found, that is, those quantities, from which the differentials originate. It is known from what I have said that dx is the differential of x, $x\,dx$ the differential of $\frac{1}{2}xx$ or $\frac{1}{2}xx$ + or − a constant quantity, $xx\,dx$ the differential of $\frac{1}{3}x^3$ + or − a constant quantity, and $x^3\,dx$ the differential of $\frac{1}{4}x^4$ + or − a constant quantity. Equally:

$$
\begin{array}{lllll}
a\,dx & \text{is the differential of } ax & & & \text{etc.} \\
ax\,dx & \text{,, \quad ,,} & \text{,,} & \text{,, } \tfrac{1}{2}axx & \text{,,} \\
axx\,dx & \text{,, \quad ,,} & \text{,,} & \text{,, } \tfrac{1}{3}ax^3 & \text{,,} \\
ax^3\,dx & \text{,, \quad ,,} & \text{,,} & \text{,, } \tfrac{1}{4}ax^4 & \text{,,}
\end{array}
$$

From this the general Rule can thus be formulated:

$$ax^p\,dx \text{ is the differential of the quantity } \frac{a}{p+1}x^{p+1}.$$

---

[10] The fact that this area is infinite has been known ever since Torricelli communicated several of his results to various mathematicians during 1646–1647; see "De infinitis hyperbolis," *Opere* (3 vols.; Monfanari, Faenza, 1919), I, 191–195 (Selection IV.9). We find this theorem first published in N. Mercator, *Logarithmotechnica* (London, 1668).

[1] This refers to the first part of the lectures, which were published only in 1922; see the introduction to Selection V.8 (L'Hôpital).

If, therefore, we inquire into the integral of a differential we must primarily find out whether the given quantity is the product of a differential and a multiple of its absolute quantity[2] raised to a certain power. This is a sign that we can find the integral by the above-mentioned rule. If, for example, the integral of $dy\sqrt{(a + y)}$ has to be found, then I observe first that dy is multiplied by a multiple of its absolute quantity $(a + y)$ raised to the power $\frac{1}{2}$; then, by the above-mentioned rule I inquire into its integral and find

$$\frac{1}{\frac{1}{2} + 1}(a + y)^{\frac{1}{2}+1}, \quad \text{that is} \quad \tfrac{2}{3}(a + y)\sqrt{(a + y)}.$$

Likewise we find the integral of $x\,dx\sqrt[3]{(a^2 + x^2)}$, which is

$$\frac{\frac{1}{2}}{\frac{1}{3} + 1}(a^2 + x^2)^{\frac{1}{3}+1} = \tfrac{3}{8}(a^2 + x^2)\sqrt[3]{(a^2 + x^2)},$$

the integral of $dx : \sqrt{(a + y)}$ equal to $2\sqrt{(a + y)}$, the integral of $dx : x$ equal to

$$\frac{1}{0}x^0 = \frac{1}{0} \times 1 = \infty.^{3}$$

Note that sometimes quantities occur whose integrals at first sight cannot be found by this rule. But they are easily found after a certain change, as in the following cases.

1. If we write instead of $dx\sqrt{(a^2xx + x^4)}$ the expression $x\,dx\sqrt{(a^2 + xx)}$, then the integral is found to be $(\tfrac{1}{3}a^2 + \tfrac{1}{3}xx)\sqrt{(a^2 + xx)}$. And if instead of $dx\sqrt{(a^3 + 3a^2x + 3axx + x^3)}$ we write $(a\,dx + x\,dx)\sqrt{(a + x)}$, then the integral is found to be $\tfrac{2}{5}(a^2 + 2ax + xx)\sqrt{(a + x)}$.

2. It also happens the other way: that one or more letters have to be placed under the square-root sign before integration is possible, as in the following expression:

$$(3ax^3\,dx + 4x^4\,dx)\sqrt{(ax + xx)}.$$

This, it seems, cannot be integrated by our rule. But if we place one x under the square-root sign, then the expression becomes

$$(3axx\,dx + 4x^3\,dx)\sqrt{(ax^3 + x^4)},$$

and the integral of this differential can be found by our rule. It is

$$\tfrac{2}{3}(ax^3 + x^4)\sqrt{(ax^3 + x^4)}.$$

---

[2] This term "absolute quantity" is explained by the examples; thus in $x\,dx\sqrt[3]{a^2 + xx}$ the "absolute quantity" is $z = a^2 + xx$, since the integrand is a multiple of $dz^{4/3}$.

[3] $\int dx : x$ had already been correctly discussed in Leibniz's first paper of 1684 (Selection V.1). Bernoulli later corrected himself; see note 5.

3. If a fraction occurs, of which the denominator is a square, a cube, or another power, then we must assume its square root to be the absolute quantity. So in

$$\frac{x\,dx}{a^4 + 2a^2xx + x^4}$$

we choose $a^2 + xx$ as the absolute quantity, and we get $-1 : (2a^2 + 2xx)$. If we had chosen $a^4 + 2a^2xx + x^4$ as the absolute quantity, then we could not have obtained the integral of the fraction.

4. If for two quantities the integrals cannot be found separately then it may happen that the integral of their combination can be found. Take, for instance,

$$\frac{a\,dx}{\sqrt{(2ax + xx)}} + \frac{x\,dx}{\sqrt{(2ax + xx)}}.$$

Neither of the two terms yields an integral. Their sum, however, $\dfrac{a\,dx + x\,dx}{\sqrt{(2ax + xx)}}$, gives $\sqrt{(2ax + xx)}$ for its integral.[4]

5. A fraction sometimes seems not to have an integral; but it may happen that multiplication of numerator and denominator by the same quantity will give an integral that can easily be obtained. Take, for instance,

$$(a\,dx + x\,dx) : \sqrt{(3a + 2x)}.$$

Multiplication of numerator and denominator by x gives

$$(ax\,dx + xx\,dx) : \sqrt{(3axx + 2x^3)},$$

and its integral is $\frac{1}{3}\sqrt{(3axx + 2x^3)}$.

6. Then again, by dividing numerator and denominator by the same quantity we may obtain an integral. For example $axx\,dx : \sqrt{(a^2xx + x^4)}$; by dividing both terms by x we get $ax\,dx : \sqrt{(a^2 + xx)}$; and its integral is $a\sqrt{(a^2 + xx)}$ according to the rule.

After another section, no. 7, Bernoulli adds a *Monitum*:

However, it is as difficult to find the integral of any differential as it is easy to find the differential of any quantity. Sometimes it is not even certain whether or not we can find the integral of a given quantity. I dare say at any rate that

---

[4] This idea proved very helpful in tackling questions pertaining to elliptic integrals. See Selection V.18.

any quantity, integral or rational, that is multiplied or divided by $x^p \sqrt{(a^2 - xx)}$, $x^p \sqrt{(ax - xx)}$, $x^p \sqrt{(a^2 + xx)}$ can be integrated or its integration can be reduced to a quadrature of a circle or of a hyperbola.[5] We shall show this in what follows.

We thus have in general to ascertain carefully whether the given quantity to be integrated can, by multiplication or division or the taking of the root, be reduced to a quantity that is the product of one of the roots multiplied by a rational or integral quantity. If we can accomplish this, then we can promptly find either the integral of the given quantity or else that it depends on and can be reduced to the quadrature of the circle or the hyperbola. For example, let

$$(a^3 + axx - x^3)\, dx \sqrt{\frac{a + x}{x}}$$

be given to be integrated. At first it seems that its integral cannot be computed and has no relation to the squaring of the circle. If, however, we assume for the absolute quantity the expression following the $\sqrt{}$ sign, which is the fraction $(a + x) : x$, then its differential is also a fraction, so that according to the rule nothing can be derived from it. To avoid this, I multiply the numerator and denominator of the irrational fraction by the numerator, then connect the product of the numerator by itself to the rational part of the quantity, so that I get a fraction of which the numerator is rational and the denominator irrational; in fact, we obtain

$$(a^3 + axx - x^3)\, dx \sqrt{\frac{a + x}{x}} = \frac{(a^4 + a^2xx + a^3x - x^4)\, dx}{\sqrt{(ax + xx)}}.$$

This gives me the clue to the nature of this integral, whether it is obtainable or reducible to the quadrature of the hyperbola. We shall now show how this can be determined and carried out.

Bernoulli goes on to give examples of integration by substitution, and applications to quadratures. Here, in Lecture X, we find the case of the area of the hyperbola between the curve and an asymptote, as follows:

Let it be required to find the nature of the curve BDC [Fig. 1] of which the subtangent is always equal to the constant a. According to this assumption,

[5] Leibniz, in his *Geometria recondita* of 1686 (Selection V.2), had suggested that such integrals as $\int \sqrt{a^2 - x^2}\, dx$, $\int \sqrt{a^2 + x^2}\, dx$ should be taken as the representation of certain "transcendental" quantities (soon to be called functions). We now express the thought of Leibniz and Bernoulli by saying that every integral of the form $\int R(x, \sqrt{ax^2 + bx + c})\, dx$, where R is a rational function, can be expressed in terms of algebraic, cyclometric, and logarithmic functions.

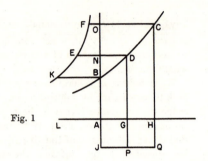

Fig. 1

$dy:dx = y:a$, hence $y\,dx = a\,dy$ and $dx = a\,dy:y$. If we multiply both sides by a, we obtain

$$a\,dx = \frac{aa\,dy}{y}.$$

Hence, in order to construct the curve we have to draw the unlimited perpendiculars BJ, LH and apply to all AB, AN, AO, etc., which are equal to y, the corresponding BK, NE, OF, etc., equal to ad/y. Then we obtain the hyperbola KEF of which AO, AL, are the asymptotes. Now apply to AG, AH, etc. the perpendiculars GP, HQ, etc. equal to a; then PQ will be a line parallel to AH. If, therefore, we take the hyperbolic area KN equal to the rectangle AP and KO equal to AQ, then the points of intersection D, C, etc. will lie on the required curve. Hence, when AG, AH, etc. are arithmetically proportional, then AB, AN, AO, etc. will be geometrically proportional, because the areas KN, EO, etc. must be equal. The curve BDC is therefore a logarithmic curve.[6]

---

11 TAYLOR. THE TAYLOR SERIES

The unsatisfactory way in which Newton introduced his fluxions and moments led to criticism, and to attempts at improvement. Brook Taylor (1685–1731), educated at St. John's College, Cambridge, was for a while secretary of the Royal Society, and in his *Methodus incrementorum directa et inversa* (London, 1715) attempted a systematic calculus of finite differences. This "method of increments" has a "direct" and an "inverse" part, just as the method of fluxions has. The book is mainly known because it contains "Taylor's series." This was the first publication of this theorem, although it was known before; it can be found in the papers of James Gregory (see Selection V.4). Taylor's book also opens the mathematical investigation of the vibrating string (see Selection V.16).

On Taylor see H. Auchter, *Brook Taylor, der Mathematiker und Philosoph* (Dissertation, Marburg, Würzburg, 1937).

[6] This is very much like Leibniz's discussion of 1684, and Bernoulli has corrected himself. See Selection V.1.

Here follows the section of the *Methodus* (pp. 21–23) that contains the Taylor series. We have changed the notation somewhat to make the paper more legible.[1]

Proposition VII. *Theorem* III. Let z and x be two variable quantities, of which z increases uniformly with given increments Δz.[2] Let $n \Delta z = v$, $v - \Delta z = \dot{v}$, $\dot{v} - \Delta z = \ddot{v}$, etc. Then I say that when z grows into $z + v$, then x grows into

$$z + \Delta x \frac{v}{1 . \Delta z} + \Delta^2 x \frac{v \dot{v}}{1 . 2 (\Delta z)^2} + \Delta^3 x \frac{v \dot{v} \ddot{v}}{1 . 2 . 3 (\Delta z)^3} + \cdots .$$

Demonstration [Fig. 1]

x	Δx	$\Delta^2 x$	$\Delta^3 x$	$\Delta^4 x$	etc.
$x + \Delta x$	$\Delta x + \Delta^2 x$	$\Delta^2 x + \Delta^3 x$	$\Delta^3 x + \Delta^4 x$	etc.	
$x + 2\Delta x + \Delta^2 x$	$\Delta x + 2\Delta x + \Delta^3 x$	$\Delta^2 x + 2\Delta^3 x$ $+ \Delta^4 x$	etc.		
$x + 3\Delta x + 3\Delta^2 x + \Delta^3 x$	$\Delta x + 3\Delta^2 x$ $+ 3\Delta^2 x + \Delta^4 x$	etc.			
$x + 4\Delta x + 6\Delta^2 x$ $+ 4\Delta^3 x + \Delta^4 x$	etc.				

The successive values of x, collected by continued addition, are x, $x + \Delta x$, $x + 2\Delta x + \Delta^2 x$, $x + 3\Delta x + 3\Delta^2 x + \Delta^3 x$, etc., as we see from the operation expressed in the table. But the numerical coefficients of the terms x, Δx, $\Delta^2 x$, etc. for these values of x are formed in the same way as the coefficients of the corresponding terms in the binomial expansion [*in dignitate binomii*]. And if n is the exponent of the expansion [*dignitatis index*], then the coefficients (according to Newton's theorem) will be $1, \frac{n}{1}, \frac{n}{1} \frac{n-1}{2}, \frac{n}{1} \frac{n-1}{2} \frac{n-2}{3}$, etc. When, therefore, z grows into $z + n\Delta z$, that is, into $z + v$, then x will be equal to the series

$$x + \frac{n}{1} \Delta x + \frac{n}{1} \frac{n-1}{2} \Delta^2 x + \frac{n}{1} \frac{n-1}{2} \frac{n-2}{3} \Delta^3 x + \text{etc.}$$

[1] Taylor used a complicated notation with dots and primes (*lineolae*) used as superscripts and subscripts, and the primes in both the *accent grave* and *accent aigu* position. We have kept his notation, except that instead of the increments z, z, etc. we have written Δz, $\Delta^2 z$, etc., and for the v with subscript *accents aigus* we have written $v_1, v_{11}, \ldots$. Taylor's notation also has its advantage. In his notation x'', x', x, $\dot{x}$, $\ddot{x}$ represent a sequence of functions of which each is the fluxion of the previous one; whereby Taylor remarks that the *lineolae* in x', x'' can be regarded as negative dots—an anticipation, if we like, of our modern operational notation D^{-2}, D^{-1}, D^0, D^1, D^2, and for the same purpose. Compare Taylor's notation to that of Newton, Selection V.7, art. 13, p. 307.

[2] Since z flows uniformly, Δz is constant, so that $\Delta^2 z$, $\Delta^3 z$, etc. are all zero. Here $\Delta^2 z$ is the increment of Δz, $\Delta^3 z$ that of $\Delta^2 z$, etc.

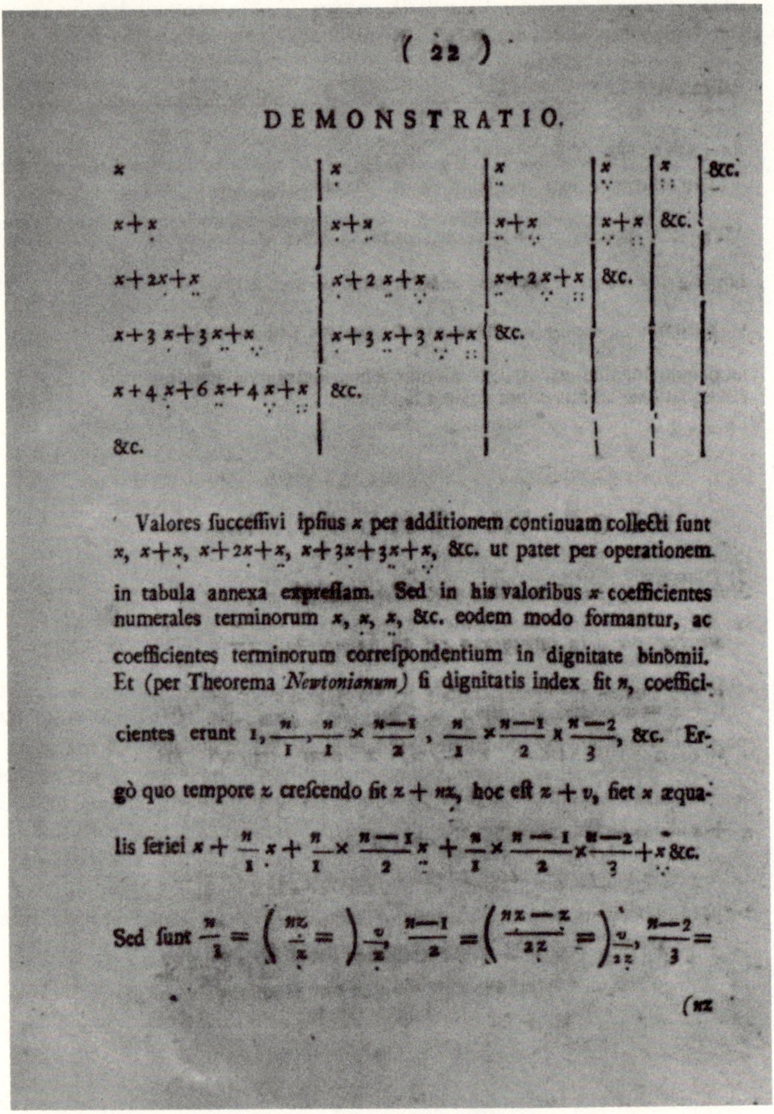

(22)

DEMONSTRATIO.

x	$\dot{x}$	$\ddot{x}$	$\dddot{x}$	x	&c.
$x+x$	$x+x$	$x+x$	$x+x$	&c.	
$x+2x+x$	$x+2\,x+x$	$x+2x+x$	&c.		
$x+3\,x+3\,x+x$	$x+3\,x+3\,x+x$	&c.			
$x+4\,x+6\,x+4\,x+x$	&c.				
&c.					

Valores fucceffivi ipfius x per additionem continuam collecti funt x, $x+x$, $x+2x+x$, $x+3x+3x+x$, &c. ut patet per operationem in tabula annexa expreffam. Sed in his valoribus x coefficientes numerales terminorum x, x, x, &c. eodem modo formantur, ac coefficientes terminorum correfpondentium in dignitate binomii. Et (per Theorema *Newtonianum*) fi dignitatis index fit n, coefficientes erunt $1, \dfrac{n}{1}, \dfrac{n}{1} \times \dfrac{n-1}{2}, \dfrac{n}{1} \times \dfrac{n-1}{2} \times \dfrac{n-2}{3}$, &c. Ergò quo tempore z crefcendo fit $z + nz$, hoc eft $z + v$, fiet x æqualis feriei $x + \dfrac{n}{1} x + \dfrac{n}{1} \times \dfrac{n-1}{2} x + \dfrac{n}{1} \times \dfrac{n-1}{2} \times \dfrac{n-2}{3} x + x$ &c.

Sed funt $\dfrac{n}{1} = \left(\dfrac{nz}{z} = \right) \dfrac{v}{z}$, $\dfrac{n-1}{2} = \left(\dfrac{nz - z}{2z} = \right) \dfrac{v}{2z}$, $\dfrac{n-2}{3} =$

(nz

Fig. 1

(23)

$\left(\dfrac{\text{II}z - 2\dot{z}}{3\dot{z}} =\right) \dfrac{\overset{\cdot\cdot}{v}}{\dot{z}}$, &c. Proinde quo tempore z crefcendo fit $z + v$,

eodem tempore x crefcendo fiet $x + x \dfrac{v}{1\dot{z}} + x \dfrac{\overset{\cdot}{v}v}{1.2\dot{z}^2} + x \dfrac{\overset{\cdot\cdot}{v}\overset{\cdot}{v}v}{1.2.3\dot{z}^3} + $

$+$ &c.

COROLL. I.

Et ipfis z, x, $\overset{\cdot}{x}$, $\overset{\cdot\cdot}{x}$, &c. iifdem manentibus, mutato figno ipfius v,

quo tempore z decrefcendo fit $x - v$, eodem tempore x decrefcen-

do fiet $x - x \dfrac{v}{1\dot{z}} + x \dfrac{\overset{\cdot}{v}v}{1.2\dot{z}^2} - x \dfrac{\overset{\cdot\cdot}{v}\overset{\cdot}{v}v}{1.2.3\dot{z}^3}$ &c. vel juxta notatio-

nem noftram $x - x \dfrac{v}{1\dot{z}} + x \dfrac{\overset{\cdot}{v}v}{1.2\dot{z}^2} - x \dfrac{\overset{\cdot\cdot}{v}\overset{\cdot}{v}v}{1.2.3\dot{z}^3}$ &c. ipfis $\overset{\cdot}{v}$, $\overset{\cdot\cdot}{v}$, &c.

converfis in $- \overset{\cdot}{v}$, $- \overset{\cdot\cdot}{v}$, &c.

COROLL. II.

Si pro Incrementis evanefcentibus fcribantur fluxiones ipfis pro-

portionales, faltis jam omnibus $\overset{\cdot}{v}$, $\overset{\cdot\cdot}{v}$, $\overset{\cdot\cdot\cdot}{v}$, $\overset{\cdot\cdot\cdot\cdot}{v}$, $\overset{\cdot}{v}$, &c. æqualibus

quo tempore z uniformiter fluendo fit $z + v$ fiet x, $x + x \dfrac{v}{1\dot{z}} +$

$x \dfrac{v^2}{1.2\dot{z}^2} + x \dfrac{v^3}{1.2.3\dot{z}^3}$ &c. vel mutato figno ipfius v, quo tem-

pore z decrefcendo fit $z - v$, x decrefcendo fiet $x - x \dfrac{v}{1\dot{z}} +$

$x \dfrac{v^2}{1.2\dot{z}^2} - x \dfrac{v^3}{1.2.3\dot{z}^3}$ + &c.   G   PROP.

Fig. 2

But

$$\frac{n}{1} = \left(\frac{n\Delta z}{\Delta z}\right) = \frac{v}{\Delta z}, \quad \frac{n-1}{2} = \left(\frac{n\Delta z - \Delta z}{2\Delta z}\right) = \frac{\dot{v}}{2\Delta z}, \quad \frac{n-2}{3} = \left(\frac{n\Delta z - 2\Delta z}{3\Delta z}\right) = \frac{\ddot{v}}{3\Delta z},$$

etc. Hence [Fig. 2] in the time that z grows into $z + v$, x grows into[3]

$$x = x + \Delta x \frac{v}{1.\Delta z} + \Delta^2 x \frac{v\dot{v}}{1.2(\Delta z)^2} + \Delta^3 x \frac{v\dot{v}\ddot{v}}{1.2.3(\Delta z)^3} + \text{etc.}$$

Corollary I. If the Δz, Δx, $\Delta^2 x$, $\Delta^3 x$ remain the same, but the sign of v is changed so that z decreases and becomes $z - v$, then x decreases at the same time and becomes

$$x - \Delta x \frac{v}{1.\Delta z} - \Delta^2 x \frac{v\dot{v}}{1.2(\Delta z)^2} - \Delta^3 x \frac{v\dot{v}\ddot{v}}{1.2.3(\Delta z)^3} - \text{etc.}$$

or[4]

$$x - \Delta x \frac{v}{1.\Delta z} + \Delta^2 x \frac{vv_1}{1.2(\Delta z)^2} - \Delta^3 x \frac{vv_1 v_{11}}{1.2.3(\Delta z)^3} + \text{etc.}$$

with $\dot{v}$, $\ddot{v}$, etc. converted into $-v_1$, $-v_{11}$, etc.

Corollary II. If we substitute for evanescent increments the fluxions proportional to them, then all $\ddot{v}$, $\dot{v}$, v, v_1, v_{11} become equal. When z flows uniformly into $z + v$, x becomes[5]

$$x + \dot{x}\frac{v}{1.\dot{z}} + \ddot{x}\frac{v^2}{1.2\dot{z}^2} + \dddot{x}\frac{v^3}{1.2.3\dot{z}^3} + \text{etc.},$$

or with v changing its sign, when z decreases to $z - v$, x becomes

$$x - \dot{x}\frac{v}{1.\dot{z}} + \ddot{x}\frac{v^2}{1.2\dot{z}^2} - \dddot{x}\frac{v^3}{1.2.3\dot{z}^3} + \text{etc.},$$

---

[3] This is Newton's well-known interpolation formula (*Principia*, Book III, Lemma 5); see also Newton, *Methodus differentialis* (London, 1711); *James Gregory tercentenary memorial volume*, ed. H. W. Turnbull (Bell, London, 1939), 119; H. W. Turnbull, *The mathematical discoveries of Newton* (Blackie, Glasgow, 1945), 46. The *Methodus differentialis* was reprinted in the *Opera Newtoni*, ed. S. Horsley, I (London, 1779), 519–528, and in D. C. Fraser, *Newton's interpolation formula* (Layton, London, 1927). There is a German translation in A. Kowalewski, *Newton, Cotes, Gauss, Jacobi. Vier grundlegende Abhandlungen über Interpolation und genäherte Quadratur* (Veit, Leipzig, 1917).

[4] The v_{11}, v_1, v, $\dot{v}$, $\ddot{v}$ form a sequence of increments, so that $v_1 - \Delta z = v$, $v_{11} - \Delta z = v_1$, or $v_1 = (n+1)\Delta z$, $v_{11} = (n+2)\Delta z$.

[5] This is the classical Taylor series, since in the Leibniz notation $\dot{x}/\dot{z} = dx/dz$, $x/(\dot{z})^2 = d^2x/dz^2$, etc. Taylor therefore obtained his series from Newton's interpolation formula by taking $\Delta x = 0$, $n = \infty$. Felix Klein has called Taylor's step "a transition to the limit of extraordinary audacity"; see *Elementary mathematics from an advanced standpoint*, trans. E. R. Hedrick and C. A. Noble, I (Dover, New York, 1924), 233. Although we shall not belittle this statement we must also take into account that Taylor's theorem had been "in the air" ever since James Gregory had it in a manuscript of 1671 (*Gregory tercentenary memorial volume*, pp. 123, 173, 356). See also A. Pringsheim, "Zur Geschichte des Taylorschen Lehrsatzes," *Bibliotheca mathematica* (3) *1* (1900), 433–479; G. Eneström, "Zur Vorgeschichte der Entdeckung des Taylorschen Lehrsatzes," *ibid.*, 12 (1911–12), 333–336.

Here follows one of Taylor's applications of the theorem.

Proposition VIII. *Problem* V. Given an equation which contains, apart from a uniformly increasing z, a certain number of other variables x. To find the value of x from given z by a series of an infinite number of terms.

Find all increments, to infinity, of the proposed equation by means of Proposition I. If $\Delta^n x$ be the infinite increment of x in the proposed equation, then by means of these equations will be given all increments $\Delta^n x$ and those with higher n expressed by means of increments of lower n. Let a, c, c_1, c_2, c_3, etc. be certain arbitrary values corresponding to z and x, Δx, $\Delta^2 x$, $\Delta^3 x$, etc.; then by means of these equations all terms c_n, c_{n+1}, and the following can be expressed in terms of the terms preceding c_n. Hence if we write $a + v$ for z, then x will be given by means of

$$x = c + c_1 \frac{v}{1.\Delta z} + c_2 \frac{v\dot{v}}{1.2(\Delta z)^2} + c_3 \frac{v\dot{v}\ddot{v}}{1.2.3(\Delta z)^3} + \text{etc.}$$

(according to Proposition VII). Here the coefficients c, c_1, c_2, etc. of the terms whose number is n are given by the same number of conditions imposed on the problem.[6]

---

12 BERKELEY. *THE ANALYST*

Among the critics of Newton's theories (and of those of Leibniz) were Jonathan Swift (1667–1745) and George Berkeley (1685–1753), both deans in the Church of England in Ireland. Swift's attack, directed at much of what the Royal Society was doing (and Newton was its president from 1703 to his death), can be found in the Laputa section of *Gulliver's travels* (1726); see M. Nicolson and N. M. Mohler, *Annals of Science 2* (1937), 405–430. Berkeley, who became Bishop of Cloyne (County Cork), in 1734, attacked Newton's fluxions in its weak spot: the infinitesimals. His lampooning of infinitesimals as "ghosts of departed

---

[6] This proposition and several others give information on the number of arbitrary constants in difference and differential equations. Taylor, on page 27 of his book, shows how the differential equation $\ddot{x} - \dot{x}z - 2x = 0$ can be solved by means of $x = A + Bz + Cz^2 + Dz^3 + Ez^4 + \cdots$, and by substituting this series as well as $\dot{x} = B + 2Cz + \cdots$, $\ddot{x} = 2C + 6Dz + \cdots$ he gets a series of recursion equations, from which he derives the solution

$$x = A + Bz + Az^2 + \tfrac{1}{3}Bz^3 + \tfrac{1}{3}Az^4 + \cdots$$

with two arbitrary constants. These series are what our textbooks often call "Maclaurin" series (see Selection V.13).

Taylor has no discussion of convergence.

See also R. Reiff, *Geschichte der unendlichen Reihen* (Laupp, Tübingen, 1889), 81. There exists a facsimile edition of the second edition of Taylor's book (London, 1717), published by Friedlander, Berlin, 1862.

quantities" was to show that he who is willing to accept the mysteries of the calculus (which nevertheless lead to true results) need not hesitate to accept the mysteries of religion. The foundation of the attack was laid in Berkeley's *Principles of human knowledge* (Dublin, 1710), Articles 130–132. The real onslaught came in *The analyst* of 1734, directed against "an infidel mathematician," supposedly Edmund Halley (1656–1742), the Astronomer Royal. The text is in English, and can be found in editions of Berkeley's works; see, for instance, *The works of George Berkeley*, IV (Nelson, London, 1951), 65–102. The original title is *The analyst, or a discourse addressed to an infidel mathematician* (London, 1734). Berkeley's full text is easily available; we only select a few pages to give the flavor. See further F. Cajori, *A history of the conceptions of limits and fluxions in Great Britain from Newton to Woodhouse* (Open Court, Chicago, London, 1919), and "Indivisibles and 'ghosts of departed quantities' in the history of mathematics," *Scientia 37* (1925), 303–306; see also *American Mathematical Monthly 24* (1917), 145–154.

---

3 The Method of Fluxions is the general key by help whereof the modern mathematicians unlock the secrets of Geometry, and consequently of Nature. And, as it is that which hath enabled them so remarkably to outgo the ancients in discovering theorems and solving problems, the exercise and application thereof is become the main if not sole employment of all those who in this age pass for profound geometers. But whether this method be clear or obscure, consistent or repugnant, demonstrative or precarious, as I shall inquire with the utmost impartiality, so I submit my inquiry to your own judgment, and that of every candid reader. Lines are supposed to be generated by the motion of points, planes by the motion of lines, and solids by the motion of planes. And whereas quantities generated in equal times are greater or lesser according to the greater or lesser velocity wherewith they increase and are generated, a method hath been found to determine quantities from the velocities of their generating motions. And such velocities are called fluxions: and the quantities generated are called flowing quantities. These fluxions are said to be nearly as the increments of the flowing quantities, generated in the least equal particles of time; and to be accurately in the first proportion of the nascent, or in the last of the evanescent increments. Sometimes, instead of velocities, the momentaneous increments or decrements of undetermined flowing quantities are considered, under the appellation of moments.

4 By moments we are not to understand finite particles. These are said not to be moments, but quantities generated from moments, which last are only the nascent principles of finite quantities. It is said that the minutest errors are not to be neglected in mathematics:[1] that the fluxions are celerities, not proportional to the finite increments, though ever so small; but only to the moments or nascent increments, whereof the proportion alone, and not the magnitude, is considered. And of the aforesaid fluxions there be other fluxions, which fluxions of fluxions are called second fluxions. And the fluxions of these second fluxions are called third fluxions: and so on, fourth, fifth, sixth, &c. *ad infinitum.* Now,

[1] We recognize the statement made by Newton in his *Quadratura curvarum* (Selection V.7).

as our sense is strained and puzzled with the perception of objects extremely minute, even so the imagination, which faculty derives from sense, is very much strained and puzzled to frame clear ideas of the least particles of time, or the least increments generated therein: and much more so to comprehend the moments, or those increments of the flowing quantities in *statu nascenti*, in their very first origin or beginning to exist, before they become finite particles. And it seems still more difficult to conceive the abstracted velocities of such nascent imperfect entities. But the velocities of the velocities, the second, third, fourth, and fifth velocities, &c., exceed, if I mistake not, all human understanding. The further the mind analyseth and pursueth these fugitive ideas the more it is lost and bewildered; the objects, at first fleeting and minute, soon vanishing out of sight. Certainly in any sense, a second or third fluxion seems an obscure mystery. The incipient celerity of an incipient celerity, the nascent augment of a nascent augment, *i.e.*, of a thing which hath no magnitude: take it in what light you please, the clear conception of it will, if I mistake not, be found impossible; whether it be so or no I appeal to the trial of every thinking reader. And if a second fluxion be inconceivable, what are we to think of third, fourth, fifth fluxions, and so on without end?

5 The foreign mathematicians are supposed by some, even of our own, to proceed in a manner less accurate, perhaps, and geometrical, yet more intelligible. Instead of flowing quantities and their fluxons, they consider the variable finite quantities as increasing or diminishing by the continual addition or subduction of infinitely small quantities. Instead of the velocities wherewith increments are generated, they consider the increments or decrements themselves, which they call differences, and which are supposed to be infinitely small. The difference of a line is an infinitely little line; of a plane an infinitely little plane. They suppose finite quantities to consist of parts infinitely little, and curves to be polygons, whereof the sides are infinitely little, which by the angles they make one with another determine the curvity of the line. Now to conceive a quantity infinitely small, that is, infinitely less than any sensible or imaginable quantity, or than any the least finite magnitude is, I confess, above my capacity. But to conceive a part of such infinitely small quantity that shall be still infinitely less than it, and consequently though multiplied infinitely shall never equal the minutest finite quantity, is, I suspect, an infinite difficulty to any man whatsoever; and will be allowed such by those who candidly say what they think; provided they really think and reflect, and do not take things upon trust.

6 And yet in the *calculus differentialis*, which method serves to all the same intents and ends with that of fluxions, our modern analysts are not content to consider only the differences of finite quantities: they also consider the differences of those differences, and the differences of the differences of the first differences. And so on *ad infinitum*. That is, they consider quantities infinitely less than the least discernible quantity; and others infinitely less than those infinitely small ones; and still others infinitely less than the preceding infinitesimals, and so on without end or limit. Insomuch that we are to admit an infinite succession of infinitesimals, each infinitely less than the foregoing, and infinitely greater than the following. As there are first, second, third, fourth, fifth, &c. fluxions, so there are differences, first, second, third, fourth, &c., in an infinite

progression towards nothing, which you still approach and never arrive at. And (which is most strange) although you should take a million of millions of these infinitesimals, each whereof is supposed infinitely greater than some other real magnitude, and add them to the least given quantity, it shall never be the bigger. For this is one of the modest *postulata* of our modern mathematicians, and is a corner-stone or ground-work of their speculations.

.   .   .   .

9 Having considered the object, I proceed to consider the principles of this new analysis by momentums, fluxions, or infinitesimals; wherein if it shall appear that your capital points, upon which the rest are supposed to depend, include error and false reasoning; it will then follow that you, who are at a loss to conduct your selves, cannot with any decency set up for guides to other men. The main point in the method of fluxions is to obtain the fluxion or momentum of the rectangle or product of two indeterminate quantities. Inasmuch as from thence are derived rules for obtaining the fluxions of all other products and powers; be the coefficients or the indexes what they will, integers or fractions, rational or surd. Now, this fundamental point one would think should be very clearly made out, considering how much is built upon it, and that its influence extends throughout the whole analysis. But let the reader judge. This is given for demonstration. Suppose the product or rectangle AB increased by continual motion: and that the momentaneous increments of the sides A and B are a and b. When the sides A and B were deficient, or lesser by one half of their moments, the rectangle was $\overline{A - \frac{1}{2}a} \times \overline{B - \frac{1}{2}b}$ i.e., $AB - \frac{1}{2}aB - \frac{1}{2}bA + \frac{1}{4}ab$. And as soon as the sides A and B are increased by the other two halves of their moments, the rectangle becomes $\overline{A + \frac{1}{2}a} \times \overline{B + \frac{1}{2}b}$ or $AB + \frac{1}{2}aB + \frac{1}{2}bA + \frac{1}{4}ab$. From the latter rectangle subduct the former, and the remaining difference will be $aB + bA$. Therefore the increment of the rectangle generated by the intire increments a and b is $aB + bA$. Q.E.D. But it is plain that the direct and true method to obtain the moment or increment of the rectangle AB, is to take the sides as increased by their whole increments, and so multiply them together, $A + a$ by $B + b$, the product whereof $AB + aB + bA + ab$ is the augmented rectangle; whence, if we subduct AB the remainder $aB + bA + ab$ will be the true increment of the rectangle, exceeding that which was obtained by the former illegitimate and indirect method by the quantity ab. And this holds universally be the quantities a and b what they will, big or little, finite or infinitesimal, increments, moments, or velocities. Nor will it avail to say that ab is a quantity exceeding small: since we are told that *in rebus mathematicis errores quam minimi non sunt contemnendi*.[2]

10 Such reasoning as this for demonstration, nothing but the obscurity of the subject could have encouraged or induced the great author of the fluxionary method to put upon his followers, and nothing but an implicit deference to authority could move them to admit. The case indeed is difficult. There can be nothing done till you have got rid of the quantity ab. In order to this the notion of fluxions is shifted: It is placed in various lights: Points which should be clear

---

[2] Newton's statement again, this time in Latin.

as first principles are puzzled; and terms which should be steadily used are ambiguous. But notwithstanding all this address and skill the point of getting rid of *ab* cannot be obtained by legitimate reasoning. If a man, by methods not geometrical or demonstrative, shall have satisfied himself of the usefulness of certain rules; which he afterwards shall propose to his disciples for undoubted truths; which he undertakes to demonstrate in a subtile manner, and by the help of nice and intricate notions; it is not hard to conceive that such his disciples may, to save themselves the trouble of thinking, be inclined to confound the usefulness of a rule with the certainty of a truth, and accept the one for the other; especially if they are men accustomed rather to compute than to think; earnest rather to go on fast and far, than solicitous to set out warily and see their way distinctly.

The subject of the next sections can be summed up in the following argument. If $(x + c)^n - x^n = nx^{n-1}0 + \dfrac{n(n-1)}{1.2}x^{n-2}0^2 + \cdots$, and we divide by 0, we can get nx^{n-1}, the fluxion of x^n, only by first supposing that $0 \neq$ zero, then $0 =$ zero. "All which seems a most inconsistent way of arguing, and such as would not be allowed of in Divinity" (Sec. 14). Then follows, somewhat later:

---

35 I know not whether it be worth while to observe, that possibly some men may hope to operate by symbols and suppositions, in such sort as to avoid the use of fluxions, momentums, and infinitesimals, after the following manner. Suppose x to be an absciss of a curve, and z another absciss of the same curve. Suppose also that the respective areas are xxx and zzz: and that $z - x$ is the increment of the absciss, and $zzz - xxx$ the increment of the area, without considering how great or how small these increments may be. Divide now $zzz - xxx$ by $z - x$, and the quotient will be $zz + zx + xx$: and, supposing that z and x are equal, this same quotient will be $3xx$, which in that case is the ordinate, which therefore may be thus obtained independently of fluxions and infinitesimals. But herein is a direct fallacy: for, in the first place, it is supposed that the abscisses z and x are unequal, without which supposition no one step could have been made; and in the second place, it is supposed they are equal; which is a manifest inconsistency, and amounts to the same thing that hath been before considered. And there is indeed reason to apprehend that all attempts for setting the abstruse and fine geometry on a right foundation, and avoiding the doctrine of velocities, momentums, &c. will be found impracticable, till such time as the object and end of geometry are better understood than hitherto they seem to have been. The great author of the method of fluxions felt this difficulty, and therefore he gave into those nice abstractions and geometrical metaphysics without which he saw nothing could be done on the received principles; and what in the way of demonstration he hath done with them the reader will judge. It must, indeed, be acknowledged that he used fluxions, like the scaffold of a

building, as things to be laid aside or got rid of as soon as finite lines were found proportional to them. But then these finite exponents are found by the help of fluxions. Whatever therefore is got by such exponents and proportions is to be ascribed to fluxions: which must therefore be previously understood. And what are these fluxions? The velocities of evanescent increments? And what are these same evanescent increments? They are neither finite quantities, nor quantities infinitely small, nor yet nothing. May we not call them the ghosts of departed quantities? [3]

---

13 MACLAURIN. ON SERIES AND EXTREMES

Berkeley's criticism stung, and during the eighteenth century many attempts were made to place the calculus on a rigorous foundation. For a report on these attempts, as far as Great Britain is concerned, see F. Cajori's works quoted in the introduction to Selection V.12. One of the most distinguished attempts was made by the Edinburgh professor Colin Maclaurin (1698–1746) in his *Treatise of fluxions* (Edinburgh, 1742). Maclaurin started, like Barrow and Newton, from the concepts of space, time, and motion. But Maclaurin's book also contains other contributions. Best known is his introduction of Taylor's series in a way that has remained familiar in elementary textbooks. He gave the method for deciding between a maximum and a minimum by investigating the sign of a higher derivative. Here follow, in the original text, some of the articles of the *Treatise* that contain these contributions. Maclaurin also considered questions of convergence in series. See H. W. Turnbull, *Bi-centenary of the death of Colin Maclaurin* (University Press, Aberdeen, 1951), also "Colin Maclaurin," *American Mathematical Monthly* 54 (1947), 318–322, and our Selection III.10, note 3.

---

751. The following theorem is likewise of great use in this doctrine. Suppose that y is any quantity that can be expressed by a series of this form $A + Bz + Cz^2 + Dz^3 +$ &c. where A, B, C, &c. represent invariable coefficients as usual, any of which may be supposed to vanish. When z vanishes, let E be the value of y, and let $\dot{E}$, $\ddot{E}$, $\dddot{E}$, &c. be then the respective values of $\dot{y}$, $\ddot{y}$, $\dddot{y}$, &c. z being supposed to flow uniformly. Then $y = E + \dfrac{\dot{E}z}{\dot{z}} + \dfrac{\ddot{E}z^2}{1 \times 2\dot{z}^2} + \dfrac{\dddot{E}z^3}{1 \times 2 \times 3\dot{z}^3}$

$+ \dfrac{\ddddot{E}z^4}{1 \times 2 \times 3 \times 4\dot{z}^4} +$ &c. the law of the continuation of which series is manifest. For since $y = A + Bz + Cz^2 + Dz^3 +$ &c. it follows that when $z = o$,

---

[3] We may think here of the many arguments involved in the Zeno paradoxes, which also played a role in the eighteenth-century discussions concerning the foundations of the calculus; see Cajori, *History of the conceptions of limits and fluxions*, quoted in the introduction to this selection, and his nine articles, "History of Zeno's arguments on motion," *American Mathematical Monthly* 22 (1915).

A is equal to y; but (by the supposition) E is then equal to y; consequently $A = E$. By taking the fluxions, and dividing by $\dot{z}, \dfrac{\dot{y}}{\dot{z}} = B + 2Cz + 3Dz^2 + $ &c. and when $z = o$, B is equal to $\dfrac{\dot{y}}{\dot{z}}$, that is to $\dfrac{\dot{E}}{\dot{z}}$. By taking the fluxions again, and dividing by $\dot{z}$, (which is supposed invariable) $\dfrac{\ddot{y}}{\dot{z}^2} = 2C + 6Dz + $ &c. Let $z = o$, and substituting $\ddot{E}$ for $\ddot{y}, \dfrac{\ddot{E}}{\dot{z}^2} = 2C$, or $C = \dfrac{\ddot{E}}{2\dot{z}^2}$. By taking the fluxions again, and dividing by $\dot{z}, \dfrac{\dddot{y}}{\dot{z}^3} = 6D + $ &c. and by supposing $z = o$, we have $D = \dfrac{\dddot{E}}{6\dot{z}^3}$. Thus it appears that $y = A + Bz + Cz^2 + Dz^3 + $ &c. $= E + \dfrac{\dot{E}z}{\dot{z}} + \dfrac{\ddot{E}z^2}{1 \times 2\dot{z}^2}$

$+ \dfrac{\dddot{E}z^3}{1 \times 2 \times 3\dot{z}^3} + \dfrac{\ddddot{E}z^4}{1 \times 2 \times 3 \times 4\dot{z}^4} + $ &c. This proposition may be likewise deduced from the binomial theorem. Let BD [Fig. 1], the ordinate of the figure

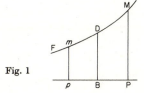

Fig. 1

FDM at B, be equal to E, $BP = z$, $PM = y$, and this series will serve for resolving the value of PM, or y, (some particular cases being excepted, as when any of the coefficients $E, \dfrac{\dot{E}}{\dot{z}}, \dfrac{\ddot{E}}{\dot{z}^2}$, &c. become infinite) into a series, not only in such cases as were described in the preceding articles, but likewise when the relation of y and z is determined by an affected equation, and in many cases when their relation is determined by a fluxional equation. This theorem was given by Dr. Taylor, *method. increm.* By supposing the fluxion of z to be represented by BP, or $\dot{z} = z$, we have $y = E + \dot{E} + \dfrac{\ddot{E}}{2} + \dfrac{\dddot{E}}{6} + \dfrac{\ddddot{E}}{24} + $ &c. (as was observed in Art. 255)[1] and hence it appears at what rate the fluxion of y of each order contributes to produce the increment or decrement of y, since $y - E = \dot{E} + \dfrac{\ddot{E}}{2} + \dfrac{\dddot{E}}{6} + \dfrac{\ddddot{E}}{24}$ + &c. If Bp be taken on the other side of B equal to BP, then $pm = A - Bz + Cz^2 - Dz^3 + $ &c. $= $ (the same quantities being represented by $\dfrac{\dot{E}}{\dot{z}}, \dfrac{\ddot{E}}{\dot{z}^2}$, &c., as

---

[1] Maclaurin's book is divided into two parts. Book I is geometrical, Book II is computational. Our selection is from Book II. Articles 255 and 261 (to which he refers below) deal with the same matter in a geometrical way.

before, or the base being supposed to flow the same way,) $E - \dfrac{\dot{E}z}{\dot{z}} + \dfrac{\ddot{E}z^2}{1 \times 2\dot{z}^2}$

$- \dfrac{\dddot{E}z^3}{1 \times 2 \times 3\dot{z}^3} + \dfrac{\ddddot{E}z^4}{1 \times 2 \times 3 \times 4\dot{z}^4} -$ &c. consequently $PM + pm = 2E +$

$\dfrac{2\ddot{E}z^2}{1 \times 2\dot{z}^2} + \dfrac{2\ddddot{E}z^4}{1 \times 2 \times 3 \times 4\dot{z}^4} +$ &c. . . .

Then, in Arts. 858–861, Maclaurin gives his criterion for maxima and minima.

---

858. When the first fluxion of the ordinate vanishes, if at the same time its second fluxion is positive, the ordinate is then a *minimum*, but is a *maximum* if its second fluxion is then negative; that is, it is less in the former, and greater in the latter case than the ordinates from the adjoining parts of that branch of the curve on either side. This follows from what was shewn at great length in *Chap. 9. B.* I, or may appear thus. Let the ordinate $AF = E$, $AP = x$ [Fig. 2],

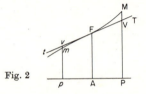

Fig. 2

and the base being supposed to flow uniformly, the ordinate $PM =$ (Art. 751)

$E + \dfrac{\dot{E}x}{\dot{x}} + \dfrac{\ddot{E}x^2}{2\dot{x}^2} + \dfrac{\dddot{E}x^3}{6\dot{x}^3} +$ &c. Let Ap be taken on the other side of A equal to

AP, then the ordinate $pm = E - \dfrac{\dot{E}x}{\dot{x}} + \dfrac{\ddot{E}x^2}{2\dot{x}^2} - \dfrac{\dddot{E}x^3}{6\dot{x}^3} +$ &c. Suppose now $\dot{E} = o$,

then $PM = E * + \dfrac{\dot{E}x}{\dot{x}} + \dfrac{\ddot{E}x^2}{2\dot{x}^2}$ &c. and $pm = E * + \dfrac{\ddot{E}x^2}{2\dot{x}^2} -$ &c. Therefore if the

distances AP and Ap be small enough, PM and pm will both exceed the ordinase AF when $\ddot{E}$ is positive; but will be both less than AF if $\ddot{E}$ be negative. But if $\ddot{E}$ vanish as well as $\dot{E}$, and $\dddot{E}$ does not vanish, one of the adjoining ordinates PM or pm shall be greater than AF, and the other less than it; so that in this case the ordinate is neither a *maximum* nor *minimum*. We always suppose the expression of the ordinate to be positive.

859. In general, if the first fluxion of the ordinate, with its fluxions of several subsequent orders, vanish, the ordinate is a *minimum* or *maximum*, when the number of all those fluxions that vanish is 1, 3, 5, or any odd number. The ordinate is a *minimum*, when the fluxion next to those that vanish is positive; but a *maximum* when this fluxion is negative. This appears from Art. 261, or by comparing the values of PM and pm in the last article. But if the number of all the fluxions of the ordinate of the first and subsequent successive orders that

vanish be an even number, the ordinate is then neither a *maximum* nor *minimum*.

860. When the fluxion of the ordinate y is supposed equal to nothing, and an equation is thence derived for determining x, if the roots of this equation are all unequal, each gives a value of x that may correspond to a greatest or least ordinate. But if two, or any even number of these roots be equal, the ordinate that corresponds to them is neither a *maximum* nor *minimum*. If an odd number of these roots be equal, there is one *maximum* or *minimum* that corresponds to these roots, and one only. Thus if $\frac{\dot{y}}{\dot{x}} = x^4 + ax^3 + bx^2 + cx + d$, then supposing all the roots of the equation $x^4 + ax^3 + bx^2 + cx + d = 0$ to be real, if the four roots are equal there is no ordinate that is a *maximum* or *minimum*; if two or three of the roots only are equal, there are two ordinates that are *maxima* or *minima*; and if all the roots are unequal there are four such ordinates.

861. To give a few examples of the most simple cases. Let $y = a^2 x - x^3$, then $\dot{y} = a^2\dot{x} - 3x^2\dot{x}$ and $\ddot{y} = -6x\dot{x}^2$. Suppose $\dot{y} = 0$, and $3x^2 = a^2$ or $x = \dfrac{a}{\sqrt{3}}$, in which case $\ddot{y} = \dfrac{-6a\dot{x}^2}{\sqrt{3}}$. Therefore $\ddot{y}$ being negative, y is a *maximum* when $x = \dfrac{a}{\sqrt{3}}$, and its greatest value is $\dfrac{2a^3}{3\sqrt{3}}$. If $y = aa + 2bx - xx$, then $\dot{y} = 2b\dot{x} - 2x\dot{x}$, and $\ddot{y} = -2\dot{x}^2$; consequently y is a *maximum* when $2b - 2x = 0$, or $x = b$. If $y = aa - 2bx + xx$ then $\dot{y} = -2b\dot{x} + 2x\dot{x}$, and $\ddot{y} = 2\dot{x}^2$; consequently y is now a *minimum* when $x = b$, if a be greater than b.

Maclaurin also considers the cases in which $\ddot{y}, \dot{\ddot{y}}, \ddot{\ddot{y}}, \ldots$ vanish.

14 D'ALEMBERT. ON LIMITS

Among the mathematicians who seriously tried to come to an understanding of the foundations of the calculus (the "metaphysics of the calculus") was Jean LeRond D'Alembert (1717–1783), long the *secrétaire perpétuel* of the French Academy and with Denis Diderot the leading spirit of the famous *Encyclopédie ou dictionnaire raisonné des sciences, des arts et des métiers* (28 vols.; Paris, 1751–1772). In this *Encyclopédie* D'Alembert wrote a number of articles,[1] and in the article entitled "Différentiel" (vol. 4, 1754) he came to the expression of the derivative as the limit of a quotient of increments, that is, of what we now write $dy/dx = \lim \Delta y/\Delta x, \Delta x \to 0$ (already, though not in a very clear way, expressed by Newton). This leading idea, however, was not followed up immediately, either by D'Alembert himself or by others. One of the difficulties that prevented acceptance of the limit concept in this case was of the same nature as the Zeno paradoxes: how can a limit be reached if the process of coming to it consists of an infinite number of steps? Only with Cauchy in the early

---

[1] For an account of several of them see G. Loria in the *Actes . . . du 3ᵉ congrès international d'histoire des sciences, tenu au Portugal en 1934* (Lisbon, 1935), 15 pp.

twenties of the nineteenth century was this difficulty overcome, so that we can ascribe the acceptance of the definition (and the way of writing, too)

$$\frac{dy}{dx} = \lim \frac{\Delta y}{\Delta x} = \lim \frac{f(x + h) - f(x)}{h}, \quad \Delta x = h \to 0$$

to the work of Cauchy: *Cours d'analyse* (Paris, 1821).

DIFFERENTIALS

...What concerns us most here is the metaphysics of the *differential* calculus.

This metaphysics, of which so much has been written, is even more important and perhaps more difficult to explain than the rules of this calculus themselves: various mathematicians, among them Rolle,[2] who were unable to accept the assumption concerning infinitely small quantities, have rejected it entirely, and have held that the principle was false and capable of leading to error. Yet in view of the fact that all results obtained by means of ordinary Geometry can be established similarly and much more easily by means of the *differential* calculus, one cannot help concluding that, since this calculus yields reliable, simple, and exact methods, the principles on which it depends must also be simple and certain.

Leibniz was embarrassed by the objections he felt to exist against infinitely small quantities, as they appear in the *differential* calculus; thus he preferred to reduce infinitely small to merely incomparable quantities. This, however, would ruin the geometric exactness of the calculations; is it possible, said Fontenelle,[3] that the authority of the inventor would outweight the invention itself? Others, like Nieuwentijt,[4] admitted only *differentials* of the first order and rejected all others of higher order. This is impossible; indeed, considering an infinitely small chord of first order in a circle, the corresponding abcissa or versed sine is infinitely small[5] of second order; and if the chord is of the second order, the abscissa mentioned will be of the fourth order, etc. This is proved easily by elementary geometry, since the diameter of a circle (taken as a finite quantity) is always to the chord as the chord to the corresponding abscissa.[6] Thus, if one admits the infinitely small of the first order, one must admit all the others, though in the end one can rather easily dispense with all this metaphysics of the infinite in the *differential* calculus, as we shall see below.

[2] Michel Rolle (1652–1719), member of the French Academy, is best known for the theorem in the theory of equations called after him. In 1700 he took part in a debate in the French Academy on the principles of the calculus; see C. Boyer, *The history of the calculus* (Dover, New York, 1949), 241. See also Selection II.13, note 5.

[3] Bernard le Bovier de Fontenelle (1657–1757) was a predecessor of D'Alembert as *secrétaire perpétuel* of the Academy. See Boyer, *History*, 241–242.

[4] Bernard Nieuwentijt (1654–1718), a physician-burgomaster of Purmerend, near Amsterdam, opposed Leibniz's concept of the calculus; see Selection V.1.

[5] Versed sin $\alpha = 1 - \cos \alpha = \alpha^2/2! - \alpha^4/4! + \cdots$ (D'Alembert still takes the dimension to be that of a chord, hence his vers α is really our R vers α).

[6] $2R : 2R \sin \alpha/2 = 2R \sin \alpha/2 : R(1 - \cos \alpha)$.

Newton started out from another principle; and one can say that the metaphysics of this great mathematician on the calculus of fluxions is very exact and illuminating, even though he allowed us only an imperfect glimpse of his thoughts.

He never considered the *differential* calculus as the study of infinitely small quantities, but as the method of first and ultimate ratios, that is to say, the method of finding the limits of ratios. Thus this famous author has never differentiated quantities but only equations; in fact, every equation involves a relation between two variables and the differentiation of equations consists merely in finding the limit of the ratio of the finite differences of the two quantities contained in the equation. Let us illustrate this by an example which will yield the clearest idea as well as the most exact description of the method of the *differential* calculus.

Let AM [Fig. 1] be an ordinary parabola, the equation of which is $yy = ax$; here we assume that $AP = x$ and $PM = y$, and a is a parameter. Let us draw

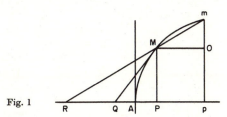

Fig. 1

the tangent MQ to this parabola at the point M. Let us suppose that the problem is solved and let us take an ordinate pm at any finite distance from PM; furthermore, let us draw the line mMR through the points M, m. It is evident, *first*, that the ratio[7] MP/PQ of the ordinate to the subtangent is greater than the ratio MP/PR or mO/MO which is equal to it because of the similarity of the triangles MOm, MPR; *second*, that the closer the point m is to the point M, the closer will be the point R to the point Q, consequently the closer will be the ratio MP/PR or mO/MO to the ratio MP/PQ; finally, that the first of these ratios approaches the second one as closely as we please, since PR may differ as little as we please from PQ. Therefore, the ratio MP/PQ is the limit of the ratio of mO to OM. Thus, if we are able to represent the ratio mO/OM in algebraic form, then we shall have the algebraic expression of the ratio of MP to PQ and consequently the algebraic representation of the ratio of the ordinate to the subtangent, which will enable us to find this subtangent. Let now $MO = u$, $Om = z$; we shall have $ax = yy$, and $ax + au = yy + 2yz + zz$. Then in view of $ax = yy$ it follows that $au = 2yx + zz$ and $z/u = a/(2y + z)$.

This value $a/(2y + z)$ is, therefore, in general the ratio of mO to OM, wherever one may choose the point m. This ratio is always smaller than $a/2y$; but the smaller z is, the greater the ratio will be and, since one may choose z as small as

[7] D'Alembert writes $\dfrac{MP}{PQ}$.

one pleases, the ratio $a/(2y + z)$ can be brought as close to the ratio $a/2y$ as we like. Consequently $a/2y$ is the limit of the ratio $a/(2y + z)$, that is to say, of the ratio mO/OM. Hence $a/2y$ is equal to the ratio MP/PQ, which we have found to be also the limit of the ratio of mO to Om, since two quantities that are the limits of the same quantity are necessarily equal to each other. To prove this, let X and Z be the limits of the same quantity Y. Then I say that $X = Z$; indeed, if they were to have the difference V, let $X = Z \pm V$: by hypothesis the quantity Y may approach X as closely as one may wish; that is to say, the difference between Y and X may be as small as one may wish. But, since Z differs from X by the quantity V, it follows that Y cannot approach Z closer than the quantity V and consequently Z would not be the limit of Y, which is contrary to the hypothesis.

From this it follows that MP/PQ is equal to $a/2y$. Hence $PQ = 2yy/a = 2x$. Now, according to the method of the *differential* calculus, the ratio of MP to PQ is equal to that of dy to dx; and the equation $ax = yy$ yields $a\,dx = 2y\,dy$ and $dy/dx = a/2y$. So dy/dx is the limit of the ratio of z to u, and this limit is found by making $z = 0$ in the fraction $a/(2y + z)$.

But, one may say, is it not necessary also to make $z = 0$ and $u = 0$ in the fraction $z/u = a/(2y + z)$, which would yield $\frac{0}{0} = a/2y$? What does this mean? My answer is as follows. First, there is no absurdity involved; indeed $\frac{0}{0}$ may be equal to any quantity one may wish: thus it may be $= a/2y$. Secondly, although the limit of the ratio of z to u has been found when $z = 0$ and $u = 0$, this limit is in fact not the ratio of $z = 0$ to $u = 0$, because the latter one is not clearly defined; one does not know what is the ratio of two quantities that are both zero. This limit is the quantity to which the ratio z/u approaches more and more closely if we suppose z and u to be real and decreasing. Nothing is clearer than this; one may apply this idea to an infinity of other cases.[8]

Following the method of differentiation (which opens the treatise on the quadrature of curves by the great mathematician Newton[9]), instead of the equation $ax + au = yy + 2yz + zz$ we might write $ax + a0 = yy + 2y0 + 00$, thus, so to speak, considering z and u equal to zero; this would have yielded $\frac{0}{0} = a/2y$. What we have said above indicates both the advantage and the inconveniences of this notation: the advantage is that z, being equal to 0, disappears without any other assumption from the ratio $a/(2y + 0)$; the inconvenience is that the two terms of the ratio are supposed to be equal to zero, which at first glance does not present a very clear idea.

From all that has been said we see that the method of the *differential* calculus offers us exactly the same ratio that has been given by the preceding calculation. It will be the same with other more complicated examples. This should be sufficient to give beginners an understanding of the true metaphysics of the *differential* calculus. Once this is well understood, one will feel that the assumption made concerning infinitely small quantities serves only to abbreviate and simplify the reasoning; but that the *differential* calculus does not necessarily

[8] Here D'Alembert refers to his articles on "Limit" and "Exhaustion" in the same *Encyclopédie*.

[9] See our Selection V.7.

suppose the existence of those quantities; and that moreover this calculus merely consists in *algebraically determining the limit of a ratio, for which we already have the expression in terms of lines, and in equating those two expressions. This will provide us with one of the lines we are looking for.* This is perhaps the most precise and neatest possible definition of the *differential* calculus; but it can be understood only when one is well acquainted with this calculus, because often the true nature of a science can be understood only by those who have studied this science.

In the preceding example the known geometric limit of the ratio of z to u is the ratio of the ordinate to the subtangent; in the *differential* calculus we look for the algebraic limit of the ratio z to u and we find $a/2y$. Then, calling s the subtangent, one has $y/s = a/2y$; hence $s = 2yy/a = 2x$. This example is sufficient to understand the others. It will, therefore, be sufficient to make oneself familiar with the previous example concerning the tangents of the parabola, and, since the whole *differential* calculus can be reduced to the problem of the tangents, it follows that one could always apply the preceding principles to various problems of this calculus, for instance to find *maxima and minima*, points of inflection, cusps, etc. . . . [10]

What does it mean, in fact, to find a maximum or a minimum ? It consists, it is said, in setting the difference[11] dy equal to zero or to infinity; but it is more precise to say that it means to look for the quantity dy/dx which expresses the limit of the ratio of finite dy to finite dx, and to make this quantity zero or infinite. In this way all the mystery is explained; it is not dy that one makes = to infinity: that would be absurd, since dy is taken as infinitely small and hence cannot be infinite; it is dy/dx: that is to say, one looks for the value of x that renders the limit of the ratio of finite dy to finite dx infinite.

We have seen above that in the *differential* calculus there are really no infinitely small quantities of the first order; that actually those quantities called u are supposed to be divided by other supposedly infinitely small quantities; in this state they do not denote either infinitely small quantities or quotients of infinitely small quantities; they are the limits of the ratio of two finite quantities. The same holds for the second-order differences and for those of higher order. There is actually no quantity in Geometry such as $d\,dy$; whenever $d\,dy$ occurs in an equation it is supposed to be divided by a quantity dx^2, or another of the same order. What now is $d\,dy/dx^2$? It is the limit of the ratio $d\,dy/dx$ divided by dx; or, what is still clearer, it is the limit of dz/dx, where $dy/dx = z$ is a finite quantity.

---

15   EULER. TRIGONOMETRY

Euler's *Introductio in analysin infinitorum* (2 vols., Lausanne, 1748; *Opera omnia*, ser. I, vols. 8, 9), written in 1745, is one of his great textbooks, from which whole generations have learned their analysis, especially their knowledge of infinite series and functions defined with

[10] Here D'Alembert refers to his articles on these subjects.
[11] D'Alembert makes little distinction between *différence* and *différentiel*.

their aid. The first volume begins with numerical and power series, recurrent series and continued fractions. The Taylor series comes into its own. Logarithmic and exponential functions, both defined by means of power series, are seen at last in their inverse character: if $y = e^x$, $x = \log_e y$. Both series are obtained by limiting operations from the binomial series. Then follows trigonometry, which is the subject of our selection. The book also contains much number theory; and the second volume contains an analytic geometry of the plane and of space, with discussion of algebraic and transcendental curves as well as quadric surfaces. The method of exposition, however, is still quite different from that presented in our present texts as analytic geometry.

The *Introductio* created order in the still somewhat uncertain field of mathematical notation; Euler's notation (with few exceptions) is our notation. The symbols sin, cos, e, π, although occasionally used before, from now on received general acceptance. The book, however, does not deal with differential and integral calculus, to which subjects Euler later devoted his *Institutiones calculi differentialis* (Saint Petersburg, 1755) and *Institutiones calculi integralis* (3 vols.; Saint Petersburg, 1768–1770); *Opera omnia*, ser. I, vols. 10–13.

The spirit of the book may somewhat be gathered from chapter 8 of the first volume (*Opera omnia*, ser. I, vol. 8), in which Euler introduces the trigonometric functions, now for the first time regarded systematically as ratios, hence as dimensionless quantities. In the two preceding chapters he had introduced exponentials and logarithms, with examples on population increase and investment, and with such series as those for log $(1 + x)$ and a^x. The symbol e, already used by Euler in 1727,[1] is used for the sum of the series

$$1 + \frac{1}{1} + \frac{1}{1.2} + \frac{1}{1.2.3} + \text{etc.}$$

This sum is given to 23 decimals. With the series

$$e^z = 1 + \frac{z}{1} + \frac{z^2}{1.2} + \frac{z^3}{1.2.3} + \text{etc.,}$$

in which z may be real, imaginary, or complex, chapter 7 comes to an end. Here follows chapter 8.

ON TRANSCENDENTAL QUANTITIES WHICH CAN BE OBTAINED FROM THE CIRCLE

126. After logarithms and exponential quantities we shall investigate circular arcs and their sines and cosines, not only because they constitute another type of transcendental quantities, but also because they can be obtained from these very logarithms and exponentials when imaginary quantities are involved.

---

[1] Euler, *Meditatio in experimenta explosione tormentorum nuper instituta*, written in 1727 but first published in 1864. The symbol e appeared in print for the first time in Euler's textbook, *Mechanica, sive motus scientia analytice exposita* (Saint Petersburg, 1736), *Opera omnia*, ser. II, vol. 1.

Let us therefore take the radius of the circle, or its sinus totus, $= 1$. Then it is obvious that the circumference of this circle cannot be exactly expressed in rational numbers; but it has been found that the semicircumference is by approximation

$$= 3.14159.26535.89793\ldots$$

[127 decimal places are given[2]] for which number I would write for short

$$\pi,$$

so that π is the semicircumference of the circle of which the radius $= 1$, or π is the length of the arc of 180 degrees.[3]

127. If we denote by z an arbitrary arc of this circle, of which I always assume the radius $= 1$, then we usually consider of this arc mainly the sine [*sinus*] and cosine [*cosinus*]. I shall denote the sine of the arc z in the future in this way

$$\text{sin.}\ A.z, \quad \text{or only} \quad \text{sin.}\ z$$

and the cosine accordingly

$$\text{cos.}\ A.z, \quad \text{or only} \quad \text{cos.}\ z.$$

Hence we shall have, since π is the arc of $180°$,

$$\text{sin.}\ 0 = 0, \quad \text{cos.}\ 0 = 1$$

and

$$\text{sin.}\ \tfrac{1}{2}\pi = 1, \quad \text{cos.}\ \tfrac{1}{2}\pi = 0 \cdots$$

Now follows a whole set of trigonometric formulas including the definitions $\text{tang.}\,z = \dfrac{\text{sin.}\ z}{\text{cos.}\ z}$, $\text{cot.}\ z = \dfrac{\text{cos.}\ z}{\text{sin.}\ z}$, the addition formulas, and identities such as

$$\text{tang.}\,\frac{a+b}{2} = \frac{\text{sin.}\ a + \text{sin.}\ b}{\text{cos.}\ a + \text{cos.}\ b}.$$

[2] Euler took this value from T. G. de Lagny, "Mémoire sur la quadrature du cercle," *Histoire de l'Académie Royale, Paris, 1719* (1727), 1° partie, 176–189, who computed π to 127 decimal places by means of a series for $\tan^{-1} 30°$.

[3] The symbol π was never used in Antiquity; it seems first to have been used by William Jones (the editor of Newton's *Analysis per aequationes*, London, 1711) in his *Synopsis palmariorum matheseos* (London, 1706), p. 243. See D. E. Smith, *History of mathematics* (Ginn, New York, 1925), II, 312. Euler used π in his *Mechanica* (1736); see note 1. See E. W. Hobson, *Squaring the circle* (Cambridge University Press, Cambridge, England, 1913). Euler, using the term *sinus totus* for the radius of the circle, adheres for the last time to the old terminology, in which the sine is a segment. See Selection IV.13, note 9.

Hereafter we omit the period after sin and cos and write i for $\sqrt{-1}$, as Euler also did in later work.[4]

---

132. Since

$$(\sin z)^2 + (\cos z)^2 = 1,$$

we shall have by factorization

$$(\cos z + i \sin z)(\cos z - i \sin z) = 1,$$

which factors, although imaginary [*etsi imaginarii*], still are of great use in combining and multiplying sines and cosines.

Now comes De Moivre's theorem[5] (though the name is not mentioned), from which follows, in §133:

$$\cos nz = \frac{(\cos z + i \sin z)^n + (\cos z - i \sin z)^n}{2}$$

and

$$\sin nz = \frac{(\cos z + i \sin z)^n - (\cos z - i \sin z)^n}{2i}.$$

When we develop these binomials in a series we shall get

$$\cos nz = (\cos z)^n - \frac{n(n-1)}{1.2}(\cos z)^{n-2}(\sin z)^2 + \text{etc.}$$

and

$$\sin nz = \frac{n}{1}(\cos z)^{n-1}\sin z - \frac{n(n-1)(n-2)}{1.2.3}(\cos z)^{n-3}(\sin)^2 + \text{etc.}$$

134. Let the arc z be infinitely small; then we get $\sin z = z$ and $\cos z = 1$; let now n be an infinitely large number, while the arc nz is of finite magnitude.

[4] "In the following I shall denote the expression $\sqrt{-1}$ by the letter i so that $ii = -1$": Euler, *De formulis differentialibus angularibus*, presented to the Saint Petersburg Academy, 1777; published in the posthumous vol. IV of the *Institutiones calculi integralis* (1794), 183–194; *Opera omnia*, ser. I, vol. 19, 129–140, p. 130.

[5] This theorem, now usually written $(\cos \varphi + i \sin \varphi)^n = \cos n\varphi + i \sin n\varphi$, appears at the opening of A. de Moivre, *Miscellanea analytica* (London, 1730), but in a different, more geometrical, form.

Take $nz = v$; then since $\sin z = z = v/n$ we shall have

$$\cos v = 1 - \frac{v^2}{1.2.3} + \frac{v^4}{1.2.3.4} - \cdots + \text{etc.}$$

and

$$\sin v = v - \frac{v^3}{1.2.3} + \frac{v^5}{1.2.3.4.5} - \cdots + \text{etc.}$$

Then, by writing $v = \frac{m}{n} \cdot \frac{\pi}{2}$ Euler obtains a series for $\sin \frac{m}{n} 90°$ with terms up to $\frac{m^{29}}{n^{29}}$, and a series for $\cos \frac{m}{n} 90°$ with terms up to $\frac{m^{30}}{n^{30}}$, the coefficients given to 28 decimals; these are followed by series for the tangent and the cotangent. He shows that it is only necessary to know the numerical values of these quantities for the values from $0°$ to $30°$ to be able to find them all by identities such as $\sin (30 + z) = \cos z - \sin (30 - z)$. Here cosec. z and sec. z are introduced.

---

138. Let us now take in the formulas of §133 the arc z infinitely small and let n be an infinitely small number ϵ [Euler writes i] such that ϵz will take the finite value v. We thus have $\epsilon z = v$ and $z = v/\epsilon$, hence $\sin z = v/\epsilon$ and $\cos z = 1$. After substituting these values we find

$$\cos v = \frac{\left(1 + \frac{vi}{\epsilon}\right)^\epsilon + \left(1 - \frac{vi}{\epsilon}\right)^\epsilon}{2},$$

$$\sin v = \frac{\left(1 + \frac{vi}{\epsilon}\right)^\epsilon - \left(1 - \frac{vi}{\epsilon}\right)^\epsilon}{2i}.$$

In the previous chapter we have seen that

$$\left(1 + \frac{z}{\epsilon}\right)^\epsilon = e^z,$$

where by e we denote the base of the hyperbolic logarithms; if we therefore write for z first iv, then $-iv$, we shall have

$$\cos v = \frac{e^{iv} - e^{-iv}}{2}$$

and

$$\sin v = \frac{e^{iv} - e^{-iv}}{2i}.$$

From these formulas we can see how the imaginary exponential quantities can be reduced to the sine and cosine of real arcs. Indeed, we have

$$e^{iv} = \cos v + i \sin v,$$

$$e^{-iv} = \cos v - i \sin v.$$

Then follow in §139 some formulas for the logarithms leading up to

$$z = \frac{1}{2i} l \frac{\cos z + i \sin z}{\cos z - i \sin z},$$

where l indicates logarithm.

140. Since $\dfrac{\sin z}{\cos z} = \tan g\, z$, the arc z can be expressed by its tangent in such a way that we have

$$z = \frac{1}{2i} l \frac{1 + i \tan g\, z}{1 - i \tan g\, z}.$$

Now we have seen above (§123) that

$$l \frac{1 + x}{1 - x} = \frac{2x}{1} + \frac{2x^3}{3} + \frac{2x^5}{5} + \frac{2x^7}{7} + \text{etc.}$$

We now put $x = i \tan g\, z$ and shall obtain

$$z = \frac{\tan g\, z}{1} - \frac{(\tan g\, z)^3}{1} + \frac{(\tan g\, z)^5}{5} + \frac{(\tan g\, z)^7}{7} + \text{etc.}$$

If we therefore put $\tan g\, z = t$, so that z is the arc of which the tangent is t, which we shall indicate by A. tang. t [our $\tan^{-1} t$], we shall have

$$z = A. \text{tang}.\, t.$$

Therefore, for known t, the corresponding arc will be

$$z = \frac{t}{1} - \frac{t^3}{3} + \frac{t^5}{5} - \frac{t^7}{7} + \frac{t^9}{9} - \text{etc.}$$

Therefore, if the tangent t is equal to the radius 1, the arc $z = 45°$ or $z = \pi/4$, and we shall have

$$\frac{\pi}{4} = 1 - \frac{1}{3} + \frac{1}{5} - \frac{1}{7} + \text{etc.},$$

which is the series first found by Leibniz to express the value of the circumference of the circle.[6]

The chapter ends with some other series for π that converge more rapidly.

16   D'ALEMBERT, EULER, DANIEL BERNOULLI. THE VIBRATING STRING AND ITS PARTIAL DIFFERENTIAL EQUATION

Among the problems that Brook Taylor discusses in the second part of his *Methodus incrementorum* (London, 1715; see Selection V.11) there are two (Nos. 17 and 18) that deal with the vibrating string. They had already been discussed in Taylor's paper "De motu nervi tensi" (On the motion of a tense string, or taut sinew), *Philosophical Transactions 28* (1713), 26–32 (published in 1714; translated in the abridged edition, vol. 6, pp. 14–17). The first problem is "To determine the motion of a tense string"; the second is "Given the length and weight of the string, as well as the stretching weight, to find the time of vibration." Taylor concludes that at any point of the arc the normal acceleration is proportional to the curvature. This means that for small vibrations Taylor has in principle the equation which we write

$$\sigma \frac{\partial^2 y}{\partial t^2} = T \frac{\partial^2 y}{\partial x^2},$$

σ being the mass per unit length and T the tension of the string, but there is no evidence that he had any notion of partial derivatives. But he did find that the motion of an arbitrary point is that of a simple pendulum, and thus found its time of vibration. He took the form of the curve to be sinusoidal.

In 1727 Johann Bernoulli suggested to his son Daniel that he take up Taylor's problem again: "Of a musical string, of given length and weight, stretched by a given weight, to find its vibrations" (*Opera omnia*, III, p. 125). But Johann could not wait and published his own "Meditations on vibrating strings with little weights at equal distances" in the *Commentarii Academiae Scientiarum Petropolitanae 3* (1728, *Opera omnia*, III, 198–210), where he set up difference equations by studying the forces working on each of n little weights along a string, and then, passing to the limit of a continuous string, concluded that

---

[6] Leibniz published this series in "De vera proportione circuli," *Acta eruditorium 1* (1682), 41 (*Mathematische Schriften*, Abth 2, vol. 1, p. 118). He had already mentioned it in letters of 1673. Before him the series appears in the writings of James Gregory (letter to Collins, 1671; see Selection V.4), and before this in a Sanskrit text of *c.* 1500 by Nīlakantha; see C. T. Rajagopal and T. V. Vedamurthi Aiyar, *Scripta Mathematica 17*, 1951, 65–74; J. E. Hofmann, *Math. phys. Semesterberichte 3* (1953), 194–206.

the form of the string must be the *trochoides socia*, the "companion of the cycloid," a name used since Roberval for the sine curve. He obtained Taylor's time of vibration.

With D'Alembert's paper, "Recherches sur la courbe que forme une corde tendue mise en vibration," *Histoire de l'Académie Royale, Berlin, 3, 1747* (1749), 214–219, followed by "Suite des recherches," *ibid.*, 220–249, begins the search for the general shape of the vibrating string, a search based on the solution of the partial differential equation indicated (in his special notation) by Taylor. This paper opened a series of contributions on the subject by D'Alembert, Euler, and Daniel Bernoulli, in which the authors came to different conclusions on the nature of an "arbitrary" function and its expansion in trigonometric functions, a controversy brought to a conclusion only in the nineteenth century by Fourier, Cauchy, Dirichlet, and Riemann.

The papers, of which excerpts follow, are: (1) D'Alembert (1747), already mentioned; (2) Euler, "Sur la vibration des cordes," *Histoire de l'Académie Royale, Berlin, 4, 1748* (1750), 69–85 (French), *Opera omnia*, ser. II, vol. 10, 63–77; (3) D'Alembert, "Addition au mémoire sur la courbe que forme une corde tendue, mise en vibration," *Histoire de l'Académie Royale, Berlin, 6, 1750* (1752), 355–360; (4) D. Bernoulli, "Réflexions et éclaircissemens sur les nouvelles vibrations des cordes exposées dans les mémoires de l'Académie de 1747 et 1748," *ibid.*, *9, 1753* (1755), 147–172; (5) Euler, "Remarques sur les mémoires precédents de M. Bernoulli," *ibid.*, 196–222, *Opera omnia*, ser. II, vol. 10, 232–254.

For further information we refer to the exemplary exposition by C. Truesdell in the introduction to Euler's *Opera omnia*, ser. II, vol. 11, part 2: "The rational mechanics of flexible or elastic bodies, 1638–1788" (1960), 435 pp. An older exposition is to be found in H. Burkhart, "Entwicklungen nach oscillierenden Funktionen und Integration der Differentialgleichungen der mathematischen Physik," *Jahresberichte der deutschen mathematischen Vereinigung 10* (1908), xii+894 pp.

---

(1) D'ALEMBERT (1747)

I. I propose to show in this paper that there exist an infinity of curves different from the elongated cycloid[1] which satisfy the problem under consideration. I shall always suppose that

1° The excursions or vibrations of the string are very small so that the arcs AM [Fig. 1] of the curve that is formed can always be taken as reasonably equal to the corresponding abscissa AP;

Fig. 1

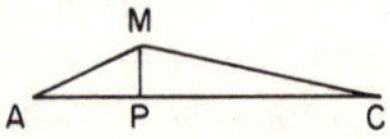

2° The string is of uniform thickness in its whole length;

3° That the force F of the tension is to the weight of the string in constant ratio, that is, as $m:l$, from which it follows that when p is gravity, and l the length of the curve, we can suppose $F = pml$;

4° That if AP or AM is called s and PM y, and if ds is taken constant, then the accelerating force at the point M along MP is $-F(d\,dy/ds^2)$ if the curve is

---

[1] This is the "compagne de la cycloide" of Roberval, hence the sine curve.

concave toward AC, or $F(d\,dy/ds^2)$ if it is convex. See Taylor, *Meth. incremen-torum*.[2]

II. This being so, let us imagine that Mm, mn [Fig. 2] are two consecutive sides of the curve at an arbitrary instant, and that $Pp = p\pi$, that is, let ds be constant. Let t be the time elapsed from the moment when the string started to vibrate: it is certain that the ordinate PM can only be expressed by a function of the time t and of the abscissa or the corresponding arc s or AP.[3] Let, there-fore, $PM = \varphi(t, s)$, that is, let it be equal to an unknown function of t and s. We shall write $d[\varphi(t, s)] = p\,dt + q\,ds$, p and q being equally functions of t and s. Now it is evident from the theorem of Mr. Euler (*Mém. de Pétersbourg* 7, p. 177) that the coefficient of ds in the differential of p must be equal to the coefficient of dt in the differential of q.[4] If, therefore, $dp = \alpha\,dt + \nu\,ds$, then $dq = \nu\,dt + \beta\,ds$, α, ν, β being again unknown functions of t and s.

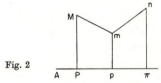

Fig. 2

III. It follows from this that, since the sides Mm, mn belong to the same curve, $pm - PM$ will be equal to the differential of $\varphi(t, s)$ when only s varies, hence that $pm - PM = q\,ds = ds.q$ and that the quantity which above we have indicated by $d\,dy$, that is, the second differential of PM taken varying only s, will be $ds.\beta\,ds$. Hence $F(d\,dy/ds^2) = F\beta$.

[2] This formula is now usually written as force $= T(\partial^2 y/\partial x^2)$.

[3] D'Alembert uses the term "function" in the way we are accustomed to use it. By 1747 this was not unusual. See our "Note on the emergence of the concept of function," below.

[4] This is the theorem that, when $F = F(x, y)$, $\partial^2 F/\partial x\ \partial y = \partial^2 F/\partial y\ \partial x$. Euler published it in "De infinitis curvis eiusdem generis," *Commentarii Academiae Scientiarum Petro-politanae* 7, *1734–35* (1740), 184–200 (*Opera omnia*, ser. I, vol. 22, 57–75). He demonstrated it as follows:

"Let A be a function [*functio*] of these t and u, let A pass into B when instead of t we put $t + dt$, and let A pass into C if $u + du$ is substituted for u. And let A pass into D when at the same time $t + dt$ is substituted for t and $u + du$ for u. From this it is clear that if in B we write $u + du$ instead of u we get D, and in the same way we also get D when in C we write $t + dt$ instead of t. This being so, let A be differentiated with t constant, and we obtain $C - A$, since when A passes into C when $u + du$ is substituted for u, its differential is $C - A$. If now in $C - A$ we substitute $t + dt$ for t we obtain $D - B$, so that the differential will be

$$D - B - C + A.$$

"Now let us change the order, and by substituting $t + dt$ for t change A into B. Then the differential of A with only t as variable will be $B - A$. If now we substitute $u + du$ for u this differential passes into $D - C$, so that its differential will be

$$D - B - C - A,$$

which agrees with the differential obtained by the previous operation." Hence, in our notation, $\partial^2 A/\partial u\ \partial t = \partial^2 A/\partial t\ \partial u$. Euler repeats the theorem and proof in his *Institutiones calculi differentialis* (Saint Petersburg, 1755), *Opera omnia*, ser. I, vol. 10, 153–157.

IV. Let us now imagine that the points M, m, n [Fig. 3] move to M', m', n'. Then it is certain that the excess of PM' over PM will be equal to the differential of $\varphi(t, s)$ taken when only t varies, that is, that $PM' - PM = p\,dt = dt.p$, and that the second differential of PM taken when only t varies, that is, the differential of MM' (or, what is the same, the space traversed by the point M under the accelerating force that animates it), is $\alpha\,dt$.

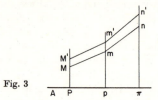

Fig. 3

V. This being so, let a be the space that a heavy body animated by gravity p would traverse in a given constant time θ; then it is evident that (Newton, *Princ. Math.*[5])

$$\alpha\,dt^2 : 2a = F\beta\,dt^2 : p\theta^2,$$

hence

$$\alpha = \frac{2aF\beta}{p\theta^2} = \frac{2apml\beta}{p\theta^2} = \beta\,\frac{2aml}{\theta^2}.\ ^6$$

VI. We shall first notice that we can represent the given time θ by a constant line of such magnitude as we like: we must only take care that we take (in order to express the parts of time that are variable and indetermined) lines t that are to the line that we take to represent θ in the ratio of these variable parts of the time to the given constant time (during which a heavy body traverses the space a). We can therefore suppose θ to be such that $\theta^2 = 2aml$, so that in this case $\alpha = \beta$. Hence, since $dp = \alpha\,dt + \nu\,ds$, dq or $\nu\,dt + \beta\,ds$ must be $= \nu\,dt + \alpha\,ds$.[7]

VII. In order to determine, by means of these conditions, the quantities α and ν, we notice that, since $dp = \alpha\,dt + \nu\,ds$ and $dq = \nu\,dt + \alpha\,ds$, we have $dp + dq = (\alpha + \nu) \cdot (dt + ds)$ and $dp - dq = (\alpha - \nu) \cdot (dt - ds)$.

---

[5] This is the place in the *Principia*, I, Sec. X, Prop. LII, where Newton, following Huygens, discusses the vibrations of the mathematical pendulum, including the cycloidal motion. Taylor had referred to it in his discussion of the vibrating string.

[6] In our notation:

$$\frac{\partial^2 \varphi}{\partial t^2}\,dt^2 : 2a = T \cdot \frac{\partial^2 \varphi}{\partial x^2} : \theta^2 g,$$

or

$$\frac{\partial^2 \varphi}{\partial t^2} = \frac{2aml}{\theta^2}\,\frac{\partial^2 \varphi}{\partial x^2}.$$

[7] By choice of the unit of time we can take $\partial^2 \varphi / \partial x^2 = \partial^2 \varphi / \partial t^2$.

From this it follows:

1° That $\alpha + \nu$ is a function of $t + s$, and that $\alpha - \nu$ is equal to a function of $t - s$;

2° That therefore we shall have $p = \dfrac{\varphi(t + s) + \Delta(t - s)}{2}$ or simply $p = \varphi(t + s) + \Delta(t - s)$, and $q = \varphi(t + s) - \Delta(t - s)$. From this it follows that (since $d\varphi = p \, dt + q \, ds$) $PM = \Psi(t + s) + \Gamma(t - s)$, $\Psi(t + s)$ and $\Gamma(t - s)$ expressing as yet unknown functions of $t + s$ and $t - s$.[8]

The general equation of the curve is therefore

$$y = \Psi(t + s) + \Gamma(t - s).$$

VIII. It is easy to see that this equation includes [*renferme*] an infinity of curves. To show it, let us take here only a special case, namely, that where $y = 0$ when $t = 0$.

We shorten the further argument: since the curve has to pass through A and B ($s = 0$, $y = 0$; $s = l$, $y = 0$), D'Alembert gets $\Gamma(-s) = -\Psi(s)$,[9] $\Gamma(t - s) = -\Psi(t - s)$, hence $y = \Psi(t + s) - \Psi(t - s)$. Therefore, since $\Psi(-s) = \Gamma(s) = -\Psi(s)$, we find that $\Psi(s)$ is an even function of s. But $\Psi(t + l) - \Psi(t - l) = 0$, hence $\Psi(t + s)$ must be found in such a way that $\Psi(s) - \Psi(-s) = 0$ and $\Psi(t + l) - \Psi(t - l) = 0$. To obtain it, the curve tOT [Fig. 4] is constructed with $TR = u = \Psi(z)$, $z = QR$, and it is found that this curve is periodic,

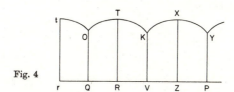

Fig. 4

[8] Since

$$d(\varphi_t + \varphi_s) = \varphi_{tt} \, dt + \varphi_{st}(dt + ds) + \varphi_{ss} \, ds,$$

and

$$\varphi_{ss} = \varphi_{tt},$$

we have

$$d(\varphi_t + \varphi_s) = (\varphi_{tt} + \varphi_{st})(dt + ds).$$

Also

$$d(\varphi_t - \varphi_s) = (\varphi_{tt} - \varphi_{st})(dt - ds).$$

Hence $\varphi_{tt} + \varphi_{st}$ is a function of $t + s$ and $\varphi_{tt} - \varphi_{st}$ is a function of $t - s$; hence $\varphi_t + \varphi_s$ is a function of $t + s$, $\varphi_t - \varphi_s$ of $t - s$. Result:

$$\varphi_t = f(t + s) + g(t - s), \quad \varphi_s = f(t + s) - g(t - s).$$

[9] D'Alembert here has no parentheses; he writes $\Gamma - s = -\Psi s$, and so forth, but he does write $\Gamma(t - s)$ and so forth.

repeating endlessly the part OTK with $QV = 2l$. D'Alembert shows geometrically how such a curve can be constructed and compares it with the cycloid. Once OTK is given, then for any t the curve y formed by the string can be found. The final paragraph points out that the initial velocity, expressed by $\Delta(s) - \Delta(-s)$, given to each point of the string must be an odd function, since otherwise the problem would be impossible.

In the second article, D'Alembert calls the curve OTK (Fig. 4) the generating curve [*courbe génératrice*], giving different types and their time of vibration. Then returning to Taylor's result, he concludes that if $-(d\,dy/ds^2):(1/R) = y:A$, where R is the radius of curvature of the curve at the end of its largest ordinate A, then

$$ds = \frac{A\,dy}{\sqrt{AA - yy}}\,\frac{\sqrt{R}}{\sqrt{A}}$$

(this implies that, for $y = A$, $dy/ds = 0$). When l is the length of the string, $R = ll/n\pi A$, and when y' is the largest ordinate at time t, then ($\theta = 2aml$):

$$\frac{dy'}{\sqrt{A^2 - y'^2}} = \frac{\pi\,dt}{\theta}\,\frac{\sqrt{2am}}{\sqrt{l}},$$

hence

$$\frac{y'}{A'} = \frac{e^p - e^{-p}}{2\sqrt{-1}}, \quad p = \frac{\pi t\sqrt{2am}\sqrt{-1}}{\theta\sqrt{l}}.$$

From this D'Alembert obtains

$$\Psi(t + s) = A\,\frac{e^q + e^{-q}}{-4}, \quad q = \frac{\pi\sqrt{-1}}{l}\,(t + s),$$

and a similar expression for $\Psi(t - s)$, except that he writes c instead of e and n instead of π. He concludes (Art. XXI):

This is the equation of the vibrating string on the general hypothesis that it will be in a straight line at the beginning of its motion, and then, under convenient impulse, it takes the form of an extremely elongated sine curve [*compagne de la cycloide extrêmement allongée*].

D'Alembert's paper continues to Art. XLVII. In Art XXII he writes that it is easy to see that

$$\Psi(t + s) - \Psi(t - s) = \Delta(t),$$

$\Gamma(s)$, Δ, and Γ expressing functions of t and of s respectively. He gets

$$\Gamma(s) = \frac{e^{Ms\sqrt{-1}} - e^{-Ms\sqrt{-1}}}{2\sqrt{-1}}$$

and

$$\Delta t = \sin Mt \text{ or } \cos Mt.$$

We see that d'Alembert was familiar with Euler's formula, $e^{ix} = \cos x + i \sin x$. His paper was followed by a paper of Euler's in the next volume of the *Histoire* of the Berlin Academy, which we give here in abstract.

(2) EULER (1749)

1. Although all that Messrs. Taylor, Bernoulli, and some others have said and discovered up to now on the subject of the vibratory motion of strings seems to have exhausted the matter, yet there remains a double limitation of such restrictive power that there exists hardly any case in which the true motion of a vibrating string can be determined. For firstly, they have supposed that the stretched strings make only almost infinitesimally small vibrations, so that in this motion the string—whether it is a straight line or a curve—can always be considered of the same length. The other limitation consists in the fact that they have supposed that all vibrations are regular, claiming that in every vibration the entire string, and this once, is stretched directly, and, looking for its curved figure starting from this situation, they have found it to be a trochoid prolonged to infinity.

Euler discards the first limitation, but the second one cannot be defined, on the ground that, "even if vibrations are not regular at the beginning of the motion, after a short time they will be regular and form a prolonged trochoid."

4. From this there arises therefore the following question, in which the whole research is comprised.

If a string of given length and mass is stretched by a given force or weight, and if instead of the straight lines we give it an arbitrary figure which, however, differs by only an infinitesimal amount from a straight line, and if then it is at once released—then to determine the total vibratory motion with which it will be agitated.

Euler then, in Arts. 5–29, derives the differential equation of the motion of the string AB of length a and mass M, acted upon by a force F. Then if $dy = p\,dx + q\,dt$, $dp = r\,dx + s\,dt$, $dq = s\,dx + u\,dt$, he finds for the ordinate $PM = y$ as function of $AP = x$ the equation:

$$P = \text{accelerating force at an arbitrary point } M$$

$$= -\frac{Far}{M} = -\frac{2\,d\,dy}{dt}.$$

Then, writing $Fa/2M = b$, $x + t\sqrt{b} = v$, $x - t\sqrt{b} = u$, he finds

$$dq + dp\sqrt{b} = dv(s + r\sqrt{b}),$$
$$dq - dp\sqrt{b} = du(s - r\sqrt{b}),$$

which leads him to the expression

$$y = f(x + t\sqrt{b}) + \varphi(x - t\sqrt{b})$$

[Euler writes $f:(x + t\sqrt{b})$]; by virtue of the initial conditions, this becomes

$$y = \tfrac{1}{2}f(x + t\sqrt{b}) + \tfrac{1}{2}f(x - t\sqrt{b}).$$

---

30. Thus having found the general solution, let us now consider some special cases of it, in which the eellike curve [*courbe anguiforme*] of Fig. 1 is a continuous curve, of which the parts are related by virtue of the law of continuity, in such

Fig. 1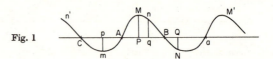

a way that its nature can be expressed by an equation [*de manière que sa nature puisse être comprise par une équation: ut eius natura aequatione comprehendi possit*]. And first: these curves always will be transcendental, since they are cut by the axis in an infinity of points. If the length of the string $AB = a$, and an arbitrary abscissa $AP = u$, and if $1 : \pi$ as the diameter of the circle to the circumference, then it is clear that the following equation, expressed by sines, provides a curve of the required form:

$$PM = \alpha \sin\frac{\pi u}{a} + \beta \sin\frac{2\pi u}{a} + \gamma \sin\frac{3\pi u}{a} + \delta \sin\frac{4\pi u}{a} + \text{etc.}$$

Indeed, if instead of u we take a, or $2a$, or $3a$, or $4a$, etc., the ordinate PM vanishes, and if u is taken negative, the ordinate itself changes into its negative.

If therefore the curve AMB were the primitive figure of the string, then after a time t, in which the heavy body descends by an altitude $= z$, and setting $v = \sqrt{(2Faz/M)}$, to the abscissa x in the figure of the string will correspond the ordinate y such that

$$y = \tfrac{1}{2}\alpha \sin \frac{\pi}{a}(x+v) + \tfrac{1}{2}\beta \sin \frac{2\pi}{a}(x+v) + \tfrac{1}{2}\gamma \sin \frac{3\pi}{a}(x+v) + \text{etc.}$$

$$+ \tfrac{1}{2}\alpha \sin \frac{\pi}{a}(x-v) + \tfrac{1}{2}\beta \sin \frac{2\pi}{a}(x-v) + \tfrac{1}{2}\gamma \sin \frac{3\pi}{a}(x-v) + \text{etc.}$$

31. Now, since $\sin(a+b)$ and $\sin(a-b) = 2\sin a \cos b$ [Euler writes sin. a. cos. b.], this equation will be transformed into this form:

$$y = \alpha \sin \frac{\pi x}{a} \cos \frac{\pi v}{a} + \beta \sin \frac{2\pi x}{a} \cos \frac{2\pi v}{a} + \gamma \sin \frac{3\pi x}{a} \cos \frac{3\pi v}{a} + \text{etc.}$$

and the primitive figure of the string will be expressed by this equation:

$$y = \alpha \sin \frac{\pi x}{a} + \beta \sin \frac{2\pi x}{a} + \gamma \sin \frac{3\pi x}{a} + \text{etc.},$$

which repeats itself every time that v becomes either $2a$, or $4a$, or $6a$, etc. But when v is either a, or $3a$, or $5a$, etc. the figure of the string will be

$$y = -\alpha \sin \frac{\pi x}{a} + \beta \sin \frac{2\pi x}{a} - \gamma \sin \frac{3\pi x}{a} + \text{etc.},$$

where we must observe that when $\beta = 0$, $\gamma = 0$, $\delta = 0$, etc. we obtain the case that is usually believed to be the only one in the vibration of strings, namely, $y = \alpha \sin \frac{\pi x}{a} \cos \frac{\pi v}{a}$, in which the curvature of the string is perpetually the line of the sines, or a trochoid prolonged to infinity. But if only the term β, or γ, or δ, etc., occurs, then this forms cases where the time of vibration is less, either by the double, or the triple, or the quadruple, etc.

This paper was followed by another one by D'Alembert in the *Histoire* of the Berlin Academy. Again we give it in abstract.

---

(3) D'ALEMBERT (1750)

1. In article XXII of this paper (*Mém. Berlin*, 1747) I have found by a very indirect method that if $\Psi(t+s) - \Psi(t-s) = \Delta(t)\Gamma(s)$, then $\Gamma(s) = \sin Ms$ and $\Delta(t) = \sin Mt$ or $\cos Mt$. This proposition is true and exact from the point of view from which I saw it at that time, but having had occasion to consider it

from a more general point of view, I have found a direct way of solving this problem, which gives an opportunity for some observations.

This new solution starts by differentiating the equation

$$\Psi(t + s) - \Psi(t - s) = \Delta(t)\Gamma(s)$$

with respect to t and to s:

$$\Gamma(t + s) - \Gamma(t - s) = \frac{d\,\Delta(t)}{dt}\,\Gamma(s),$$

$$\Gamma(s) = \frac{d\Psi}{ds},$$

$$\Gamma(t + s) + \Gamma(t - s) = \Delta(t)\,\frac{d\Gamma(s)}{ds},$$

from which by renewed differentiation it follows that

$$\frac{d\,d\,\Delta(t)}{\Delta(t)\,dt^2} = \frac{d\,d\Gamma(s)}{ds^2\Gamma(s)},$$

hence $d\,d\,\Delta(t) = A\,dt^2\,\Delta(t)$, $d\,d\Gamma(s) = A\,ds^2\Gamma(s)$, A being a constant. After some computation this leads to:

$$\Delta(t) = Me^{t\sqrt{A}} + ge^{-t\sqrt{A}}, \quad \Gamma(s) = Me^{s\sqrt{A}} - Me^{-s\sqrt{A}}$$

Here e is used for D'Alembert's c; M and g are constants determined by boundary conditions. If $\Psi(t + s) - \Psi(t - s) = 0$ for $s = 0$ and $s = l$, as in Art. XXII, $\sqrt{A}$ must be imaginary. Consequences of this process of differentiating lead further to $d^n y = ky\,dx^n$, which had been integrated by Euler and by D'Alembert himself. Then:

In the *Mémoires* of 1748 Mr. Euler has discussed the vibrating string by a method quite similar to mine as far as the essential part of the problem is concerned, but only (it seems to me) a little beyond this. This great geometer remarks (as I have done), that the curve formed by the string at the beginning of its motion is the same that I have called the *generating curve*. But I believe that a warning should here be in place—for fear that some readers may get a wrong sense of these words—that in order to have this generating curve it is not sufficient to transport the initial curve alternately above or below the axis. It is also necessary that this curve should satisfy the conditions that I have expressed in my paper. These conditions are that, if we suppose $y = \Sigma$ for the equation of the initial curve, then Σ must be an odd function of s, and that in general the ordinates that are distant from each other by the quantity $2l$ must be equal. And this cannot be so unless the curve is not mechanical and such as I have determined it in my *Mémoire*. In any other case the problem cannot be solved, at least by my method, and I am not certain whether it will not surpass the power of known analysis. Indeed, it seems to me that we cannot express y analytically in a more general way than by supposing it a function of t and s. But under this supposition we find the solution of the problem only for the cases in which all the different figures of the vibrating string can be comprehended by one and the same equation. It seems to me that in all other cases it will be impossible to give y a general form.

D'Alembert then makes some remarks on the time of vibration, which, he says, may well be independent of the form of the curve, although this will be difficult to prove.

(4) DANIEL BERNOULLI (1753)

Daniel Bernoulli (1700–1782), son of Johann, after some years in Saint Petersburg, returned in 1733 to Basel, where he became a professor in the university. He was mostly interested in applied mathematics. Through his *Hydrodynamica* (1738) he became a founder of hydrodynamics and the kinetic theory of gases.

I. Mr. Taylor was the first to obtain the number of vibrations made in a given time by a string uniformly thick, of given length and given weight, and stretched by a given force. It was not possible to determine this number without knowing in advance the curve taken by the string during the whole time that its vibration lasted; he therefore proved that this curve was always "the companion of an extremely elongated cycloid," for which the ordinates represent the sines of the arcs represented by the abscissas. I think that only in this form can the vibrations become regular, simple, and isochronous despite the inequality of the deviations [*excursions*]. Since I always had this idea I could only be surprised to see in the *Mémoires* [of the Berlin Academy] of the years 1747 and 1748 an infinity of other curves claimed to be endowed with the same property. I really needed the great names of Messrs. D'Alembert and Euler, whom I could not suspect of any carelessness, to make me examine whether there would not be anything in this aggregate of curves that conflicted [*équivoque*] with those of Mr. Taylor, and in what sense they could be admitted. I immediately saw that this multitude of curves could be admitted only in quite an improper sense. I do not the less esteem the calculations of Messrs. D'Alembert and Euler, which certainly contain all that analysis can have at its deepest and most sublime, but which show at the same time that an abstract analysis which is accepted without any synthetic examination of the question under discussion is liable to surprise rather than enlighten us. It seems to me that we have only to pay attention to the nature of the simple vibrations of the strings to foresee without any calculation all that these two great geometers have found by the most thorny and abstract calculations that the analytical mind can perform.

II. Let us first observe that, according to Mr. Taylor's theory, a stretched string can perform uniform vibrations in an infinity of ways, physically speaking different from each other, but geometrically speaking amounting to the same, since in every one of them only the unit that serves as measure is changed. These different ways are characterized by the number of loops [*ventres*] that the string can form during its vibration. When there is only one loop [Fig. 1], then the vibrations are the slowest, and they produce the fundamental tone; when there are two loops, and one node [*noeud*] in the middle of the axis [Fig. 2], then the vibrations are doubled, and they produce the octave of the fundamental tone; when the string forms three, four, or five loops, with two, three, or four nodes, at equal distances, as in Figs. 3, 4, 5, then the vibrations are multiplied by three,

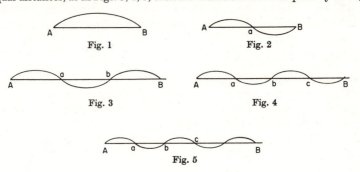

Fig. 1   Fig. 2

Fig. 3   Fig. 4

Fig. 5

four, or five, and produce the twelfth, the double octave, or the major third of the double octave relative to the fundamental tone. In every type of these vibrations the total displacements can be large, or small, at discretion, provided that the largest must be considered as extremely small. The nature of these vibrations is such that not only does each point begin and end every simple vibration at the same instant, but also all the points place themselves after every simple half-vibration in the position of the axis AB. We must regard all these conditions as essential, and then we have at once the curves described by Mr. Taylor as satisfying the problem.

But although, separating these conditions, we can find an infinity of curves which separately satisfy some condition, I shall show how little reason there is in this case to call the vibrations isochronous for every point. It is with these vibrations as with the reciprocal motions of bodies that descend and ascend alternately on a curve: if we require that all these descents, as well as the ascents, large or small, be isochronous, the curve can only be the cycloid. However, if one simply wants the entire vibrations to be isochronous with respect to each other, then one can give as many curves as one likes that satisfy this problem. This is so because I have shown in the *Mémoires* of Petersburg that, whatever curve of descent may be given, it is always possible to determine the curve of ascent such that the two times used for descent and ascent, taken together, are constant, whatever inequality there may be between the amplitudes of the deviations.

In Sec. III there is experimental proof that sounds can have an infinity of simple vibrations, shown for trumpets, flutes, and so forth, together with other remarks on music. No finite formula by which to express them can be given. Reference is made to a paper of 1749 by Euler, where he has given a solution like Bernoulli's.

IV. My conclusion is that all the sounding bodies contain potentially an infinity of tones, and an infinity of corresponding ways of performing their regular vibrations—in short, that in every different kind of vibration the inflections of the parts of the sounding body are made in a different way. However, it is not of this multitude of vibrations applied to stretched strings that Messrs. D'Alembert and Euler claim to speak; this was not unknown to Mr. Taylor. They multiply to infinity every kind of vibration, create an infinity of curves according to every interval between two neighboring nodes, such that every point begins and ends its vibration at the same instant—while, according to the theory of Mr. Taylor, every one of these intervals must necessarily assume the only curve possible, that of the companion of the extremely elongated cycloid. The apparent contradiction between such great geometers seems to me to ask for some clarification.

Then, in Secs. V–XI, Bernoulli claims that all the new curves of D'Alembert and Euler are comprised in his construction.

XII. Let us see if all the new curves found by Mr. Euler are contained in our discussion. For this purpose we must give an equation for all Taylorian curves, of which our five figures are as many examples. I shall use the notation of Mr. Euler. Let the length of the string AB be $= a$, let π be the semicircumference of the circle with unit radius, let the largest ordinate at the middle of every loop for the first figure be $= x$, for the second $= \beta$, for the third $= \gamma$, for the fourth $= \delta$, and finally let x be an arbitrary abscissa, and y the ordinate for the abscissa; then we shall have, according to Mr. Taylor:

for the first figure $\qquad y = \alpha \sin \dfrac{\pi x}{a},$

for the second figure $\qquad y = \beta \sin \dfrac{2\pi x}{a},$

for the third figure $\qquad y = \gamma \sin \dfrac{3\pi x}{a},$

for the fourth figure $\qquad y = \delta \sin \dfrac{\sin 4\pi x}{a},$

and so forth.

Hence, combining all these curves, according to Fig. 6 (for which we have combined only the first two figures),[10] we shall have in the general case the following equation for the same abscissa x:

$$y = \alpha \sin\frac{\pi x}{a} + \beta \sin\frac{2\pi x}{a} + \gamma \sin\frac{3\pi x}{a} + \delta \sin\frac{4\pi x}{a} + \text{etc.,}$$

in which the quantities α, β, γ, δ, etc. are positive [*affirmatives*] or negative quantities.

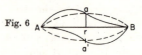

Fig. 6

XIII. Here we have therefore found this infinity of curves without any computation, and our equation is the same as that of Mr. Euler.

This refers to Euler's paper of 1749, but in Euler's paper it is a special case of the eellike curve, namely, a curve whose nature might be expressed by an equation [*puisse être comprise par une équation*].

He does not treat this infinite multitude as being the general case, and he gives it only in §30 as one of the particular cases, but this is certain—although I am not yet quite clear about it—: if there are still other curves, then I do not see in what sense they can be admitted.

Then follow among others some special cases of the equation, and a mechanical derivation of Bernoulli's equation by integration of Hooke's law for every term.

This paper was followed by one of Euler's, part of which is given now.

(5) EULER (1753)

Euler begins by stating that it is to Bernoulli that we owe the felicitous idea that the same string can produce at the same time several different sounds, an idea that can be extended to various other types of sounding bodies. Then he continues:

---

2. Mr. Bernoulli draws all his excellent reflections uniquely from the investigations made by the late Mr. Taylor on the motion of strings, and maintains

---

[10] Figure 6 has been somewhat simplified.

against Mr. D'Alembert and me that the solution of Taylor is sufficient for the explanation of all motions to which a string can be subjected, in such a way that the curves which a string takes during its motion are always either a simple elongated trochoid, or a mixture of two or more curves of the same kind. Now, although such a mixture can no longer be regarded as a trochoid, and the possibility of combining several of Mr. Taylor's curves already makes his solution insufficient in other respects as well, the motion of a curve could be such that it would be impossible to reduce it to the type of Taylor trochoids.

3. If all curves that the string could form during its motion were described by the equation

$$y = \alpha \sin \frac{\pi x}{a} + \beta \sin \frac{2\pi x}{a} + \gamma \sin \frac{3\pi x}{a} + \delta \sin \frac{4\pi x}{a} + \text{etc.,}$$

then Mr. Bernoulli's opinion would be justified, since by taking every term separately we see that such an equation as $y = \mu \sin mx/a$ always gives one of the trochoids indicated by Mr. Taylor, and our equation will be formed by several trochoids. However, even if the number of terms in this equation becomes infinite, I doubt whether we can say that the curve is composed of an infinity of trochoids. The infinite number seems to destroy the nature of such a composition. But I agree that Mr. Bernoulli could have come to the discovery of all these curves by the reasoning based on the composition of Taylorian trochoids alone and that the equation mentioned, even if continued to infinity, is a very natural result of this reasoning.

4. However, there are cases where this equation, if continued to infinity, can be reduced to a finite equation, and then it would be very improper to say that the curve is formed by an infinity of trochoids, since the equation itself would provide an idea and a much simpler construction for it. For instance, if the coefficients α, β, γ, δ, etc. form a geometric progression, then the infinite equation is reduced to this finite equation

$$y = \frac{c \sin \pi x/a}{1 - n \cos \pi x/a},$$

which without doubt determines curves that can fit the motion of a string, in agreement with Mr. Bernoulli, provided that n is a number less than unity. This string could therefore well produce at the same time an infinity of sounds, of which the higher would become weaker and weaker; but the equation offers us a much simpler idea of this curve than if we were to say that it is composed of an infinity of Taylorian trochoids.

5. But there is more: I have give this equation,

$$y = \alpha \sin \frac{\pi x}{a} + \beta \sin \frac{2\pi x}{a} + \gamma \sin \frac{3\pi x}{a} + \delta \sin \frac{4\pi x}{a} + \text{etc.,}$$

only as a particular solution of the formula which in general contains all the curves that a string in motion can assume and there are an infinity of other curves that could not be expressed by this equation. If Mr. Bernoulli would agree with this, then he could not have maintained that all curves of a vibrating string could result uniquely from a combination of two or three Taylorian curves, and he would have recognized that the reasoning based on this combination is not sufficient to provide a complete solution to the problem in question. Neither would he have regarded the method used by Mr. D'Alembert and me as too cumbersome to arrive at a general solution, which could be obtained from a simple physical consideration. The principal question which I must face is therefore: are all curves of a string set in motion contained in this formula, or are they not?

6. If the Bernoulli solution is not the general one, then the Euler–D'Alembert solution is better:

7. Now it seems to me that this fact is beyond doubt, when we consider that we can begin by giving the string an arbitrary shape. Indeed, let us conceive that the string, before it is released, has been given a figure not contained in the form $y = \alpha \sin \pi x/a + \beta \sin 2\pi x/a + $ etc. There is no doubt that the string, after being suddenly released, will be forced into a certain motion. It is also certain that the figure which it will assume after the first instant will be very different from that described by this equation, and, even if one were to maintain that after several instants it will assume a figure contained in this expression, it cannot be denied that before this happens the motion will be quite different from that which is contained in the reasoning of Mr. Bernoulli. Since this first motion is therefore certainly not in agreement with the laws derived from Taylor's theory, it seems to me that this fact alone is quite sufficient to show that this theory is not able to explain to us all motions to which a string is susceptible.

Euler continues this argument, pointing out that the initial figure "often cannot be expressed by any equation, be it algebraic or transcendental, and is not even included [*renfermée*] in any law of continuity." Then he defines the problem:

11. Having at the beginning given to the string an arbitrary figure, either algebraic, or transcendental, or even mechanical, to determine the motion of the string after its release. Stating the problem thus, it is quite clear that the solution obtained from the combination of two chords can only be regarded as a very special one. But, it will be asked, is a general solution possible? I believe that the solution which I have given is in no respect limited; at any rate I cannot discover any fault in it, and nobody has yet shown the insufficiency. It is

true that Mr. D'Alembert, although he has rebuked me, saying that my solution was not different from his, has claimed—but without giving any proof—that my solution does not extend to all possible figures that the string can have at the start. It is the same sentiment that Mr. Bernoulli seems to share when he maintains that the motion of a string cannot be determined unless its initial figure is contained in the equation to which I have already referred several times.

Then Euler tackles the problem again and discusses his solution: $y = \varphi(x + ct) + \Psi(x - ct)$, φ, Ψ "two arbitrary functions." This part is of a certain interest, because Euler here approaches the modern notation for partial differentiation. When $y = P\,dx + Q\,dy + R\,du$, he writes

$$P = \left(\frac{dy}{dx}\right), \quad Q = \left(\frac{dy}{dt}\right), \quad R = \left(\frac{dz}{dt}\right);$$

the differential equation is written

$$\left(\frac{d\,dy}{dt^2}\right) = cc\left(\frac{d\,dy}{dx^2}\right).$$

Note on the emergence of the concept of function. The concept of function appeared gradually in such works as Oresme's latitude of forms, Galilei's study of the dependence of velocity on time, and Descartes's graphical representation of algebraic expressions. In the tabulation of chords and of sines, the function concept can even be traced to antiquity; the many tables published during the sixteenth century constituted an added preparation. Newton may have been the first to formulate a descriptive term, using the word *genita* for a quantity obtained from other quantities by means of the four species (Selection V.6). The first appearance in print of the term *functio* may have been in Leibniz's "De linea ex lineis numero infinitis ordinatim ductis," *Acta Eruditorum* (1692), *Mathematische Schriften*, Abth. 2, Band I, 266–269, a paper in which he shows how to find the evolute of a family of curves. Here we read of the "tangent line and some other functions depending on it," for example, the perpendiculars to the tangent drawn from the curve to an axis. (This paper contains a number of other terms now commonly used; see Selection V.1, note 1.) We find the same idea in another of Leibniz's papers: "Considérations sur la différence qu'il y a entre l'analyse ordinaire et le nouveau calcul des transcendantes," *Journal des Sçavans* (1694), *Mathematische Schriften*, Abth. 2, Band I, 306–308; here he calls the abscissa, the ordinate, the chord, the tangent, and the normal (both drawn from the curve to an axis), and other related segments "functions" of a curve. The term was then taken over by the Bernoullis. In July 1698 Leibniz wrote to Johann Bernoulli, "I am pleased that you use the term function in my sense." Bernoulli replied from Groningen in August 1698: "To denote a function of some indeterminate quantity x, I like to use the corresponding capital letters X or Greek ξ, so that we can see at the same time on which indeterminate quantity the function depends." In the same letter he used the symbols $X = x$ and $X = \sqrt{x}$ (*Mathematische Schriften*, Abth. 1, Band III, 531–532). Gradually the function concept lost its immediate geometric character

Johann Bernoulli later defined the term as follows: "We here denote by *function* of a variable quantity a quantity composed in some way or other of this variable quantity and constants" (*Opera omnia*, II, 241). In this paper Bernoulli then used the term "function" quite freely in enunciating his theorems. Euler took it over and in a paper in *Commentarii Academiae Scientiarum Petropolitanae 7, 1734–35* (1740), 184–200 (*Opera omnia*, ser. I, vol. 22, 57–75) introduced the notation $f\left(\dfrac{x}{a} + c\right)$ for "an arbitrary function of $\dfrac{x}{a} + c$." In Chapter I of his *Introductio* of 1748 (see Selection V.15) Euler repeated Bernoulli's statement, adding the word "analytic," and continued, "Therefore every analytic expression in which apart from a variable quantity z all quantities that compose this expression are constants is a function of this z, such as $a + 3z$, $az - 4zz$, $az + baa - zz - c^2$, etc."

Euler then classified functions, using the terms "algebraic" and "transcendental," "single-valued" and "multiple-valued." In the second volume of the *Introductio* he discussed curved lines, and wrote (Chapter I):

"A continuous curve is of such a nature that it can be expressed by one definite function of x. But if a curved line is of such a nature that various parts of it, BM, MD, DN, etc., are expressed by various functions of x such that, after the part BM has been defined with the aid of one function, the part MD is described by another function, then we call such curved lines discontinuous or mixed and irregular, because they are not formed according to one constant law and are composed of parts of various continuous curves."

In his *Institutiones calculi differentialis* (Saint Petersburg, 1755), Euler returned to these statements in the *Introductio* and then showed how to differentiate these functions. It is clear, therefore, that in Euler's opinion (and in that of his contemporaries and pupils), a function was a relation to be expressed by some analytical expression, as a polynomial, a sine, a logarithm, or even an integral of such expressions.

It was the exchange of opinions among Euler and some of his colleagues due to the vibrating-string discussion that brought about a certain feeling of disturbance among those who used the concept of function in this way. As we have seen in Selection V.16, Taylor had shown that there are sinusoidal solutions. D'Alembert found the solution in the form $z = f(at + x) + f(at - x)$, with $f(x)$ an "arbitrary function," but was not sure that this "analytic way," as he called it, of expressing a solution was sufficient to describe all forms of the string in motion; in other words, he was not sure whether any continuous curve could be given by an expression $y = f(x)$. Euler thought that this could be done. But Daniel Bernoulli derived the solution in the form of an infinite trigonometric series and gave it as his opinion that this combination of "Taylorian" functions could give the general solution, something Euler doubted. Euler's conclusion was (Selection V.16(5)) that his trigometric solution was only a particular solution of the formula which in general contains all the curves that the string in motion can assume, and there are an infinity of other curves that cannot be expressed by this equation.

For Euler, "arbitrary functions" were able to represent all "curves of the string" and conversely. Later (1759) Lagrange argued that an arbitrary function in great generality can be expressed by a trigonometric series. On Lagrange's definition of function see Selection V.19.

As we have said, the concept of function was clarified in the nineteenth century by the work of Fourier, Cauchy, Dirichlet, and Riemann; see, for example, P. E. B. Jourdain, "The origins of Cauchy's conceptions of a definite integral and of the continuity of a function," *Isis 1* (1913), 661–703; A. Pringsheim in *Encyklopädie der mathematischen Wissenschaften* (Teubner, Leipzig), II (1899), 1–53.

17   LAMBERT. IRRATIONALITY OF π

By 1750 the number π had been expressed by infinite series, infinite products, and infinite continued fractions, its value had been computed by infinite series to 127 places of decimals (see Selection V.15), and it had been given its present symbol. All these efforts, however, had not contributed to the solution of the ancient problem of the quadrature of the circle; the question whether a circle whose area is equal to that of a given square can be constructed with the sole use of straightedge and compass remained unanswered. It was Euler's discovery of the relation between trigonometric and exponential functions that eventually led to an answer. The first step was made by J. H. Lambert, when, in 1766–1767, he used Euler's work to prove the irrationality not only of π, but also of e.

Johann Heinrich Lambert (1728–1777) was a Swiss from Mülhausen (then in Switzerland). Called to Berlin by Frederick the Great, he became a member of the Berlin Academy and thus a colleague of Euler and Lagrange. His name is also connected with the introduction of hyperbolic functions (1770), with perspective (1759, 1774), and with the so-called Lambert projection in cartography (1772).

Lambert published his proof of the irrationality of π in his "Vorläufige Kenntnisse für die, so die Quadratur und Rectification des Circuls suchen," *Beyträge zum Gebrauche der Mathematik und deren Anwendung 2* (Berlin, 1770), 140–169, written in 1766, and in more detail in the "Mémoire sur quelques propriétés remarquables des quantités transcendentes circulaires et logarithmiques," *Histoire de l'Académie, Berlin, 1761* (1768), 265–322, presented in 1767. They have been reprinted in the *Opera mathematica*, ed. A. Speiser (2 vols.; Füssli, Zurich, 1946, 1948), I, 194–212, II, 112–159. The following text is a translation from pp. 132–138 of vol. II. Lambert writes tang where we write tan. See also F. Rudio, *Archimedes, Huygens, Lambert, Legendre. Vier Abhandlungen über die Kreismessung* (Teubner, Leipzig, 1892).

---

37. *Now I say that this tangent* [$\tan \varphi/\omega$] *will never be commensurable to the radius, whatever the integers ω, φ may be.*[1]

[1] In the previous sections Lambert expands tan v, v an arbitrary arc of a circle of radius 1, into a continued fraction, and gets for $v = 1/w$

$$\tan v = \cfrac{1}{w - \cfrac{1}{3w - \cfrac{1}{5w - 1}}} \quad \text{etc.}$$

Investigating the partial fractions and their residues, he finds infinite series like

$$\tan v = \frac{1}{w} + \frac{1}{w(3w^2 - 1)} + \frac{1}{(3w^2 - 1)(15w^3 - 6w)} + \cdots$$

and shows (in §34) that these series converge more rapidly than any decreasing geometric series. Then, if $w = \omega{:}\varphi$, ω, φ being relatively prime integers, he finds for the partial fractions of tan v (§36):

$$\frac{\varphi}{\omega}, \quad \frac{3\omega\varphi}{3\omega^2 - \varphi^2}, \quad \frac{15\omega^2\varphi - \varphi^3}{15\omega^3 - 6\varphi^2\omega}, \quad \frac{105\omega^3\varphi - 10\omega\varphi^3}{105\omega^4 - 45\omega^2\varphi^2 + \varphi^2}, \quad \text{etc.,}$$

and (§37):

$$\tan \frac{\varphi}{\omega} = \frac{\varphi}{\omega} + \frac{\varphi^3}{\omega(3\omega^2 - \varphi^2)} + \frac{\varphi^5}{(3\omega^2 - \varphi^2)(15\omega^3 - 6\omega^2\varphi)} + \text{etc.}$$

Then follows the text which we reproduce.

38. To prove this theorem, let us write

$$\tan\frac{\varphi}{\omega} = \frac{M}{P},$$

such that M and P are quantities expressed in an arbitrary way, even, if you like, by decimal sequences, which always can happen, even when M, P are integers, because we have only to multiply each of them by an irrational quantity. We can also, if we like, write

$$M = \sin\frac{\varphi}{\omega}, \quad P = \cos\frac{\varphi}{\omega},$$

as above. And it is clear that, even if $\tan\varphi/\omega$ were rational, this would not necessarily hold for $\sin\varphi/\omega$ and $\cos\varphi/\omega$.

39. Since the fraction M/P exactly expresses the tangent of φ/ω, it must give all the quotients w, $3w$, $5w$, etc., which in the present case are

$$+\frac{\omega}{\varphi}, \quad -\frac{3\omega}{\varphi}, \quad +\frac{5\omega}{\varphi}, \quad -\frac{7\omega}{\varphi}, \quad \text{etc.}$$

40. Hence, if the tangent of φ/ω is rational, then clearly M will be to P as an integer μ is to an integer π, such that, if μ, π are relatively prime, we shall have

$$M:\mu = P:\pi = D,$$

and D will be the greatest common divisor of M, P. And since reciprocally

$$M:D = \mu, \quad P:D = \pi,$$

we see that, since M, P are supposed to be irrational quantities, their greatest common divisor will be equally an irrational quantity, which is the smaller, the larger the quotients μ, π are.

41. Here are therefore *the two suppositions of which we must show the incompatibility*. Let us first divide P by M, and the quotient must be $\omega:\varphi$. But since $\omega:\varphi$ is a fraction, let us divide φP by M, and the quotient ω will be the φ-tuple of $\omega:\varphi$. It is clear that we could divide it by φ if we wished to do so. This is not necessary, since it will be sufficient that ω be an integer. Having thus obtained ω by dividing φP by M, let the residue be R'. This residue will equally be the φ-tuple of what it would have been, and that we have to keep in mind. Now, since $P:D = \pi$, an integer, we still have $\varphi P:D = \varphi\pi$, an integer. Finally, $R':D$ will also be an integer. Indeed, since

$$\varphi P = \omega M + R',$$

we shall have

$$\frac{\varphi P}{D} = \frac{\omega M}{D} + \frac{R'}{D}.$$

But $\varphi P : D = \varphi\pi$, $\omega M : D = \omega\mu$; hence

$$\varphi\pi = \omega\mu + \frac{R'}{D},$$

which gives

$$\frac{R'}{D} = \varphi\pi - \omega\mu = \text{integer},$$

which we shall call r', so that $R'/D = r'$. The residue of the first division will therefore still have the divisor D, the greatest common divisor of M, P.

42. Now let us pass to the second division. The residue R' being the φ-tuple of what it would have been if we had divided P instead of φP, we must take this into account by the second division, where we divide φM, instead of M, by R' in order to obtain the second quotient, which $= 3\omega : \varphi$. However, in order to avoid the fractional quotient here also, let us divide $\varphi^2 M$ by R', in order to have the quotient 3ω, an integer. Let the residue be R'', and we shall have

$$\varphi^2 M = 3\omega R' + R'';$$

hence, dividing by D,

$$\frac{\varphi^2 M}{D} = \frac{3\omega R'}{D} + \frac{R''}{D}.$$

But

$$\frac{\varphi^2 M}{D} = \varphi^2 m = \text{integer},$$

$$\frac{3\omega R'}{D} = 3\omega r' = \text{integer};$$

hence

$$\varphi^2 m = 3\omega r' + \frac{R''}{D},$$

which gives $R''/D = \varphi^2 m - 3\omega r' = $ an integer number, which we shall write $= r''$, so that

$$\frac{R''}{D} = r''.$$

Hence the greatest common divisor of M, P, R' is still of the second residue R''.

43. Let the next residues be R''', $R^{\text{iv}}, \ldots, R^n, R^{n+1}, R^{n+2}, \ldots$ which correspond to the φ-tuple quotients $5\omega, 7\omega, \ldots, (2n-1)\omega, (2n+1)\omega, (2n+3)\omega, \ldots$, and we have to prove in general that if two arbitrary residues R^n, R^{n+1}, in

immediate succession, still have D as divisor, the next residue R^{n+2} will have it too, so that, if we write

$$R^n : D = r^n,$$
$$R^{n+1} : D = r^{n+1},$$

where r^n and r^{n+1} are integers, we shall also have

$$R^{n+2} : D = r^{n+2},$$

an integer. This is the demonstration.

We omit this proof in §44, since the reasoning follows that of §42.

---

45. Now we have seen that r', r'' are integers (§§41, 42), hence also r''', $r^{iv}, \ldots, r^n, \ldots$ to infinity will be integers. Hence any one of the residues R', R'', $R''', \ldots, R^n, \ldots$ to infinity will have D as common divisor. Let us now find the value of these residues expressed in M, P.

46. Every division provides us with an equation for this purpose, since we have

$$R' = \varphi P - \omega M,$$
$$R'' = \varphi^2 M - 3\omega R',$$
$$R''' = \varphi^2 R' - 5\omega R'', \quad \text{etc.}$$

But let us observe that in the existing case the quotients ω, 3ω, 5ω, etc. are alternately positive and negative and that the signs of the residues succeed each other in the order $- - + +$. These equations can therefore be changed into

$$R' = \omega M - \varphi P,$$
$$R'' = 3\omega R' - \varphi^2 M,$$
$$R''' = 5\omega R'' - \varphi^2 R',$$
$$\cdot \quad \cdot \quad \cdot \quad \cdot \quad \cdot \quad \cdot$$

or in general

$$R^{n+2} = (2n - 1)R^{n+1} - \varphi^2 R^n.$$

From this we see that every residue is related to the two preceding in the same way as the numerators and denominators of the fractions that approximate the value of $\tan \varphi/\omega$ (§36).

47. Let us make the substitutions indicated by these equations in order to express all these residues by M, P. We shall have

$$R' = \omega M - \varphi P,$$

$$R'' = (3\omega^2 - \varphi^2)M - 3\omega\varphi P,$$

$$R''' = (15\omega^3 - 6\omega\varphi^2)M - (15\omega^2\varphi - \varphi^3)P, \quad \text{etc.}$$

And since these coefficients of M, P are the denominators and numerators of the fractions we found above for $\tan \varphi/\omega$ (§36), we see also that we shall have

$$\frac{M}{P} - \frac{\varphi}{\omega} = \frac{R'}{\omega P},$$

$$\frac{M}{P} - \frac{3\omega\varphi}{3\omega^2 - \varphi^2} = \frac{R''}{(3\omega^2 - \varphi^2)P},$$

$$\frac{M}{P} - \frac{15\omega^2\varphi - \varphi^3}{15\omega^3 - 6\omega\varphi^2} = \frac{R'''}{(15\omega^3 - 6\omega\varphi^2)P}, \quad \text{etc.}$$

48. But we have

$$\frac{M}{P} = \tan\frac{\varphi}{\omega};$$

hence (§§37, 34)

$$\frac{M}{P} - \frac{\varphi}{\omega} = \frac{\varphi^3}{\omega(3\omega^2 - \varphi^2)} + \frac{\varphi^5}{(3\omega^2 - \varphi^2)(15\omega^2 - 6\omega\varphi^2)} + \text{etc.,}$$

$$\frac{M}{P} - \frac{3\omega\varphi}{3\omega^2 - \varphi^2} = \frac{\varphi^5}{(3\omega^2 - \varphi^2)(15\omega^3 - 6\omega\varphi^2)} + \text{etc.;}$$

hence

$$\frac{R'}{\omega P} = \frac{\varphi^3}{\omega(3\omega^2 - \varphi^2)} + \frac{\varphi^5}{(3\omega^2 - \varphi^2)(15\omega^3 - 6\omega\varphi^2)} + \text{etc.,}$$

$$\frac{R''}{(3\omega^2 - \varphi^2)P} = \frac{\varphi^5}{(3\omega^2 - \varphi^2)(15\omega^3 - 6\omega\varphi^2)} + \text{etc.,}$$

$$\frac{R'''}{(15\omega^3 - 6\omega\varphi^2)P} = \frac{\varphi^7}{(15\omega^3 - 6\omega\varphi^2)(105\omega^4 - 45\omega^2\varphi^2 + \varphi^4)} + \text{etc.}$$

Thus all the residues can be found by means of the sequence of differences (§37)

$$\tan\frac{\varphi}{\omega} = \frac{\varphi}{\omega} + \frac{\varphi^3}{\omega(3\omega^2 - \varphi^2)} + \frac{\varphi^5}{(3\omega^2 - \varphi^2)(15\omega^3 - 6\omega\varphi^2)}$$

$$+ \frac{\varphi^7}{(15\omega^3 - 6\omega\varphi^2)(105\omega^4 - 45\omega^2\varphi^2 + \varphi^4)} + \text{etc.}$$

by omitting 1, 2, 3, 4, etc. of the first terms and multiplying the sum of the following terms by the first factor of the denominator of the first term that is retained and by P.

49. Now, this sequence of differences is more convergent than a decreasing geometric progression (§§34, 35). Hence the residues R', R'', R''', etc. decrease in such a way that they become smaller than any assignable quantity. And as every one of these residues, having D as common divisor, is a multiple of D, it follows that this common divisor D is smaller than any assignable quantity, which makes $D = 0$. Consequently $M : P$ is a quantity incommensurable with unity, hence irrational.

50. Hence *every time that a circular arc* $= \varphi/\omega$ *is commensurable with the radius* $= 1$, *hence rational, the tangent of this arc will be a quantity incommensurable with the radius, hence irrational. And conversely, every rational tangent is the tangent of an irrational arc.*

51. Now, since the tangent of 45° is rational, and equal to the radius, the arc of 45°, and hence also the arc of 90°, 180°, 360°, is incommensurable with the radius. Hence *the circumference of the circle does not stand to the diameter as an integer to an integer*. Thus we have here this theorem in the form of a corollary to another theorem that is infinitely more universal.

52. Indeed, it is precisely this absolute universality that may well surprise us.

Lambert then goes on to draw consequences from his theorem concerning arcs with rational values of the tangent. Then he draws an analogy between hyperbolic and trigonometric functions and proves from the continued fraction for $e^u + 1$ that e and all its powers with integral exponents are irrational, and that all rational numbers have irrational natural logarithms. He ends with the sweeping conjecture that "no circular or logarithmic transcendental quantity into which no other transcendental quantity enters can be expressed by any irrational radical quantity," where by "radical quantity" he means one that is expressible by such numbers as $\sqrt{2}$, $\sqrt{3}$, $\sqrt[4]{4}$, $\sqrt{2 + \sqrt{3}}$, and so forth. Lambert does not prove this; if he had, he would have solved the problem of the quadrature of the circle. The proof of Lambert's conjecture had to wait for the work of C. Hermite (1873), and F. Lindemann (1882). See, for instance, H. Weber and J. Wellstein, *Encyklopädie der Elementar-Mathematik* (3rd ed.; Teubner, Leipzig, 1909), I, 478–492; G. Hessenberg, *Transzendenz von e und π* (Teubner, Leipzig, Berlin, 1912); U. G. Mitchell and M. Strain, "The number *e*," *Osiris 1* (1936), 476–496.

18 FAGNANO AND EULER. ADDITION THEOREM OF ELLIPTIC INTEGRALS

Count Giulio Carlo de'Toschi di Fagnano (1682–1766), Spanish consul in his home town of Sinigaglia (Italy) and an amateur mathematician, published in the *Giornali de'letterati d'Italia* for the years 1714–1718 a series of papers on the summation of the arcs of certain curves, a problem induced by a paper of Johann Bernoulli's of 1698.[1] These papers of Fagnano are reproduced in his *Opere mathematiche* (2 vols.; Albrighi, Segati & Co., Milan, Rome, Naples), II (1911), from which our selection has been translated. In vol. 19 of the *Giornali* Fagnano posed the following problem (*Opere*, II, 271):

Problem. Let a biquadratic primary parabola, which has as its constituent equation $x^4 = y$, and also a portion of it, be given. We ask that another portion of the same curve be assigned such that the difference of the two portions be rectifiable.

It had already been recognized by the brothers Bernoulli that what would be called elliptic arcs are not rectifiable, but that sums or differences might be representable by arcs of circles or straight lines. Fagnano gave a solution of his own problem, and generalized it to a number of cases, all involving elliptic integrals. One of his conclusions, sometimes called Fagnano's theorem, dates from 1716 and is found in the paper entitled "Teorema da cui si deduce una nuova misura degli archi elittici, iperbolici, e cicloidali," *Giornali 26* (*Opere*, II, 287–292).

Theorem. In the two polynomials below, X and Z, and in equation (1) the letters h, l, f, g represent arbitrary constant quantities.

I say, in the first place, that if in equation (1) the exponent s expresses the positive unity [$s = +1$], then the integral of the polynomial $X - Z$ is equal to $-hxz/\sqrt{-fl}$.

I say, in the second place, that if in the same equation (1) the exponent s expresses the negative unity [$s = -1$], then the integral of

$$X + Z = \frac{xz\sqrt{-h}}{\sqrt{g}}.$$

Here

$$X = \frac{dx\sqrt{hx^2 + l}}{\sqrt{fx^2 + g}},$$

$$Z = \frac{dz\sqrt{hz^2 + l}}{\sqrt{fz^2 + g}},$$

(1) $$(fhx^2z^2)^s + (flx^2)^s + (flz^2)^s + (gl)^s = 0.$$

---

[1] An account of the contributions of Fagnano to this problem can be found in Cantor, *Geschichte*, III (2nd. ed., 1901), 465–472. Johann Bernoulli's paper, entitled "Theorema universale rectificationi linearum curvarum inserviens" (Universal theorem useful for the rectification of curved lines), appeared in the *Acta Eruditorum* of October 1698 (*Opera omnia*, I, 249–253); in it he asked whether there are curves with arcs that are not rectifiable, but are such that sums or differences of arcs are rectifiable. He claims that the parabola $3a^2y = x^3$ has that property. See Selection V.10, note 4.

The first part of the theorem Fagnano applies to the difference of arcs of an ellipse and of a cycloid, the second part to the sum of arcs of a hyperbola.

Then, in another article, "Metodo per misurare la lemniscata," *Giornali 29* (1718; *Opere*, II, 293–313), he applied his considerations to the lemniscate, a curve discovered by Jakob Bernoulli in 1694.[2] After a reference to the two brothers Bernoulli, Fagnano continues:

Let the lemniscate be $CQACFC$ [Fig. 1], its semiaxis $CA = a$; then it is known that if we take the origin of the abscissa (x) at the center C and call (y) the ordinates [*le ordinate*] normal to the axis, then the nature of the lemniscate is

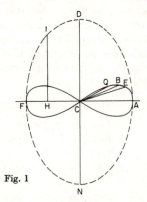

Fig. 1

expressed by this equation: $x^2 + y^2 = a\sqrt{x^2 - y^2}$. It is also known that if we call z the indeterminate chord $CQ = \sqrt{x^2 + y^2}$, then the direct arc

$$CQ = \int \frac{a^2\,dz}{\sqrt{a^4 - z^4}},$$

and the inverse arc

$$QA = \text{arc. } CA - \text{arc. } CQ = \int \frac{a^2\,dz}{\sqrt{a^4 - z^4}}.^3$$

[2] The lemniscate was introduced by Jakob Bernoulli in an article entitled "Constructio curvae accessus et recessus aequabilis" in the *Acta Eruditorum* of September 1694 (*Opera*, II, 608–612) dealing with elastic curves. Here he discusses the curve with equation $xx + yy = a\sqrt{(xx - yy)}$, which curve "of four dimensions" has, as he says, a form "jacentis notae octonari ∞, seu complicatae in nodum fasciae, sive lemnisci, d'un noeud de ruban Gallis" (like a lying eightlike figure, folded in a knot of a bundle, or of a lemniscus, a knot of a French ribbon), *lemniskos* being a knot in the form of an eight. The curve was soon known as a lemniscate.

[3] It was not yet customary to indicate the limits of the integral at the bottom and top of the integral sign, so that the integrals for arcs CQ and QA look alike. Our modern notation $\int_a^b$ is due to J. Fourier; see his *Théorie analytique de la chaleur* (Didot, Paris, 1822), 237–238.

Take the ellipse $ADFNA$, of which the minor semiaxis is $CF = a$, and the major semiaxis $CD = a\sqrt{2}$, and call z the indeterminate abscissa CH, with its origin in the center C of the ellipse, and equal to the chord CQ of the lemniscate, and draw the ordinate HI parallel to the major axis. Then it is already known that the direct arc DI of this ellipse has as its expression

$$\int dz \frac{\sqrt{a^2 + z^2}}{\sqrt{a^2 - z^2}},$$

and the inverse arc

$$IF = \text{arc. } DF - \text{arc. } DI = \int -dz \frac{\sqrt{a^2 + z^2}}{\sqrt{a^2 - z^2}}.$$

Finally, take the equilateral hyperbola LMP with semiaxis $SM = a$ [Fig. 2]. If we call t the indeterminate radius [*applicata*] SO, then it is known that if we take the arc MO starting from the center M this arc is expressed as follows:

$$\int \frac{t^2 \, dt}{\sqrt{t^4 - a^4}}.$$

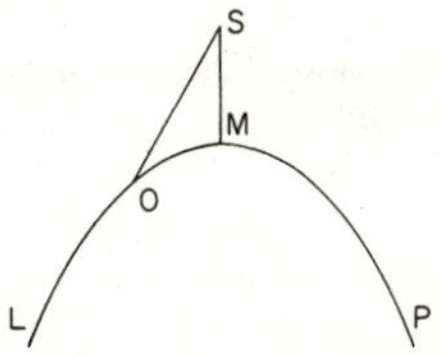

Fig. 2

Theorem I. *Let the two equations written below be* (1) *and* (2); *then I say that if we take the first of them, then also the other is valid:*

(1)
$$t = a \frac{\sqrt{a^2 + z^2}}{\sqrt{a^2 - z^2}},$$

(2)
$$\int \frac{a^2 \, dz}{\sqrt{a^4 - z^4}} = \int dz \frac{\sqrt{a^2 + zz}}{\sqrt{a^2 - az}} + \int \frac{t^2 \, dt}{\sqrt{t^4 - a^4}} - \frac{zt}{a}.$$

The truth of this theorem can be shown by differentiation, and substituting for t and dt their values in terms of z and dz taken from equation (1).

Corollary. If in the lemniscate the chord $CQ = z$, and in the ellipse the abscissa CH is also $= z$, and in the equilateral hyperbola LMP the central radius $SO = t$, and if we assign to t its value expressed in equation (1) and substitute in equation (2) the arcs of the curves in terms of their expressions already indicated in the statements above, we obtain

$$\text{arc. } CQ = \text{arc. } DI + \text{arc. } MO = \frac{zt}{a}.$$

Fagnano has in Theorem II another substitution which leads (see Figs. 1 and 2) to

$$\text{arc. } QA = \text{arc. } IF + \text{arc. } ML - \frac{1}{z}\sqrt{a^4 - z^4},$$

and then goes on to

Theorem III. *If we consider equation (7) and equation (8) below, then I say that, given the first one, the other also is valid:*

$$(7) \qquad\qquad u = a\frac{\sqrt{a^2 - z^2}}{\sqrt{a^2 + z^2}},$$

$$(8) \qquad\qquad \int \frac{a^2\,dz}{\sqrt{a^4 - z^4}} = \int - \frac{a^2\,du}{\sqrt{a^4 - u^4}}.$$

From these equations Fagnano again derives some expressions for the arc. Other pairs are (*Opere*, II, 304–309):

$$x = \frac{\sqrt{1 \mp \sqrt{1 - z^4}}}{2}, \qquad\qquad \frac{\pm\,dz}{\sqrt{1 - z^4}} = \frac{dx\sqrt{2}}{\sqrt{1 + x^4}},$$

$$x = \frac{\sqrt{1 \mp z}}{\sqrt{1 \pm z}}, \qquad\qquad \frac{\mp\,dz}{\sqrt{1 - z^4}} = \frac{dx\sqrt{2}}{\sqrt{1 + x^4}},$$

$$(9) \quad \frac{u\sqrt{2}}{\sqrt{1 - u^4}} = \frac{1}{z}\sqrt{1 - \sqrt{1 - z^4}}, \qquad \frac{dz}{\sqrt{1 - z^4}} = \frac{2\,du}{\sqrt{1 - u^4}}, \quad (10)$$

$$(12) \quad \frac{\sqrt{1 - t^4}}{t\sqrt{2}} = \frac{1}{z}\sqrt{1 - \sqrt{1 - z^4}}, \qquad \frac{dz}{\sqrt{1 - z^4}} = \frac{-2\,dt}{\sqrt{1 - t^4}}. \quad (13)$$

The last equations allow Fagnano to duplicate an arc of the lemniscate, and so to divide the quadrant of the lemniscate into three equal parts; $t = z$ then gives $z = \sqrt[4]{-3 + 2\sqrt{3}}$.

He also shows how to divide the quadrant into five equal parts.

Two more sets of equations show how to duplicate an arc of the lemniscate. Fagnano concludes that he can divide the quadrant of the lemniscate therefore into 2×3^m, 3×2^m, 5×2^m equal parts. "And this is a new and singular property of my curve."

Much later, Fagnano republished his papers in his *Produzioni matematiche* (Pesaro, 1750; reprinted as vol. II of the *Opere matematiche*). When this book reached the Berlin Academy in 1751, Euler, who was asked to express an opinion on it, quickly grasped the importance of Fagnano's transformations for the integration of a number of differential equations of a particular kind, involving radicals. In his "Observationes de comparatione arcuum curvarum ellipticarum," *Novi Commentarii Academiae Scientiarum Petropolitanae 6, 1756–57*

(1761), 58–84 (*Opera omnia*, ser. I, vol. 20, 80–107), he took up, in his own way, Fagnano's investigations on the arcs of the ellipse, the hyperbola, and the lemniscate. In chapter I he sets up formulas on sums and differences of the arcs of the ellipse, in chapter II of the hyperbola, then in chapter III he takes up analogous problems for the case of the lemniscate $(xx + yy)^2 = xx - yy$.

Theorem 4. If, in the lemniscatic curve that we have described here [Fig. 3] we draw a chord CM = z and another one besides which is[4]

$$CN = u = \sqrt{\frac{1 - zz}{1 + zz}},$$

then the arc CM is equal to the arc AN, or also: the arc CN is equal to the arc AM.

Fig. 3

The demonstration is like that of Fagnano in a similar case. In Corollary 1 Euler writes $CN = CA\sqrt{\dfrac{CA^2 - CM^2}{CA^2 + CM^2}}$, in Corollary 2 he changes $u = \sqrt{\dfrac{1 - zz}{1 + zz}}$ into $z = \sqrt{\dfrac{1 - uu}{1 + uu}}$ and those expressions into $uuzz + uu + zz = 1$, "hence the points M and N can be interchanged, from which it follows that arc CM = arc AN as well as arc CN = arc AM."

Corollary 3 states that, since CQ, the abscissa of N, is equal to $u\sqrt{\dfrac{1 + uu}{2}}$ and QN, its ordinate, to $u\sqrt{\dfrac{1 - uu}{2}}$, therefore $CQ = \dfrac{u}{1 + zz}$, $QN = \dfrac{uz}{1 + zz}$, and hence $QN/CQ = z$, and $AT = z = CM$ (AT is the tangent at A).

Corollary 6 points out that the point O, which divides the whole quadrant CA into two equal parts, also divides all arcs MN into two equal parts.

Theorem 5. If in a lemniscatic curve with axis CA = 1 we construct [Fig. 4] one chord CM = z and another arc besides which is

$$CM^2 = u = \frac{2z\sqrt{1 - z^4}}{1 + z^4},$$

then the arc CM² subtended by this chord u is twice the arc subtended by chord CM.

---

[4] Compare Fagnano's case (7).

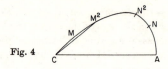

Fig. 4

The demonstration leads, via $uu = \dfrac{4zz - 4z^6}{1 + 2z^4 + z^8}$, through

$$\sqrt{(1 - uu)} = \frac{1 - 2zz - z^4}{1 + z^4}, \quad \sqrt{(1 + uu)} = \frac{1 + 2zz - z^4}{1 + z^4}, \quad \sqrt{(1 - u^4)} = \frac{1 - 6z^4 + z^8}{(1 + z^4)^2}.$$

and

$$du = \frac{2\,dz(1 - 6z^4 + z^8)}{(1 + z^4)^2\,\sqrt{(1 - z^4)}}$$

to

$$\frac{du}{\sqrt{1 - u^4}} = \frac{2\,dz}{\sqrt{1 - z^4}},\ ^5$$

or, since arc $CM = \displaystyle\int \frac{dz}{\sqrt{1 - z^4}}$, arc $CM^2 = \displaystyle\int \frac{du}{\sqrt{1 - u^4}}$,

$$\text{arc } CM^2 = 2 \text{ arc } CM + \text{const.};$$

but, since $z = 0$ gives $u = 0$, the constant is zero, so that

$$\text{arc } CM^2 = 2 \text{ arc } CM.\ ^6$$

In Corollary 1 to Theorem 5 it is pointed out that, if

$$CN = \sqrt{\frac{1 - zz}{1 + zz}},$$

$$CN^2 = \frac{1 - 2zz - z^4}{1 + 2zz - z^4} = \sqrt{\frac{1 - uu}{1 + uu}},$$

then arc $AN = $ arc CM, arc $AN^2 = $ arc CM^2, arc $AN^2 = 2$ arc AN.

[5] Compare Fagnano's case (10), interchanging the letters u and z.

[6] See C. L. Siegel, "Zur Vorgeschichte des Eulerschen Additionstheorems," *Sammelband zu Ehren des 250. Geburtstages Leonhard Eulers*, ed. K. Schröder (Akademie Verlag, Berlin, 1959), 315–317.

In Corollary 4 it is pointed out that when M and N^2 coincide the arc $CMNA$ is divided into three equal parts. This leads to a fifth-degree equation,

$$(1 + z)(1 - \mu z + zz)(1 + \mu z + zz) = 0,$$

with $\mu = 1 + \sqrt{3}$, hence $CM = \dfrac{1 + \sqrt{3} - \sqrt{2\sqrt{3}}}{2}$, $CN = \sqrt{\dfrac{2\sqrt{3}}{1 + \sqrt{3}}}$.

Other corollaries give formulas for half a given arc and the fifth part of a quadrant; the number of equal parts that can be computed is $2^m(1 + 2^n)$.

Theorem 6. If the chord of a simple arc CM is z and the chord of the n-fold arc $CM^2 = u$, then the chord of the $(n + 1)$-fold arc is

$$CM^{n+1} = \frac{z\sqrt{\dfrac{1 - uu}{1 + uu}} + u\sqrt{\dfrac{1 - zz}{1 + zz}}}{1 - uz\sqrt{\dfrac{(1 - uu)(1 - zz)}{(1 + uu)(1 + zz)}}}.$$

In the paper "De integratione aequationis differentialis," *Novi Commentarii Academiae Scientiarum Petropolitanae 6, 1756–57* (1761), 37–57 (*Opera omnia*, ser. I, 20, 58–79), printed in front of the previous paper but written somewhat later, Euler returned, in his own way, to the principle expressed in "Fagnano's theorem," and thereby clarified its character. The full title of the paper reads in translation:

On the integration of the differential equation

$$\frac{m\,dx}{\sqrt{1 - x^4}} = \frac{n\,dy}{\sqrt{1 - y^4}},$$

---

comparing the case first with that of $\dfrac{m\,dx}{\sqrt{1 - x^2}} = \dfrac{n\,dy}{\sqrt{1 - y^2}}$, which leads to $m \sin^{-1} x = n \sin^{-1} y + C$ (Euler writes A sin for $\sin^{-1}$).

Theorem. I therefore say that of the differential equation

$$\frac{dx}{\sqrt{1 - x^4}} = \frac{dy}{\sqrt{1 - y^4}}$$

the complete integral equation is

$$xx + yy + ccxxyy = cc + 2xy\sqrt{1 - c^4}.$$

Demonstration. When we take this equation its differential will be

$$x\,dx + y\,dy + ccxy(x\,dy + y\,dx) = (x\,dy + y\,dx)\sqrt{1 - c^4},$$

from which we obtain

$$dx[x + ccxyy - y\sqrt{(1 - c^4)}] + dy[y + ccxxy - x\sqrt{(1 - c^4)}] = 0.$$

Solving the same equation we obtain

$$y = \frac{x\sqrt{(1 - c^4)} + c\sqrt{(1 - x^4)}}{1 + ccxx} \quad \text{and} \quad x = \frac{y\sqrt{(1 - c^4)} - c\sqrt{(1 - y^4)}}{1 + ccyy}.$$

If we now assign to the radical $\sqrt{(1 - x^4)}$ the sign $+$, we must assign to the radical $\sqrt{(1 - y^4)}$ the sign $-$, so that the value $x = 0$ gives in both cases the value $y - c$. Therefore we have

$$x + ccxyy - y\sqrt{(1 - c^4)} = -c\sqrt{(1 - y^4)},$$

$$y + ccxxy - x\sqrt{(1 - c^4)} = c\sqrt{(1 - x^4)}.$$

When we substitute these values in the differential equation, we obtain

$$-c\,dx\sqrt{(1 - y^4)} + c\,dy\sqrt{(1 - x^4)} = 0,$$

or

$$\frac{dx}{\sqrt{(1 - x^4)}} = \frac{dy}{\sqrt{(1 - y^4)}}.$$

The integral of this differential equation is therefore

$$xx + yy + ccxxyy = cc + 2xy\sqrt{(1 - c^4)}.$$

and, since it contains the arbitrary constant c, it is the complete integral.
Q.E.D.

10. If, therefore, we have the equation

$$\frac{dx}{\sqrt{(1 - x^4)}} = \frac{dy}{\sqrt{(1 - y^4)}},$$

then the complete value of the integral in x is

$$x = \frac{y\sqrt{(1 - c^4)} \pm c\sqrt{(1 - y^4)}}{1 + ccyy},$$

which passes into $x = y$ if the constant c vanishes, and if we place $c = 1$ we obtain

$$x = \pm \frac{\sqrt{(1 - y^4)}}{1 + yy} = \sqrt{\frac{1 - yy}{1 + yy}},$$

which are both particular values already found above [in §9]. From here we obtain other particular values, but which lead to imaginaries. Thus if we take $c = 0$ we obtain

$$x = \frac{\sqrt{-1}}{y},$$

and if we take $cc = -1$ we obtain

$$x = \sqrt{\frac{yy + 1}{yy - 1}},$$

which also satisfy the equation in question.

---

19 EULER, LANDEN, LAGRANGE. THE METAPHYSICS OF THE CALCULUS

Many eighteenth-century mathematicians tried to give a solid foundation to the calculus. We present here three of these attempts. Euler, in his *Institutiones calculi differentialis* (Saint Petersburg, 1755; *Opera omnia*, ser. I, vol. 10), gave his theory of the zeros of different orders, dx being, he said, equal to 0. John Landen (1719–1790), an English surveyor and land agent, best remembered because of his contributions to the theory of elliptic integrals, defined his derivative by the "residue" $\left[\frac{f(x_1) - f(x_0)}{x_1 - x_0}\right]_{x_0 = x_1}$, expanding $f(x)$ in a power series in x (concentrating on the binomial theorem). We find this in the *Discourse concerning the residual analysis* (London, 1758). A few years later, Lagrange, in his "Note sur la métaphysique du calcul infinitésimal," *Miscellanea Taurinensia 2* (1760–61), reprinted in *Oeuvres*, V (1877), 597–599, gave what he thought to be an improvement on Landen's "algebraic" method, basing his whole comprehensive reevaluation of the principles of the calculus on the Taylor expansion. Lagrange later gave a full exposition in his *Théorie des fonctions analytiques* (Paris, 1797), of which the second edition, revised (1813), is reprinted in *Oeuvres*, IX (1881).

Euler's method has long been rejected, often with a kind of shoulder shrugging indicating that even the great Euler sometimes slept. A more appreciative note has recently been struck by A. P. Juschkewitch, "Euler und Lagrange über die Grundlagen der Analysis,"

Sammelband zu Ehren des 250. Geburtstages Leonhard Eulers, ed. K. Schröder (Akademie Verlag, Berlin, 1959).

Here follows a section of Euler's text, from *Opera omnia*, ser. I, vol. X, 69–72.

---

(1) EULER

83. This doctrine of the infinite, however, will be better explained when we explain what the infinitely small of the mathematicians is. There is no doubt that any quantity can be diminished until it vanishes and is transformed into nothing. But an infinitely small quantity is nothing else but a vanishing quantity and, therefore, actually will be $= 0$. This definition of the infinitely small is in agreement with the other which states that it is smaller than any given quantity. Indeed, if a quantity were so small that it is smaller than any given one, then it certainly could not be anything else but zero; for if it were not $= 0$, then a quantity equal to it could be shown, which is against the hypothesis. To those who ask what the infinitely small quantity in mathematics is, we answer that it is actually $= 0$. Hence there are not so many mysteries hidden in this concept as there are usually believed to be. These supposed mysteries have rendered the calculus of the infinitely small quite suspect to many people. Those doubts that remain we shall thoroughly remove in the following pages, where we shall explain this calculus.

84. In order to show that the infinitely small quantity is really zero we must first meet the objection: why do we not always characterize the infinitely small quantities by the same sign 0, instead of using particular symbols to designate them? Since all zeros are equal among themselves, it seems superfluous to discriminate among them by means of different signs. It is true that any two zeros are equal in such a way that their difference is zero, yet, since there are two methods of comparison, one arithmetic, the other geometric, we see this difference between them (depending on the origin of the quantities to be compared): the arithmetic ratio of two arbitrary zeros is equality, but not the geometric one. This can best be understood from the geometric proportion $2:1 = 0:0$, in which the fourth term $= 0$ as well as the third. It is in the nature of a proportion that when the first term is twice the second, then the third term must also be twice the fourth.

85. This, however, is also clear in ordinary arithmetic. It is known that a zero multiplied by an arbitrary number gives zero and that $n \cdot 0 = 0$ as well as $n:1 = 0:0$. From this it seems possible that two quantities, whatever their geometric ratio may be, will always be equal if we look at them from the arithmetic point of view. Hence if two zeros can have an arbitrary ratio, then I judge that different signs should be applied, especially when we have to consider a geometric ratio of different zeros. The calculus of the infinitely small is therefore nothing but the investigation of the geometric ratio of different infinitely small quantities. This enterprise will be thrown into the greatest confusion unless we use different signs to indicate these infinitely small quantities. No other method can be efficient.

86. Hence, if we introduce into the infinitesimal calculus a symbolism in which we denote by dx an infinitely small quantity, then $dx = 0$ as well as $a\,dx = 0$ (a an arbitrary finite quantity). Notwithstanding this, the geometric ratio $a\,dx : dx$ will be finite, namely, $a : 1$, and this is the reason that these two infinitely small quantities dx and $a\,dx$ (though both $= 0$) cannot be confused with each other when their ratio is investigated. Similarly, when different infinitely small quantities dx and dy occur, their ratio is not fixed though each of them $= 0$. And in an investigation of a ratio of two such infinitely small quantities we need all the power of the differential calculus. The use of this comparison, though at first sight it looks quite small, will more and more be appreciated and will then shine forth in the open.

87. When, therefore, the infinitely small is indeed nothing, it is clear that a finite quantity neither increases nor decreases, if an infinitely small quantity is either added to it or subtracted from it. Let a be a finite quantity and dx be infinitely small, then $a + dx$ as well as $a - dx$ and $a + n\,dx$ in general $= a$. Then whether we take for $a \pm n\,dx$ and a an arithmetic or a geometric relation, in both cases we obtain equality. Indeed, the arithmetic ratio of equality is obvious, since, as $n\,dx = 0$, we have

$$a \pm n\,dx - a = 0.$$

This clearly gives the geometric ratio of equality, which is

$$\frac{a \pm n\,dx}{a} = 1.$$

From this follows the rule, accepted by most people, that *infinitely small quantities vanish in comparison with finite ones, and thus can be rejected in so far as those finite quantities are concerned.*

The objection that the analysis of the infinites neglects mathematical rigor disappears therefore automatically, since nothing else is rejected but what is nothing at all. Hence we can in good right affirm that in this exalted science we can maintain the highest mathematical rigor as well as we find it in the ancient books.

88. Since an infinitely small quantity dx is indeed $= 0$, its square dx^2, cube dx^3, and any other power with positive exponent will also be $= 0$, and they will therefore equally vanish when compared with finite quantities. And so will also an infinitely small quantity dx^2 vanish when compared with dx, because $dx \pm dx^2$ stands to dx in the ratio of equality, whether the comparison is carried out in an arithmetic or a geometric way. There is no doubt about the former; as to geometric comparison, we obtain

$$(dx \pm dx^2) : dx = \frac{dx \pm dx^2}{dx} = 1 \pm dx = 1.$$

We shall equally find that $dx \pm dx^3 = dx$ and in general that $dx \pm dx^{n-1} = dx$, provided n is greater than zero, for the geometric ratio will be

$$(dx \pm dx^{n+1}) : dx = 1 \pm dx^n,$$

and this ratio will be that of equality since $dx^n = 0$. If, therefore, as is customary with powers, we call dx infinitely small of the first order, dx^2 of the second order, dx^3 of the third order, and so on, then it is manifest that the infinitely small of higher order will vanish with those of the first order.

———————————

Now comes a section of Landen's text (with notation slightly modernized).

———————————

(2) LANDEN

In the application of the *Residual Analysis*, a geometrical or physical problem is naturally reduced to another purely algebraical; and the solution is then readily obtained, without any supposition of motion, and without considering quantities as composed of infinitely small particles.

It is by means of the following theorem, *viz.*

$$\frac{x^{m/n} - v^{m/n}}{x - v} = x^{(m/n)-1} \times \frac{1 + \dfrac{v}{x} + \left(\dfrac{v}{x}\right)^2 + \left(\dfrac{v}{x}\right)^3 \qquad (m)}{1 + \left(\dfrac{v}{x}\right)^{m/n} + \left(\dfrac{v}{x}\right)^{2m/n} + \left(\dfrac{v}{x}\right)^{3m/n} \quad (n)},$$

(where m and n are any integers), that we are enabled to perform all the principal operations in our said Analysis; and I am not a little surprized, that a theorem so obvious, and of such vast use, should so long escape the notice of algebraists!

I have no objection against the truth of the method of fluxions, being fully satisfied, that even a problem purely algebraical may be very clearly resolved by that method, by bringing into consideration lines, and their generation by motion. But I must own, I am inclined to think, such a problem would be more naturally resolved by pure algebra, without any such consideration of lines and motion.—Suppose it required to investigate the binomial theorem; i.e., to expand $(1 + x)^{m/n}$ into a series of terms of x, and known coefficients. To do this by the method of fluxions, we first assume

$$(1 + x)^{m/n} = 1 + ax + bx^2 + cx^3 + dx^4 \quad \&c.$$

We, to proceed with perspicuity, are next to conceive x, and each term of that assumed equation, to be denoted by some line, and that line to be described by the motion of a point: Then, supposing $\dot{x}$ to be the velocity of the point describing the line x, and taking, by the rules taught by those who have treated of the said method, the several contemporary velocities of the other describing points, or the fluxions of the several terms in the said equation, we get

$$\frac{m}{n} \times (1 + x)^{(m/n)-1} \times \dot{x} = a\dot{x} + 2bx\dot{x} + 3cx^2\dot{x} + 4dx^3\dot{x} \quad \&c.$$

because, when the space described by a motion is always equal to the sum of the spaces described in the same time by any other motions, the velocity of the first motion is always equal to the sum of the velocities of the other motions.

From which last equation, by dividing by $\dot{x}$, or supposing $\dot{x}$ equal to unity, we have

$$\frac{m}{n} \times (1 + x)^{(m/n)-1} = a + 2bx + 3cx^2 + 4dx^3 \quad \&c.$$

Consequently, multiplying by $1 + x$, we have

$$\frac{m}{n} \times (1 + x)^{m/n}, \quad \text{or its equal} \quad \frac{m}{n} + \frac{m}{n}ax + \frac{m}{n}bx^2 + \frac{m}{n}cx^3 \quad \&c.$$

$$= a + \frac{2b}{a}\Big\}x + \frac{3c}{2b}\Big\}x^2 + \frac{4d}{3c}\Big\}x^3 \quad \&c.$$

From whence, by comparing the homologous terms, the coefficients a, b, c, &c. will be found.

The same theorem is investigated by the *Residual Analysis*, in the following manner.

Assuming, as above,

$$(1 + x)^{m/n} = 1 + ax + bx^2 + cx^3 \quad \&c.$$

we have

$$(1 + y)^{m/n} = 1 + ay + by^2 + cy^3 \quad \&c.$$

and, by subtraction,

$$(1+x)^{m/n} - (1+y)^{m/n} = a \cdot (x - y) + b \cdot (x^2 - y^2) + c \cdot (x^3 - y^3) + d \cdot (x^4 - y^4) \quad \&c.$$

If, now, we divide by the residual $x - y$, we shall get

$$(1 + x)^{(m/n)-1} \times \frac{1 + \dfrac{1+y}{1+x} + \left(\dfrac{1+y}{1+x}\right)^2 + \left(\dfrac{1+y}{1+x}\right)^3 \quad (m)}{1 + \left(\dfrac{1+y}{1+x}\right)^{m/n} + \left(\dfrac{1+y}{1+x}\right)^{2m/n} + \left(\dfrac{1+y}{1+x}\right)^{3m/n} \quad (n)}$$

$$= a + b \cdot (x + y) + c \cdot (x^2 + xy + y^2) + d \cdot (x^3 + x^2y + xy^2 + y^3) \quad \&c.$$

which equation must hold true let y be what it will: From whence, by taking y equal to x, we find, as before,

$$\frac{m}{n} \times (1 + x)^{(m/n)-1} = a + 2bx + 3cx^2 + 4dx^3 \quad \&c.$$

The rest of the operation will therefore be as above specified.

Now, as to either of these methods of investigation, I shall not take upon me to say any thing in particular; it is submitted to the reader to compare one with the other, and judge which of the two is most natural.

Finally, we present a section from Lagrange's *Théorie des fonctions analytiques*, from *Oeuvfes*, X, 20ff. He opens his book with a definition of function, primitive function, and derivative function.

---

(3) LAGRANGE

We define as a *function* of one or several quantities any mathematical expression in which those quantities appear in any manner, linked or not with some other quantities that are regarded as having given and constant values, whereas the quantities of the function may take all possible values. Thus in a function we consider only the quantities which are supposed to be variables without regard to the constants it may contain.

The term *function* was used by the first analysts in order to denote in general the powers of a given quantity. Since then the meaning of this term has been extended to any quantity formed in any manner from any other quantity. Leibniz and the Bernoullis were the first to use it in this general sense, which is nowadays the accepted one.[1]

Let us assign to the variable of a function some increment by adding to this variable an arbitrary quantity; we can, if the function is algebraic, expand it in terms of the powers of this quantity by using the familiar rules of algebra. The first term of the expansion will be the given function, which will be called the primitive function; the following terms will be formed of various functions of the same variable multiplied by the successive powers of the arbitrary quantity. These new functions will depend only on the primitive function from which they are derived and may be called the *derivative* functions. Generally speaking, whether the primitive function is algebraic or not, it can always be expanded in the same manner, and in this way it will give rise to the derivative functions. The functions considered from this point of view lead to an analysis superior to the ordinary one because of its generality and of its numerous applications and we shall see in this work that the analysis that is commonly called *transcendental* or *infinitesimal* is, in fact, not different from that of the primitive and derivative functions; and that the differential and integral calculus is also, properly speaking, nothing else but the calculus of those very same functions.

Lagrange then criticizes the foundation of the calculus by Newton and others on infinitesimals based on motion, with which, he says, even Maclaurin had difficulty. He praises the

[1] See Selection V.16.

approach of "a capable English mathematician" (Landen) although it is cumbersome. Then Lagrange gives his own method:

Now let us consider a function $f(x)$ of a variable x. If we replace x by $x + i$, i being any arbitrary quantity, it will become $f(x + i)$ and, by the theory of series, we can expand it in a series of the form

$$f(x) + pi + qi^2 + ri^3 + \cdots,$$

in which the quantities $p, q, r, \ldots$, the coefficients of the powers of i, will be new functions of x, which are derived from the primitive function of x, and are independent of the quantity i.

But, in order to prove what we claim, we shall examine the actual form of the series representing the expansion of a function $f(x)$ when we substitute $x + i$ for x, which involves only positive integral powers of i.

This assumption is indeed fulfilled in the cases of various known functions; but nobody, to my knowledge, has tried to prove it a priori—which seems to me to be all the more necessary since there are particular cases in which it is not satisfied. On the other hand, the differential calculus makes definite use of this assumption, and the exceptional cases are precisely those in which objections have been made to the calculus.

I will first prove that in the series arising by the expansion of the function $f(x + i)$ no fractional power of i can occur except for particular values of x.

After having accomplished this, Lagrange continues in Chapter II, entitled "Derived functions, their notation and algorithm," as follows:

We have seen that the expansion of $f(x + i)$ generates various other functions $p, q, r, \ldots$, all of them derived from the original function $f(x)$, and we have given the method for finding these functions in particular cases. But in order to establish a theory concerning these kinds of functions we must look for the general law of their derivation.

For this purpose, let us take once more the general formula $f(x + i) = f(x) + pi + qi^2 + ri^3 + \cdots$, and let us suppose that the undetermined quantity x is replaced by $x + o$, o being any arbitrary quantity independent of i. Then $f(x + i)$ will become $f(x + i + o)$, and it is clear that we shall obtain the same result by simply substituting $i + o$ for i in $f(x + i)$. The result must also be the same whether we replace the quantity i by $i + o$ or x by $x + o$ in the expansion $f(x)$.

The first substitution yields

$$f(x) + p(i + o) + q(i + o)^2 + r(i + o)^3 + \cdots,$$

or, expanding the powers of $i + o$ and writing out for the sake of simplicity no more than the first two terms of each power (since the comparison of these terms will be sufficient for our purpose):

$$f(x) + pi + qi^2 + ri^3 + si^4 + \cdots + po + 2qio + 3ri^2o + 4si^3o + \cdots.$$

In order to carry out the other substitution, we note that we obtain $f(x) + f'(x)o + \cdots, p + p'o + \cdots, q + q'o + \cdots, r + r'o + \cdots$ when we replace x by $x + o$ in the functions $f(x), p, q, r, \ldots$, respectively; here we retain in the expansion only the terms that include the first power of o. It is clear that the same expression will become $f(x) + pi + qi^2 + ri^3 + si^4 + \cdots + f'(x)o + p'io + q'i^2o + r'i^3o + \cdots$.

Since these two results must be identical whatever the values of i and o may be, comparison of the terms involving $o, io, i^2o, \ldots$, will give:

$$p = f'(x), \quad 2q = p', \quad 3r = q', \quad 4s = r', \ldots$$

Now it is clear that in the same way that $f'(x)$ is the first derived function of $f(x)$, p' is the first derived function of p, q' the first derived function of q, r' the first derived function of r, and so on. Therefore, if, for the sake of greater simplicity and uniformity, we denote by $f'(x)$ the first derived function of $f(x)$, by $f''(x)$ the first derived function of $f'(x)$, by f''' the first derived function of $f''(x)$, and so on, we have

$$p = f'(x), \quad \text{and hence} \quad p' = f''(x);$$

consequently

$$r = \frac{q'}{2} = \frac{f''(x)}{2}, \quad \text{hence} \quad q' = \frac{f'''(x)}{2},$$

consequently

$$r = \frac{q'}{3} = \frac{f'''(x)}{2 \cdot 3}, \quad \text{hence} \quad r' = \frac{f^{\mathrm{iv}}(x)}{2 \cdot 3},$$

consequently

$$s = \frac{r'}{4} = \frac{f^{\mathrm{iv}}(x)}{2 \cdot 3 \cdot 4}, \quad \text{hence} \quad s' = \frac{f^{\mathrm{v}}(x)}{2 \cdot 3 \cdot 4},$$

and so on.

Then by substituting these values in the expansion of the function $f(x + i)$, we obtain

$$f(x + i) = f(x) + f'(x)i + \frac{f''(x)}{2}i^2 + \frac{f'''(x)}{2\cdot 3}i^3 + \frac{f^{\mathrm{iv}}(x)}{2\cdot 3\cdot 4}i^4 + \cdots.$$

This new expression has the advantage of showing how the terms of the series depend on each other and above all how we can form all the derived functions involved in the series provided that we know how to form the first derived function of any primitive function.

We shall call the function $f(x)$ the *primitive function* with respect to the functions $f'(x), f''(x), \ldots$ that are derived from it; these functions are called the *derived functions* with respect to the former one. Moreover, we shall call the first derived function $f'(x)$ the *first function*, the second derived function the *second function*, the third derived function the *third function*, and so on. In the same way, if y is supposed to be a function of x, we denote its derived function by $y', y'', y''', \ldots$, respectively, so that, y being the primitive function, y' will be its *first function*, y'' its *second function*, y''' its *third function*, and so on.

Consequently, if x is replaced by $x + i$, y will become

$$y + y'i + \frac{y''i^2}{2} + \frac{y'''i^3}{2\cdot 3} + \cdots.$$

Thus, provided that we have a method of computing the first function of any primitive function, we can obtain, by merely repeating the same operation, all the derived functions, and consequently all the terms of the series that result from expanding the primitive function.

Finally, only a little knowledge of the differential calculus is necessary to recognize that the derived functions $y', y'', y''' \ldots$ of x coincide with the expressions

$$\frac{dy}{dx}, \quad \frac{d^2y}{dx^2}, \quad \frac{d^3y}{dx^3}, \quad \text{respectively.}$$

Lagrange then continues with examples of functions and their derivatives.

The reader interested in this discussion of the foundations of the calculus should further consult C. B. Boyer, *The history of the calculus* (Dover, New York, 1959). On Lagrange see also Judith V. Grabiner, "The calculus as algebra, J. L. Lagrange, 1733–1813," dissertation, Harvard University, 1966.

20 JOHANN AND JAKOB BERNOULLI. THE BRACHYSTOCHRONE

Johann Bernoulli, *Acta Eruditorum 6* (June 1696), 269 (*Opera omnia*, I, 161), challenged the learned world to solve the following "problema novum":

"Let two points A and B be given in a vertical plane. To find the curve that a point M, moving on a path AMB, must follow such that, starting from A, it reaches B in the shortest time under its own gravity."

The interesting thing was, he added, that this curve is not a straight line, but a curve well known to geometers.

After another, more explicit explanation of the problem, published at Groningen in January 1697 (*Opera omnia*, I, 166–169), he then gave the solution himself in the *Acta Eruditorum* of May 1697, pp. 206–211 (*Opera omnia*, I, 187–193), under a title beginning "Curvatura radii in diaphanis non uniformibus" (The curvature of a ray in nonuniform media). He showed that the required curve, which he called a brachystochrone, is a cycloid. In the same issue, pp. 211–217, was also published the solution of his older brother Jakob: "Solutio problematum fraternorum ... una cum propositione reciproca aliorum" (Solution of problems of my brother ... together with the proposition of others in turn; *Opera*, I, 768–778). We present both papers, from which we can see that the one by Jakob contains a general principle, namely, that a curve which constitutes a maximum or a minimum as a whole must also possess this property in the infinitesimal. Johann's paper also contains a general principle, that of the parallel between (geometric) optics and (point) mechanics, which would lead to the work of W. R. Hamilton in the 1830's. In the same *Acta Eruditorum* of 1697 are also contributions to the same problem by Leibniz, L'Hôpital, Tschirnhaus, and Newton. All except L'Hôpital find the cycloid as the solution.

These papers open the history of a new field, the calculus of variations. The papers of the Bernoullis exist in a German translation by P. Stäckel, Ostwald's *Klassiker*, No. 46 (Engelmann, Leipzig, 1894).

---

(1) THE SOLUTION OF JOHANN BERNOULLI

The curvature of a ray in nonuniform media, and the solution of the problem to find the brachystochrone, that is, the curve on which a heavy point falls from a given position to another given position in the shortest time, as well as on the construction of the synchrone or the wave of the rays.

...We have a just admiration for Huygens, because he was the first to discover that a heavy point on an ordinary *Cycloid* falls in the same time [*tautochronos*], whatever the position from which the motion begins.[1] But the reader will be greatly amazed [*an non obstupescus plane*], when I say that exactly this *Cycloid*, or *Tautochrone of Huygens*, is our required *Brachystochrone*. I reached this understanding in two ways, one indirect and one direct. When I pursued the first, I discovered a wondrous agreement between the curved path of a light ray in a continuously varying medium and our *Brachystochrone*. I also found other rather mysterious things [*in quibus nescio quid arcani subest*] which might be useful in dioptric investigations. It is therefore true, as I claimed when I proposed

---

[1] Huygens, *Horologium oscillatorium* (Paris, 1673), Proposition XXV: In a cycloid with vertical axis and with its vertex down, the times of descent in which a mobile particle, starting from rest at an arbitrary point of the curve, reaches the lowest point are equal among themselves, and have to the time of the vertical fall along the total axis of the cycloid a ratio equal to that of the semicircumference of a circle to its diameter. *Oeuvres complètes*, XVIII (1934), 185. See Selection IV.18.

the problem, that *it is not just naked speculation,* but *also very useful for other branches of knowledge,* namely, for dioptrics. But in order to confirm my words by the deed, let me here give the first mode of proof!

Fermat, in a letter to De la Chambre,[2] has shown that a light ray passing from a thin to a more dense medium, is bent toward the perpendicular in such a way that, under the supposition that the ray moves continuously from the light to the illuminated point, it follows the path that requires the shortest time. With the aid of these principles he showed that the sine of the angle of incidence and the sine of the angle of refraction are in inverse proportion to the densities of the media, hence directly as the velocities with which the light ray penetrates these media. Later Leibniz, in the *Acta Eruditorum,* 1682, pp. 185 sequ., and soon afterward the famous Huygens in his *Treatise on light,* p. 40,[3] have demonstrated this more comprehensively and, by most valid arguments, have established the physical, or better the metaphysical, principle which Fermat seems to have abandoned at the insistence of Clerselier, remaining satisfied with his geometric proof and giving up his rights all too lightly.

Now we shall consider a medium that is not homogeneously dense, but consists of purely parallel horizontally superimposed layers, of which each consists of diaphanous matter of a certain density decreasing or increasing according to a certain law. It is then manifest that a ray which we consider as a particle will not be propagated in a straight line, but in a curved path. This has already been considered by Huygens in his above-mentioned *Treatise on Light,* but he did not determine the nature of this minimizing curve such that the particle, whose velocity increases and decreases depending on the density of the medium, will pass from point to point in the shortest time. We know that the sines of the angles of refraction at the separate points are to each other inversely as the densities of the media or directly as the velocities of the particles, so that the brachystochrone curve has the property that the sines of its angles of inclination with respect to the vertical are everywhere proportional to the velocities. But now we see immediately that the brachystochrone is the curve that a light ray would follow on its way through a medium whose density is inversely proportional to the velocity that a heavy body acquires during its fall. Indeed, whether the increase of the velocity depends on the constitution of a more or less resisting medium, or whether we forget about the medium and suppose that the acceleration is generated by another cause according to the same law as that of gravity, in both cases the curve is traversed in the shortest time. Who prohibits us from replacing the one by the other?

In this way we can solve the problem for an arbitrary law of acceleration, since it is reduced to the determination of the path of a light ray through a medium of arbitrarily varying density. Hence let *FGD* [Fig. 1] be the medium bounded by the horizontal line *FG* on which the luminous point *A* is situated.

---

[2] Fermat's letters to Martin Cureau de la Chambre are of 1657 and 1662 (*Oeuvres,* II, 354–359, 457–463). The law of refraction was published by Descartes in his *Dioptrique* (1637). Fermat first opposed it, but then reestablished it by a maximum-minimum principle.

[3] Huygens, *Traité de la lumière* (Leiden, 1690), 40; *Oeuvres complètes,* XIX (1737), 489. On Leibniz, see Selection V.1, note 13, p. 279.

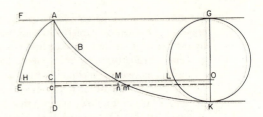

Fig. 1

Let the curve AHE, with vertical axis AD, be given, its ordinates HC determining the densities of the medium at altitude AC or the velocities of the light rays or particles at M. Let the curved line of the light ray, which we wish to determine, be ABM. Let us write for AC, x; for CH, t; for CM, y; and for the differentials Cc, dx; diff. $mn = dy$; diff. $Mm = dz$, finally, let a be an arbitrary constant. Then Mm is the total sine, mn the sine of the angle of refraction or the angle of inclination of the curve with respect to the vertical. As we have said before, the ratio of mn to CH is constant, hence

$$dy : t = dz : a,$$

so that

$$a\, dy = t\, dz,$$

or

$$aa\, dy^2 = tt\, dz^2 = tt\, dx^2 + tt\, dy^2.$$

This gives a general differential equation for the required curve ABM:

$$dy = t\, dx : \sqrt{(aa - tt)}.$$

In this way I have solved at one stroke two important problems—an optical and a mechanical one—and have achieved more than I have demanded from others: I have shown that the two problems, taken from entirely separate fields of mathematics, have the same character.

Now let us take a special case, namely the common hypothesis first introduced by Galilei, who proved that the velocities of falling bodies are to each other as the square roots [*in ratione subduplicata*] of the altitudes traversed—then this is really the given problem. Under this assumption the given curve AHE is a parabola $tt = ax$, hence $t = \sqrt{ax}$. If this value is substituted in the original equation, we obtain

$$dy = dx \sqrt{\frac{x}{a - x}},$$

from which I conclude that the *Brachystochrone* is the ordinary *Cycloid*. For when the circle GLK of radius a rolls on AG and the rolling starts at A, the point K describes a cycloid, of which the differential equation is exactly

$$dy = dx \sqrt{\frac{x}{a - x}},$$

if $AC = x$, $CM = y$.

Bernoulli then shows this analytically by writing

$$dx\sqrt{\frac{x}{a-x}} = \frac{1}{2}\frac{a\,dx}{\sqrt{ax-x^2}} - \frac{1}{2}\frac{a\,dx - 2x\,dx}{\sqrt{ax-x^2}},$$

which integrated gives

$$CM = \text{arc } GL - LO,$$

from which, since $MO = CO - \text{arc } GL + LO = \text{arc } LK + LO$, it follows that $ML = \text{arc } LK$.[4]

To solve the problem completely he then shows that from a given point as vertex a cycloid can be described that passes through a second given point.

Before I end I must voice once more the admiration that I feel for the unexpected identity of Huygens' tautochrone and my brachystochrone. I consider it especially remarkable that this coincidence can take place only under the hypothesis of Galilei, so that we even obtain from this a proof of its correctness. Nature always tends to act in the simplest way, and so it here lets one curve serve two different functions, while under any other hypothesis we should need two curves, one for tautochronic oscillations, the other for the most rapid fall. If, for example, the velocities were as the altitudes, then both curves would be algebraic, the one a circle, the other one a straight line.[5]

Bernoulli then introduces the *synchrone*: the curve PB [Fig. 2] in a vertical plane such that a heavy body falling from A along this curve reaches the points B in the same time as a heavy body falling on the cycloid AB. Referring to Huygens, he concludes that PB is

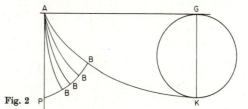

Fig. 2

---

[4] This gives the equation of the cycloid in the form

$$x = \frac{a}{2}(1 - \cos t), \quad y = \frac{a}{2}(t - \sin t), \quad t = \pi - \varphi, \quad \text{arc } LK = a\varphi.$$

The differential equation can already be found in Leibniz's first paper on the integral calculus of 1686 (see Selection V.2).

[5] The cases mentioned are $t = ax$ and $t = ax^{1/3}$.

also a cycloid intersecting all cycloids with initial point A at a right angle. He ends by suggesting that other orthogonal trajectories of given families of curves be found.[6]

(2) THE SOLUTION OF JAKOB BERNOULLI

Bernoulli begins by saying that, though he did not care for his brother's challenge, he tackled his problem at the invitation of Leibniz and solved it in a few weeks, finding what he calls the oligochrone. Then he begins with the following:

Lemma. Let $ACEDB$ [Fig. 3] be the desired curve along which a heavy point falls from A to B in the shortest time, and let C and D be two points on it as close together as we like. Then the segment of arc CED is among all segments of arc with C and D as end points the segment that a heavy point falling from A traverses in the shortest time. Indeed, if another segment of arc CFD were traversed in a shorter time, then the point would move along $ACFDB$ in a shorter time than along $ACEDB$, which is contrary to our supposition.

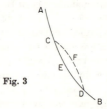

Fig. 3

Hence in a plane arbitrarily inclined to the horizon (the plane need not be horizontal), take ACB [Fig. 4] as the required curve, on which a heavy point from A reaches B in a shorter time than on any other curve in this plane. Take on it two points C and D infinitesimally close together and draw the horizontal line AH, the vertical CH, and DF normal to it. Take E halfway between C and F and complete the parallelogram DE by means of the line EI. On EI we now must determine a point G such that the time of fall through CG + the time of fall through GD is a minimum. I denote this by

$$t_{CG} + t_{GD};^7$$

we have to keep in mind that the fall begins at the altitude of A. If we now take on the line EI another point L [Fig. 4a] such that GL is incomparably small as compared to EG, and if we draw CL and DL, then, according to the nature of a minimum:

$$t_{CL} + t_{LD} = t_{CG} + t_{GD}$$

[6] Johann Bernoulli does not yet use the term "orthogonal trajectories." The concept played an important role in the work of Leibniz and Bernoulli in those days. The connection with Huygens's theory of light was clear. The term "trajectory" dates from an article by Johann Bernoulli in the *Acta Eruditorum* of 1698 (*Opera omnia*, I, 266).

[7] We write t_{CG} instead of Bernoulli's tCG.

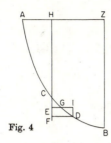

Fig. 4

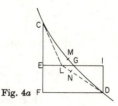

Fig. 4a

and hence

$$t_{CG} - t_{CL} = t_{LD} - t_{GD}.$$

I now reason as follows. According to the nature of the fall of heavy bodies,

$$CE:CG = t_{CE}:t_{CG},$$

$$CE:CL = t_{CE}:t_{CL},$$

hence

$$CE:(CG - CL) = t_{CE}:(t_{CG} - t_{CL}).^8$$

If we take a point M on CG such that $CG - CL = MG$, then we have, because of the similarity of the triangles MLG and CEG,

$$CE:GL = EG \times t_{CE}:CG \times (t_{CG} - t_{CL}).$$

In the same way we find, according to the nature of the fall of heavy bodies,

$$EF:GD = t_{EF}:t_{GD}$$

$$EG:LD = t_{EF}:t_{LD},$$

hence

$$EF:(LD - GD) = t_{EF}:(t_{LD} - t_{GD}).$$

If we take on DL the point N such that $LD - GD = LN$, then we have, because of the similarity of the triangles LNG and GID,

$$LN:LG = GI:GD,$$

hence

$$EF:LG = GJ \times t_{EF}:GD \times (t_{LD} - t_{GD}).$$

---

[8] Bernoulli uses no parentheses.

By comparison we obtain

$$EG \times t_{CE} : CG \times (t_{CG} - t_{CL}) = GI \times t_{EF} : GD \times (t_{LD} - t_{GD}) = CG : GD,$$

and by permutation

$$EG \times t_{CE} : GI \times t_{EF} = CG \times (t_{CG} - t_{CL}) : GD \times (t_{LD} - t_{GD}) = CG : GD,$$

because there is a minimum.

But according to the law of gravity we have

$$EG \times t_{CE} : GI \times t_{EF} = \frac{EG}{\sqrt{HC}} : \frac{GI}{\sqrt{HE}},$$

and therefore finally:

$$\frac{EG}{\sqrt{HC}} : \frac{GI}{\sqrt{HE}} = CG : GD.$$

By the way, please let Mr. Nieuwentyt take notice of the use of second differentials [*differentio-differentiales*], which he wrongly neglects.[9] Indeed, we were forced to suppose that the part GL of the infinitesimally small segments EG, GI is infinitesimally small with respect to them, and I fail to see how without it a solution of the problem can be obtained.

Now EG and GI are elements of the abscissa AH, CG and GD are elements of the curve, HC and HE their ordinates, and CE and EF elements of the ordinate. The problem can therefore be reduced to the purely geometric one of determining the curve of which the line elements are directly proportional to the elements of the abscissa and indirectly proportional to the square roots of the ordinates. I find that this property belongs to the *Isochrone* of Huygens, which therefore is also the *Oligochrone*, namely the cycloid, well known to the geometers.

Bernoulli gives a geometric proof, but we can readily verify that the equation $ds = a\,dx/\sqrt{y}$, if $AH = x$, $HC = y$, leads to the same equation

$$dy\sqrt{\frac{y}{a-y}} = dx$$

that Johann Bernoulli arrived at. Jakob then shows, like Johann, how to find a cycloid with horizontal base AH passing through the given points A and B.

[9] On Bernard Nieuwentijt see Selection V.1.

Bernoulli then indicates some other problems that can be solved by his method. They are:

(1) To find on which of the infinitely many cycloids (or circles, parabolas, etc.) passing through A with the same base AH a heavy point can fall from A to the vertical line ZB in the shortest time.

(2) To find the path of a particle moving in a medium of varying density, which curve is the same as the refraction curve studied by Huygens and himself.

(3) To find isoperimetric figures of different kinds;[10] he especially challenges his brother Johann to solve the following problem: Among all isoperimetric figures on the common base BN [Fig. 5], to find the curve BFN which—though not having itself the largest area— is such that this property belongs to another curve BZN of which the ordinate PZ is pro-portional to a power or a root of the segment PF or the arc BF. Johann will get 50 ducats from a gentleman known to Jakob if he solves this problem before the end of the year.

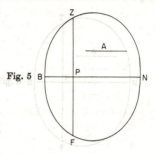

Fig. 5

21   EULER. THE CALCULUS OF VARIATIONS

After having mastered the methods that the Bernoullis had developed in the study of iso-perimetric problems, Euler began to develop his own approach shortly before 1732. Where the Bernoullis had only solved specific problems, Euler began to look for a general theory. This theory, which began to take shape after 1740, appeared finally in the majestic volume entitled *Methodus inveniendi lineas curvas maximi minimive proprietate gaudentes sive solutio problematis isoperimetrici latissimo sensu accepti* (A method for discovering curved lines having a maximum or minimum property or the solution of the isoperimetric problem taken in its widest sense; Lausanne, Geneva, 1744; *Opera omnia*, ser. I, vol. 25, 1952). The book consists of six chapters with two appendices. It does not yet present the calculus of variations in the form in which we know it—that was Lagrange's work, the importance of which Euler immediately understood when it appeared. Euler's method still has a geometric character, but Euler understood its nonessential nature: in chap. I, §32, he remarks: "It is thus possible to reduce problems of the theory of curves to problems belonging to pure analysis. And conversely, every problem of this kind proposed in pure analysis can be con-sidered and solved as a problem of the theory of curves." Euler, however, preferred to deal with such problems in a geometric way, because by this means the method is "wonderfully

[10] Here the isoperimetric problems enter into the calculus. The theorem that of all figures of the same perimeter the circle has the largest area is ascribed to Zenodorus, who lived between 200 B.C. and A.D. 100. Some of his theorems can be found in Pappus' "Collection." See T. L. Heath, *A manual of Greek mathematics* (Clarendon Press, Oxford, 1931), 382–383.

aided and brought nearer to the intellect" (*mirifice adiuvetur atque intellectu facilis red-datur*).

Chapter I deals mainly with the type of questions that occur in the calculus of variations (the term *calculus variationum* does not appear in the book, being first employed by Euler in a paper of 1760 (1766), to indicate Lagrange's algorithm which uses δx, δy). Euler makes a difference between absolute and relative maxima and minima. In Chapter II we begin to meet the many special problems that give the book its charm. In chapter III he discusses the case in which certain other indetermined quantities occur under the integral. Chapter IV contains more special problems, chapter V discusses the relative method, and chapter VI gives more problems. The first appendix deals with elastic curves. The book abounds in examples.

The book was republished as ser. I, vol. 25 of the *Opera omnia* with a 55-page German introduction by C. Carathéodory (containing a classification of Euler's examples). There exist a partial German translation by P. Stäckel in Ostwald's *Klassiker*, No. 46 (Engelmann, Leipzig, 1894) and a complete Russian translation (Moscow and Leningrad, 1934).

We begin with a section of chapter I.

Hypothesis I. The abscissa is denoted by x, the ordinate [*applicata*] by y; further, $dy = p\,dx$, $dp = q\,dx$, $dq = r\,dx$, $dr = s\,dx$, and so on. The integral under consideration is $\int Z\,dx$, where Z must be such that $Z\,dx$ cannot be integrated; Z can be a function [*functio*] not only of x and y, but also of p, q, r, ...

Then the principle, which Jakob Bernoulli had established, is announced in

---

Proposition II. *Theorem.* If *amz* [Fig. 1] is a curve in which the value of the formula $\int Z\,dx$ is a maximum or a minimum, and Z is an algebraic or a determined function of $x, y, p, q, r, \ldots$, then every portion *mn* of this curve has the special property that, if it is referred to the abscissa MN, the value of $\int Z\,dx$ is also a maximum or minimum.

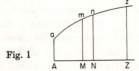

Fig. 1

The proof follows essentially the reasoning of Jakob Bernoulli (see Selection V.20(2)). One of the corollaries points out that the reasoning does not hold when in Z there appear indeterminate integrals, as $\int y\,dx$. Then follows

Proposition III. *Theorem.* If amz is a curve, corresponding to the abscissa AZ, for which $\int Z\,dx$ is a maximum or a minimum, while Z contains indefinite integral expressions, then the property of a maximum or a minimum does not hold for any arbitrary part of the curve, but belongs to the whole curve corresponding to the abscissa AZ.

After the proof of this theorem, and certain corollaries, comes

Hypothesis II. When the abscissa AZ [Fig. 2] of a curve is divided into innumerable infinitely small elements $IK, KL, LM, \ldots$, all equal to one another, and some portion AM is denoted by X, to which some variable function F

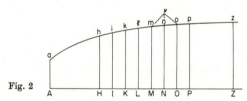

Fig. 2

corresponds, then we shall denote the values of the function F for the following points of the abscissa N, O, P, Q, and for the preceding points $L, K, I, H, \ldots$ by $F', F'', F''', \ldots$ for $N, O, P, \ldots$, and $F_{\prime}, F_{\prime\prime}, F_{\prime\prime\prime}, \ldots$ for $L, K, I, \ldots$. Thus we can indicate in an easy way, without prolix writing of differentials, the value of a subscript prime variable function at any point of the abscissa.

There follow five corollaries, which express the following identities:

$$F' = F + dF, \qquad F = F_{\prime} + dF_{\prime},$$
$$F'' = F' + dF', \qquad F_{\prime} = F_{\prime\prime} + dF_{\prime\prime},$$
$$F''' = F'' + dF'', \qquad F_{\prime\prime} = F_{\prime\prime\prime} + dF_{\prime\prime\prime},$$

and so forth, and, when the ordinates $Mm, Nn, Oo, Pp, \ldots$ are indicated by $y, y', y'', y''', \ldots$ and $Ll, Kk, Ii, \ldots$ by $y_{\prime}, y_{\prime\prime}, y_{\prime\prime\prime}, \ldots$ then, since $p = \dfrac{dy}{dx} = \dfrac{Nn - Mm}{dx}$,

$$p = \frac{y' - y}{dx}, \quad p' = \frac{y'' - y'}{dx}, \quad p'' = \frac{y''' - y''}{dx}, \quad p''' = \frac{y^{\mathrm{iv}} - y'''}{dx},$$

$$p_{\prime} = \frac{y - y_{\prime}}{dx}, \quad p_{\prime\prime} = \frac{y_{\prime} - y_{\prime\prime}}{dx}, \quad p_{\prime\prime\prime} = \frac{y_{\prime\prime} - y_{\prime\prime\prime}}{dx}, \quad \text{etc.}$$

$$q = \frac{dp}{dx} = \frac{p' - p}{dx} = \frac{y'' - 2y' + y}{dx^2},$$

$$q' = \frac{y''' - 2y'' + y'}{dx^2}, \quad \text{etc.,}$$

$$r = \frac{y''' - 3y'' + 3y' - y}{dx^3}, \quad \text{etc.}$$

Corollaries VI–VIII. If $\int Z \, dx$ is referred to the abscissa $AM = x$, then the value corresponding to the next element $MN = dx$ is $Z \, dx$. In a similar way we shall indicate the values of $\int Z \, dx$ belonging to the elements $MN, MO, OP, \ldots$ by $Z \, dx, Z' \, dx, Z'' \, dx, \ldots$. Then if the expression $\int Z \, dx$ is referred to the abscissa $AM = x$, the value belonging to the abscissa AZ is

$$\int Z \, dx + Z \, dx + Z' \, dx + Z'' \, dx + \cdots,$$

until we arrive at point Z.

When therefore we must find the curve for which, for the given abscissa, the value of $\int Z \, dz$ is the largest or smallest, we must obtain a maximum or minimum of this expression $\int Z \, dx + Z \, dx + Z' \, dx + Z'' \, dx +$ etc.

Proposition IV. *Theorem.* When the expression $\int Z \, dx$ has a maximum or minimum for the curve *amnoz* [Fig. 2] referred to the given abscissa AZ, and we conceive another curve *amvoz* which differs from the first one only by an infinitely small amount, then the value of $\int Z \, dz$ is the same for both curves.

After demonstration and several corollaries there follows

Definition V. The *differential value* of a given expression for a maximum or minimum is the difference of the values which this expression receives on the required curve and on the curve that results from it by an infinitely small change.

One of the corollaries points out that for a maximum or minimum of $\int Z \, dx$ the differential value vanishes.

CHAPTER II

Proposition I. *Problem.* When in a curve *amz* [Fig. 2] some ordinate Nn is augmented by an infinitely small segment nv, then we must find the increase or decrease of the separate quantities determined by the curve.

To obtain the solution all quantities depending on y' are changed; the others remain fixed. For instance, $p = (y' - y)/dx$ increases by the particle $n\nu/dx$, and $p' = (y'' - y')/dx$ decreases by the particle $n\nu/dx$. Reasoning in a similar way, we find the following table of quantities that change:

Quantity:	y'	p	p'	$q_{,}$	q	q'	$r_{,,}$	$r_{,}$	r	r'
Change:	$+n\nu$	$+\dfrac{n\nu}{dx}$	$-\dfrac{n\nu}{dx}$	$+\dfrac{n\nu}{dx}$	$-\dfrac{2n\nu}{dx}$	$+\dfrac{n\nu}{dx}$	$+\dfrac{n\nu}{dx^3}$	$-\dfrac{3n\nu}{dx^3}$	$+\dfrac{3n\nu}{dx}$	$-\dfrac{n\nu}{dx^3}$

Among the corollaries we find one stating that from the changes in the primary quantities all the changes in the quantities that are composed of them can be found. These changes can in a sense be considered their differentials. From the ordinary differential of, say, $y'\sqrt{(1 + p^2)}$, which is $dy'\sqrt{(1 + p^2)} + y'p\,dp/\sqrt{(1 + p^2)}$, we can therefore find as the change of the function

$$+n\nu\sqrt{(1 + p^2)} + \frac{y'pn\nu}{dx\sqrt{(1 + p^2)}}.$$

Proposition II. *Problem.* When Z is a determined function of x and y alone, to find the curve az for which the value of the expression $\int Z\,dx$ is a maximum or minimum.

When $dz = M\,dx + N\,dy$, the required curve is given by $N\,dxn\nu = 0$, or $N = 0$.
Among the corollaries is the case in which Z is a function of z only, when all curves having the same axis are all solutions. When Z as a function of x and y is algebraic, the solution is algebraic. A maximum or minimum may also occur when $N = \infty$. Several examples follow; one is to find the curve for which for all curves corresponding to the same abscissa $\int (ax - yy)y\,dx = 0$ has a maximum or minimum. Answer: $ax - 3yy = 0$. Euler then discusses whether this is a maximum or a minimum, and finds the value of the integral.

Proposition III. *Problem.* When Z is a determined function of x, y, and p, so that

$$dZ = M\,dx + N\,dy + P\,dp,$$

to find among all curves corresponding to the same abscissa the curve for which $\int Z\,dx$ is a maximum or minimum.
Solution. Let amz be the required curve, and imagine the ordinate $Nn = y'$ augmented by a particle $n\nu$. Then the differential value of the expression $\int Z\,dx$, or of the equivalent expression $Z\,dx + Z'\,dx + Z''\,dx +$ etc., together with $Z_{,}\,dx + Z_{,,}\,dx + Z_{,,,}\,dx +$ etc., must be $= 0$. We obtain the differential value

of the whole quantity $\int Z \, dz$, resulting from the translation of the point n to ν, when we look for the differential values of the separate terms, insofar as they have been affected by the translation, and combine them into a sum. But as a result of the translation of the point n to ν only those terms are changed that contain the quantities y', $p_{\cdot}$, and p', hence only the terms $Z \, dx$ and $Z' \, dx$; since just as Z depends on x as well as on y and p, so Z' is a function of y' and p'. We must therefore differentiate those members, and substitute in their differentials for dy', dp, and dp' the above-mentioned values $+n\nu$, $+n\nu/dx$, and $-n\nu/dx$. But just as $dZ = M \, dx + N \, dy + P \, dp$, so $dZ' = M' \, dx + N' \, dy' + P' \, dp'$. The differential value of Z is therefore $P(n\nu/dx)$, that of Z' is equal to $N' \cdot n\nu - P' \cdot n\nu/dx$, and that of $Z \, dx + Z' \, dx$, hence also of the whole expression $\int Z \, dx$, is equal to $n\nu \cdot (P + N' \, dx - P')$. But $P' - P = dP$ and for N' we may write N, so that the differential value will be $= n\nu \cdot (N \, dx - dP)$. And since we obtain the equation of the required curve by equating the differential value of the expression $\int Z \, dx$ to zero, we obtain $0 = N \, dx - dP$ or $N - dP/dx = 0$, which equation expresses the nature of the required curve. Which is what we have to find. Q.E.I.

Corollaries point out that $N - dP/dx = 0$ is always a differential equation of the second order [*gradus*], unless there is no p in P. There are therefore two constants, so that two points on the curve may be prescribed. A number of special cases are discussed. The co-ordinates x and y may be interchanged. Among the examples we find nos. 33, 34, 36, 38:

$Z = \sqrt{(1 + pp)}$, $y = a + nx$ (the straight line as shortest distance between two points);

$$Z = \frac{\sqrt{(1 + pp)}}{\sqrt{x}}, \quad y = \int dx \sqrt{\frac{x}{a - x}}, \quad \text{the cycloid;}[1]$$

$$Z \, dy = \frac{y \, dy^3}{dx^2 + dy^2};[2]$$

$$x = \frac{a}{2}\left(\frac{3}{4p^4} + \frac{1}{pp} + 1 + \log p\right),$$

from which the curve can be constructed, using logarithms (Euler writes lp for our $\log p$), $Z = (xx + yy)^n\sqrt{(1 + pp)}$, many cases, depending on n; for example, $n = \frac{1}{2}$ gives

$$x^2 - y^2 = 2kxy + C.$$

---

[1] This is the brachystochrone; see Selection V.20.

[2] This is the problem found in Newton's *Principia*, Book II, Sect. 7, Prop. 34, Scholium: to find the shape of a volume of rotation moving in a fluid with uniform velocity parallel to its axis under a pressure perpendicular to the surface and proportional to the square of the velocity in the direction of the normal to this surface. Newton without proof gave the differential equation in a geometric form; proofs were given in the *Acta Eruditorum* 5 (1697), 8 (1699), and *11* (1699) by N. Fatio de Duillier, L'Hôpital, and Johann Bernoulli. The differential equation is $y \, dx \, dy^3 = a \, ds^3$; see Johann Bernoulli, *Opera omnia* (Geneva, 1744), 307–315.

Corollary III, Art. 39, finishes with the remark:

From this we obtain the following rule for the solution of problems in which the curve with a maximum or minimum of $\int Z \, dz$ is desired, where

$$dZ = M \, dx + N \, dy + P \, dp:$$

differentiate Z, place zero instead of $M \, dx$ in the differentials $M \, dx + N \, dy + P \, dp$, keep $N \, dy$ unchanged, and write $-p \, dP$ instead of $P \, dp$. Then in this way we obtain $N \, dy - y \, dP = 0$, an equation which because of $dy = p \, dx$ passes exactly into $N - dP/dx = 0$, which is the one we have already found. A method free from a geometric solution is therefore desired, from which it will be clear that in such an investigation of maxima and minima instead of $P \, dp$ we must write $-p \, dP$.[3]

Proposition IV. *Problem.* When Z is a function of x, y, p, and q, so that

$$dZ = M \, dx + N \, dy + P \, dp + Q \, dq,$$

to find among all curves corresponding to the same abscissa the curve for which $\int Z \, dz$ is a maximum or minimum.

The solution, by a reasoning along the same lines as in Prop. III, but now using

$$dy' = +n\nu, \qquad dp' = \frac{n\nu}{dx}, \qquad dq' = +\frac{n\nu}{dx^2},$$

$$dy = 0, \qquad dp = \frac{n\nu}{dx}, \qquad dq = -\frac{2n\nu}{dx^2},$$

$$dy_{,} = 0, \qquad dp_{,} = 0, \qquad dq_{,} = +\frac{n\nu}{dx^2},$$

leads to

$$n\nu \cdot dx\left(N' - \frac{P'}{dx} + \frac{P}{dx} + \frac{Q'}{dx^2} - \frac{2Q}{dx^2} + \frac{Q_{,}}{dx^2}\right) = n\nu \cdot dx\left(N' - \frac{dP}{dx} + \frac{d \, dQ_{,}}{dx^2}\right)$$

$$= n\nu \cdot dx\left(N - \frac{dP}{dx} + \frac{d \, dQ}{dx^2}\right),$$

so that the required equation is

$$N - \frac{dP}{dx} + \frac{d \, dQ}{dx^2} = 0.$$

[3] This is the paragraph to which Lagrange refers; see Selection V.22, p. 407.

Again many special cases and examples are given. Example II is: To find the curve Am [Fig. 3] wich with its evolute AR and the radius of curvature mR at every point has the smallest area ARm. The answer is a cycloid. In Proposition V Euler derives for the case $dZ = M\,dx + N\,dy + P\,dp + Q\,dq + R\,dr + S\,ds + T\,dt + \cdots$ the condition

$$0 = N - \frac{dP}{dx} + \frac{d\,dQ}{dx^2} + \frac{d^3 R}{dx^3} + \frac{d^4 S}{dx^4} - \frac{d^5 T}{dx^5} + \cdots.$$

Fig. 3

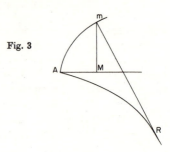

In later chapters we find problems that belong to the isoperimetric type. For instance, chapter V, 41 solves the problem:

To find among all curves of the same length, connecting the points a and z [Fig. 4], the curve that encloses the largest or smallest area $aAZz$. Answer: the circle. Similarly, chapter V, 45: To find among all curves enclosing the same area $aAZz$ the curve that by rotation about the axis AZ gives the surface of smallest area. Answer: a curve of the third order, belonging to type 68 of Newton, $9b(x - c)^2 = (2b - y)^2(2y - b)^4$

Fig. 4

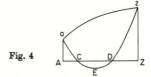

The first appendix exists in an English translation by W. A. Oldfather, C. A. Ellis, and D. M. Brown, "Leonhard Euler's elastic curves," *Isis 20* (1933), 72–160; there is a German translation by H. Linsenbarth in Ostwald's *Klassiker*, No. 175 (Engelmann, Leipzig, 1910). The second appendix contains Euler's first publication of the principle of least action.

22   LAGRANGE. THE CALCULUS OF VARIATIONS

Lagrange had already studied Euler's papers when he was in his teens, and Euler's book of 1744 in particular. As a young professor at the Artillery School in Turin he began to correspond with Euler on this subject as early as 1755, when he was 21 years of age. Euler

---

4 *Enumeratio linearum tertii ordinis* (1706); see Selection III.8.

encouraged him, and the first results of Lagrange's work were published in the "Essai d'une nouvelle méthode pour déterminer les maxima et les minima des formules intégrales indéfinies" (Attempt at a new method for determining the maxima and minima of indefinite integral formulas), *Miscellanea Taurinensia 2* (1760–61, 173–195; *Oeuvres*, I (1867), 355–362). Lagrange's aim was to present the "calculus of variations" in a purely analytic form. A German translation by P. Stäckel can be found in Ostwald's *Klassiker*, No. 47 (Engelmann, Leipzig, 1894). We give here the principal part of his paper.

The first problem of this kind solved by the geometers is that of the *Brachystochrone*, or line of most rapid descent, which Mr. Jean Bernoulli proposed toward the end of the last century. It was solved only for particular cases, and it was not until some time later, on the occasion of the investigations on *Isoperimetrics*, that the great geometer whom we mentioned and his illustrious brother Mr. Jacques Bernoulli gave some general rules for solving several other problems of the same kind. But since these rules were not general enough, all these investigations were reduced by the famous Mr. Euler to a general method, in a work entitled *Methodus inveniendi...*, an original work which everywhere radiates a deep knowledge of the calculus. But, however ingenious and fertile his method may be, we must recognize that it does not have all the simplicity that might be desired in a subject of pure analysis. The author has made us aware of this in Article 39 of Chapter II of his book, by the words, "A method free from a geometric solution is therefore required..."

Now here is a method that demands only a very simple application of the principles of the differential and integral calculus, but first of all I must warn you that, since this method demands that the same quantities vary in two different manners, I have, in order not to confuse these variations, introduced into my calculations a new characteristic δ. Thus δZ will express a difference of Z that will not be the same as dZ, but that nevertheless, will be formed by means of the same rules; so that when we have an equation $dZ = m\,dx$ we might just as well have $\delta Z = m\delta x$, and other expressions in the same way.

This being settled, I come first to the following problem.

I

Problem I. Given an indefinite integral expression represented by $\int Z$, where Z indicates a given arbitrary function of the variables x, y, z and their differentials [*différences*] dx, dy, dz, d^2x, d^2y, d^2z, ..., to find the relation among these variables so that the formula $\int Z$ become a maximum or a minimum.

Solution. According to the known method *de maximis et minimis* we shall have to differentiate the proposed $\int Z$, and, regarding the quantities x, y, z, dx, dy, dz, d^2x, d^2y, d^2z, ... as variables, make the resulting differential [*différentielle*] equal to zero. When, therefore, we indicate these variations by δ, we shall have first, for the equation of the maximum or minimum,

$$\delta \int dZ = 0,$$

or, what is equivalent to it,

$$d \int \delta Z = 0.$$

Now, let Z be such that

$$
\begin{aligned}
\delta Z = n\delta x &+ p\delta\, dx + q\delta\, d^2x + r\delta\, d^3x + \cdots \\
&+ N\delta y + P\delta\, dy + Q\delta\, d^2y + R\delta\, d^3y + \cdots \\
&+ \nu\delta z + \pi\delta\, dz + \chi\delta\, d^2z + \rho\delta\, d^3z + \cdots;
\end{aligned}
$$

then we obtain from it the equation

$$
\begin{aligned}
\int n\delta x &+ \int p\delta\, dx + \int q\delta\, d^2x + \int r\delta\, d^3x + \cdots \\
&+ \int N\delta y + \int P\delta\, dy + \int Q\delta\, d^2y + \int R\delta\, d^3y + \cdots \\
&+ \int \nu\delta z + \int \pi\delta\, dx + \int \chi\delta\, d^2z + \int \rho\delta\, d^3z + \cdots = 0,
\end{aligned}
$$

but it is easily understood that

$$\delta\, dx = d\delta x, \quad \delta\, d^2x = d^2\delta x,$$

and the others in the same way; moreover, we find by the method of integration by parts,

$$
\int p\, d\delta x = p\delta x - \int dp\delta x,
$$

$$
\int q\, d^2\delta x = q\, d\delta x - dq\delta x + \int d^2q\delta x,
$$

$$
\int r\, d^3\delta x = r\, d^2\delta x - dr\, d\delta x + d^2r\delta x - \int d^3z\delta x,
$$

and the others in a similar way. The preceding equation will therefore be changed into the following:

$$
\begin{aligned}
\int (n &- dp + d^2q - d^3r + \cdots)\delta x \\
&+ \int (N - dP + d^2Q - d^3R + \cdots)\delta y \\
&+ \int (\nu - d\pi + d^2\chi - d^3\rho + \cdots)\delta z \\
&+ (p - dq + d^2r - \cdots)\delta x + (q - dr + \cdots)\, d\delta x \\
&\qquad\qquad\qquad\qquad + (r - \cdots)\, d^2\delta x + \cdots \\
&+ (P - dQ + d^2R - \cdots)\delta y + (Q - dR + \cdots)\, d\delta y \\
&\qquad\qquad\qquad\qquad + (R - \cdots)\, d^2\delta y + \cdots \\
&+ (\pi - d\chi + d^2\rho - \cdots)\delta z + (\chi - d\rho + \cdots)\, d\delta z \\
&\qquad\qquad\qquad\qquad + (\rho - \cdots)\, d^2\delta z + \cdots = 0,
\end{aligned}
$$

(A)

from which we obtain first the indefinite equation

(B)
$$(n - dp + d^2q - d^3r + \cdots)\delta x$$
$$+ (N - dP + d^2Q - d^3R + \cdots)\delta y -$$
$$+ (\nu - d\pi + d^2\chi - d^3\rho + \cdots)\delta z = 0,$$

and then the determinate equation

(C)
$$(p - dq + d^2r - \cdots)\delta x + (q - dr + \cdots)\, d\delta x$$
$$+ (r - \cdots)\, d^2\delta x + \cdots$$
$$+ (P - dQ + d^2R - \cdots)\delta y + (Q - dR + \cdots)\, d\delta y$$
$$+ (R - \cdots)\, d^2\delta y + \cdots$$
$$+ (\pi - d\chi + d^2\rho - \cdots)\delta z + (\chi - d\rho + \cdots)\, d\delta z$$
$$+ (\rho - \cdots)\, d^2\delta z + \cdots = 0.$$

This equation refers to the last part of the integral $\int Z$; but we must observe that, since each of its terms, such as $p\delta x$, depends on an integration by parts of the formula $\int p\, d\delta x$, we may add to or subtract from it a constant quantity. The condition by which this constant must be determined is that $p\delta x$ must vanish at the point where the integral $\int p\, d\delta x$ begins; we must therefore take away from $p\delta x$ its value at this point. From this we obtain the following rule. Let us express the first part of equation (C) generally by M, and let the value of M at the point where the integral $\int Z$ begins be indicated by $'M$, and at the point where this integral ends, by M'; then we have $M' - 'M = 0$ for the complete expression of equation (C). Now, in order to free the equations obtained from the undetermined differentials $\delta x, \delta y, \delta z, d\delta x, d\delta y, \ldots$, we must first examine whether, by the nature of the problem, there exists some given relation among them, and then, having reduced them to the smallest number possible, we must equate to zero the coefficient of each of those that remain. If they are absolutely independent of each other, then equation (B) will give us immediately the three following:

$$n - dp + d^2q - d^3r + \cdots = 0,$$
$$N - dP + d^2Q - d^3R + \cdots = 0,$$
$$\nu - d\pi + d^2\chi - d^3\rho + \cdots = 0.$$

Next follows the example

$$\int \frac{\sqrt{dx^2 + dy^2 + dz^2}}{\sqrt{x}},$$

which is the brachystochrone in empty space and leads (a) to the result that the curve is plane, and (b) to $dt = \sqrt{x}\, dx/\sqrt{c - x}$. The case of the brachystochrone on a surface is also discussed; here the relation $\delta z = p\delta x + q\delta y$ has to be taken into consideration. Lagrange takes the cases in which the end points are fixed, as well as those in which they are subjected to certain other conditions. This, says Lagrange, makes his method more general than that of Euler, since Euler keeps the end points fixed; moreover, he lets only y vary in Z.

Problem II. To make the expression $\int Z$ a maximum or minimum, under the supposition that Z is an arbitrary algebraic function of the variables x, y, z with their differences dx, dy, dz, d^2x, d^2y, ... and of the quantity $\Pi = \int Z'$, Z' being another arbitrary algebraic function of the variables $x, y, z, \ldots$ and their differentials dx, dy, dz, d^2x, d^2y, ... alone.

Solution. Let us have, by differentiating with respect to δ,

$$\delta Z = L\delta\Pi + n\delta x + p\delta\, dx + q\delta\, d^2x + \cdots$$
$$+ N\delta y + P\delta\, d^2y + Q\delta\, d^2y + \cdots$$
$$+ \nu\delta z + \pi\delta\, dz + \chi\delta\, d^2z + \cdots$$

and

$$\delta Z' = n'\delta x + p'\delta\, dx + q'\delta\, d^2x + \cdots$$
$$+ N'\delta y + P'\delta\, dy + Q'\delta\, d^2y + \cdots$$
$$+ \nu'\delta z + \pi'\delta\, dz + \chi'\delta\, d^2z + \cdots$$

then we shall have, by hypothesis,

$$\delta\Pi = \delta\int Z' = \int \delta Z' = \int (n'\delta x + p'\delta\, dx + q'\delta\, d^2x + \cdots),$$

hence

$$\delta\int Z = \int \delta Z = \int (n\delta x + p\delta\, dx + q\delta\, d^2x + \cdots)$$
$$+ \int L \int (n'\delta x + p'\delta\, dx + q'\delta\, d^2x + \cdots).$$

The first section can be reduced, as in Problem I, to

$$\int (n - dp + d^2q - \cdots)\delta x + (p - dq + \cdots)\delta x + (q - \cdots)\, d\delta x + \cdots.$$

As to the second one, we shall transform it into

$$\int L \times (n'\delta x + p'\delta\, dx + q'\delta\, d^2x + \cdots) - \int \left[\left[\int L \times (n'\delta x + p'\delta\, dx + q'\delta\, d^2x) \right].$$

Now let the total value of the integral $\int L$ be represented by H. If we take this quantity H as a constant, then the preceding transformed equation will be reduced to this one:

$$\int\left[\left(H - \int L\right)(n'\,\delta x + p'\,\delta\,dx + q'\,\delta\,d^2x + \cdots),\right.$$

which can easily be transformed by partial integration into

$$\int\left[n'\left(H - \int L\right) - dp'\left(H - \int L\right) + d^2q'\left(H - \int L\right) + \cdots\right]\delta x$$

$$+ \left[p'\left(H - \int L\right) - dq'\left(H - \int L\right) + \cdots\right]\delta x$$

$$+ \left[q'\left(H - \int L\right) - \cdots\right]d\,\delta x + \cdots.$$

If, therefore, we write, for short,

$$n + n'\left(H - \int L\right) = (n),$$

$$p + p'\left(H - \int L\right) = (p),$$

$$q + q'\left(H - \int L\right) = (q),$$

and equally

$$N + N'\left(H - \int L\right) = (N),$$

$$P + P'\left(H - \int L\right) = (P),$$

$$Q + Q'\left(H - \int L\right) = (Q),$$

and so forth, as well as

$$\nu + \nu'\left(H - \int L\right) = (\nu),$$

$$\pi + \pi'\left(H - \int L\right) = (\pi),$$

$$\chi + \chi'\left(H - \int L\right) = (\chi),$$

we shall have, in general,

$$\delta \int Z = \int [(n) - d(p) + d^2(q) - \cdots] \delta x$$
$$+ \int [(N) - d(P) + d^2(Q) - \cdots] \delta y$$
$$+ \int [(\nu) - d(\pi) + d^2(\chi) - \cdots] \delta z$$
$$+ [(p) - d(q) + \cdots] \delta x - [(q) - \cdots] \delta \, dx + \cdots$$
$$+ [(P) - d(Q) + \cdots] \delta y - [(Q) - \cdots] \delta \, dy + \cdots$$
$$+ [(\pi) - d(\chi) - \cdots] \delta z - [(\chi) - \cdots] \delta \, dz$$
$$= 0,$$

an equation reduced to the form of equation (A) of the preceding problem, hence, etc.

In a corollary this method is applied to the case in which Z' contains another indefinite integral function $\Pi' = \int Z''$.

Problem III. To find the equation of the maximum or the minimum of the formula $\int Z$, if Z is simply given by a differential equation that does not contain other differentials of Z than the first.

This is the case in which we can write

$$\delta \, dZ + T\delta Z = n\delta x + p\delta \, dx + \cdots + N\delta y + P\delta \, dy + \cdots + \nu\delta z + \pi\delta \, dz,$$

which is then solved as a linear differential equation in δZ, taking $\delta \, dZ = d\delta Z$.

There are two appendices. In the first we find (a) the problem of the surface of least area among all surfaces with the same given perimeter:

$$\delta \int \int dx \, dy \sqrt{1 + p^2 + q^2} = 0, \qquad p = \left(\frac{dz}{dx}\right), \qquad q = \left(\frac{dz}{dy}\right),$$

which leads to the condition that both $p \, dx + q \, dy$ and $\dfrac{p \, dy - q \, dx}{\sqrt{1 + p^2 + q^2}}$ have to be exact differentials,[1] and (*b*) the problem of the surface of least area among all surfaces of equal volume:

$$\delta\left(\int\int z\,dx\,dy\right) = 0, \qquad \delta\left(\int\int dx\,dy\sqrt{1 + p^2 + q^2}\right) = 0,$$

which leads to the condition that both $p\,dx + q\,dy$ and $\dfrac{p\,dy - q\,dx}{\sqrt{1 + p^2 + q^2}} + kx\,dy$ (k an arbitrary coefficient) must be exact differentials. This is verified for the sphere.

In the second appendix we find the problem of the polygon of largest area among all polygons of the same given number of sides. It is shown that this polygon is inscribed in a circle, a theorem proved geometrically by Cramer (*Histoire de l'Académie Royale, Berlin,* 1752). If only the sum of the sides is given, the polygon is regular.

Lagrange's paper was followed in the same number of the *Miscellanea Taurinensis*, pp. 196–298, by a longer one: "Application de différents problèmes de dynamique" (*Oeuvres*, I, 365–468).

These investigations of Euler and Lagrange on the calculus of variations were supplemented by Legendre in a paper in the *Histoire de l'Académie Royale, Paris, 1786* (1788), 7–37, in which he studied the second variation. A German translation appears in the same volume 47 of Ostwald's *Klasskiker* in which we find Euler's and Lagrange's research.

23 MONGE. THE TWO CURVATURES OF A CURVED SURFACE

Where Lagrange can be called the first great mathematician who was an analyst (hence, by his own philosophy, an algebraist), Gaspard Monge (1746–1818) can be called the first geometer. He first taught at the military academy in Mézières near Sedan, after which he went to Paris, where he played a leading political role in the Revolution, and helped in founding the École Polytechnique (1795), of which he became the leading spirit until his dismissal by the Bourbons in 1815. It was primarily through his teaching that descriptive geometry, analytical geometry, and differential geometry were established as special fields. Of his *Géométrie descriptive* (Paris, 1795) there exists an English extract by J. F. Heather (Lockwood, London, 1809; Weale, London, 1851). His lessons in differential geometry were first published in *Feuilles d'analyse appliquée a la géométrie a l'usage de l'École Polytechnique* (Paris, 1795), later elaborated in *Application de l'algèbre a la géométrie* (5th ed.; Bachelier, Paris, 1850). The following selection is from *Feuille* XV, almost equivalent with chap. XV, pp. 124–139, of the 1850 edition, entitled "Des deux courbures d'une surface courbe." In previous *Feuilles* Monge had discussed conical surfaces, surfaces of revolution, canal surfaces, developable surfaces, and other surfaces, using that combination of differential calculus and geometrical reasoning typical of his way of thinking. On Monge see R. Taton, *L'Oeuvre scientifique de Monge* (Presses Universitaires, Paris, 1951).

We begin with a part of Section I of the paper.

---

[1] Two examples of these "minimal surfaces," the catenoid and the right helicoid, were found by Jean-Baptiste Meusnier, a pupil of Monge's, in the *Mémoires des savants étrangers de l'Académie 10* (Paris, 1785). He also interpreted here Lagrange's analytic condition geometrically as indicating that the mean curvature is zero. The catenoid had already appeared in chap. V, 44 of Euler's *Methodus inveniendi*, but not as a minimal surface.

If we represent by x, y, z the coordinates of an arbitrary point of a curved surface, and by x', y', z' those of the surface of a sphere, then the equation of the sphere that will have its center at the point of the curved surface, and for its radius the quantity R, will be

(A) $$(x - x')^2 + (y - y')^2 + (z - z')^2 = R^2.$$

We have seen[1] that when x', y', z' are regarded as constants in this equation, and if we differentiate it successively regarding first x, then y as sole variables, the two equations obtained

(B) $$x - x' + (z - z')p = 0,$$

(C) $$y - y' + (z - z')q = 0,[2]$$

are, in x', y', z' those of the two normal planes to the curved surface passing through the point under consideration on the surface. They are perpendicular, the one to the xz-plane, the other to the yz-plane. Hence, these two equations are those of the two projections of the normal to the curved surface passing through the same point of the surface. In these two equations x', y', z' are the variables of the normal, and the five quantities x, y, z, p, q, which belong to the point of the surface through which passes the normal, are constants for the same normal, and vary in value when we pass from one normal to another.

If, from the first point of the surface under consideration, we pass in a certain direction to another point at an infinitesimal distance, then the five quantities x, y, z, p, q will increase by their respective differentials dx, dy, dz, dp, dq and among these five differentials exist the following three equations:

$$dr = p\,dx + q\,dy, \quad dp = r\,dx + s\,dy, \quad dq = s\,dx + t\,dy;[3]$$

and the value of the quantity dy/dx will determine, in the plane of the x, y, the projection of the direction along which we pass from the first point to the second.

This being established, if we conceive through the second point a new normal to the curved surface, and if this normal is in the same plane as the first and hence intersects it somewhere at a point, then this point of intersection will be that of the first normal for which the three coordinates x', y', z' do not change

---

[1] In chap. I (p. 5 of the 1850 ed.), where Monge discusses tangent planes and normals to curved surfaces $f(x, y, z) = 0$.

[2] On p. 48, Monge introduces the now common notation $p = dz/dx$, $q = dz/dy$ for the partial derivatives of z with respect to x and y. The notation $\partial z/\partial x$ appears in the nineteenth century, at the suggestion of Jacobi in his theory of determinants (*Crelle's Journal für die reine und angew. Mathem. 22* (1841),3 19, and received for a long time only a gradual acceptance. It seems to have first been proposed by Legendre, *Histoire de l'Académie Royale, Paris, 1786* (1788), p. 8.

[3] On p. 71 Monge introduces r, s, t as the partial derivatives of the second order, which he calls *différences partielles du second ordre*.

when x and y change their values. Hence, if we differentiate equations (B) and (C), regarding x', y', z' as constants, we obtain

$$dx + p^2\, dx + pq\, dy + (z - z')(r\, dx + s\, dy) = 0,$$

$$dy + pq\, dx + q^2\, dy + (z - z')(s\, dx + t\, dy) = 0.$$

Eliminating first dy/dx and then $(z - z')$, we obtain two equations equivalent to the two preceding ones:

(D) $\qquad (z - z')^2(rt - s^2) + (z - z')[(1 + q^2)r - 2pqs + (1 + p^2)t]$
$$+ 1 + p^2 + q^2 = 0,$$

(E) $\qquad \dfrac{dy^2}{dx^2}[(1 + q^2)s - pqt] + \dfrac{dy}{dx}[(1 + q^2)r - (1 + p^2)t]$
$$- (1 + p^2)s + pqr = 0.$$

The four equations (B), (C), (D), (E) will belong to the point of intersection of the two consecutive normals. However, the first three of these equations suffice to determine the three coordinates x', y', z' of this point. The fourth equation (E), which does not contain any of the coordinates, is therefore a condition that has to be satisfied, and that, by determining the value of dy/dx, indicates the direction according to which we must pass from the first point of the surface to the second, so that the new normal would be in the same plane as the first one, and would have a point in common with it.

· · · · · · · ·

II

Since equation (E) is of the second algebraic degree with respect to dy/dx, and provides two values for this quantity, it follows that, having drawn a normal through an arbitrary point of a curved surface, we can always pass on the surface in two different directions from this point to another, at an infinitely small distance, for which the normal is in the same plane as the first. These two directions are, in general, the only ones for which this result can occur, so that, except in the very particular case for which equation (E) is always satisfied whatever the value of dy/dx may be, if we pass from the first point to the second along any other direction, the new normal will not be in the same plane with the first, and will have no point in common with it.

The two directions in question have between them a very remarkable property, and that is that they are at right angles. Indeed, whatever may be the surface on which we operate, and whatever be the point of this surface that we consider, we can always suppose that the three rectangular planes of projection, whose position was first arbitrary, are chosen in such a way that the tangent plane to the surface at this point is parallel to the xy-plane. The quantities p, q, under this assumption are both equal to zero, and equation (E) becomes

$$\frac{dy^2}{dx^2} + \frac{dy}{dx}\left(\frac{r - t}{s}\right) - 1 = 0.$$

Now, if we represent by m and m' the two values of dy/dx which this equation provides, then we shall have

$$mm' + 1 = 0;$$

hence the projections of the two directions on the xy-plane are at right angles. But these directions themselves, since they are in the tangent plane, are parallel to their projections, hence they are also at right angles.[4]

III

Equation (D) being also of the second algebraic degree, with respect to $z - z'$, it is clear that the three equations (B), (C), (D) give two values for each of the quantities x', y', z', and that these double values will be those that correspond respectively to the two points of intersection of the first normal with the two other normals that are in the same plane with it.

Operating on each of these points of intersection in particular, and to begin with on the first of them, then, if we think of the sphere whose center is at this point and whose surface passes through the point of the curved surface, it is evident that the two normals to the curved surface, which intersect at the center, will also be normals to the sphere. The curved surface and that of the sphere will therefore have two consecutive normals in common, and consequently, will have two consecutive tangent planes in common; they will thus have the same curvature in the direction of the plane that passes through the two normals, that is to say, in the direction determined by the corresponding value of dy/dx, and the center of this curvature will be nothing else but the point where the two normals meet, hence the center of the sphere itself.

In the same way Monge finds a second sphere corresponding to the other value of dy/dx.

Hence, every curved surface has at every one of its points two curvatures whose directions are in two normal planes perpendicular to one another, and whose centers are on the same normal.

The three quantities x, y, z being the coordinates of the point of the surface, and the other three x', y', z' being the coordinates of the center of curvature, it is evident that the distance of these two points, that is to say, the value of the

---

[4] These directions were first discovered, in a rather complicated way, by Euler, "Recherches sur la courbure des surfaces," *Hist. Acad. Roy. des Sciences, Berlin* (1760; publ. 1767), 119–143, *Opera omnia*, ser. I, vol. 28 (1955), 1–22; he asked for the maximum and minimum values of the normal curvature of plane sections through the normal at a point of the surface. Monge introduced them in his "Mémoire sur la théorie des déblais et des remblais," *Histoire de l'Académie Royale, Paris, 1781* (1784), 666–704, where he investigates the developable surfaces formed by congruences of straight lines. This approach is directly related to that of our Selection from the *Feuilles*.

radius of curvature, is nothing else but the quantity R of equation (A). Hence, if between the four equations (A), (B), (C), (D) the three quantities $x - x'$, $y - y'$, $z - z'$ are eliminated, then we shall have an equation of the second degree which will give, in p, q, r, s, t, the two values of R, hence those of the two radii of curvature.

If we write for short

$$g = rt - s^2,$$

$$h = (1 + q^2)r - 2pqs + (1 + p^2)t,$$

$$k^2 = 1 + p^2 + q^2,$$

then the result of this elimination is

$$gR^2 + hkR + k^2 = 0,$$

from which it follows that the expression for the two radii of curvature is

(F) $$R = \frac{k}{2g}[-h + \sqrt{h^2 - 4k^2g}] = \frac{-2k^3}{h + \sqrt{h^2 - 4k^2g}}.$$

IV

Since every normal to a curved surface always meets two other infinitely close normals placed in two normal places perpendicular to each other, let us imagine that from the normal at the first point of the surface we pass to one of the two infinitely close normals that intersect it, that subsequently we pass from this second normal, in the same sense, to the one that intersects it, that from this third one we pass, in the same sense, to the one that intersects this one, and so on along the whole extent of the surface. It is evident that we shall pass through a developable surface which will everywhere be perpendicular to the curved surface, and which will intersect it in a curved line whose elements will all be directed along one of the curvatures of the surface; this curve will thus be a line of the first curvature. Performing the same operation, and in the same sense, for all other points of the surface, we shall have the family [*la suite*] of all the lines of the first curvature, which will divide the curved surface into zones of variable width.

In the same way Monge derives the lines of the second curvature, each perpendicular to all those of the first curvature, and conversely.

These two families of curves will divide the curved surface into elements that can be regarded as rectangular. [As examples we get the meridians and parallels on a surface of revolution.]

We have seen that equation (E) expresses the relation that must exist between dy/dx and the five quantities p, q, r, s, t, so that two consecutive normals intersect; it is therefore the equation of the projection of the line of curvature on the xy-plane. If we thus have differentiated twice the given equation of the curved surface to obtain in x, y the values of p, q, r, s, t, and if we substitute these values into (E), then we shall have an ordinary differential equation in x, y, dy/dx, which will be that of the lines of curvature. But this equation is of the second degree with respect to dy/dx, and hence, when we shall have integrated it and completed it with an arbitrary constant, which we shall represent by A, this constant will be of the second degree, and the integral will be in general of the form

$$A^2 + Ag(x, y) + f(x, y) = 0,^5$$

in which the two functions g, f will be given by the integration.

Monge then shows how to find the A for the specific lines of curvature passing through a point $x = a$, $y = b$ on the surface.

If in the equation (E) of the lines of curvature we substitute for r, t their values taken from $dp = r\,dx + s\,dy$, $dq = s\,dx + t\,dy$, the quantity s disappears at the same time; and, with the aid of $dz = p\,dx + q\,dy$, this equation takes the form

(E) $$dp(dy + q\,dz) = dq(dx + p\,dz),$$

under which we shall see that it often presents itself.

In Section V Monge discusses the developables formed by the normals to the surface, and in Section VI the surfaces formed by the edges of regression of these developables, hence the locus of the centers of principal curvature of the surface, consisting of two sheets. One of his results is as follows:

If, on the sheet of the centers of one of the curvatures, we consider any one of the edges of regression of which it is the locus, then this edge will be, between any two of its points, the shortest line that can be drawn on the sheet. Indeed,

---

[5] Monge uses two differently shaped f's, which we replace by g and f.

the osculating plane of this edge,[6] that is to say, the plane that passes through two of its consecutive tangents, is tangent to the developable surface to which the edge belongs, and which is the locus of its tangents; it is therefore tangent to the sheet of the centers of the other curvature, and consequently normal to the first sheet at the point of osculation. But the line whose osculating plane is normal to the surface at the point of osculation is the shortest that can be traced, on this surface, between any two of its points, or, what amounts to the same, it is the line that would be traced by a thread extended between these two points.

This relation is then explained by a mechanical consideration.

In the next *Feuille*, No. XVI, Monge applies his theory to the determination of the lines of curvature of the ellipsoid. At present we usually determine these lines with the aid of Dupin's theorem on triply orthogonal systems, published by Monge's pupil Charles Dupin in *Développements de géométrie* (Courcier, Paris, 1813), 239.

---

[6] The term *osculation* was introduced by Leibniz in his "Meditatio nova de natura anguli contactus et osculi" (New meditation on the nature of the angle of contact and of kiss), *Acta Eruditorum* (June 1686), 289–292 (*Mathematische Schriften*, Abth. 2, Band III, 326–329), where he introduced the circle of osculation. A weakness in his characterization was corrected by Jakob Bernoulli, *Acta Eruditorum* (March 1692), 110–118 (*Opera*, I, 473–481), who also advanced the terminology: "Quod si omnes intersectiones quibus alias datae curvae se mutuo secare possunt, in unum punctum confluant, oritur coitus, qui est consummatissimus earum congressus . . . " Then, in 1694, he found the formula for the radius of curvature, $ds^3 : dx\, d\, dy$; see Selection IV.15 (Huygens), note 6. The name "osculating plane" (*planum osculans*) of a curve appears first in Johann Bernoulli, *Opera omnia*, IV, 113, 115, but the plane itself had already been introduced in the work done by Leibniz and the Bernoullis on geodesic lines on surfaces (*Acta Eruditorum*, 1697, 1698).

INDEX

Page numbers in italics indicate, for persons, places with biographical data, and for concepts, the places where they are introduced.